T0336499

CARBONATE SYSTEMS DURING THE OLIGOCENE–MIOCENE CLIMATIC TRANSITION

Other publications of the International Association of Sedimentologists

SPECIAL PUBLICATIONS

41 Perspectives in Carbonate Geology
A Tribute to the Career of Robert Nathan Ginsburg
Edited by Peter K. Swart, Gregor P. Eberli and Judith A. McKenzie
2009, 387 pages, 230 illustrations

40 Analogue and Numerical Modelling of Sedimentary Systems
From Understanding to Prediction
Edited by P. de Boer, G. Postma, K. van der Zwan, P. Burgess and P. Kukla
2008, 336 pages, 172 illustrations

39 Glacial Sedimentary Processes and Products
Edited by M.J. Hambrey, P. Christoffersen, N.F. Glasser and B. Hubbard
2007, 416 pages, 181 illustrations

38 Sedimentary Processes, Environments and Basins
A Tribute to Peter Friend
Edited by G. Nichols, E. Williams and C. Paola
2007, 648 pages, 329 illustrations

37 Continental Margin Sedimentation
From Sediment Transport to Sequence Stratigraphy
Edited by C.A. Nittrouer, J.A. Austin, M.E. Field, J.H. Kravitz, J.P.M. Syvitski and P.L. Wiberg
2007, 549 pages, 178 illustrations

36 Braided Rivers
Process, Deposits, Ecology and Management
Edited by G.H. Sambrook Smith, J.L. Best, C.S. Bristow and G.E. Petts
2006, 390 pages, 197 illustrations

35 Fluvial Sedimentology VII
Edited by M.D. Blum, S.B. Marriott and S.F. Leclair
2005, 589 pages, 319 illustrations

34 Clay Mineral Cements in Sandstones
Edited by R.H. Worden and S. Morad
2003, 512 pages, 246 illustrations

33 Precambrian Sedimentary Environments
A Modern Approach to Ancient Depositional Systems
Edited by W. Altermann and P.L. Corcoran
2002, 464 pages, 194 illustrations

32 Flood and Megaflood Processes and Deposits
Recent and Ancient Examples
Edited by I.P. Martini, V.R. Baker and G. Garzón
2002, 320 pages, 281 illustrations

31 Particulate Gravity Currents
Edited by W.D. McCaffrey, B.C. Kneller and J. Peakall
2001, 320 pages, 222 illustrations

30 Volcaniclastic Sedimentation in Lacustrine Settings
Edited by J.D.L. White and N.R. Riggs
2001, 312 pages, 155 illustrations

29 Quartz Cementation in Sandstones
Edited by R.H. Worden and S. Morad
2000, 352 pages, 231 illustrations

28 Fluvial Sedimentology VI
Edited by N.D. Smith and J. Rogers
1999, 328 pages, 280 illustrations

27 Palaeoweathering, Palaeosurfaces and Related Continental Deposits
Edited by M. Thiry and R. Simon Coinçon
1999, 408 pages, 238 illustrations

26 Carbonate Cementation in Sandstones
Edited by S. Morad
1998, 576 pages, 297 illustrations

25 Reefs and Carbonate Platforms in the Pacific and Indian Oceans
Edited by G.F. Camoin and P.J. Davies
1998, 336 pages, 170 illustrations

24 Tidal Signatures in Modern and Ancient Sediments
Edited by B.W. Flemming and A. Bartholomä
1995, 368 pages, 259 illustrations

23 Carbonate Mud-mounds
Their Origin and Evolution
Edited by C.L.V. Monty, D.W.J. Bosence, P.H. Bridges and B.R. Pratt
1995, 543 pages, 330 illustrations

REPRINT SERIES

4 Sandstone Diagenesis: Recent and Ancient
Edited by S.D. Burley and R.H. Worden
2003, 648 pages, 223 illustrations

3 Deep-water Turbidite Systems
Edited by D.A.V. Stow
1992, 479 pages, 278 illustrations

2 Calcretes
Edited by V.P. Wright and M.E. Tucker
1991, 360 pages, 190 illustrations

Special Publication Number 42 of the International Association of Sedimentologists

Carbonate Systems During the Oligocene–Miocene Climatic Transition

EDITED BY

Maria Mutti

Institut für Erd- und Umweltwissenschaften Universität Potsdam
Postfach 60 15 53 D-14415 Potsdam
Germany

Werner Piller

Institute for Earth Sciences (Geology & Paleontology)
University of Graz, Heinrichstrasse 26
A-8010 Graz, Austria

Christian Betzler

Geologisch-Palaeontologisches Institut
Bundesstr. 55, D-20146 Hamburg, Germany

SERIES EDITOR

Ian Jarvis

School of Geography, Geology and the Environment
Centre for Earth and Environmental Science Research
Kingston University London
Penrhyn Road
Kingston upon Thames KT1 2EE
UK

SERIES CO-EDITOR

Tom Stevens

Department of Geography
Royal Holloway, University of London
Egham, Surrey
TW20 0EX
UK

A John Wiley & Sons, Ltd., Publication

Blackwell Publishing was acquired by John Wiley & Sons in February 2007. Blackwell's publishing program has been merged with Wiley's global Scientific, Technical and Medical business to form Wiley-Blackwell.

Registered office
John Wiley & Sons Ltd, The Atrium, Southern Gate, Chichester, West Sussex, PO19 8SQ, UK

Editorial offices
9600 Garsington Road, Oxford, OX4 2DQ, UK
The Atrium, Southern Gate, Chichester, West Sussex, PO19 8SQ, UK
111 River Street, Hoboken, NJ 07030-5774, USA

For details of our global editorial offices, for customer services and for information about how to apply for permission to reuse the copyright material in this book please see our website at www.wiley.com/wiley-blackwell

Library of Congress Cataloguing-in-Publication Data

Carbonate systems during the Oligocene–Miocene climatic transition / edited by
M. Mutti, W. Piller, and C. Betzler.
p. cm. – (Special publication ... of the international association of
sedimentologists ; no. 42)
Includes bibliographical references and index.
ISBN 978-1-4443-3791-4 (hardcover : alk. paper)
1. Rocks, Carbonate. 2. Geology, Stratigraphic–Oligocene. 3. Geology,
Stratigraphic–Miocene. 4. Paleoclimatology–Oligocene. 5.
Paleoclimatology–Miocene. I. Mutti, M. (Maria) II. Piller, Werner E. III.
Betzler, Christian.
QE471.15.C3C376 2010
552′.58–dc22

2010017502

A catalogue record for this book is available from the British Library.

Set in 10/12 pt Melior by Thomson Digital, Noida, India
Printed and bound in Malaysia by Vivar Printing Sdn Bhd

1 2010

Contents

Int. Assoc. Sedimentol. Spec. Publ. (2010) **42**, vii–xii

Miocene carbonate systems: an introduction

MARIA MUTTI, WERNER PILLER and CHRISTIAN BETZLER

The Oligocene and Miocene comprise the most important phases in the long-term, step-wise global cooling that, during the Cenozoic, led from a greenhouse to an icehouse Earth. Recent major advances in the understanding and time-resolution of climate events taking place at this time, as well as the proliferation of studies on Oligocene and Miocene shallow-water/neritic carbonate systems, invite us to re-evaluate the significance of these carbonate systems in the context of changes in climate and Earth surface processes. These changes can be traced by the palaeoecology and palaeobiogeography of neritic and pelagic biota. Additionally, the skeletons of these biota represent archives for proxy data, such as stable isotopes. Stable isotopes can be used to trace steps of climate change and track associated changes in Earth surface processes such as continental weathering and runoff, sea-level changes, current circulation patterns, and surface water temperatures. As has been demonstrated during the past decades, carbonate systems, because of a wide dependence on the ecological requirements of organisms producing the sediment, are sensitive recorders of changes in environmental conditions on the Earth surface.

This Special Publication is based on scientific contributions presented at an international workshop entitled "Evolution of Carbonate Systems during the Oligocene–Miocene Climatic Transition" that took place in Potsdam in February 2005. The workshop, sponsored by the European Science Foundation (ESF) and the Deutsche Forschungsgemeinschaft (DFG), had the objective of addressing the dynamic evolution of carbonate systems deposited during the Oligocene and Miocene in the context on climatic and Earth surfaces processes specific for this time interval. In order to shed new light on this subject, this volume addresses a number of key issues including: climatic trends and controls over deposition; temporal changes in carbonate producers and palaeoecology; carbonate terminology; facies; processes and environmental parameters (including water temperature and production depth profiles); carbonate producers and their spatial and temporal variability; and tectonic controls over architecture.

CLIMATIC TRENDS AND CONTROLS OVER DEPOSITION

The main controls over shallow-water carbonate assemblages on shorter time scales are ambient water temperature, nutrient content, and salinity of the water masses. Palaeotemperature reconstructions, based on integrated oxygen-isotope records and Mg/Ca ratios of foraminifera tests show that important temperature variations took place during the Oligocene and Miocene, and that some are out of phase with the main glaciation events. Their magnitude is large enough to cause shifts in benthic communities.

The paper by **Billups** and **Scheidereich** discusses the application of Mg/Ca in foraminifera for separating the effect of temperature versus ice volume on oxygen isotope records. They illustrate examples from the Oligocene/Miocene boundary and the Middle Miocene. Of great relevance to carbonate studies, their results indicate that intervals of oxygen-isotope excursions are mostly out of phase with the reconstructed palaeotemperature changes. This implies that a simple correlation between carbonate facies changes and the oxygen-isotope curve is a flawed approach. Furthermore, the temperature fluctuations recorded by Ca/Mg variations at one single location are higher than previously assumed, and can reach several degrees Celsius in amplitude. This is sufficient to trigger significant faunal turnovers and the shifting of an environment from the tropical to the temperate domain.

It should be noted that the time resolution available in most shallow-water carbonate records is not adequate enough to be compared with climatic events as recorded by deep-sea records. In part, there is an intrinsic limitation regarding achievable age resolution of shallow-water strata. However, additional efforts should be made to achieve the best possible resolution, by integrating different dating schemes. This is necessary to resolve the entire climatic transition and its detailed effects on carbonate deposition.

TEMPORAL CHANGES IN CARBONATE PRODUCERS

Based on a literature compilation, Kiessling (2005) introduced the apparent paradox of expanding reef habitats during the Cenozoic at a time of longer term cooling. The greatest increase in recorded reef sites was from the Late Oligocene to the Early Miocene, whereas preserved reef volumes increased most strongly from the Early to Middle Miocene. Average reef diversity peaked in the Late Oligocene and then declined towards the Late Miocene, whereas the global diversity of scleractinian corals steadily increased through the Oligocene and Miocene, and there were almost no extinctions at the genus level across the Oligocene–Miocene boundary. Kiessling (2005) argued that neither climatic data nor any other quantified physico-chemical parameter are able to explain the large expansion of reefs across the Oligocene–Miocene boundary, thus implying a predominantly biological control.

In this volume, **Perrin** and **Kiessling** show that most Cenozoic buildups occur within a latitudinal belt broadly centred on the tropical regions and slightly shifted to the north. During the Cenozoic, the reef belt shows both gradual latitudinal shifts and latitudinal contraction/expansion. In particular, the latitudinal width of the reef belt seems to have been reduced near the Eocene–Oligocene boundary and increased again after the Rupelian. It was wider than today during most of the Miocene, its widest extension occurring during the Middle Miocene.

Bosellini and **Perrin** analyze patterns of generic richness and inferred palaeotemperatures for each stage of the Oligocene–Miocene time interval and compare it with global palaeoclimatic curves based on stable oxygen isotopes. Except for the Mid-Miocene Climatic Optimum, which is not recorded in the generic richness of the Mediterranean z-coral communities, their coral richness-derived palaeotemperatures correlate with the palaeoclimatic trends based on the isotopic curves. Their results show a gradual increase of temperature from the Early to the Late Rupelian and a gradual widening of the temperature range after the Burdigalian by progressively adapting to a larger temperature range from the mid-Miocene onwards to the Messinian.

CARBONATE TERMINOLOGY

Researchers with different focuses and expertise tend to use different terminology to describe carbonate sediments. The general consensus is that this is a problem affecting communication and identification of the real scientific problems. Furthermore, using different terminology considerably reduces the possibility to compare different study areas. Currently, there are two major terminologies in use for the description of neritic carbonates. One is based on the observation of the most important sediment-forming components (chlorozoan, foramol: Lees & Buller, 1972; Lees, 1975), the other takes into account the feeding priorities of the carbonate-producing biota (heterozoan, photozoan: James, 1997). None of the systems ultimately allows the reconstruction of controlling factors, such as water temperature or nutrient content. It can be argued that only descriptive terms should be used and an interpretative terminology should be avoided. In this volume, **Kindler** and **Wilson** provide a review of carbonate grain associations, and stress the need for a purely descriptive nomenclature to define regionally important groupings of carbonate deposits.

ENVIRONMENTAL CONTROLS OVER DEPOSITIONAL FACIES AND PROCESSES

A major compilation at the Mediterranean scale published in 1996 (*SEPM Concepts in Sedimentology and Paleontology*, Volume 5), mainly focused on reefal (scleractinian) carbonate systems, but also brought attention to the existence of distinct systems referred to as "non-tropical" or "cool-water" carbonates. Over the last decade, a number of studies have focused on these carbonates and have generated debate over the environmental and climatic parameters controlling the occurrence of these carbonate systems.

The main controls over carbonate assemblages and the resulting depositional facies on shorter time scales are temperature, nutrients and salinity of the water masses in which the biota live, and on longer time scales are biological evolution, geodynamics and changes in palaeogeography. It is essential to understand the degree to which carbonate facies assemblages record the effects of temperature changes versus nutrients or other environmental changes. This issue is of relevance with regard to how changes in carbonate facies can be related to the climatic changes specific to the time interval. It has been observed that local parameters may play an overriding role that can mask the global signal. Therefore it is necessary to understand regional settings before making global implications. Topography and geomorphology of an area are important concerns that may strongly affect depositional systems.

Martin, Braga, Sanchez-Almazo and **Aguirre** give an overview of Miocene carbonate platforms in the western Mediterranean. Lithofacies changes of carbonate rocks in the stratigraphic record at the margins of the Neogene basins in southern Spain were mainly the result of temperature variations during the Late Neogene. These variations promoted the alternation of non-tropical and tropical carbonate deposition during the last 10 Myr in the region. The authors state that heterozoan carbonates accumulated on ramps during sea-level lowstands, whereas photozoan carbonates accumulated during rising sea-level and highstands. Changes in sea-surface temperatures, reconstructed on the basis of fossil assemblages and stable isotope analyses, are considered to have been the major control.

Ruchonnet and **Kindler** present a study based on the Ragusa Platform in south-eastern Sicily and show how two different facies models must be used to describe the internal architecture of this carbonate edifice of latest Serravallian to Tortonian age. During the Late Serravallian, a heterozoan system dominated the inner ramp and was characterized by biota of the foramol association, with rhodoliths dominating the mid-ramp. During the Early Tortonian, a photozoan association dominated, with chlorozoan biota in the inner ramp and branching corallinaceans and rhodoliths in the middle ramp. This change in biotic assemblages had profound consequences for the overall geometry of the platform, which evolved from a Late Serravallian distally steepened ramp to a flat-topped platform in the Early Tortonian.

Brandano, Westphal and **Mateu-Vicens** discuss a case of Miocene carbonates from the central Apennines and analyze this record with regard to the problem of the sensitivity of a foramol-rhodalgal carbonate ramp to sea-level changes. They argue that the middle and outer ramp deposits are remarkably monotonous, implying that they do not record high-frequency sea-level fluctuations.

Brandano, Corda and **Castorina** integrate data from sedimentological and microfacies analyses to reconstruct the depositional model of a carbonate platform outcropping in the central Apennines. They recognize a narrow inner zone dominated by red algae and rhodoliths, a middle zone dominated by rhodoliths and larger foraminifera, and an outer ramp with three different subzones. These consist of a proximal zone dominated by bryozoan colonies, bivalves and echinoids, an intermediate zone with benthic and planktic foraminifera, echinoids and bivalves, and a distal zone with marls with silica-sponge spicules and calcarenites with mollusc, bryozoan and echinoid debris. The stratigraphic architecture reflects two second-order sequences.

Benisek, Marcano, Betzler and **Mutti** present a case study from the Sedini Limestone in northern Sardinia, where both heterozoan and photozoan assemblages occur, offering an ideal opportunity to demonstrate: (1) the importance of different biotic assemblages in determining the depositional architecture; and (2) the role of differential early diagenesis in stabilizing depositional facies, thus affecting depositional geometries. They recognize two depositional sequences. The lower one is a homoclinal ramp characterized by heterozoan assemblages. The upper sequence documents a progressive steepening of the carbonate platform slope and is dominated by photozoan assemblages. The differential distribution and amount of early cements in both sequences support the observed changes in overall depositional geometry. Early diagenetic features are rare in the lower sequence but rich and diverse in the upper sequence, allowing facies stabilization and the development of steep progradational slopes.

CARBONATE PRODUCERS AND PALAEOECOLOGY

The participation of a wide range of specialists with different expertise and professional backgrounds at the workshop provided a unique perspective into approaching the significance of carbonate facies in a broader palaeoecological and palaeoclimatic context. First, it is clear that the palaeoecological information that can be extracted from biotic assemblages is too often ignored by physical sedimentologists. For example, in addition to corals, molluscs, red algae, larger benthic foraminifera and echinoids all have a very clear palaeoecological signal, providing information on water depth, temperature and salinity, which are critical for sedimentological interpretations. However, this information is available in most cases only if the proper taxonomic identifications are made. Acquiring this information calls for very close co-operation across expertise boundaries. It was also noticed that bryozoa are the group of biocalcifiers least understood in terms of palaeoecological requirements. However, the discussion showed that palaeoecological reconstruction of carbonate sedimentary systems based on taxonomy also has its limits. Whereas Miocene and younger assemblages are very similar to Holocene associations, Oligocene and older assemblages have different compositions on the generic and species level.

In this volume, **Brandano** and **Piller** present a detailed palaeoecological study of a 7 m thick rhodolith interval from the middle ramp of the Lower Miocene Latium-Abruzzi carbonate platform in central Italy. They describe in detail the morphology and the taxonomy of the corallines and provide criteria for a palaeodepth reconstruction. Sedimentological and biogenic criteria indicate a low energy environment. Therefore the movement of rhodoliths cannot be generated by palaeocurrents, but by biogenic activities, particularly those of regular echinoids.

Braga, Bassi and **Piller** provide a review of the palaeoenvironmental significance of coralline red algae occurring both in reef-related carbonates and as the main components in shallow-water heterozoan settings from temperate regions. They show how the known distribution of corallines in the Oligocene does not suggest any palaeogeographical differentiation. In contrast, for the Miocene, the occurrence of taxa still living today allows the rough differentiation of the following palaeobiogeographic regions: (a) a tropical region (characterized by thick *Hydrolithon* plants and *Aethesolithon*); (b) a subtropical Mediterranean (with common *Spongites* and *Neogonolithion* species); and (c) a temperate region with shallow-water assemblages dominated by *Lithophyllum*. Furthermore, it is possible to identify changes in assemblages and growth forms occurring with depth, proving a very useful palaeobathymetric tool.

Harzhauser and **Piller** point out how molluscs are generally strongly underrated when interpreting carbonate systems in the circum-Mediterranean area, because of taphonomic loss and because taxonomic interpretation in thin-sections is difficult. They argue that this affects mostly small-sized gastropod species that, however, display a high diversity within modern carbonate systems. Mollusc shells may contribute up to 80% of the sediment within nearshore settings and especially around Miocene oolite shoals. Nevertheless, in most micro-facies studies this highly indicative and extremely species-rich group is just referred to as "gastropod", "bivalve" or, at the most accurate, as "oyster" or "pectinid". Obviously, such identifications are meaningless in terms of ecology or biogeography. They emphasize the importance of a classical, taxonomy-oriented palaeontology in any facies-analysis.

Kroh and **Nebelsick** show that echinoderms and their debris can contribute considerable sediment to Oligo-Miocene carbonate systems. They review the morphology of the echinoderm skeleton of all five extant echinoderm classes and discuss reproduction and growth of echinoderms, the chemistry of the skeleton, and the crystallography and diagenesis of echinoderm ossicles. The composition of the echinoderm fauna in carbonate environments and their abundance is also controlled by climate. Many groups are currently restricted to certain climate zones, and in particular the shallow water forms are sensitive to temperature changes. Additionally, echinoderms are useful tools for palaeoenvironmental reconstruction as they are sensitive to water agitation, currents, depth and type of substrate.

GEODYNAMIC CONTROLS OVER ARCHITECTURE

Geodyamic, by affecting topography and the subsidence regime of an area is a fundamental control that may strongly affect depositional systems. **Iryu, Inagaki, Suzuki** and **Yamamoto** present the results of investigations on a drill core on Kita-daito-jima, a carbonate island, lying ~350 km east of Okinawa-jima (southwestern Japan). The reconstructed Oligocene to Pliocene age-depth section of Kita-daito-jima shows that: (1) reef formation on Kita-daito-jima was controlled by combined effects of sea-level changes and tectonic movements (subsidence and uplift); and (2) two types of reef formation may be recognized, the growth of which kept up with the subsidence of the island and the rapid reef formation which commenced at sea-level falls.

Vigorito and, **Murru** and **Simone** present sedimentary and architectural patterns of Miocene carbonate successions deposited in the Sardinia

Rift Basin in three different physiographic settings: (1) narrow rift-related submerged valleys; (2) isolated fault-blocks in the axial portion of the rift; and (3) small basins located at the edge of the rift itself. Palaeophysiography and in turn pre- and synsedimentary tectonics appear to exert a major control over facies distribution and depositional architectures as well as over the location and morphology of the carbonate factories.

FUTURE INITIATIVES

Many possible future research topics were discussed at the workshop:

- It is crucial to note the importance of sea-level changes and the necessity to enhance time-resolution to achieve a better fit and understanding of forcing mechanisms.
- One very interesting idea was to compare the role of nutrients and light adaptation to evolutionary patterns. This could be applied to the average depth of coral growth and to how the size of larger foraminifera (*Lepidocyclina*) changes with respect to depth of the photic zone (small-shallow, large-deeper).
- It is necessary to "bring some order" into the spatial and temporal distribution of these carbonate systems, to develop maps for different time slices, and to recognize spatial changes with latitude, as well as temporal changes associated with climatic evolution.
- Before such maps can be compiled, it is necessary to rationalize the terminology used, largely because the different terminologies and study approaches in use hinder the possibility of a direct comparison of existing data.
- Finally, studies on neritic carbonates should be better compared to basinal, pelagic records, where complementary information on climate and oceanographic parameters is recorded.

We would like to thank all the reviewers who have helped to shape this volume, including Torsten Bickert, Dirk Nürnberg, Nancy Budd, Katharina Billups, Jochen Halfar, Jeff Lukasik, Cedric John, Toni Simo, John Reijmer, Andre Freiwald, Lucia Simone, Andrea Knoerich, Phil Bassant, Georg Warrlich, Giovanna della Porta, Wolf-Christian Dullo, James Nebelsick, Martyn Pedley, Thilo Bechstädt, Gregor Eberli, Finn Surlyk and Arndt Peterhänsel.

REFERENCES

Kiessling, W. (2005) Long-term relationships between ecological stability and biodiversity in Phanerozoic reefs. *Nature*, **433**, 410–413.

James, N.P. (1997) The cool-water carbonate depositional realm. In: *Cool-water Carbonates* (Eds N.P. James and J.A.D. Clarke). *SEPM Spec. Publ.*,**56**, 1–20.

Lees, A. (1975) Possible influence of salinity and temperature on modern shelf carbonate sedimentation. *Mar. Geol.*, **19**, 59–198.

Lees, A. and **Buller, A.T.** (1972) Modern temperate-water and warmwater shelf carbonate sediments contrasted. *Mar. Geol.*, **13**, 67–73.

Int. Assoc. Sedimentol. Spec. Publ. (2010) **42**, 1–16

A synthesis of Late Oligocene through Miocene deep sea temperatures as inferred from foraminiferal Mg/Ca ratios

KATHARINA BILLUPS and KATHLEEN SCHEIDERICH[1]

School of Marine Science and Policy, University of Delaware, 700 Pilottown Road, Lewes, DE 19958, USA (E-mail: kbillups@udel.edu)

ABSTRACT

Published benthic foraminiferal Mg/Ca records have been compiled that span the latest Oligocene through Miocene, including new data for the South Atlantic. This synthesis, the first such of Mg/Ca data, necessitates consideration of uncertainties and limitations and provides a general perspective on the evolution of deep-sea temperatures over this period. Published Mg/Ca records show temperature patterns through the Miocene that are consistent with those first synthesized by Kennett (1985) utilizing isotope and other data. Accordingly, the early Miocene was an interval of relative warmth culminating in a climatic optimum at ~16 Ma that was characterized by the warmest (Mg-derived) temperatures of the past 20 million years. After the climatic optimum, palaeotemperatures dropped by 3–4 °C during the second major advance of Antarctic ice between ~15 Ma and 13 Ma. For the late Miocene, between 11 and 8.5 Ma, a distinct increase in benthic foraminiferal Mg/Ca ratios at two Atlantic sites provides evidence for deep to intermediate water circulation changes. Thereafter, temperatures close to modern are recorded at all sites. Assuming constant seawater Mg/Ca ratios through time, it can be concluded that the early Miocene climate was generally warmer than today and that by the late Miocene temperatures approached modern values.

Keywords Oligocene, Miocene, palaeotemperatures, palaeoclimate, palaeoceanography, foraminifera.

INTRODUCTION

The Miocene, a time of climatic extremes ranging from an early Miocene climatic optimum at about 16 Ma to the mid-Miocene period of ice growth on Antarctica between about 15 and 13 Ma, has long been of particular interest to palaeoceanographers. A volume synthesizing Miocene climate was published in 1985 in the context of the Cenozoic Paleoceanography Project (Kennett, 1985). Quantitative micropalaeontological data and stable-isotope records provided a view of the evolution of Miocene climate primarily based on results from the Deep Sea Drilling Project. Accordingly, multi-species planktonic oxygen-isotope records from throughout the world ocean suggested that during the early Miocene, surface-water temperatures were warm and meridional surface-water temperature gradients were small, with modern ocean conditions developing after the mid-Miocene expansion of the Antarctic ice sheet (Savin *et al.*, 1985). Benthic foraminiferal faunal assemblages from the Pacific Ocean also indicated warm deep-waters and a less well-stratified water column, in comparison to the modern ocean (Woodruff *et al.*, 1985). Many of the modern forms of benthic foraminifera first appeared during the mid-Miocene, indicative of deep water cooling (Woodruff *et al.*, 1985). In short, these earlier studies documented a relatively warm early Miocene, followed by cooling of high-latitude surface and Pacific deep waters concurrent with mid-Miocene ice sheet expansion.

Much of what is known about past climate change (including the Miocene) comes from the oxygen-isotopic composition of benthic foraminifera. Although this proxy outlines large-scale climate change, the absolute magnitude and the relationship between ice growth and decay and ocean temperature changes cannot be determined uniquely. Foraminiferal $\delta^{18}O$ values reflect the $\delta^{18}O$ value of seawater due to global storage of ice at the poles and regional changes such as

[1]Present address: Department of Geology, University of Maryland, College Park, MD20742, USA.

evaporation and precipitation at the sea surface, as well as the water temperature in which the organism calcifies. Thus, certain assumptions are required in order to interpret the records with respect to any of the palaeoenvironmental indicators (e.g. ice volume versus palaeotemperature). This inherently limits the degree to which past changes in temperature and ice-volume can be quantified.

The recent development of foraminiferal Mg/Ca ratios as a proxy for palaeotemperatures provides an opportunity to improve our understanding of past climate change. Several studies have explored the potential of this approach to resolve questions of longer term climate change. In particular, Lear *et al.* (2000) constructed a record of benthic foraminiferal Mg/Ca ratios to constrain first-order changes in Cenozoic deep-water temperature and, in conjunction with the foraminiferal $\delta^{18}O$ values, the evolution of the isotopic composition of seawater. Billups & Schrag (2002) then developed higher-resolution records focused on the mid-Miocene and the Eocene, both periods of ice-sheet expansion. Other subsequent studies have further improved the temporal resolution of pre-Pleistocene Mg/Ca records, affording a more detailed view of Cenozoic climate change (Lear *et al.*, 2003, 2004; Shevenell *et al.*, 2004). These studies have elucidated the nature of the relationship between the temperature of seawater and its oxygen-isotopic composition, leading to new insights into the mechanisms of climate change.

The objective of this study is to compile published late Oligocene through Miocene Mg/Ca-derived palaeotemperature records. Collectively, these studies provide a continuous view of the evolution of Miocene palaeotemperatures in different parts of the world's oceans (Fig. 1). For the late Miocene, sufficient published records now exist to constrain deep and intermediate water temperatures in the Atlantic, Southern Ocean, and Pacific. To augment this array of data, a new benthic foraminiferal Mg/Ca record from the South Atlantic is presented (Ocean Drilling Program Site 1088, Fig. 1). It will be shown that, taken together, these geochemical records provide a perspective on Miocene climate history that is not substantially different from those of earlier reconstructions presented in Kennett (1985).

Mg/Ca palaeothermometry

Foraminiferal Mg/Ca ratios depend primarily on the temperature and the Mg/Ca ratio of the

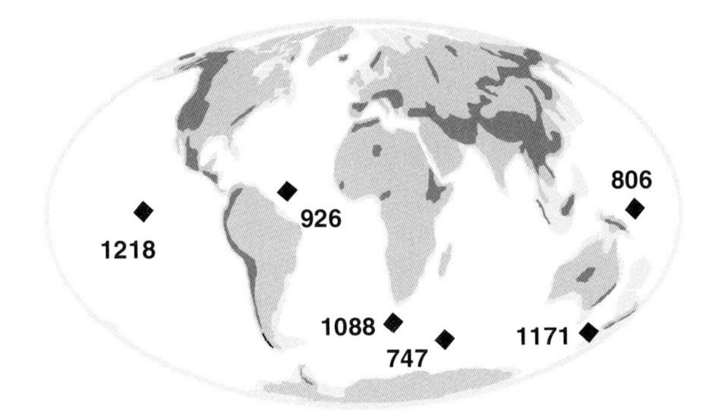

Fig. 1. General locations of sites discussed in this study. Table 1 summarizes the position, modern water depth and palaeodepth (as far as available), and the original citation. In stratigraphic order, Site 1218 in the eastern Pacific spans the Eocene through Miocene, Site 747 in the Southern Ocean reflects the latest Oligocene through Miocene, Site 1171 spans the mid-Miocene, and Sites 926, 806, and 1088 span the late Miocene. Note that Site 1171 reflects a surface water record.

seawater in which the foraminifera calcify. Thus, if seawater Mg/Ca ratios can be constrained, or if it is assumed that these ratios are invariant on the time scale under consideration, then foraminiferal Mg/Ca ratios provide a quantitative proxy for water temperature. On the relatively short time scales of the Quaternary, when secular changes in seawater Mg and Ca concentrations are not significant because of the long residence times of these cations (~13 Myr and 1 Myr, respectively; e.g. Broecker & Peng, 1982), the tool has been widely used to reconstruct absolute sea-surface temperatures (e.g. Hastings *et al.*, 1998; Elderfield & Ganssen, 2000; Lea *et al.*, 2000; Stott *et al.*, 2004). Currently, a number of temperature calibrations exist for planktonic (Nürnberg *et al.*, 1996; Lea *et al.*, 1999; Dekens *et al.*, 2002; Rosenthal & Lohmann, 2004; Anand *et al.*, 2003) and benthic (Lear *et al.*, 2002; Martin *et al.*, 2002) foraminiferal species, offering a multitude of opportunities for quantifying palaeoclimatic change.

General limitations associated with the Mg/Ca ratio proxy include calcite dissolution, diagenesis, insufficient removal of clays, and seawater carbon chemistry. Dissolution removes those portions of a foraminiferal shell that contain higher Mg/Ca ratios because Mg destabilizes the crystal lattice (Brown & Elderfield, 1996; Dekens *et al.*, 2002; Fehrenbacher *et al.*, 2006). Diagenesis adds inorganic calcite which, owing to a larger Mg distribution coefficient than biotic calcite, has comparatively higher Mg/Ca ratios than biogenic calcite (Morse & Bender, 1990; Carpenter & Lohmann, 1992).

Incomplete removal of high-Mg aluminosilicate clays from a test may yield abnormally high Mg/Ca ratios (Barker *et al.*, 2003). In fact, the foraminiferal cleaning protocol may affect how effectively contaminants such as clays and manganese crusts are removed from the tests (e.g. Rosenthal *et al.*, 2002). Lastly, recent studies have shown that at low seawater carbonate-ion concentrations there is a 'carbonate-ion effect' on benthic foraminiferal Mg/Ca ratios (Dekens *et al.*, 2002; Martin *et al.*, 2002; Elderfield *et al.*, 2006).

It is possible to mitigate some of these factors. Sites can be chosen above a certain water depth to minimize carbonate-ion concentration effects on Mg incorporation into a test and post-depositional dissolution effects on Mg removal from a test. Scanning electron microscopy may show evidence of euhedral inorganic calcite crystals on a test. An inverse relationship between Mg/Ca and Sr/Ca ratios in a sample can be used as an identifier of diagenesis, because the Sr/Ca ratios of inorganic calcite are lower than in biogenic calcite (Baker *et al.*, 1982; Morse & Bender, 1990; Carpenter & Lohmann, 1992). Clay contamination can be detected by measuring the Al content of samples (Barker *et al.*, 2003).

With respect to the synthesis presented here, these uncertainties limit the comparison of absolute palaeotemperatures. Sites discussed here come from different water depths relative to the lysocline, and temporal changes in the lysocline differ among the sites. Thus post-depositional dissolution cannot be ruled out as having affected foraminiferal Mg/Ca ratios and palaeotemperature reconstructions. Furthermore, as detailed below, samples were cleaned and analyzed using different methods, which is now known to introduce differences in the Mg-derived temperature values (Rosenthal *et al.*, 2002). In this regard, the temperature reconstructions presented below are intended to provide a general view of Miocene climate evolution.

Additionally, on long time scales, an important caveat arises from uncertainties in fluctuating Mg/Ca ratios of seawater through time. As recently collated by Tyrrell & Zeebe (2004), a variety of lines of evidence, including geochemical models (Hardie, 1996; Wilkinson & Algeo, 1998; Stanley & Hardie, 1998), measurements of fluid inclusions (e.g. Lowenstein *et al.*, 2001) and porefluids (Fantle & DePaolo, 2006), have been used in an attempt to constrain changes in seawater Mg/Ca ratios through time. Mg/Ca ratios tend to increase toward the modern in all cases, but the rate of change differs (Fig. 2), illustrating the complexity of Mg^{2+} and Ca^{2+} cycling through carbonate and silica reservoirs. As noted above, the residence times of Mg^{2+} and Ca^{2+} in seawater are ~13 Myr and 1 Myr, respectively. Therefore, although absolute palaeotemperatures for the Oligocene and Miocene

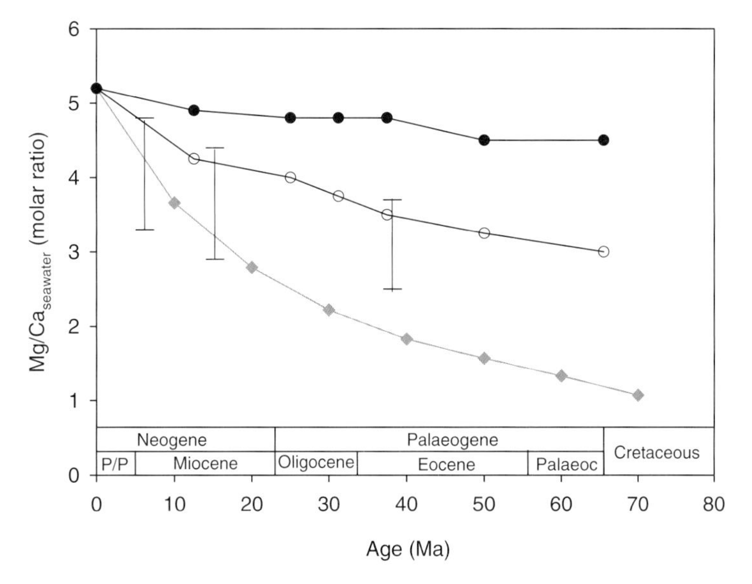

Fig. 2. Examples of Cenozoic seawater Mg/Ca ratios. Curves from the geochemical box models of Wilkinson & Algeo (1998) are marked by filled and open circles; the curve using a two end-member mixing model of Stanley & Hardie (1998) is marked by diamonds. The vertical bars indicate seawater Mg/Ca ratios determined from halite fluid-inclusions (Lowenstein *et al.*, 2001). In all cases, Mg/Ca ratios increase toward the present, but the rate of change differs.

must be interpreted with caution, but the amplitude of change occurring over less than 1 Myr is robust. It should be noted that none of the Mg-derived palaeotemperature records reviewed here include a correction for temporal changes in seawater Mg/Ca ratios. The effects of this simplification on the interpretation of the data with respect to palaeotemperatures will be discussed.

METHODS

Data compilation

Site 747

Located on the Kerguelen Plateau in the Indian Ocean Sector of the Southern Ocean at 1695 m water depth, Ocean Drilling Program Site 747 is south of the modern Polar Frontal Zone and is bathed by the Antarctic Circumpolar Current (Table 1; Fig. 1). With a relatively low resolution (100 kyr), the geochemical record spans the past ~27 Myr (Billups & Schrag, 2002). Mg/Ca ratios (and $\delta^{18}O$ values) were measured on *Cibicidoides mundulus* and *Cibicidoides wuellerstorfi*. A ~0.2 mmol mol^{-1} Mg/Ca species offset was determined, and *C. mundulus* Mg/Ca ratios were adjusted (by −0.2 mmol mol^{-1}) to those of *C. wuellerstorfi*.

Since the publication of the Site 747 study in 2002, a number of Mg/Ca temperature calibrations have become available. The genus-specific *Cibicidoides* calibration of Lear *et al.* (2002) may now be used to calculate palaeotemperatures from the Site 747 *Cibicidoides* Mg/Ca ratios. Analytical methods are described in detail by Billups & Schrag (2002) and followed the methodologies of Brown & Elderfield (1996), Hastings *et al.* (1998) and Lear *et al.* (2000), which involve repeated sonication in deionized water and methanol to remove clays as well as sonication in a hot alkaline peroxide solution to oxidize organic matter. Mg/Ca ratios were measured on an inductively coupled plasma - atomic emission spectrometer (ICP-AES) following the method outlined by Schrag (1999). Analytical precision of a standard solution was better than 0.3%. Stable isotope splits were analyzed on a Finnegan Mat252 equipped with a Kiel device at Woods Hole Oceanographic Institution with a precision of 0.07 per mille for $\delta^{18}O$.

Site 1218

Located in the northern tropical equatorial Pacific at 4828 m of water depth, Site 1218 is bathed by deep Pacific water (Lear *et al.*, 2004; Table 1; Fig. 1). Since the Eocene, the site has drifted north from its more equatorial position with a subsidence of over 1000 m (Lear *et al.*, 2004). During the Eocene through early Miocene, the site was located just above the calcium carbonate compensation depth, and dissolution has removed all but the most dissolution-resistant of the planktonic foraminiferal tests from the latest Oligocene through early Miocene assemblages (Lear *et al.*, 2004).

For the Oligocene/early Miocene interval (30–20 Ma) the temporal resolution of this record is ~40–50 kyr, while across the Oligocene/Miocene boundary it is between 20–30 kyr. Lear *et al.* (2004) measured Mg/Ca ratios on *Oridorsalis umbonatus*, while *Cibicidoides* were used for the stable-isotope record. Palaeotemperatures were derived using the *O. umbonatus* calibration of Lear *et al.* (2002). The Mg/Ca cleaning procedure was detailed in Lear *et al.* (2004); it involved oxidative as well as reductive steps and analysis by a Finnegan MAT Element Sector Field Inductively Coupled Plasma – Mass Spectrometer (ICP-MS) at Rutgers University, with a long-term precision of 1.2%. Stable-isotope ratios were measured at the Southampton Oceanography Centre using a Europa Geo 20-20 mass spectrometer with an analytical precision of better than 0.1 per mille for $\delta^{18}O$.

Table 1. Site summary indicating location, modern water depth, palaeodepth, time slice and source of data

Site	Longitude	Latitude	Water depth (m)	Palaeodepth (m)	Time Slice	Reference
747	76.5°E	54.5°S	1695	nd	Neogene	1
1218	135.5°W	9°N	4828	3700–4300	Eo–early Mio	2
1171	149°E	48.3°S	2150	1600	Mid-Miocene	3
926	42.9°W	3.7°N	3598	3698	Late Miocene	4
806	159.5°E	0.5°N	2521	2520	Late Miocene	4
1088	15°E	42°S	2082	nd	Late Miocene	5

nd = not determined, Eo = Eocene, Mio = Miocene. References: 1, Billups & Schrag (2002); 2, Lear *et al.* (2004); 3, Shevenell *et al.* (2004); 4, Lear *et al.* (2003); 5, this study.

Site 1171

Although the major focus of this synthesis is on Miocene deep-water temperatures, the planktonic foraminiferal Mg/Ca record from Site 1171 offers a temperature record spanning the mid-Miocene interval of Antarctic ice sheet growth at orbital scale resolution (Shevenell *et al.*, 2004); hence it is included here. Site 1171 is located at 2150 m water depth within the Antarctic Circumpolar Current, in the sub-Antarctic southwest Pacific (Table 1; Fig. 1). Mg/Ca ratios were measured on the mixed layer dweller *Globigerina bulloides*, for which a Mg/Ca-temperature calibration has been established (Mashiotta *et al.*, 1999). Analytical methods were detailed in Shevenell *et al.* (2004) and involved oxidative as well as reductive cleaning steps and analysis via ICP-MS, with a precision of 0.7%. Stable-isotope analyses were conducted on separate picks of *G. bulloides* and measured on a Finnegan Mat 251 with a precision of 0.1 per mille for $\delta^{18}O$.

Sites 926 and 806

Sites 926 and 806 were the subjects of a study to trace deep water masses in the Atlantic since the late Miocene (Lear *et al.*, 2003; Table 1; Fig. 1). Site 926 is located in the equatorial Atlantic on Ceara Rise at 3598 m water depth. It lies above the modern mixing zone of North Atlantic Deep Water and Antarctic Bottom Water. Conversely, Site 806 is located in the western equatorial Pacific at 2521 m water depth and should not be sensitive to changes in deep water masses. As detailed by Lear *et al.* (2003), Mg/Ca ratios were measured on *C. wuellerstorfi*, *C. mundulus* and *O. umbonatus* using species-specific calibrations to derive palaeotemperatures. Samples were cleaned using oxidative and reductive protocols and analyzed using ICP-MS with an analytical precision of 1.5–1.6%. Stable-isotope records were taken from the literature and augmented by additional analyses (*Cibicidoides*) conducted at Rutgers University using a VG Optima and at the University of Cambridge using a VG Prism with an analytical precision of better than 0.08 per mille for $\delta^{18}O$.

Site 1088

New benthic foraminiferal (*C. wuellerstorfi* and *C. mundulus*) Mg/Ca data were generated at intermediate water Site 1088 (Atlantic sector of the Southern Ocean; Fig. 1; Table 1). Site 1088, which is located on Agulhas Ridge at about 2000 m water depth, lies in the modern mixing zone of Upper Circumpolar Deep Water and North Atlantic Deep Water. Late Neogene benthic foraminiferal stable-isotope records have illustrated the sensitivity of this region to bottom-water circulation changes from the late Miocene through the early Pliocene (Billups, 2002). Thus, Mg/Ca records were generated with the aim of providing additional constrains on intermediate- and deep-water mass properties during this interval of time.

Geochemical cleaning procedures for the Site 1088 record include several sonication steps in deionized water and methanol to remove fine clays, sonication in NH_4Cl to remove exchangeable ions, sonication in hot, buffered peroxide to oxidize organic matter, and a weak acid leach. Samples were then dissolved in 2% nitric acid (trace metal grade) and analyzed by ICP-AES at the University of South Florida, St. Petersburg. Matrix effects were constrained using standards of varying Mg/Ca ratios; analytical precision (the relative standard deviation of repeat measurements of a reference solution) was better than 0.5 %. Palaeotemperatures were calculated using the genus-specific *Cibicidoides* calibration of Lear *et al.* (2002) after adjusting *C. mundulus* Mg/Ca ratios as noted above for Site 747. Stable isotope samples were analyzed at the University of California Santa Cruz using a VG Optima with a precision of better than 0.08 per mille for $\delta^{18}O$.

Age models

Comparison of the different geochemical records necessitates a common temporal reference framework. The Neogene time-scale has recently undergone significant revision with incorporation of orbitally tuned ages for biostratigraphic and magnetostratigraphic datums (Gradstein *et al.*, 2004; Ogg *et al.*, 2008). Two of the records presented here have orbitally tuned ages (Sites 1218 and 926) and are therefore already consistent with the Gradstein *et al.* (2004) time scale. No adjustments have been made to the originally published ages.

The published age model of the Site 1171 Mg/Ca record is based on biostratigraphic, magnetostratigraphic and stable-isotope control points reported on the Berggren *et al.* (1995) time scale (Shevenell *et al.*, 2004). Orbitally tuned ages, however, were recently derived for a 1-Myr long mid-Miocene portion of Site 1171 (Holbourn *et al.*, 2005), and this age model is adopted here. To derive ages outside the orbitally tuned section, constant sedimentation rates were assumed. It is noted that the ages outside of the 13.2 to 14.3 Ma tuned interval are likely to change as orbitally tuned ages become available for the entire record.

Sites 747, 806 and 1088 do not have orbitally tuned age models. For Site 747, ages were updated to the Berggren *et al.* (1995) time scale by Billups & Schrag (2002). This age model was revised further by using orbitally tuned ages of Miocene glacial events Mi 1 (23.0 Ma), Mi 3 (13.9 Ma), and Mi 4 (13.2 Ma), which are readily identifiable in the Site 747 $\delta^{18}O$ record, as depth-age control points (Shackleton *et al.*, 2000; Holbourn *et al.*, 2005, Westernhold *et al.*, 2005 respectively) to bring the ages in line with the Gradstein *et al.* (2004) time scale. Regarding Site 806, Lear *et al.* (2003) assigned orbitally tuned ages to the original shipboard biostratigraphic datums, so this record is consistent with Gradstein *et al.* (2004). In the case of Site 1088, Billups (2002) revised the age model using the benthic foraminiferal $\delta^{13}C$ record and tuned ages for the onset (7.6 Ma) and termination (6.6 Ma) of the late Miocene carbon-isotope shift (from Hodell *et al.*, 2001). As additional control points, the depth of the epoch boundaries are correlated with the corresponding ages on the Gradstein *et al.* (2004) time scale.

Although the geochemical records compiled here are consistent with Gradstein *et al.* (2004), uncertainties remain that are related to the relatively coarse temporal resolution of the records (e.g. Sites 926, 747, 806, 1088). Therefore, to minimize the impact of age model uncertainties on data interpretations, the discussion of temporal changes in deep-sea temperatures through time is limited to averages within individual Epochs (e.g. early, middle, late Miocene).

RESULTS

Neogene overview

A benthic foraminiferal Mg-derived palaeotemperature record that spans the Neogene, albeit at low temporal resolution, was published in 2002 by Billups & Schrag (Fig. 3). The $\delta^{18}O$ record (Fig. 3A)

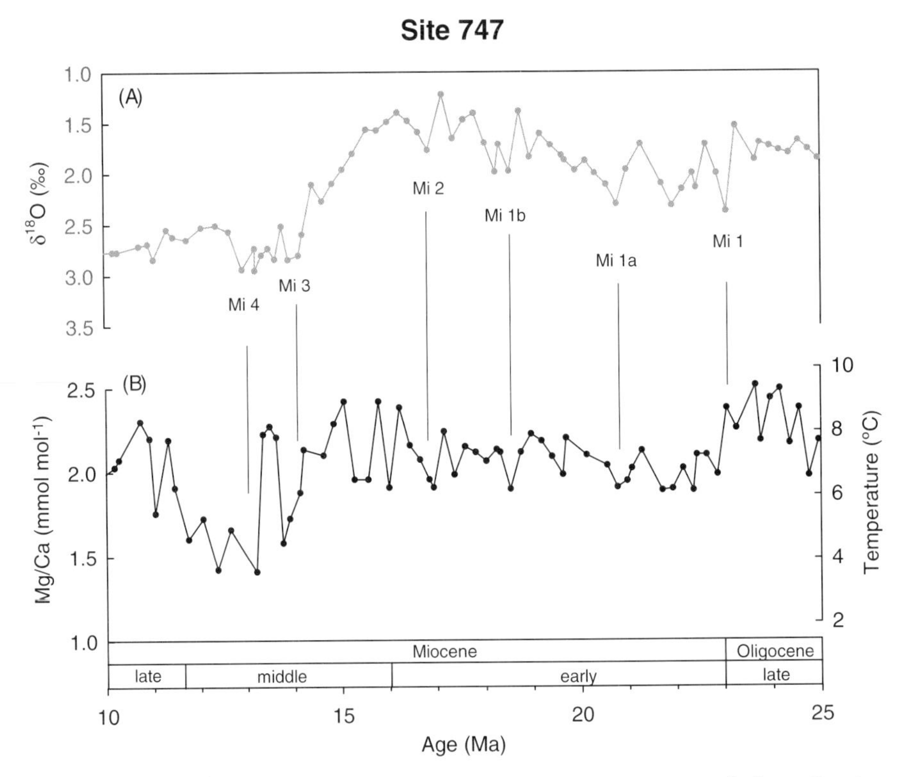

Fig. 3. (A) Benthic foraminiferal $\delta^{18}O$ and (B) Mg/Ca-derived palaeotemperature records from Southern Ocean Site 747 (Billups & Schrag, 2002). Note that the left hand y-axis in (B) indicates the Mg/Ca ratios employed to calculate the palaeotemperatures using the *Cibicidoides* equation of Lear *et al.* (2002). So-called Miocene glaciation events (Mi events after Miller *et al.*, 1991) are labelled.

displays the well-known increase at the Oligocene/Miocene boundary (the Mi1 Miocene glacial event defined in 1991 by Miller *et al.*), a general decrease toward the late early Miocene climatic optimum, and a step-wise increase during the mid-Miocene, signifying the second major ice growth phase on Antarctica.

Mg-derived palaeotemperatures (Fig. 3B) are relatively warm during the late Oligocene, cool at the Oligocene/Miocene boundary, increase by ~2 °C toward the late early Miocene climatic optimum, and then drop by 3–4 °C during the mid-Miocene, contemporaneously with increasing $\delta^{18}O$ values. These palaeotemperature changes are consistent with the conclusions of Shackleton & Kennett (1975), drawn from co-varying planktonic and benthic foraminiferal $\delta^{18}O$ records.

At the smaller scale, it becomes apparent that at the Oligocene/Miocene boundary, minimum temperatures are not reached until after the $\delta^{18}O$ maximum (Fig. 3B and A respectively). In fact, the $\delta^{18}O$ maximum is accompanied by relatively warm temperatures. The smaller-scale early Miocene glaciations (Mi1a and Mi1b), however, are concurrent with cooling events of ~1 °C. At the end of the middle Miocene, there is a brief but pronounced return to warm temperatures, and temperatures increase again during the late Miocene. As discussed by Billups & Schrag (2002), when the $\delta^{18}O$ of seawater is derived from these temperatures and the corresponding foraminiferal $\delta^{18}O$ values, excellent agreement is observed between increases in the $\delta^{18}O$ of seawater and sequence stratigraphic boundaries in the Haq *et al.* (1987) sea-level curve.

Oligocene/Miocene boundary

The benthic foraminiferal Mg/Ca record of Lear *et al.* (2004) from the eastern equatorial Pacific Site 1218 provides a highly resolved palaeotemperature history across the Oligocene/Miocene boundary (Fig. 4). As at Site 747 (discussed above), here too, $\delta^{18}O$ values are relatively low during the late Oligocene and increase rapidly at the Oligocene/Miocene boundary (Fig. 4A). However, Mg/Ca-derived palaeotemperatures do not show a longer-term increase toward the Oligocene/Miocene boundary; rather, the boundary event is marked by a decrease in Mg/Ca ratios. This appears to be part of a series of warming/cooling cycles that distinguish the late Oligocene (Fig. 4B). Temperatures cool by ~2 °C, which is equivalent to the magnitude of the temperature change observed at Site 747.

Lear *et al.* (2004) highlighted that in the deep eastern equatorial Pacific, minimum temperatures are reached prior to the $\delta^{18}O$ maximum, and that temperatures were already increasing when $\delta^{18}O$ values reached a maximum (Fig. 4C provides an expanded view of the boundary). These particular findings are opposite of those recorded at Site 747, where the $\delta^{18}O$ maximum precedes the temperature minimum (Fig. 3). It may well be that regional water-mass effects are important in determining the specific relationship between bottom-water temperatures and the $\delta^{18}O$ of the benthic foraminifera. For example, Site 747 lies at intermediate water depths in the Southern Ocean, whereas Site 1218 is a deep water site in the eastern Pacific. Alternatively, the true $\delta^{18}O$-temperature relationship may not have been captured by the relatively low temporal resolution record from Site 747.

Middle Miocene

Shevenell *et al.* (2004) published a high-resolution planktonic foraminiferal Mg/Ca-derived palaeotemperature record from sub-Antarctic Pacific Site 1171 that details the mid-Miocene transition of surface ocean cooling and ice growth on orbital time scales (Fig. 5). Although this record reflects surface-water changes, and the main focus of the present study is the evolution of bottom waters, it is included here because to-date, it is the only temperature record that shows this important climatic transition in any detail. Accordingly, during the interval of ice growth, surface waters cooled in a stepwise fashion by 6–7 °C (Fig. 5A and B). The overall magnitude is larger than in the intermediate waters at the more southwesterly Site 747, which is consistent with generally higher temperature sensitivity of the surface versus the deeper ocean. As Shevenell *et al.* (2004) discussed, this sea-surface temperature record reveals orbitally (eccentricity) forced temperature variations preceding benthic foraminiferal $\delta^{18}O$ values, and hence Antarctic ice-volume change. An expanded view focusing on the mid-Miocene interval of ice growth (Fig. 5C) illustrates how temperatures reached minimum values approximately 100 kyr before $\delta^{18}O$ values reached their maximum. Shevenell *et al.* (2004) proposed that ocean circulation changes played a major role in mid-Miocene global climate cooling, rather than greenhouse gas forcing.

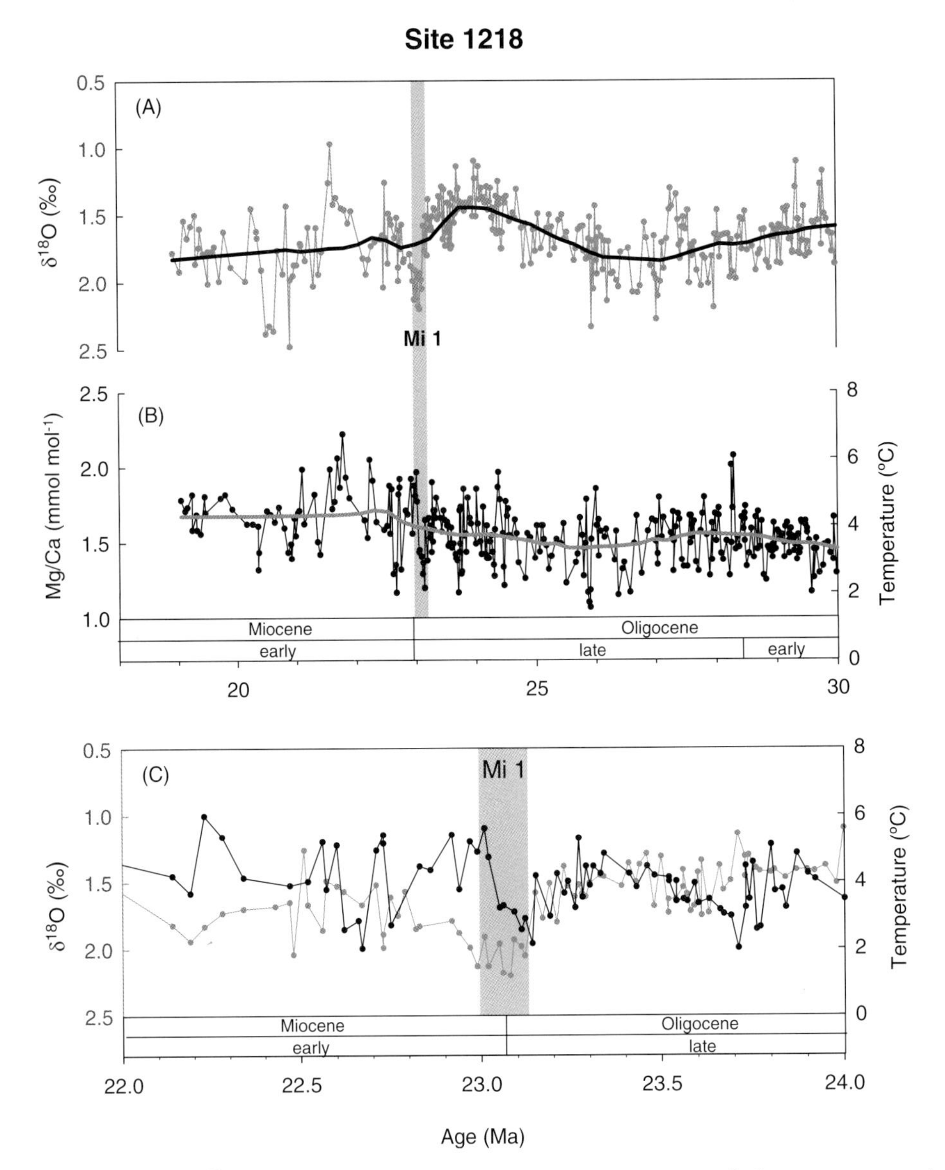

Fig. 4. (A) Benthic foraminiferal $\delta^{18}O$ and (B) Mg/Ca-derived palaeotemperature records from eastern tropical Pacific Site 1218 (Lear *et al.*, 2004). (C) An expanded view of the Oligocene/Miocene climate transition. Note that the left hand y-axis in (B) indicates the *Oridorsalis* Mg/Ca ratios used to calculate the palaeotemperatures. Records were smoothed using a weighted (10% sampling portion) Gaussian fit (heavy lines).

Late Miocene

At South Atlantic Site 1088 (Fig. 1; Table 1), results from Mg/Ca analyses presented here indicate an increase in ratios beginning at about 10 Ma, with a maximum at ~9 Ma and a decrease thereafter until ~7 Ma, whilst foraminiferal $\delta^{18}O$ values remain relatively constant (Fig. 6B and A, respectively). Mg/Ca ratios then fluctuate about a relatively constant mean during the latest Miocene through Pliocene. The increase in foraminiferal $\delta^{18}O$ values during the late Pliocene, signifying growth of ice in the Northern Hemisphere (Fig. 6A), is not accompanied by a decrease in Mg/Ca ratios that would indicate concomitant cooling of the intermediate water mass with the Northern Hemisphere (Fig. 6B).

Continuous Mg-derived palaeotemperature records for late Miocene through Pliocene sections in the western tropical Atlantic (Ocean Drilling Program Site 926) and western tropical Pacific

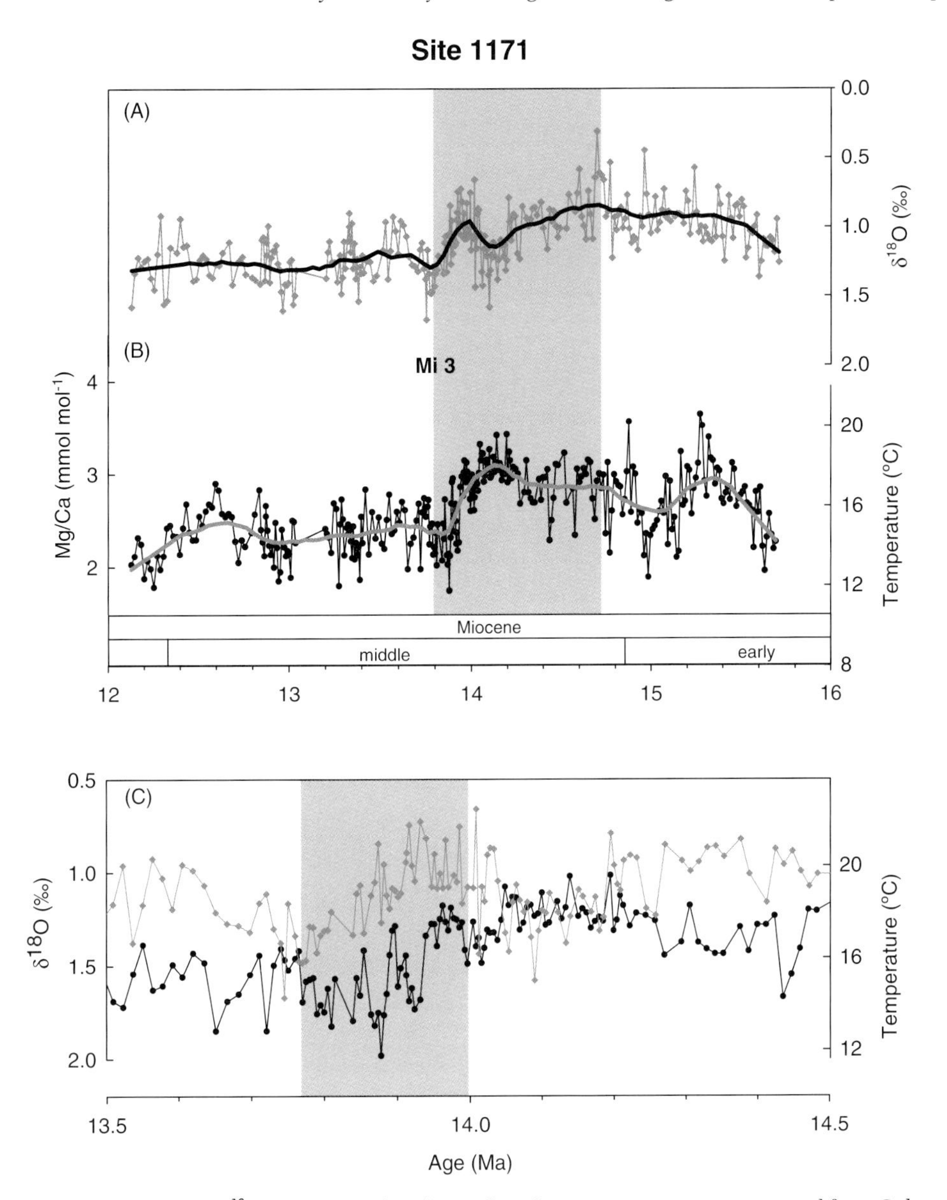

Fig. 5. (A) Planktonic foraminiferal $\delta^{18}O$ and (B) Mg/Ca-derived surface-water temperature record from Subantarctic Pacific Site 1171 (Shevenell *et al.*, 2004). (C) An expanded view of the middle Miocene climate transition. Note that the left hand y-axis in (B) indicates the *Globigerina bulloides* Mg/Ca ratios used to calculate the palaeotemperatures. Records were smoothed using a weighted (10% sampling portion) Gaussian fit (heavy lines). Grey shading highlights the interval of $\delta^{18}O$ increase and temperature decrease.

(Site 806) were published by Lear *et al.* (2003). Together with the South Atlantic data generated for this study, the Mg/Ca-derived palaeotemperature records provide a spatial view beginning in the tropical Atlantic (Fig. 7A), and extending through the South Atlantic (Fig. 7B) and into the western equatorial Pacific (Fig. 7C). Lear *et al.* (2003) noted that in the tropical Atlantic, palaeotemperatures were cool in comparison to the Pacific Ocean before ~11 Ma, and again between ~8.5 and 6 Ma. In the interim, Atlantic temperatures first increased by 4 °C, and then decreased by the same amount with maxima slightly warmer than Pacific deep waters. The Mg/Ca data from the South Atlantic (Site 1088) agrees with the findings from Site 926: a warming of ~4 °C is apparent between 10 Ma and 7 Ma. Benthic foraminiferal $\delta^{18}O$ records from each site (Fig. 7A, B and C, respectively) do not indicate any major climatic changes during this interval of time.

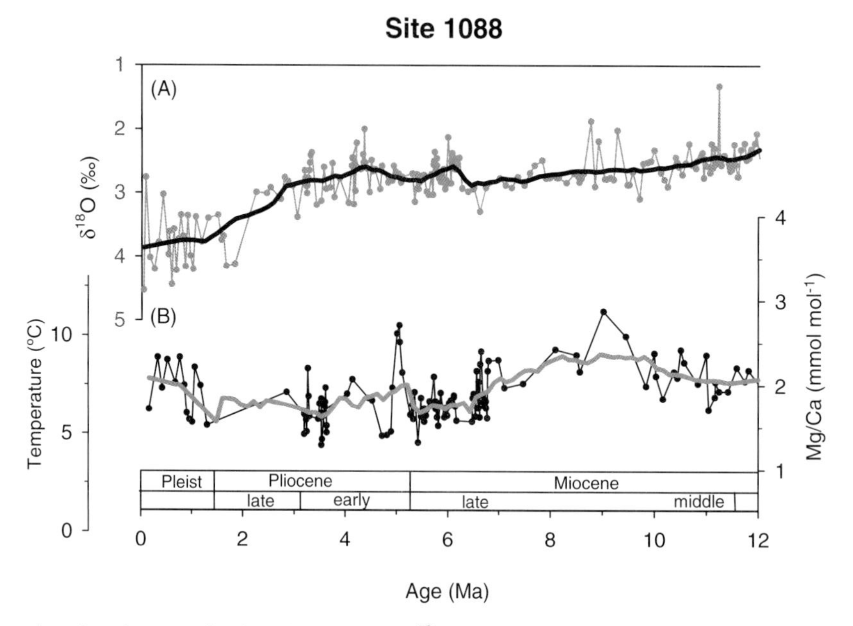

Fig. 6. (A) Site 1088 benthic foraminiferal (*Cibicidoides*) $\delta^{18}O$ (Billups, 2002) and (B) Mg/Ca data with corresponding temperature estimates (this study). Records were smoothed using a weighted (10% sampling portion) Gaussian fit (heavy lines).

DISCUSSION

Uncertainties in palaeotemperature reconstructions

Changes in seawater geochemistry

On the long time scale of interest in this review, consideration of seawater Mg/Ca ratios becomes important. Although there are large discrepancies, the available data suggest that seawater ratios may have been as much as 2.5 mol mol^{-1} lower than modern during the early Neogene (Fig. 2). Assuming that the partitioning of Mg into foraminiferal calcite from variable seawater ratios is a linear function, Mg-derived temperatures using modern seawater ratios may underestimate palaeotemperatures by a factor of 2 (e.g. equation 1, Lear *et al.*, 2000). Thus, allowing for lower seawater Mg/Ca ratios in the past results in warmer palaeotemperature estimates with the magnitude of the discrepancy decreasing toward the modern. However, these uncertainties do not call into question the direction of change; the early Miocene remains an interval of warmth in comparison to today, although how much warmer than the modern is equivocal. In the late Miocene time slice, on the other hand, the site-to-site comparison of averaged palaeotemperatures is robust, as all sites would be equally affected by changes in seawater Mg/Ca ratios.

Recent studies have demonstrated that benthic foraminiferal Mg/Ca ratios are offset from temperature calibrations when tests precipitate in waters with low carbonate-ion concentration (Dekens *et al.*, 2002; Martin *et al.*, 2002; Fehrenbacher *et al.*, 2006). This effect is different from post-depositional carbonate dissolution (Martin *et al.*, 2002; Lear *et al.*, 2004; Elderfield *et al.*, 2006). It poses an additional uncertainty to temperature reconstructions, in particular if the regional deep-water carbonate-ion concentration changed through time. At ~3700–4300 m palaeodepth, benthic foraminiferal Mg/Ca ratios from Pacific Site 1218 may be the most severely affected by contemporaneous CO_3^{-2} undersaturation (and post-depositional dissolution).

Furthermore, benthic foraminiferal Mg/Ca ratios from the Atlantic sites (926 and 1088) spanning the late Miocene may have been influenced by temporal changes in the carbonate-ion concentration. The late Miocene is known as a period of intense changes in carbonate preservation (King *et al.*, 1997). Specifically, the end of the carbonate crash in the Pacific at about 10.5 Ma (Lyle *et al.*, 1995) corresponds to an increase in carbonate preservation in the Atlantic, which can be explained by enhanced flow of relatively nutrient-depleted (CO_3^{-2} enriched) northern-sourced deep water in the Atlantic (Nisancioglu *et al.*, 2003). Thus an increase in the Mg/Ca ratios, as observed in the

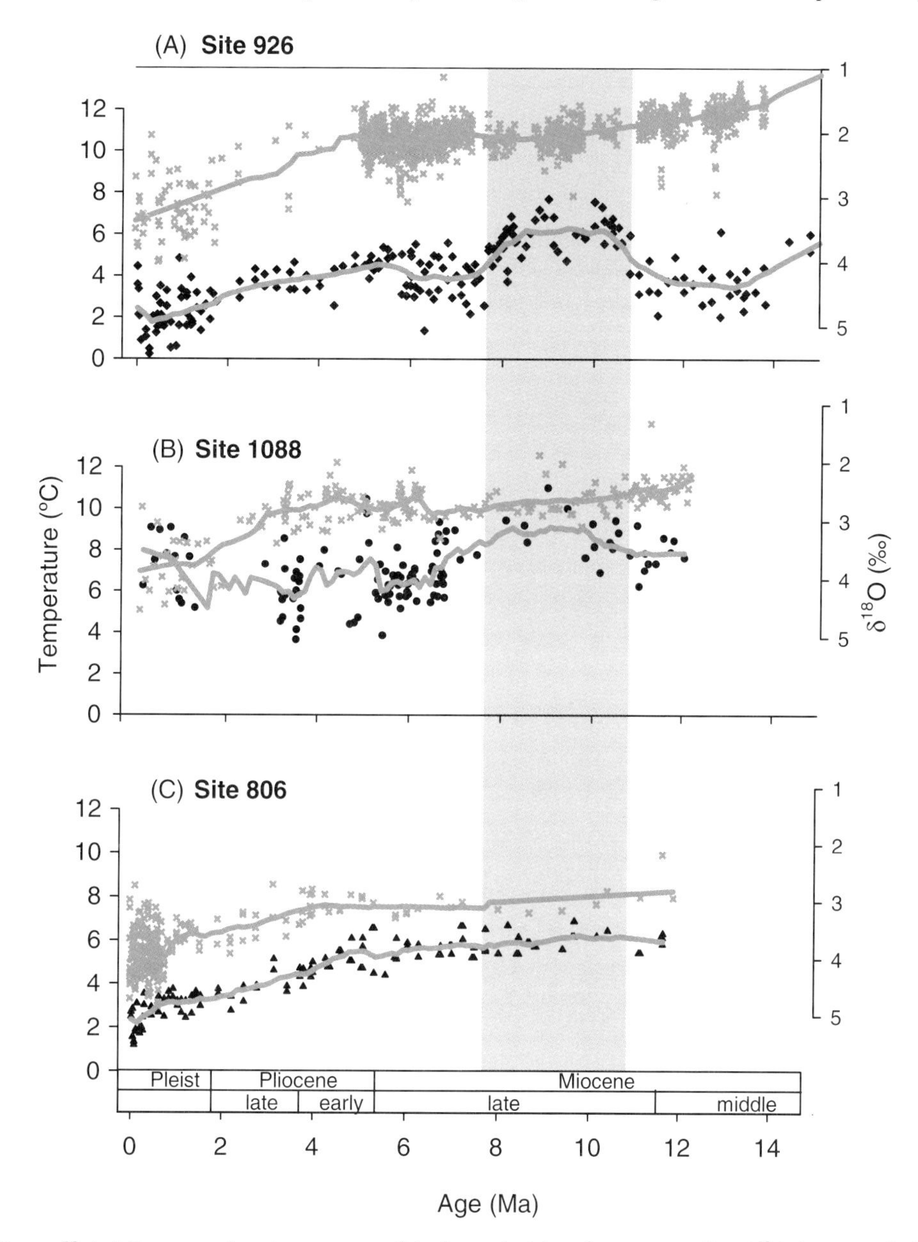

Fig. 7. Comparison of late Miocene palaeotemperature (black symbols) and corresponding $\delta^{18}O$ (grey symbols) records from the: (A) North Atlantic; (B) South Atlantic; and (C) Pacific. Records were smoothed using a weighted (10% sampling portion) Gaussian fit (heavy lines). Grey shading highlights an apparent warming of deep waters in the Atlantic between 11 Ma and 8 Ma.

two Atlantic sites at around 11 Ma, may partially reflect an increase in the CO_3^{-2} concentration associated with new deep-water circulation patterns. Although studies are emerging that seek to quantify the effect of carbonate-ion concentration on foraminiferal Mg/Ca ratios and palaeotemperatures (Dekens *et al.*, 2002; Elderfield *et al.*, 2006), little is known about the changing carbonate-ion concentrations in deep water through time (e.g. Tyrrell & Zeebe, 2004). Hence, at the magnitude of the uncertainty on the temperature reconstructions cannot be estimated.

Analytical limitations

The records compiled in this study were constructed using different cleaning methods (e.g. oxidative only, oxidative and reductive, with or without acid leaching) and analyzed using different techniques (ICP-AES versus ICP-MS). Recent studies have elucidated that Mg/Ca ratios of tests cleaned with a reductive step are lower than those cleaned using only the oxidative step (Rosenthal *et al.*, 2002; Elderfield *et al.*, 2006). Methodological differences among laboratories may introduce an uncertainty of about 2–3 °C (e.g. Rosenthal *et al.*, 2002). Consequently, only a more general discussion of similarities and differences in long-term trends among the sites will be presented here.

Palaeotemperatures through time

Modern ocean temperatures at each of the deep and intermediate water sites (Fig. 8A) reflect the relatively warm (2.2 °C) North Atlantic Deep Water bathing tropical Atlantic Site 926, relatively warm intermediate water at South Atlantic Site 1088 (2.6 °C), relatively cold circumpolar waters at Site 747 (1.6 °C), and colder waters filling the deep western (2 °C) and eastern Pacific (1 °C). In the sub-Antarctic southwest Pacific at Site 1171, surface-water temperatures have a relatively large seasonal range (8–12 °C), reflecting the position of the sub-Antarctic frontal zone (Shevenell *et al.*, 2004; Fig. 8A).

At all sites, the reconstructed late Oligocene through Miocene palaeotemperatures are higher than the temperatures in the modern ocean,

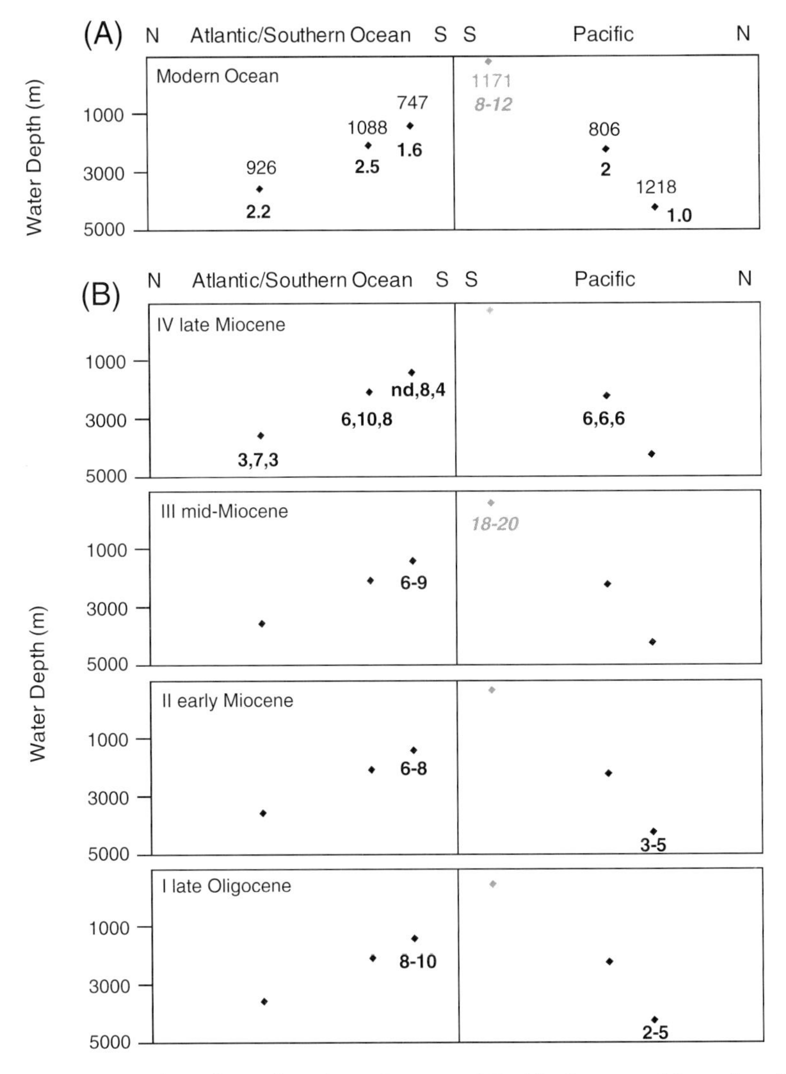

Fig. 8. (A) Relative site locations in the Atlantic/Southern Ocean and Pacific Oceans with modern bottom-water temperatures (°C; Levitus & Boyer 1994). Note that Site 1171 reflects surface water. (B) Palaeotemperature reconstructions averaged for the: (I) late Oligocene; (II) early Miocene; (III) mid-Miocene climatic optimum; (IV) late Miocene. For the late Miocene time slice (IV), the three temperatures, from left to right, reflect the average temperature after, during and prior to (in that order) the late Miocene warming event.

regardless of the specific time slice (Fig. 8B I – IV). The least deviation, with respect to the modern, is recorded in the deep eastern Pacific where late Oligocene and early Miocene temperatures are at most 4 °C warmer than today, assuming no temporal change in seawater Mg/Ca ratios. In contrast, Southern Ocean (Site 747) water temperatures are up to 8 °C warmer than today during these intervals of time (Fig. 8B). The mid-Miocene climatic optimum, which is recorded at Sites 747 and 1171, shows warmer temperatures by 7–10 °C. The magnitude of the subsequent cooling, about 6 °C at Site 747 and 7–8 °C at Site 1171, is also comparable (not shown).

For the late Miocene (Fig. 8B IV), the spatial representation of deep- and intermediate-water temperatures is reasonably good. In contrast to the modern Atlantic-to-Pacific temperature gradient defined by a colder Pacific, during the late Miocene, the western Pacific appears to have remained relatively warm. Only during the late Miocene temperature maximum do the Atlantic and Southern Ocean sites approach, or exceed, temperatures recorded in the Pacific. Lear *et al.* (2003) proposed that the relatively cool temperatures in the Atlantic flanking the 11–8 Ma warm excursion reflect the formation of cool North Atlantic Deep Water before the closure of the Central American Seaway diverted warmer and more saline waters to the source regions of deep water formation. The authors discussed that this interpretation reconciles benthic foraminiferal carbon-isotope records (a proxy for changes in the relative nutrient content of a water mass) and geophysical evidence (a seismic reflector indicating sediment erosion), which both suggest enhanced North Atlantic Deep Water formation before and after the temperature maximum of 11–8.5 Ma.

An alternate view emerges if it is assumed that this late Miocene Mg/Ca maximum at the Atlantic sites reflects a temporary incursion of warm waters, rather than a relatively high flux of cooler water before and after the event. The begin of the warming event at about 11 Ma is consistent with the timing of the closure of the eastern Mediterranean, which would result in an increase in relatively warm and saline deep- to intermediate-water flowing westward into the North Atlantic (Woodruff & Savin, 1989, 1991), where it may have contributed to Northern Component Deep Water formation (Wright *et al.*, 1992). In the modern ocean, Mediterranean overflow contributes significantly to the temperature and salinity of the North Atlantic Deep Water (e.g. Schmitz & McCartney, 1993). It may be that with continued constriction of the Straits of Gibraltar, the overflow of warm intermediate water lessened and allowed Northern Component Deep Water to cool after ~8 Ma. Temperature proxy records from the higher latitudes of the North Atlantic may help to clarify the mechanisms (e.g. warmer source water or enhanced relative flux) responsible for the late Miocene deep-water warming event in the tropical Atlantic Ocean.

A scenario of increased relative flux of North Atlantic Deep Water between ~11 Ma and 8.5 Ma (as oppose to assuming a temperature change only) would appear to conflict with the benthic foraminiferal $\delta^{13}C$ evidence for deep water circulation changes. The difference between Atlantic and Pacific $\delta^{13}C$ records is commonly used to infer the relative flux of nutrient-depleted North Atlantic Deep Water in the Atlantic, but this approach assumes that the $\delta^{13}C$ value of the source water remained constant through time (for a recent review see Hodell *et al.*, 2006). If the reduced Atlantic-to-Pacific $\delta^{13}C$ gradient between ~11 Ma and 8.5 Ma, which was originally interpreted to indicated enhanced North Atlantic Deep Water flux (e.g. Wright *et al.*, 1991 and discussed by Lear *et al.*, 2003, see above) were brought about by an increase in the $\delta^{13}C$ value of the Pacific deep water, then the relative flux of North Atlantic Deep Water in the Atlantic cannot be resolved using this approach. Other proxies for deep water flow, such as neodymium isotopes (e.g. Frank *et al.*, 2002), are needed to constrain deep water circulation changes better during this interval of time.

Lastly, as noted above, the beginning of increasing Mg/Ca-derived temperatures in the Atlantic at ~11 Ma is close in timing with the beginning of enhanced carbonate preservation in the Atlantic (King *et al.*, 1997). Recent modelling experiments by Nisancioglu *et al.* (2003) illustrate that the change in carbonate preservation can be explained by an increase in the southward flux of North Atlantic Deep Water in the Atlantic Ocean, once the Central American Seaway shoals to a depth that obstructs deeper water outflow to the Pacific. Younger, relatively nutrient-depleted North Atlantic Deep Water is characterized by a higher carbonate-ion concentration than older, nutrient-rich southern-sourced deep waters, at least in the modern ocean, so it cannot be ruled out that a portion of the apparent late Miocene warming may indeed reflect enhanced Mg incorporation in response to an increase in the carbonate-ion concentration of bottom water. If correct, this factor would not change the interpretation of the geochemical records with respect to ocean circulation changes, but it would affect the absolute magnitude of the warming associated with the northern-sourced water mass.

CONCLUSION

This Late Oligocene through Miocene Mg/Ca synthesis provides a first order view of the evolution through time of deep, intermediate, and for the middle Miocene, surface-water temperatures. The Mg-derived palaeotemperatures support the Miocene climate framework provided by the earlier, $\delta^{18}O$ and faunal-based reconstructions of Savin *et al.* (1985) and Woodruff *et al.* (1985). The early Miocene was a time of relatively warmth, and it is well after the mid-Miocene that deep and intermediate water temperatures approached a modern distribution.

Although potentially increasing seawater Mg/Ca ratios cannot be accounted for, it is noted that if they were lower, as proposed in some studies, palaeotemperature reconstructions would be even warmer with respect to today. For the late Miocene, Mg-derived palaeotemperatures trace deep and intermediate water circulation changes. However, whether the increase in late Miocene Mg/Ca ratios reflects a warming, enhanced Mg incorporation associated with higher bottom water CO_3^{-2} levels, or a combination of both, cannot currently be resolved.

ACKNOWLEDGEMENTS

We would like to thank Torsten Bickert and an anonymous reviewer for thoughtful and constructive reviews that were invaluable in revising the manuscript. We are also grateful to Carrie Lear for providing detailed suggestions to further improve the manuscript. K. Billups acknowledges the donors of the Petroleum Research Fund, administered by the American Chemical Society ACS. This research used samples by the Ocean Drilling Program (ODP). ODP is sponsored by the US National Science Foundation (NSF) and participating countries under the management of Joint Oceanographic Institutions (JOI) Inc.

REFERENCES

Anand, P., Elderfield, H. and **Conte, M.H.** (2003) Calibration of Mg/Ca thermometry in planktonic foraminifera from a sediment trap time series. *Paleoceanography*, **18**, doi: 10.1029/2002PA000846.

Baker, P. A., Gieskes, M. and **Elderfield, H.** (1982) Diagenesis of carbonates in deep-sea sediments: evidence from Sr/Ca ratios and interstitial dissolved Sr^{2+} data. *J. Sed. Petrol.*, **52**, 71–82.

Barker, S., Greaves, M. and **Elderfield, H.** (2003) A study of cleaning procedures used for foraminiferal Mg/Ca paleothermometry. *Geochem. Geophys. Geosyst.*, **4**, 10.1029/2003GC000559.

Berggren, W.A., Kent, D.V., Swisher, C.C. III and **Aubry, M.P.** (1995) A revised Cenozoic geochronology and Chronostratigraphy, In *Geochronology, Time Scales and Stratigraphic Correlation: Framework for and Historical Geology* (Eds W. A. Berggren, D.V. Kent, M.-P. Aubry and J. Hardenbol). *SEPM Spec. Publ.*, **54**, 138–144.

Billups, K. (2002) Late Miocene through early Pliocene deep water circulation and climate change viewed from the subantarctic Southern Ocean. *Palaeogeogr. Palaeoclimatol. Palaeoecol.*, **185**, 287–307.

Billups, K. and **Schrag, D.P.** (2002) Paleotemperatures and ice-volume of the past 27 myr revisited with paired Mg/Ca and stable isotope measurements on benthic foraminifera, *Paleoceanography*, **17**, 10.1029/2000PA000567.

Billups, K. and **Schrag, D.P.** (2003) Application of benthic foraminiferal Mg/Ca ratios to questions of early Cenozoic climate change. *Earth Planet. Sci. Lett.*, **209**, 181–195.

Broecker, W.S. and **Peng, T.-H.** (1982) *Tracers in the Sea.* Lamont-Doherty Earth Observatory, Palisades, NY.

Brown, S.J. and **Elderfield, H.** (1996) Variations in Mg/Ca and Sr/Ca ratios in planktonic foraminifera caused by postdepositional dissolution: Evidence of shallow Mg-dependent dissolution. *Paleoceanography*, **11**, 543–551.

Carpenter, S.J. and **Lohmann, K.C.** (1992) Sr/Mg ratios of modern marine calcite: empirical indicators of ocean chemistry and precipitation rate. *Geochim. Cosmochim. Acta*, **56**, 1837–1849.

Dekens, P.S., Lea, D.W., Pak, D.K. and **Spero, H.J.** (2002) Core top calibration of Mg/Ca in tropical foraminifera: refining paleotemperature estimation. *Geochem. Geophys. Geosyst.*, **4**, 10.1029/2001GC000200.

Elderfield, H. and **Ganssen, G.** (2000) Past temperatures and the $\delta^{18}O$ of surface ocean waters inferred from foraminiferal Mg/Ca ratios. *Nature*, **405**, 442–445.

Elderfield, H., Yu, J., Anand, P., Kiefer, T. and **Nyland, B.** (2006) Calibrations for foraminiferal Mg/Ca paleothermometry and the carbonate ion hypothesis. *Earth. Planet. Sci. Lett.*, **250**, 633–649.

Fantle, M.S. and **DePaolo, D.J.** (2006) Sr isotopes and pore fluid chemistry in carbonate sediment of the Ontong Java Plateau: calcite recrystallization rates and evidence for a rapid rise in seawater Mg over the last 10 million years. *Geochim. Cosmochim. Acta*, **70**, 3883–3904.

Fehrenbacher, J., Martin, P.A. and **Eshel, G.** (2006) Glacial deep water carbonate chemistry inferred from foraminiferal Mg/Ca: a case study from the western tropical Atlantic, *Geochem. Geophys. Geosyst.*, **7**, Q09P16.

Frank, M., Whiteley, N., Kasten, S., Hein, J.R. and **O'Nions, K.** (2002) North Atlantic Deep Water export to the

Southern ocean over the past 14 Myr: Evidence from Nd and Pb isotopes in ferromanganese crusts. *Paleoceanography*, **17**, doi: 10.1029/2000PA000606.

Gradstein, F.M., **Ogg, J.G.**, **Smith, A.G.**, **Bleeker, W.** and **Lourens, L.** (2004) A new geologic time scale, with special reference to Precambrian and Neogene. *Episodes*, **27**, 83–100.

Haq, B.U., **Hardenbol, J.** and **Vail, P.R.** (1987) Chronology of fluctuating sea levels since the Triassic. *Science*, **235**, 1156–1167.

Hardie, L.A. (1996) Secular variations in seawater chemistry: an explanation for coupled secular variation in the mineralogies of marine limestones and potash evaporates over the past 600 m.y. *Geology*, **24**, 279–283.

Hastings, D.W., **Russell, A.D.** and **Emerson, S.R.** (1998) Foraminiferal magnesium in *Globigerinoides sacculifer* as a paleotemperature proxy. *Paleoceanography*, **13**, 161–169.

Hodell, D.A., and **Venz-Curtis, K.** (2006) Late Neogene history of deepwater ventilation in the Southern Ocean. *Geochem. Geophys. Geosyst.*, **7**, doi: 10.1029/2005GC001211.

Hodell, D.A., **Curtis, J.H.**, **Sierro, F.J.** and **Raymo, M.E.** (2001) Correlation of late Miocene to early Pliocene sequences between the Mediterranean and North Atlantic. *Paleoceanography*, **16**, 164–178.

Holbourn, A., **Kuhnt, W.**, **Schulz, M.** and **Erlenkeuser, H.** (2005) Impacts of orbital forcing and atmospheric carbon dioxide on Miocene ice-sheet expansion. *Nature*, **438**, 483–487.

Kennett, J.P. (Ed.) (1985) *The Miocene Ocean: Paleoceanography and Biogeography. Geol. Soc. Am. Mem.*, **163**, 337 pp.

King, T.A., **Ellis, W.G.**, **Murray, D.W.**, **Shackleton, N.J.** and **Harris, S.** (1997) Miocene evolution of carbonate sedimentation on Ceara Rise: a multivariate data/proxy approach. *Proc. ODP Sci. Res.*, **154**, 349–366.

Lea, D.W., **Machiotta, T.A.**, and **Spero, H.J.** (1999) Controls on magnesium and strontium uptake in planktonic foraminifera determined by live culturing. *Geochim. Cosmochim. Acta*, **63**, 2369–2379.

Lea, D.W., **Pak, D.K.** and **Spero, H.J.** (2000) Climate impact of late Quaternary equatorial Pacific sea surface temperatures. *Science*, **289**, 1719–1724.

Lear, C.H., **Elderfield, H.** and **Wilson, P.A.** (2000) Cenozoic deep-sea temperatures and global ice volumes from Mg/Ca in benthic foraminiferal calcite. *Science*, **287**, 269–272.

Lear, C.H., **Rosenthal, Y.** and **Slowey, N.** (2002) Benthic foraminiferal Mg/Ca paleothermometry: a revised core-top calibration. *Geochim. Cosmochim. Acta*, **66**, 3375–3387.

Lear, C.H., **Rosenthal, Y.** and **Wright, J.D.** (2003) The closing of a seaway: ocean water masses and global climate change. *Earth Planet. Sci. Lett.*, **210**, 425–436.

Lear, C.H., **Rosenthal, Y.**, **Coxall, H.K.** and **Wilson, P.A.** (2004) Late Eocene to early Miocene ice sheet dynamics and the global carbon cycle. *Paleoceanography*, **19**, doi: 10.1029/2004PA001039.

Levitus, S. and **Boyer, T.P.** (1994). *World Ocean Atlas 1994*, *4*, *Temperature*. NOAA Atlas NESDIS 4, 129 pp. National Oceanic and Atmospheric Administration, Silver Spring, MD.

Lowenstein, T.K., **Timofeff, M.N.**, **Brennan, S.T.**, **Hardie, L.A.** and **Demicco, R.V.** (2001) Oscillations in Phanerozoic seawater chemistry: Evidence from fluid inclusions. *Science*, **294**, 1086–1088.

Lyle, M., **Dadey, K.A.** and **Farrell, J.W.** (1995) The late Miocene (11-8 Ma) eastern Pacific carbonate crash: evidence for reorganization of deep-water circulation by the closure of the Panama gateway. *Proc. ODP Sci. Res.*, **138**, 821–838.

Martin, P.A., **Lea, D.W.**, **Rosenthal, Y.**, **Shackleton, N.J.**, **Sarnthein, M.** and **Papenfuss, T.** (2002) Quaternary deep sea temperature history derived from benthic foraminiferal Mg/Ca. *Earth Planet. Sci. Lett.*, **198**, 193–209.

Mashiotta, T.A., **Lea, D.W.** and **Spero, H.J.** (1999) Glacial-interglacial changes in subantarctic sea surface temperature and the ^{18}O-water using foraminiferal Mg. *Earth Planet. Sci. Lett.*, **170**, 417–432.

Miller, K.G., **Wright, J.D.** and **Fairbanks, R.G.** (1991) Unlocking the ice-house: Oligocene-Miocene oxygen isotopes, eustacy and margin erosion. *J. Geophys. Res.*, **96**, 6829–6848.

Morse, J.W. and **Bender, M.L.** (1990) Partition coefficients in calcite: examination of factors influencing the validity of experimental results and their application to natural systems, *Chem. Geol.*, **82**, 265–277.

Nisancioglu, K.H., **Raymo, M.E.** and **Stone, P.H.** (2003) Reorganization of Miocene deep water circulation in response to the shoaling of the central American Seaway. *Paleoceanography*, **18**, doi: 10.1029/2002PA000767.

Nuernberg, D.J., **Bijma, J.** and **Hemleben, C.** (1996) Assessing the reliability of magnesium in foraminiferal calcite as a proxy for water mass temperature. *Geochim. Cosmochim. Acta.*, **60**, 803–814.

Ogg, J.G., **Ogg, G.** and **Gradstein, F.M.** (2008) *The Concise Geologic Time Scale*. Cambridge University Press, Cambridge, 177 pp.

Rosenthal, Y. and **Lohmann, G.P.** (2002) Accurate estimation of sea surface temperatures using dissolution-corrected calibrations for Mg/Ca paleothermometry. *Paleoceanography*, **17**, doi: 10.1029/2001PA000749.

Rosenthal, Y., **Perron-Cashman, S.**, **Lear, C.H.**, **Baard, E.**, **Barker, S.**, **Billups, K.**, **Bryan, M.**, **Delaney, M.L.**, **deMenocal, P.B.**, **Dwyer, G.S.**, **Elderfield, H.**, **German, C.R.**, **Greaves, M.**, **Lea, D.W.**, **Machiotto, T.M.**, **Pak, D.K.**, **Ravelo, A.C.**, **Paradis, G.L.**, **Russell, A.D.**, **Schneider, R.R.**, **Scheiderich, K.**, **Stott, L.**, **Tachikawa, K.**, **Tapa, E.**, **Thunell, R.**, **Wara, M.**, **Weldeab, S.** and **Wilson, P.A.** (2004) Inter-laboratory comparison study of Mg/Ca and Sr/Ca, measurements in planktonic foraminifera for paleoceanographic research. *Geochem. Geophys. Geosyst.*, **5**, 10.1029/2003GC000650.

Savin, S., **Abel, L.**, **Barrera, E.**, **Hodell, D.**, **Keller, G.**, **Kennett, J.P.**, **Killingley, J.**, **Murphy, M.** and **Vincent, E.** (1985) The evolution of Miocene surface and near surface marine temperatures: oxygen isotope evidence. In: *The Miocene Ocean: Paleoceanography and Biogeography* (Ed. J.P. Kennett). *Geol. Soc. Am. Mem.*, **163**, 49–82.

Schmitz, Jr., W.J. and **McCartney, M.S.** (1993) On the North Atlantic circulation. *Rev. Geophys.*, **31**, 29–49.

Schrag, D.P. (1999) Rapid analysis of high-precision Sr/Ca ratios in corals and other marine carbonates. *Paleoceanography*, **14**, 97–102.

Shackleton, N.J. and **Kennett, J.P.** (1976) Paleotemperature history of the Cenozoic and the initiation of Antarctic glaciation: oxygen and carbon isotopic analyses in DSDP Sites 277, 279, and 281. *Init. Rep. Deep Sea Drilling Proj.*, **29**, 743–755.

Shackleton, N.J., **Hall, M.A.**, **Raffi, I.**, **Tauxe, L.** and **Zachos, J.** (2000) Astronomical calibration age for the Oligocene-Miocene boundary. *Geology*, **28**, 447–450.

Shevenell, A.E., **Kennett, J.P.** and **Lea, D.W.** (2004) Middle Miocene Southern Ocean cooling and Antarctic cryosphere expansion, *Science*, **305**, 1766-L1770.

Stanley, S.M. and **Hardie, L.A.** (1998) Secular oscillations in the carbonate mineralogy or reef-building and sediment producing organisms driven by tectonically forced shifts in seawater chemistry. *Palaeogeogr. Palaeoclimatol. Palaeoecol.*, **144**, 3–19.

Stott, L., **Cannariato, K.**, **Thunell, R.**, **Haug, G.H.**, **Koutavas, A.** and **Lund, S.** (2004) Decline of surface temperature and salinity in the western tropical Pacific Ocean in the Holocene epoch. *Nature*, **431**, 56–59.

Tyrell, T. and **Zeebe, R.E.** (2004) History of carbonate ion concentration over the past 100 million years. *Geochim. Cosmochim. Acta*, **68**, 3521–3530.

Westerhold, T., **Bickert, T.** and **Rohl, U.** (2005) Middle to late Miocene oxygen isotope stratigraphy of ODP Site 1085 (SE Atlantic): new constrains on Miocene climate variability and sea level fluctuations. *Palaeogeogr. Palaeoclimatol. Palaeoecol.*, **217**, 205–222.

Wilkinson, H. and **Algeo, T.J.** (1998) Sedimentary carbonate record of calcium-magnesium cycling. *Am. J. Sci.*, **289**, 1158–1194.

Woodruff, F. (1985) Changes in Miocene deep sea benthic foraminiferal distribution in the Pacific Ocean: relationship to paleoceanography. In: *The Miocene Ocean: Paleoceanography and Biogeography* (Ed. J.P. Kennett). *Geol. Soc. Am. Mem.*, **163**, 131–176.

Woodruff, F. and **Savin, S.M.** (1989) Miocene deep water oceanography. *Paleoceanography*, **4**, 87–140.

Woodruff, F. and **Savin, S.M.**, (1991) Mid-Miocene isotope stratigraphy in the deep sea: high resolution correlations, paleoclimatic cycles, and sediment preservation. *Paleoceanography*, **6**, 755–806.

Wright, J., **Miller, K.G.** and **Fairbanks, R.G.** (1991) Evolution of modern deepwater circulation: evidence from the late Miocene Southern Ocean. *Paleoceanography*, **6**, 275–290.

Wright, J., **Miller, K.G.** and **Fairbanks, R.G.** (1992) Early and middle Miocene stable isotopes: Implications for deep-water circulation and climate. *Paleoceanography*, **7**, 357–389.

Int. Assoc. Sedimentol. Spec. Publ. (2010) **42**, 17–34

Latitudinal trends in Cenozoic reef patterns and their relationship to climate

CHRISTINE PERRIN* and WOLFGANG KIESSLING†

Laboratoire des Mécanismes et Transferts en Géologie, Université Paul Sabatier, 14 avenue Edouard Belin, 31400 Toulouse & Muséum National d'Histoire Naturelle, Paléobiodiversité et Paléoenvironnements (UMR 5143 - USM 203), Département Histoire de la Terre, 8 rue Buffon, 75005 Paris, France
(E-mail: christine.perrin@lmtg.obs-mip.fr)
†*Museum für Naturkunde, Leibniz Institute for Research on Evolution and Biodiversity at the Humboldt University Berlin, 10115 Berlin, Germany*

ABSTRACT

Reefs, especially tropical coral reefs, are commonly thought to be highly sensitive to climate change. The relationship between Cenozoic reef distribution, as recorded in the PaleoReefs database, and palaeoclimatic change, as inferred from geological and geochemical proxies, has been tested. The focus is on the Oligocene–Miocene transition, but the entire pre-Pleistocene Cenozoic reef distribution was analyzed to put the results into a broader context. It is found that reef distribution patterns are not cross-correlated with palaeoclimate change. If anything, the global cooling trend from the Eocene to the Miocene is correlated with an increase of reef carbonate production and a latitudinal expansion of the reef belt, rather than the expected opposite. This reef–climate paradox is best explained by a combination of two end-member hypotheses: (1) a macroevolutionary shift led from an extrinsic control (i.e. climate-driven) on reef development during the early Palaeogene greenhouse, to an intrinsic control (i.e. biological adaptation to new oceanographic and nutrient conditions) in the late Palaeogene–Neogene icehouse; and (2) the deleterious effects of climatic cooling on reef building were made up by changes in oceanography that led to a reduction of equatorial upwelling, and an increase of habitat area of low-nutrient shallow-water settings favourable for zooxanthellate coral reef growth.

Keywords reefs, palaeoclimate, Oligocene, Miocene, greenhouse, icehouse.

INTRODUCTION

Reefs, and especially shallow-water coral reefs, are commonly seen as excellent climatic tracers, and their occurrence at any given time or place is thought to reflect tropical to subtropical conditions. Although previous studies have challenged the general relationship between the latitudinal range of reefs and palaeoclimate (Parrish, 1998; Kiessling, 2001), Cenozoic coral reefs are still used to track palaeoclimate change or palaeolatitudinal movements (Grigg, 1997; Tarduno *et al.*, 2003). In particular, as the global distribution of present-day zooxanthellate coral reefs is limited by minimum sea-surface temperatures, a positive relationship between past reef distributions and global temperature is generally expected. However, global patterns of Phanerozoic reefs have been demonstrated to result from highly complex factors, with many interactions and potential feedbacks (see reviews in Flügel & Flügel-Kahler, 1992; Kauffman & Fagerstrom, 1993; Copper, 1994; Hallock, 1997; Kiessling *et al.*, 2002).

The aim of this paper is to analyze the relationship between reef patterns and global climate over the pre-Pleistocene Cenozoic, with a particular focus on the greenhouse–icehouse transition during the Oligocene–Miocene interval. For this purpose, three aspects were analyzed: (1) temporal changes of reef abundance and global reef distribution (especially latitude); (2) latitudinal trends of reef attributes; and (3) the link between reef properties and global physico-chemical changes inferred from geochemical proxies.

Although the main focus is on the Oligocene–Miocene transition, this global approach, which necessarily examines processes working at different time-scales, requires patterns to be analyzed over the entire pre-Pleistocene Cenozoic, in order to take into account the broader temporal context of longer-term global cooling.

DATABASE AND METHODS

The principal source of information for this paper is the PaleoReefs database. This database lists palaeontological, petrographic, environmental, and geometrical attributes of Phanerozoic reefs in a palaeogeographic framework (for a detailed description of the structure and content of the PaleoReefs database see Kiessling & Flügel, 2002. For the pre-Pleistocene Cenozoic, the database currently holds 683 reef sites, based on 327 references. An individual reef site represents data from reefs of the same age, composition and environmental setting within an area of roughly 350 km^2. Present-day coordinates of each reef site are provided together with its corresponding palaeo-coordinates calculated with the aid of the plate tectonic rotation files of Golonka (2002).

The database lists all kinds of reefs, broadly defined as laterally confined structures built by the growth and/or metabolic activity of sessile benthic organisms in an aquatic environment (Kiessling, 2003). The biotic composition of the reef-building assemblages and the characteristics of any framework are reported and categorized in PaleoReefs by four main features (Kiessling & Flügel, 2002):

(1) The nature of dominant reef-builders is recorded by the main taxonomical group at the class or phylum level;
(2) Reef-building assemblages are classified into 29 different categories for the pre-Pleistocene Cenozoic reefs;
(3) The diversity of reef-building assemblages is recorded by three main categories of species richness: low diversity (<6 species); moderate diversity (6–24 species); and high diversity (≥25 species), supplemented by actual species counts, if available;
(4) The constructional reef type is categorized as true reef, reef-mound, mud-mound and biostrome. True reefs are those buildups where the reef-building organisms form a rigid framework with a syndepositional relief. Reef-mounds are formed by abundant reef-building organisms without evidence of a rigid framework but instead a matrix-supported syndepositional relief. Mud-mounds consist predominantly of carbonate mud, skeletal organisms occurring only as minor constituents. Biostromes are characterized by a dense growth of skeletal organisms but without syndepositional relief.

In this paper, the term "reef" is used in its broader sense (i.e. a carbonate buildup), encompassing true rigid-framework reefs, reef-mounds, mud-mounds and biostromes. When "reef" is employed in its strict meaning, the expressions "true reef" or "framework reefs" have been used. Although in the Cenozoic, shallow-water coral reefs strongly predominate (422 reef sites), other reef types are also common at particular times. As these other reefs (examples are bryozoan reefs throughout the Cenozoic, deep-water coral reefs in the Danian and Pliocene, foraminiferal reefs in the Eocene, and coralline algal reefs in the Miocene) probably have quite different environmental preferences than shallow-water coral reefs, they were treated separately in the analyses.

The affinity of reefs for latitudinal zones and bathymetric settings have also been separated:

(1) Reefs similar to modern tropical coral reefs: scleractinian corals (most of them inferred to have hosted photosymbionts) are either the dominant or the second most important reef-building group. Reefs grew in shallow water (usually above fair-weather wave base, but at least in the photic zone). Reefs with unknown composition have also been included in this category if seismic or well data suggest the presence of large reef complexes, which in the Cenozoic were usually constructed by corals.
(2) Reefs with uncertain affinities: all reefal structures that have a biotic composition different from (1), and either have a broad ecological tolerance (e.g. oyster reefs, coralline algal reefs) or different ecological affinities (e.g. encrusting foraminiferal reefs). Mud-rich mound structures and reefs that grew in brackish water are included here.
(3) Reefs with a preference for high latitude or deep water; here, deep-water zooxanthellate coral reefs and bryozoan mounds are combined.

As these three "ecological" kinds of reefs may have different latitudinal distributions and may exhibit distinct relationships with climate parameters, all three are considered in the analyses to avoid *a priori* biases by the sole consideration of the conventional tropical warm shallow-water coral reef.

The variations of reef sizes in time and space were also evaluated. Reef size is recorded in PaleoReefs by three parameters: reef thickness; width; and lateral extent. These measures are recorded as categorical values and, if known, also at metric scales. The measures can be used to calculate the preserved volume of each buildup and also the globally preserved reef volume, as described by Kiessling *et al.* (2000) and Kiessling (2006).

The stratigraphic assignment of Cenozoic reefs is usually not as good as would be expected, given the wealth of high-precision stratigraphic data in offshore sequences and land mammal associations, offering an era in which global and regional correlations can be resolved to a one million year level or finer (Spencer-Cervato *et al.*, 1994; Alroy, 1994). In fact, only slightly more than half of all Cenozoic reefs are reliably assigned to a single stage. In addition, the sample size within individual stages is often too low for meaningful statistical tests. Problems with sample size are especially pronounced when an attempt is made to determine the latitudinal range of reefs. A sub-epoch level stratigraphic resolution was therefore chosen as the basic stratigraphic framework for this study (Table 1).

A more detailed regular binning of 5 Myr was occasionally chosen for assessing cross-correlations with environmental parameters. All analyses excluded reefs with imprecise stratigraphic assignments (e.g. Eocene or Miocene/Pliocene), which sum up to 19% of all Cenozoic reefs. Although the regular binning scheme often results in artificial boundaries within stages, it has the advantage of increasing sample size (13 bins as compared to 11 sub-epochs) and better complies with requirements of time-series analyses. Since the last bin in this time series contains only reefs from less than 4 Myr (5.5 to 1.8 Ma, because Pleistocene reefs were excluded from the analyses), a correction factor was applied.

Since reef data in the Southern Hemisphere are scarce, the total latitudinal range of reefs is somewhat arbitrary. Therefore the focus of this study was on the northern latitudinal limit of reef growth, which is fairly well constrained through the entire Cenozoic. The study focuses on global patterns, but also separates three important regions, the: Indo-Pacific (between 90° east and 120° west longitude); Mediterranean-Tethys (between 30° west and 90° east), and Caribbean-Western Atlantic (between 120° west and 30° west).

RESULTS

Abundance and exposed volume of reefs

The assessment of reef abundance (i.e. absolute number of reef sites per time-interval) may vary according to the temporal and spatial scales at which it is estimated. When considered at the time scale of supersequences (Golonka & Kiessling, 2002), the abundance of reef ecosystems at a global scale does not show significant fluctuations from the late Palaeocene to the early Oligocene. The global number of reef sites increased markedly during the Chattian–Aquitanian supersequence, this increasing abundance being further accentuated during the Burdigalian–Serravallian supersequence. This is followed by a decrease in the total number of reefs during the Tortonian–Messinian, and then by an important drop in the Pliocene (Perrin, 2002).

When viewed at the sub-epoch level sample resolution, the overall pattern is similar but there are some noteworthy differences, especially when

Table 1. Definition of stratigraphic intervals used in this study. The time scale is based on Gradstein *et al.* (2004). The number of reef sites refer only to those with precise age assignments

Interval	Mid-point age (Ma)	Lower boundary age (Ma)	Upper boundary age (Ma)	Number of reef sites
Danian	63.6	65.5	61.7	22
Selandian-Thanetian	58.8	61.7	55.8	33
Ypresian	52.2	55.8	48.6	20
Middle Eocene	42.9	48.6	37.2	23
Priabonian	35.6	37.2	33.9	17
Rupelian	31.2	33.9	28.4	23
Chattian	25.7	28.4	23.0	55
Early Miocene	19.5	23.0	16.0	142
Middle Miocene	13.8	16.0	11.6	131
Late Miocene	8.5	11.6	5.3	108
Pliocene	3.6	5.3	1.8	58

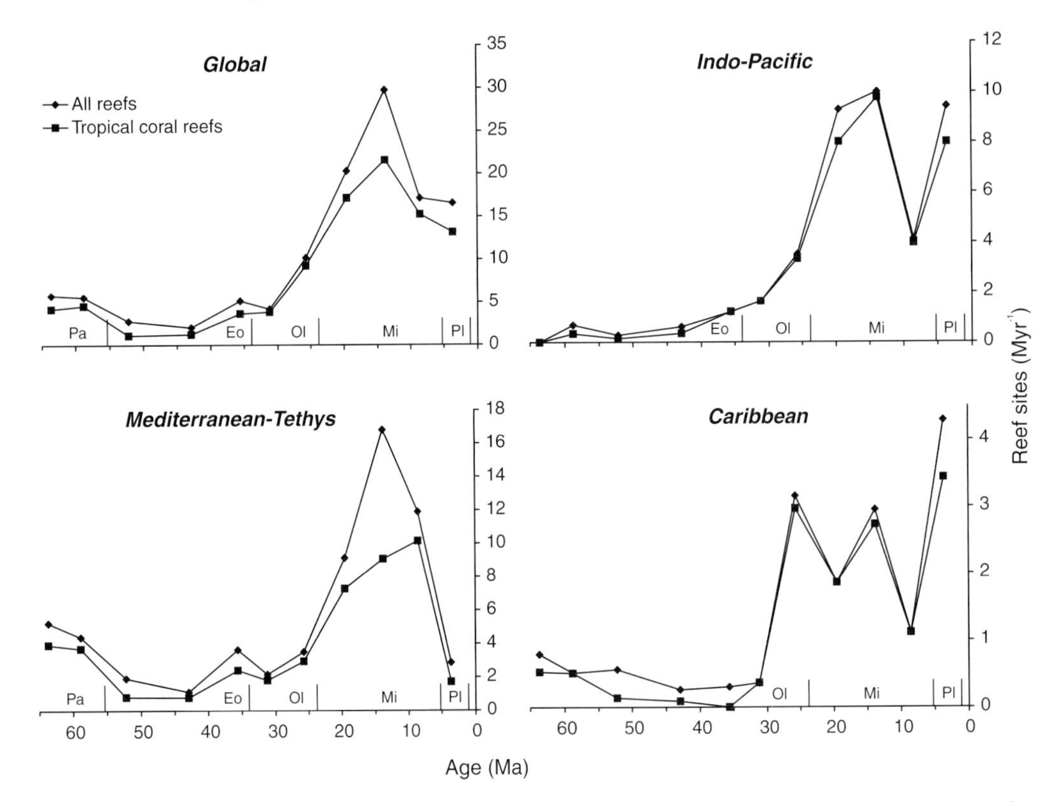

Fig. 1. Number of recorded pre-Pleistocene Cenozoic reef sites, filtered to include only reefs with at least sub-epoch level stratigraphic resolution and normalized to 1 Myr. Data are shown globally, for the Indo-Pacific, Mediterranean-Tethys, and the Caribbean. Numbers are recorded at the sub-epoch level (Table 1) for all reefs and tropical coral reefs.

the data are normalized for interval duration (Fig. 1). Palaeocene reefs were more common than Eocene reefs. A marked increase in reef numbers is observed within the Oligocene and across the Oligocene–Miocene transition. The increase culminated in a mid-Miocene peak, followed by a Late Miocene to Pliocene decline. It should be noted that the marked drop in normalized reef abundance occurred during the Late Miocene and not at the Miocene–Pliocene boundary. The pattern is quite similar when the analysis is limited to tropical-type coral reefs (Fig. 1). The only notable exception is the reduced Middle Miocene peak.

Regional patterns are markedly different. In the three areas, the number of reefs was greater during the Palaeocene than during the Eocene, with a decline across the Palaeocene–Eocene boundary. The Oligocene saw a marked increase in reef abundance in the Caribbean and Indo-Pacific regions, whereas the maximum increase in the Mediterranean happened during the Early Miocene. The Caribbean is the only region where a marked decline in reef abundance occurred from the Late Oligocene to the Early Miocene. The mid-Miocene peak, followed by a decline in the Late Miocene, is seen in all three areas. However, in the Mediterranean area the abundance of tropical-type coral reefs continued to increase in the Late Miocene. It should be noted also that the Late Miocene to Pliocene decline shown in the global curve appears to be directly related to the disappearance of most reefs in the Mediterranean region, while by contrast the two remaining reef areas are marked by an important development of reefs at that time.

The total volume of reefs was calculated for each sub-epoch, after removing the subsurface buildups (Fig. 2), which may introduce some bias in the size of reefs (Kiessling, 2002, 2006). The results differ in some important details from the simple counts. First, there is a peak in the Late Eocene, which is largely due to one large reef tract in Turkey (Kemper, 1966; Keskin, 1966). Second, there is little change across the Oligocene–Miocene transition but a pronounced increase from the Early to the Middle Miocene. Third, the Miocene reef boom is more spread out between the Middle and Late Miocene, and the Pliocene decline appears more pronounced.

Regional patterns also show little change across the Oligocene–Miocene transition except in the

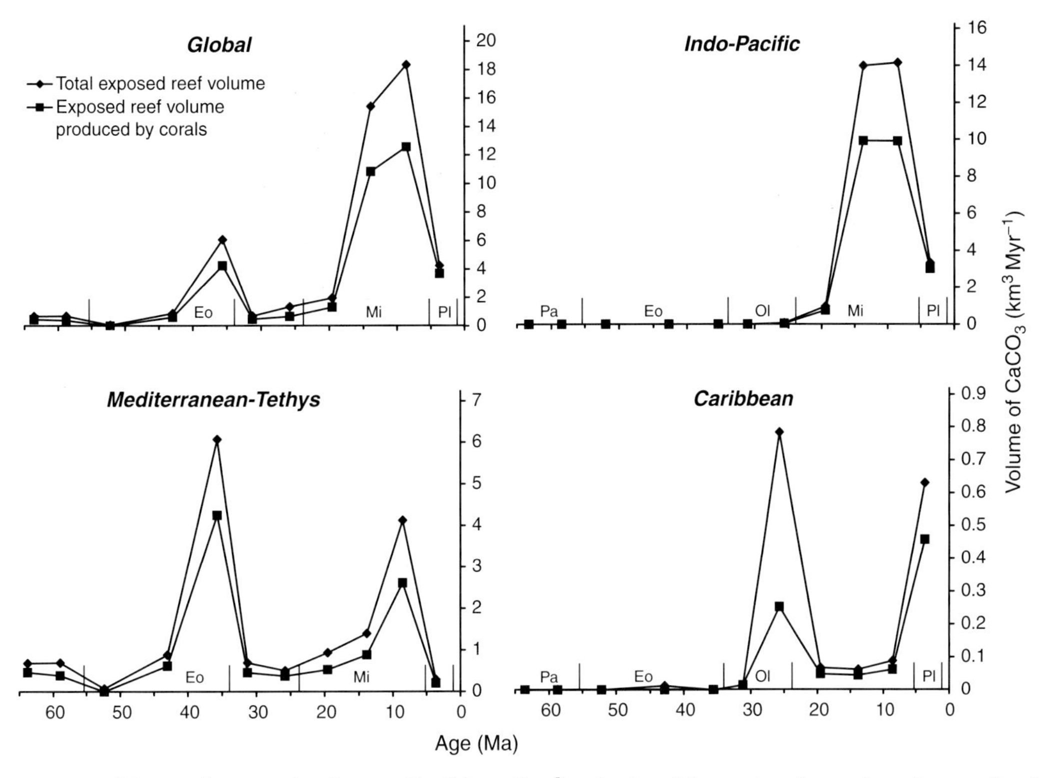

Fig. 2. Variations of the total exposed volume of buildups (km^3) calculated for each sub-epoch and normalized to 1 Myr intervals. Reefs known from the subsurface and not resolved to a sub-epoch stratigraphic resolution are excluded. Note that these estimates are very conservative (Kiessling, 2006). Data are shown globally, for the Indo-Pacific, Mediterranean-Tethys, and the Caribbean.

Caribbean region, where the reef volume per million years reached its maximum in the Late Oligocene, followed by a strong decrease in the early Miocene. The Early to Late Miocene global increase in reef volume is largely due to the reef expansion in the Indo-Pacific region and to a lesser degree in the Mediterranean. The comparison of the regional curves reveals strong regional differences in exposed volume between Miocene reefs. In particular, the acme of reef building in terms of exposed volume was reached slightly later than in terms of reef numbers, both in the Indo-Pacific and in the Mediterranean. The Caribbean region, which was isolated from the two other reef provinces from the Late Oligocene onwards, displays a very different development of reefs during the Miocene.

Fluctuations in the latitudinal extension of reef belt

Cenozoic buildups are distributed within a latitudinal belt broadly centred on the tropics but slightly shifted to the north. The latitudinal width of this belt varied through time, but less than would be expected from palaeoclimatic fluctuations, as discussed below. The maximum northward extension of reefs with an inferred tropical affinity was well beyond 40°N during most of the Cenozoic. It was only in the Pliocene that tropical coral reefs attained a latitudinal distribution comparable to the Holocene. The widest extension of the tropical reef zone occurred in the mid-Miocene (Fig. 3).

When considering only the coral-dominated buildups, a similar trend is seen in the variation of their latitudinal belt, although this is generally slightly narrower, except during the Danian and the Pliocene when some deep-water coral-dominated buildups occurred at high latitudes (Fig. 3). Similarly, the occurrence of buildups that developed below fair-weather wave base at high latitude tends to extend the width of latitudinal belt during the Danian, Lutetian, Messinian and Pliocene.

The northern margin of the reef belt is relatively well constrained in terms of bathymetry and dominant reef-builders. This is not the case for Upper Palaeocene to Lower Eocene buildups occurring in the Southern Hemisphere. In addition, the small number of buildups during the Palaeocene and

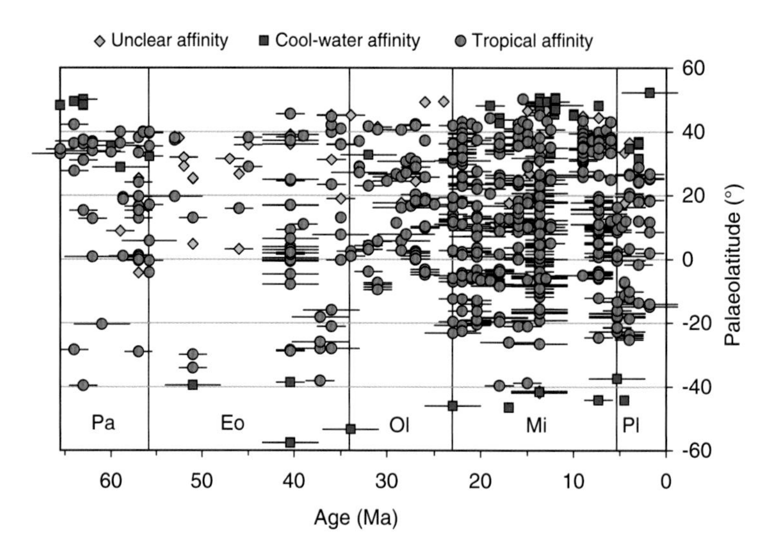

Fig. 3. Global latitudinal distribution of reefs according to their tropical or cool-water affinity. Horizontal bars around each point indicate reliability of age assignment. Long bars, poorly resolved age; short bars, moderately resolved age; no bars, precise age assignment.

Eocene may also bias views of their global distribution and hence estimates of the latitudinal extension of the reef belt for these time intervals.

When looking at the time intervals with sufficiently large numbers of reef sites (Late Eocene to Pliocene) and focussing on reefs with a tropical affinity in the Northern Hemisphere, there is only modest variation. A moderate narrowing from the Late Eocene to the Oligocene was followed by similarly moderate expansion in the Miocene, which culminated in the Middle Miocene and then declined sharply in the Pliocene. The southward shift of the northern margin of the reef belt near the Messinian–Pliocene boundary is closely related to the late Messinian disappearance of the Mediterranean coral framework reefs, which were the northernmost shallow-water coral reefs at the global scale in the Late Miocene.

Latitudinal trends in reef properties

Dominant reef builders

The nature of dominant reef-builders shows a relatively good correlation with palaeolatitude, especially during the Neogene (Fig. 4). After the Eocene, corals were by far the dominant reef builders. The proportion of coral-dominated reefs gradually increased from the Rupelian to the Pliocene. The two other important potential reef-builders are algae (mainly non-geniculate coralline algae) and, to a lesser degree, bryozoans. Other reef-building organisms such as microbes, serpulids, vermetid gastropods, hydrozoans and oysters were occasionally dominant in some reefs, but these represent only a few percent (<3%) of all buildups worldwide for this time interval. Distinct reef-building communities developed at high latitude from the Aquitanian onwards. These are particularly coralline algal and algal-bryozoan assemblages, together with less frequent coral-bivalve and coral-bryozoan communities.

From the Late Eocene onwards, high-latitude buildups were often dominated by algae or bryozoans, whereas corals tend to dominate reefs occurring in the central part of the reef belt (Fig. 4). This latitudinal distribution of dominant reef-builders is clearly seen in the Miocene of the Mediterranean and the Indo-Pacific regions (Fig. 5). A different pattern is observed in the western Atlantic region where coralline algal-dominated assemblages preferentially developed in the equatorial zones from the Chattian to mid-Miocene, while coral buildups occurred north of these areas.

Size of buildups

One of the most commonly held views is that shallow warm waters of the tropical zone provide ideal conditions for carbonate production, favouring the prolific growth of carbonate producers through easier biomineralization. Therefore, the link between the size of reefs and their palaeolatitudinal setting was tested. The size of reefs, and especially their thickness, when considered

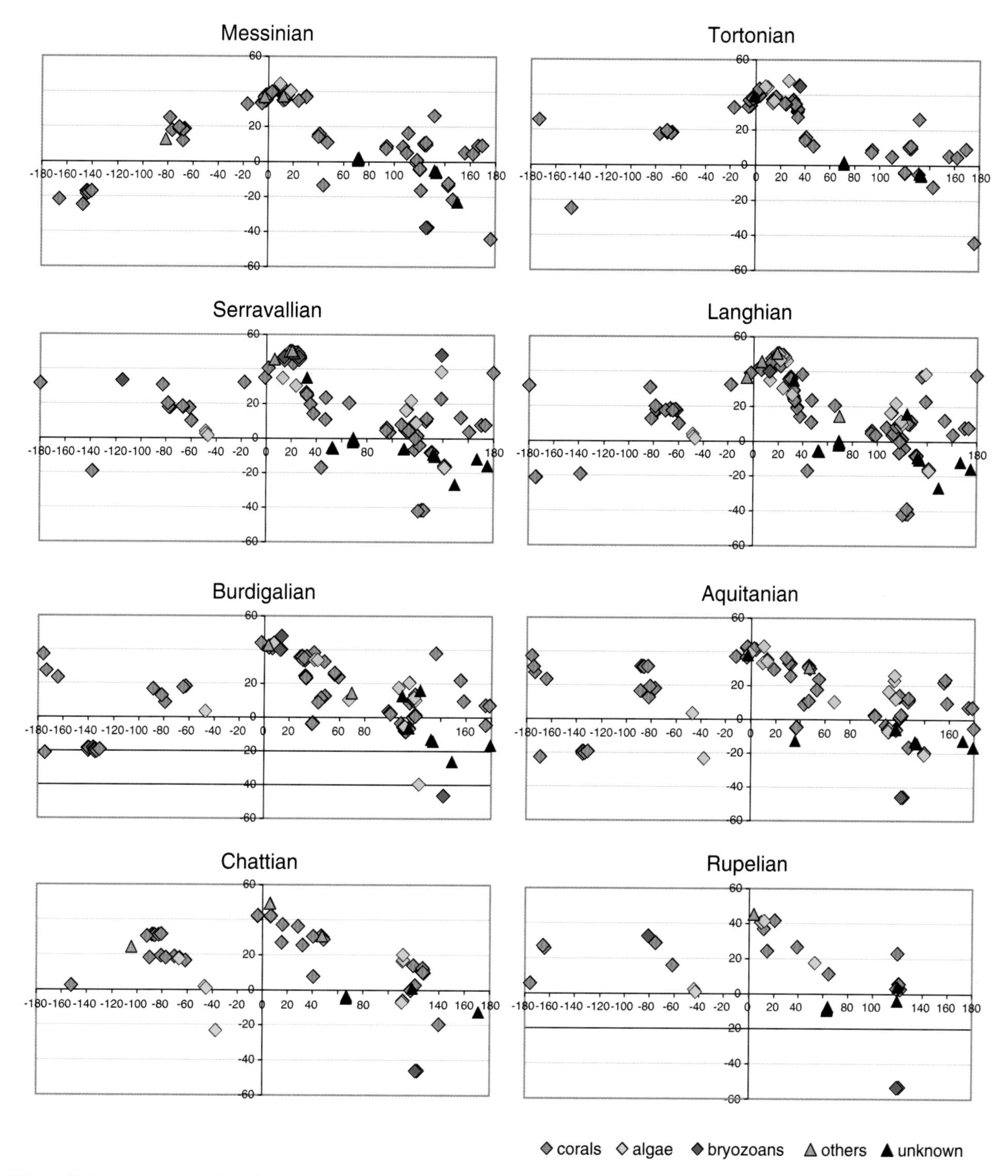

Fig. 4. Palaeogeographic distribution of reefs dominated by particular groups of organisms in individual stages of Oligocene (Rupelian, Chattian) and Miocene (Aquitanian to Messinian) age. Axes show palaeolatitude in degrees.

at a global scale, tends to be negatively correlated with palaeolatitude, independent of the time interval considered. Two parameters were used to test the relationship between the size of buildup or reef-tract and palaeolatitude: the thickness of reefs and their lateral extent.

The link between the thickness of reefs and their palaeolatitude becomes obvious from the Middle

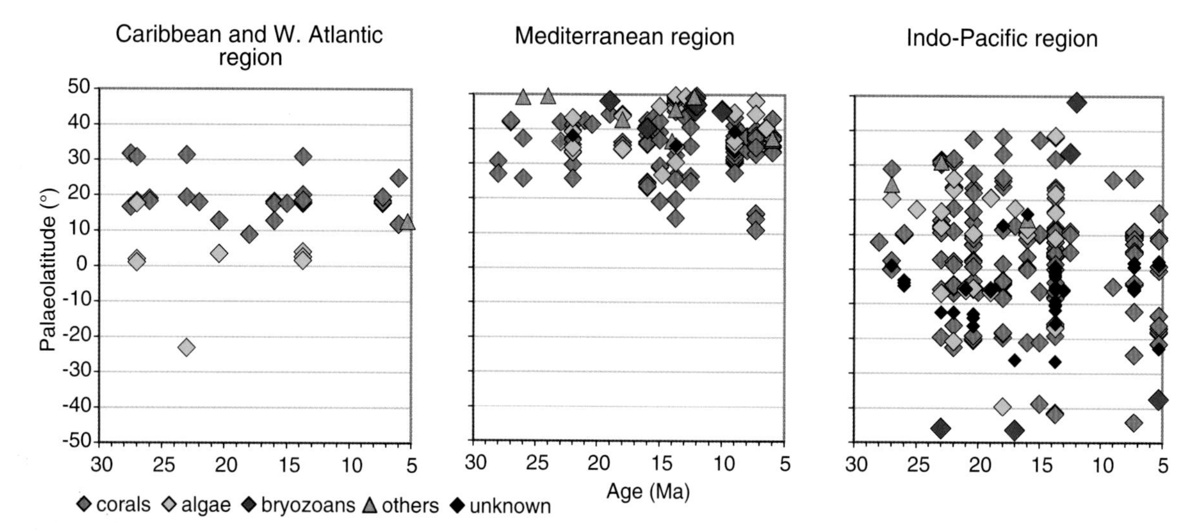

Fig. 5. Latitudinal distribution of reefs according to the type of main reef-builders in the three main regions of the Caribbean-western Atlantic, Mediterranean and Indo-Pacific. A preferential latitudinal distribution of dominant reef-builders is clearly seen in the three regions during Miocene times (5–23 Ma).

Eocene onwards, and tends to be further emphasized from the end of the Eocene to the Miocene. At the global scale, the latitudinal distribution of buildups according to their thickness shows that the large majority of reef sequences thicker than 100 m did not occur outside the tropical zone, particularly after the Oligocene–Miocene boundary. Reefs with a thickness of less than 100 m were distributed in the entire latitudinal reef-belt independent of palaeolatitude. There is an overall negative correlation between reef thickness and absolute palaeolatitude of $R = -0.43$ ($p < 0.001$) (Fig. 6).

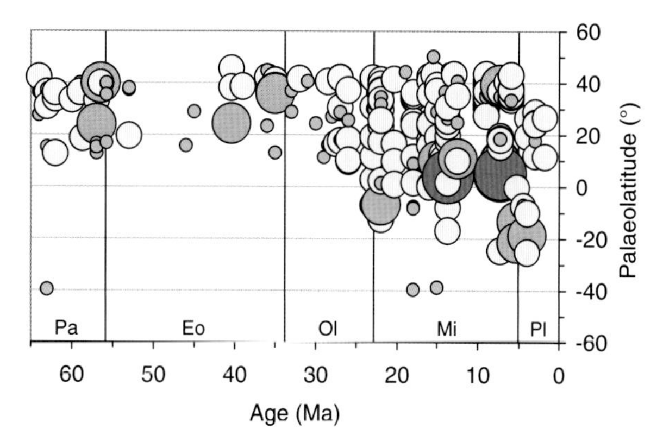

Fig. 6. Temporal variation of the global latitudinal distribution of reefs according to their thickness range. Thickness is recorded into four categories, shown by the four different diameters and colours of the circles (<10 m, small green; 10–100 m, medium yellow; 100–500 m, large orange; >500 metres, large red). Reefs with unknown thickness were excluded.

The lateral extent of reefs is less correlated with absolute palaeolatitude ($R = -0.37$, $p < 0.001$). This may be due to the often imprecise estimates of lateral reef extent in the published literature. However, some high-latitude deep-water reef-mounds and biostromes occurring on the southern margin of the reef-belt from Lutetian to Burdigalian times, and on its northern margin during the Pliocene, did actually extend over several hundred metres (Lowry, 1970; James & von der Borch, 1991; Feary & James, 1995; Henriet *et al.*, 1998). Although this tends to alter the correlation in terms of reef volume, it clearly appears that a development of reefs having a large exposed volume in the tropical zone occurred during the Oligocene–Miocene transition (Fig. 7).

Buildups known from subsurface data tend to have much greater sizes than those seen in outcrop (Kiessling, 2006). Therefore, subsurface reef sites were excluded in order to test if the link between size and palaeolatitude is still apparent with the outcrop data only. The results are basically identical, although correlation coefficients are somewhat reduced ($R = -0.24$, $p < 0.001$ for thickness; $R = -0.21$, $p < 0.001$ for lateral extent). The size of many buildups occurring in the tropics from the Selandian to the Rupelian is largely unknown, which renders it difficult to compare reef sizes before and after the Oligocene–Miocene boundary.

Diversity of reef-building assemblages

Reef diversity was measured by the average species richness of reef builders within reefs. Diversity

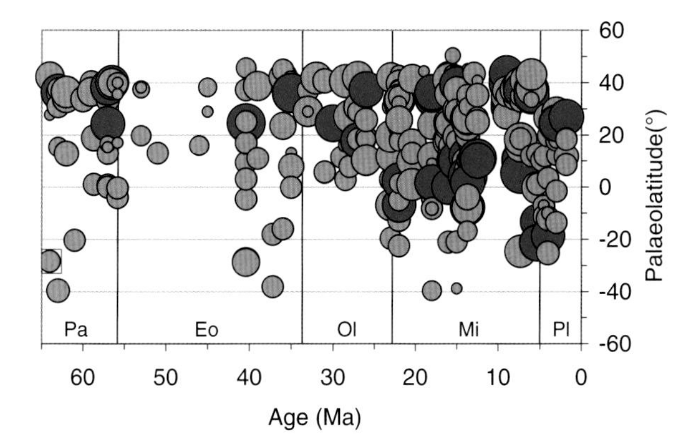

Fig. 7. Temporal variation of the global latitudinal distribution of reefs according to their calculated exposed volume. The logarithm of the exposed volume of reefs is proportional to the diameter of circles. Subsurface reefs were excluded. Major reef structures of >20 km^3 volume are highlighted in blue.

measures, imprecise as they often are, were mostly interval-scaled, just as the measures of reef sizes used here. The average diversity of reef building assemblages shows a significant increase from the Rupelian to the Chattian, which saw the Cenozoic acme of reef diversity according to Kiessling (2002, fig. 22). Average global reef diversity declined steeply through the Miocene and only increased in the Pliocene, after the disappearance of the low-diversity Mediterranean reef province.

When considering all buildups from the Danian to the Pliocene, the diversity of reef-building assemblages shows a significant inverse correlation with palaeolatitude ($R = -0.29$, $p < 0.001$ for exposed reefs and interval measures). While diverse reef-building assemblages are rare during most of the Palaeocene and Eocene, high-diversity reefs became more frequent from the Rupelian onwards. These tended to develop in the median part of the reef zone, with only a few exceptions of highly diverse bioconstructional assemblages growing at high latitudes, close to the northern margin of the reef belt. During the Miocene, the large majority of buildups occurring outside the inter-tropical regions were produced by reef-building communities having a low to moderate species richness.

The diversity pattern from the Chattian to the Messinian differs in the three main reef areas (Fig. 8). In the Indo-Pacific province, highly diverse reefs were located in the tropics, those having a moderate diversity occurred in the entire latitudinal range, while reefs of low diversity developed close to the northern and southern margins of the reef-belt. No specific distribution is seen in the Caribbean-Western Atlantic region, probably because Chattian and Miocene buildups in this area are developed only in the tropical zone. By contrast, the Mediterranean Province was mainly characterized by reefs of low to moderate diversity and several high-diversity buildups were developing at palaeolatitudes between 40° and 50°N.

There is a preferential occurrence of high-diversity reefs in the lowest latitudes, leading

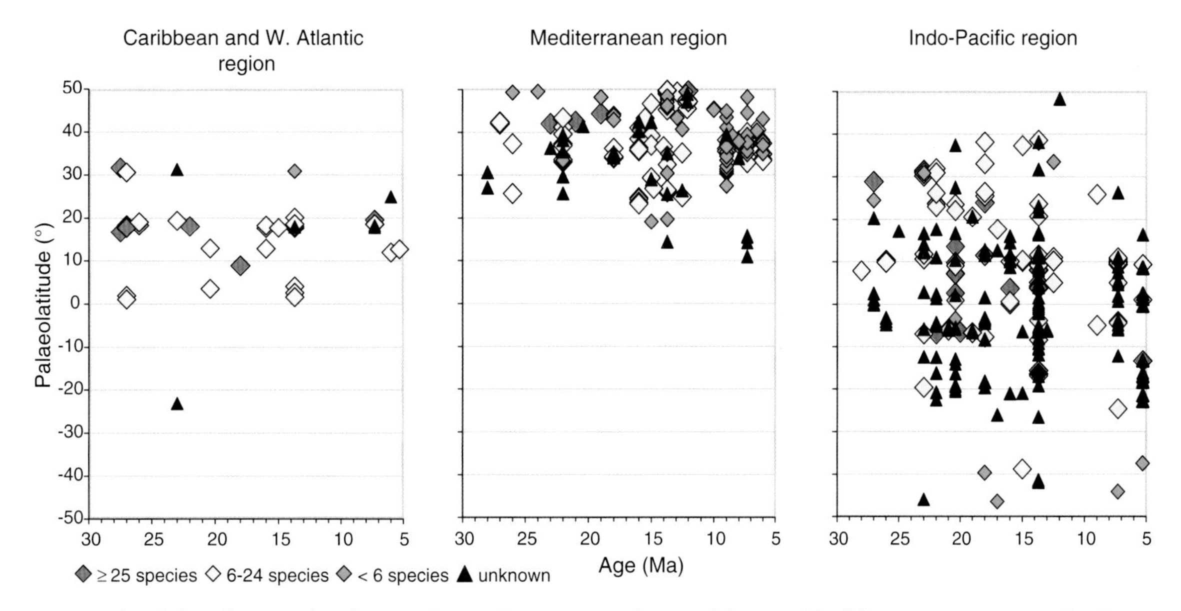

Fig. 8. Latitudinal distribution of reefs according to the species richness of their reef-building communities in the three main regions of the Caribbean-western Atlantic, Mediterranean and Indo-Pacific. The diversity pattern differs substantially in the three main reef areas.

to a statistically clear pattern at a global scale ($R = -0.26$, $p < 0.001$ for the Chattian to Pliocene interval). However, the correlation is weak, even if the dataset is restricted to coral-dominated reefs or to buildups having a tropical affinity. Although many factors may alter the link between palaeolatitude and species richness of reef-building communities, the absence of gradients may be a simple statistical artifact, related to small sample size and more detailed palaeontological studies in the richer countries (Kiessling, 2006), in which the northernmost Cenozoic reefs are located.

Testing the relationship between reef properties and global physico-chemical changes

Several statistical tests have been performed in order to reveal the degree of correlation between properties of buildups and physico-chemical parameters of seawater through time. For this purpose, variations of several seawater parameters such as temperature, Sr-isotope, C- isotope, and O-isotope composition, Mg/Ca ratio and sea-level were used from published data (Frakes *et al.*, 1992; Stanley & Hardie, 1998; Veizer *et al.*, 1999; Lear *et al.*, 2000; Zachos *et al.*, 2001; Golonka & Kiessling, 2002). Since there is a significant autocorrelation for nearly all reef attributes as well as for all Earth system parameters, detrended values rather than raw data were tested for cross-correlations. Without such treatment, autocorrelated trends may suggest correspondences that do not indicate causal relationships. Therefore it was tested whether changes in reef attributes are correlated with changes in Earth system parameters such as global temperature, or more precisely any autocorrelations were removed by using weighted first differences (see Kiessling, 2005 for details).

The analyses result in almost no significant cross-correlations. In particular, there is no correlation between changes in different physico-chemical parameters of seawater and changes in: (1) the global number of tropical reefs produced per million years (Fig. 9) or; (2) their latitudinal distribution, in particular the distribution in the Northern Hemisphere (Fig. 10). It should be noted that the palaeotemperature data of Lear *et al.* (2000) have been used for the graphic comparisons. These are based on deep-sea benthic foraminifera and thus do not necessarily correspond to the shallow-water temperatures in which reefs thrived. However, being based on Mg/Ca ratios, these temperature estimates have the advantage of not being biased by fluctuations in global ice volume, as oxygen isotopes are. The Mg/Ca ratios are thus more directly linked to the temperature evolution of the oceans than the $\delta^{18}O$ values reported in Zachos *et al.* (2001). In any case, tests have been applied to a variety of independent climate proxies and geochemical data and found the same basic results (see references above). These results suggest that global physico-chemical changes in seawater parameters did not exert a significant control on global changes of reef attributes. Therefore, the influence of biological change (ecological requirements of reef-building organisms and biotic interactions) or unmeasured oceanographic changes may have been more important.

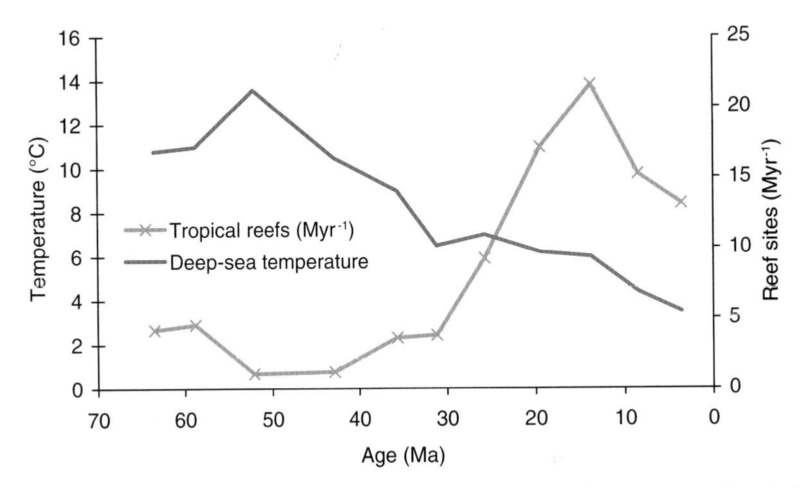

Fig. 9. Comparison of the development of global reef numbers produced per million years and variations of deep-sea water temperature (data from Lear *et al.*, 2000) through the Cenozoic. There is no significant correlation between the two time series. Similar results are obtained with the temperature curve of Zachos *et al.* (2001), based on oxygen isotopes.

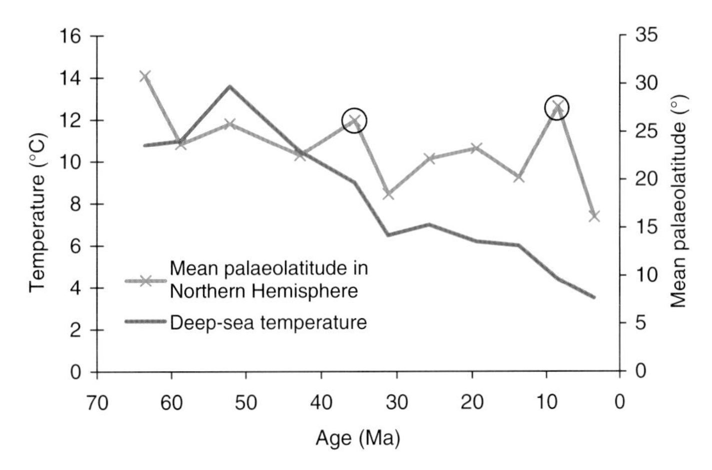

Fig. 10. Comparison of the fluctuations of the mean palaeolatitudes of reefs with tropical affinity in the Northern Hemisphere and variations of deep-sea water temperature (data from Lear *et al.*, 2000) through the Cenozoic. The absence of a significant cross-correlation is mostly due to two outliers in the Late Eocene and Late Miocene (marked with circles). Note that among all the tests performed, this graph shows the closest match between any reef attribute and an Earth system parameter.

A more detailed analysis, based on 5-Myr intervals, utilized the isotope compilation of Jan Veizer (2004). The carbon- and oxygen-isotope ratios of planktonic and benthic foraminifera were averaged to the same 5-Myr time increments as the reef data. The results show similar temporal changes between the number of buildups per Myr and the isotopic composition of foraminifera, especially benthic foraminifera, until 28 Ma (i.e. approximately the Rupelian–Chattian boundary, Fig. 11). This implies that during most of the Palaeogene, the abundance of reefs was probably linked to seawater temperature, i.e. that it was favoured by globally cooler temperatures. An important change occurred between 28 and 23 Ma, after which this link disappeared, indicating that the post-Rupelian global development of reefs was not directly related to seawater temperature.

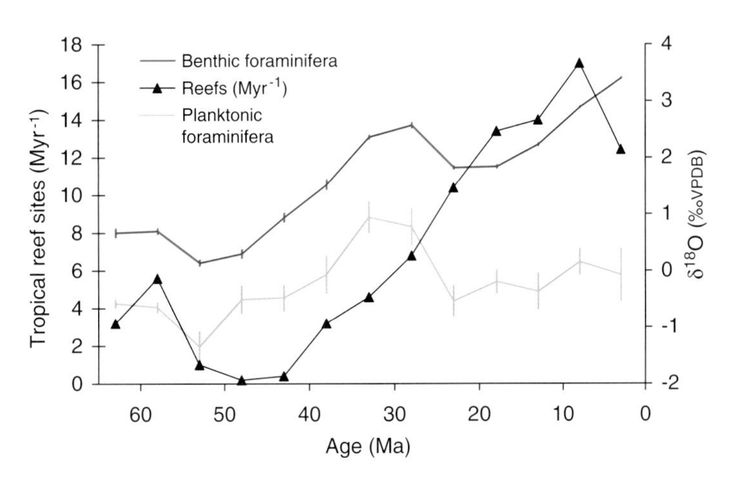

Fig. 11. Comparison between the development of tropical buildups and variations of isotopic composition of benthic and planktonic foraminifera. Equal time-slices of 5 Myr were used and reefs with poor precision in age assignments were excluded. Error bars for oxygen isotopes represent 95% confidence intervals of the mean values. Similar temporal changes occur for the number of buildups per Myr and the isotopic composition of foraminifera until 28 Ma (approximately the Rupelian–Chattian boundary), suggesting a link between the abundance of reefs and seawater temperature (note that higher $\delta^{18}O$ values indicate cooler temperatures). After 23 Ma, this link disappeared.

DISCUSSION

Significance of reef properties, potential controls and interactions

Dominant reef builders and composition of reef-building assemblages

One of the most accepted current views for explaining latitudinal gradients in the composition of shallow-marine biotas is that temperature is the main driving factor (Roy *et al.*, 1998; Valentine *et al.*, 2002). Gradual decrease of temperature with higher latitude acts together with other factors such as radiant energy, nutrients and aragonite saturation to build thresholds for the ecological requirements of reef faunas and floras (e.g. Harriott & Banks, 2002). This correlation is certainly stronger for symbiont-bearing, light-dependent reef-builders such as zooxanthellate scleractinian corals. The northern limit of zooxanthellate coral distribution in the present-day oceans is explained by a temperature threshold below which the symbiotic relationship between the zooxanthellae and the coral host does not fully function any more. The value of this limiting temperature likely varies significantly among species. When regarding major reef-building zooxanthellate coral species, it becomes a threshold for the capacity of coral communities to form framework reefs.

Independent of the global icehouse versus greenhouse climate conditions, the latitudinal distribution of both temperature and radiant energy gradients is altered to some degree by three categories of factors: (1) those producing longitudinal patterns which disturb latitudinal gradients; (2) factors acting at global and regional scale with non-exclusive latitudinal effects on reef-building communities; and (3) factors acting at a large regional scale.

The first category is directly related to global oceanographic and atmospheric circulation inducing large-scale surface oceanic currents controlling dispersal, and hence recruitment, of many major shallow-water taxa. For instance, the paucity of reef occurrence off the western coasts of continents since the early Palaeocene is thought to mainly result from the anticlockwise circulation of palaeocurrents favouring both inputs of cold waters from the South Pole and the development of upwelling along west-facing continental margins (Perrin, 2002).

Perhaps the most important factor belonging to the second category is represented by radiant energy. Although this factor has a marked latitudinal component through astronomic parameters related to Earth's geometry and movement, it depends also on atmospheric conditions, which tend to attenuate the latitudinal pattern through scattering and absorption by gases, solid and liquid particles and clouds in the atmosphere.

The third category includes much more diverse factors. One example is provided by the differentiation of equatorial and tropical carbonate platforms in southeast Asia. In this region, humid equatorial climatic conditions during the early to middle Miocene strongly influenced or even limited the development of coral reef communities by continuous inputs of siliciclastics and volcanics, high levels of terrestrial run-off in adjacent areas and potential increases in nutrient levels in comparison with the more favourable arid tropical to subtropical conditions (Wilson & Lokier 2002).

Local factors can be broadly regarded as those related to local habitat conditions. Although they are unlikely to affect markedly the global pattern, they may be responsible of some alteration at a regional scale or when the amount of reefs is particularly low in the time interval considered.

Size of buildups

The analyses presented here show that there is a significant negative correlation between the size of buildups and absolute palaeolatitude, and that this link tends to be strengthened after the Eocene and further emphasized after the Oligocene–Miocene boundary. Reef size and thickness can be regarded as reflecting a certain stability of environmental conditions suitable for reef growth. From this point of view, the tropical zone appears to provide this relative environmental stability compared with subtropical areas which are more sensitive to global and regional climatic fluctuations. Reef size itself is probably very dependent on local conditions (including variations in subsidence) which may alter the general pattern.

In addition, assessment of buildup size suffers from biases. A major problem is the lack of data for many reefs, in particular for the lateral extent of reefs. The discrepancy between the patterns obtained with all reefs and with outcropping sites only underlines the difficulty in evaluating the true sizes of reefs. When all Phanerozoic reefs are considered, it appears that buildups studied by seismic exploration tend to be thicker than their outcropping counterparts. This demonstrates that either their size at outcrop tends to be underestimated, due to erosion or burial of some reef parts for example, or that subsurface surveys tend to overestimate reef dimensions (Kiessling, 2002, 2006).

Diversity of reef-building assemblages

Testing for changes in latitudinal gradients of diversity is hampered by lack of data, especially in the Southern Hemisphere. When considering only the Chattian to Messinian time interval, the pattern looks very different in the three main reef provinces due to contextual or historic biogeographic reasons. The Indo-Pacific region is the only one which shows the preferential distribution of high-diversity buildups at low latitude and of low-diversity buildups at high latitudes. It also represents the only region where the reefs occur within a latitudinal range large enough for this pattern to be evaluated.

Diversity of reef-building communities is also positively correlated with the distribution of true reefs, as the complex internal structure of framework reefs has a larger capacity of integrating a higher number of different reef-builders by providing a larger spectrum of ecological niches (Perrin *et al.*, 1995). It should be noted that the patterns of reef diversity as analysed in this paper (diversity within individual reef structures), strongly differ from patterns of generic or species richness of particular taxa (e.g. reef corals within regions or at global scale) and represent the response of the reef ecosystems to biotic and abiotic temporal changes. This response appears decoupled from the evolutionary history (i.e. turnover, extinction and speciation) of the main reef-building group such as reef scleractinians (e.g. Budd, 2000).

At the global scale, the general pattern of reef diversity through time may also be disturbed by the fact that most Mediterranean reefs are of low to moderate diversity, with many of them occurring outside the tropical to subtropical belt. In this sense, as emphasized by Rosen (1988) for the distribution of reef corals, the expected latitudinal pattern may be altered to some degree by historical causes. Here, the progressive shift of the Mediterranean region outside the tropical belt favourable for prolific reef growth has probably induced the development of ecological adaptations by corals and coral reefs for being able to produce well-developed coral reefs with low-to-moderate diversity in that area.

Latitudinal trends in reef properties: any link to climate?

In addition to reef expansion (increase of the number of buildups and global preserved reef volume) during the Oligocene–Miocene transition, it is shown that many slight changes occurred at or close to the Rupelian–Chattian boundary. These were further accentuated at the Chattian–Aquitanian boundary and often strengthened during the Miocene. This is notably expressed in the increasingly marked latitudinal distribution of reef properties after the late Eocene, in particular those related to the biotic composition of reef-building assemblages and their capacities as carbonate producers, although the diversity of reef-builders is much more susceptible to attenuation by other interacting factors, including taxonomical bias.

These changes (Fig. 12) began with the development of specific dominant reef-builders at high latitude from the late Eocene onwards, and then continued with the dominance of corals as main reef-builders near the Priabonian–Rupelian boundary. High-diversity reef-building communities became more abundant from the Rupelian onwards and from that time tended to preferentially develop in the median part of reef-belt, while the proportion of reefs dominated by scleractinian corals continued to increase at the global scale. There was also an important increase of the number of buildups at a global scale after the Rupelian–Chattian boundary. From the early Miocene onwards, specific reef-building communities occurred at high latitudes where buildups are mainly represented by reef-mounds and biostromes.

Data analysis shows that no drastic changes can be demonstrated in the characteristics of reefs at the Oligocene–Miocene climatic transition, but instead several minor changes did occur step by step. Most changes took place during a time interval of about 10 million years between the Eocene–Oligocene to the Oligocene–Miocene boundaries. Climatic change was probably also gradual and the successive responses of reef ecosystems show that these were able to adapt progressively to climatic cooling. This gradual adaptation is expressed by latitudinal trends in some reef characteristics resulting from latitudinal gradients in ecological conditions. Although latitudinal gradients are necessarily linked in some way to radiant energy and the related surface temperature, these factors are probably not sufficient to explain why and how climatic cooling could favour reef development.

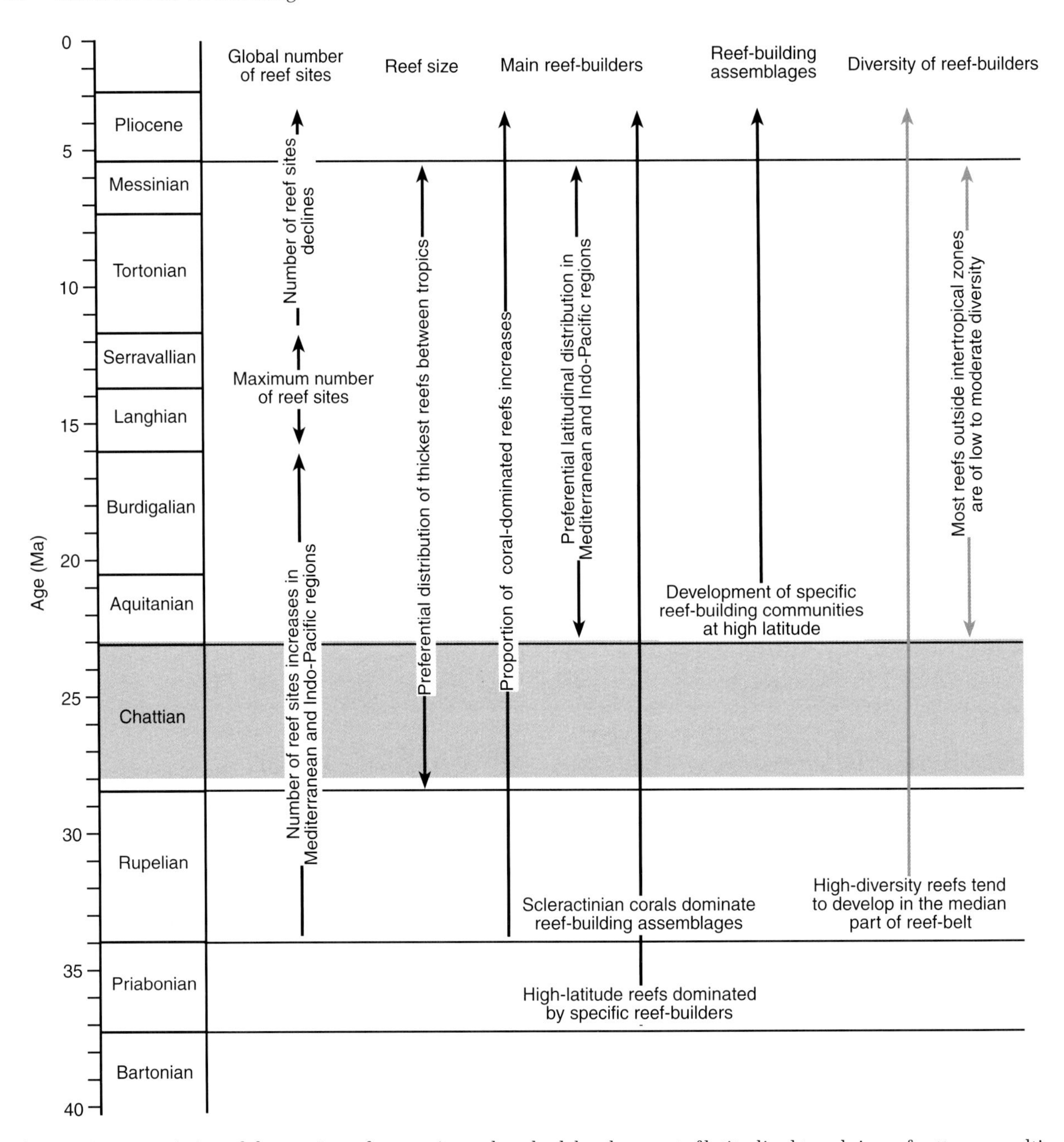

Fig. 12. Apparent timing of changes in reef properties and gradual development of latitudinal trends in reef patterns resulting from the analyses presented here. The less secure changes are shown by grey arrows. Shaded area corresponds to the major global change in seawater temperature and its decoupling with the global development of reefs between 28 and 23 Myr.

During greenhouse conditions, average surface temperatures are believed to be warmer and more uniformly distributed than today, with an extended belt of tropical conditions and increased atmospheric CO_2 (Fischer, 1984; Royer *et al.*, 2004). During icehouse conditions, the marked contrast of radiant energy levels between the equator and the poles produces a strong difference of surface temperature between high and low latitudes, which would act to enhance atmospheric and oceanic circulation. While the temperature and insolation gradients are likely to increase latitudinal patterns in the global distribution of reef-building communities, a strengthened palaeoceanographic circulation, which is additionally controlled by the global repartition of land-masses, may amplify the effects of longitudinal patterns. The strengthening of a coupled atmospheric-palaeoceanographic circulation would favour the recycling of deep water masses through coastal and equatorial upwelling and hence the input of nutrients in surface waters (Murray, 1995).

The differentiation into distinct tropical and high-latitude reef-building communities, which took place during the late Eocene and was further accentuated during the Miocene, is likely the expression of these enhanced temperature and insolation gradients during the progressive settlement of icehouse conditions. However, statistical tests have shown that no direct significant correlation between reef properties and physicochemical parameters of seawater can be established. It should be emphasized that the modest cross-correlation between reef abundance and oxygen-isotope values of both planktonic and benthic foraminifera was lost approximately at the same time as when the time series of benthic and planktonic isotope values started to diverge (Fig. 11). This suggests that oceanographic changes such as new oceanic gateways, initiation of modern style thermohaline ocean circulation, and changes in ocean heat transport (Bice *et al.*, 2000; Davies *et al.*, 2001) may have triggered reef expansion.

In any case, the link between climate and reef development is rather indirect, perhaps due to complex non-linear relationships acting at multiple spatial and temporal scales. The global distribution of continental masses, the geometry of coastal areas and the abundance of islands in the tropical belt exert a strong control on the oceanic circulation, both at global and regional scales, and are of fundamental importance for explaining the pattern of reef characteristics. In particular, the Cenozoic is characterized by long north-south oriented coasts along the margins of the oceans. We suggest that this, together with the pre-existing anticlockwise oceanic circulation, has favoured the preferential development of upwelling zones along the western margins of continents relative to equatorial upwelling, especially when icehouse conditions enhanced oceanic circulation. The preferential occurrence of coastal upwelling on the western continental coasts preserved large oceanic tropical areas where oligotrophic conditions prevailed. The development of nutrient-poor environments may have eliminated r-strategy competitors (Hallock 1988; Brasier, 1995). This probably has led to the settling and worldwide development of large coral reefs.

Although the well-established stepwise Cenozoic cooling (Abreu & Baum, 1997; Lear *et al.*, 2000; Pearson & Palmer, 2000; Zachos *et al.*, 2001; Billups & Schrag, 2003; DeConto & Pollard, 2003; Ford & Golonka, 2003) cannot be linearly correlated with any change in reef attributes, some qualitative observations can be made. Variations observed in reef patterns began to develop when global cooling was effective enough during the transitional intervals from greenhouse to icehouse conditions. Notably, the dominance of scleractinian corals in reef-building communities and the development of particular assemblages in high-latitude buildups occurred when pCO_2 concentrations were already low, and when cooling of the Southern Ocean was strong enough to permit the formation and growth of the Antarctic ice-sheet. The further development of latitudinal trends in reef properties follows the progressive development of the insolation and thermal gradients between the poles and the equator. The global development of reefs was mainly related to cooler seawater temperature until the Rupelian–Chattian boundary, although temperature was not a prime control after 28 Ma when icehouse conditions prevailed. An increase in the Mg/Ca ratio of the oceans (Stanley & Hardie, 1998) may partially explain the proliferation of coral reefs in the icehouse, although the overall correspondence between ocean chemistry and reef abundance is weak (Kiessling et al., 2008).

CONCLUSIONS

The relationship between reefs and climatic fluctuations is highly complex and results from interactions and feedbacks between multiple intrinsic and extrinsic factors acting at different temporal and spatial scales. In particular, patterns appear slightly different when examined at the scale of supersequences or at a finer temporal resolution. Analyses of pre-Pleistocene Cenozoic reefs recorded in the PaleoReefs database together with physico-chemical climatic-related fluctuations of seawater parameters has shown that:

(1) There is a gradual development of latitudinal trends in some reef properties, most of them are closely related to the biotic composition of reef-building communities;
(2) The timing of this development coincides with global palaeoclimatic changes;

(3) Although the development of reefs at global scale appears to be controlled by seawater temperature before the Rupelian–Chattian boundary, statistical analyses reveal that there is no evidence for direct relationships between characteristics of reefs and seawater parameters after 28 Ma, when atmospheric CO_2 concentrations were already relatively low and icehouse conditions prevailed;
(4) Two contrasting behaviours of reef development can be distinguished. Reef patterns appear to have been predominantly driven by extrinsic factors such as seawater temperature during greenhouse or greenhouse-to-icehouse periods, while the prime control in the Neogene icehouse period is the biological adaptation to the distribution of oligotrophic waters and a lack of nutrients. This icehouse behaviour of coral reefs is best explained by a combination of a macroevolutionary shift towards biological controls and a change in oceanographic conditions leading to a geographical change of nutrient regimes.

It has been established that low pCO_2 levels produce high saturation states of aragonite in shallow tropical seas, which facilitates the biomineralisation of aragonite skeletons at low energy cost for the organisms (Hallock, 2005). This may explain the primacy of scleractinian corals approximately after the Rupelian–Chattian boundary and their efficiency as reef-builders in tropical oceans.

REFERENCES

Abreu, V.S. and **Baum, G.** (1997) Glacio-eustasy: a global link for sequence boundaries during the Cenozoic. In: *Sedimentary Events and Hydrocarbon Systems* (Ed. B. Beauchamp), CSPG-SEPM Joint Convention 1997, Program with abstracts, p. 17. Canadian Society of Petroleum Geologists, Calgary, Canada.

Alroy, J. (1994) Appearance event ordination: a new biochronologic method. *Paleobiology*, **20**, 191–207.

Bice, K.L., **Scotese, C.R.**, **Seidov, D.** and **Barron, E.J.** (2000) Quantifying the role of geographic change in Cenozoic ocean heat transport using uncoupled atmosphere and ocean models. *Palaeogeogr. Palaeoclimatol. Palaeoecol.*, **161**, 295–310.

Billups, K. and **Schrag, D.P.** (2003) Application of benthic foraminiferal Mg/Ca ratios to questions of Cenozoic climate change. *Earth Planet. Sci. Lett.*, **209**, 181–195.

Brasier, M.D. (1995) Fossil indicators of nutrient levels. 1: Eutrophication and climate change. In: *Marine Palaeoenvironmental Analysis from Fossils* (Eds D.W.J. Bosence and P.A. Allison). *Geol. Soc. London Spec. Publ.*, **83**, 113–132.

Budd, A.F. (2000) Diversity and extinction in the Cenozoic history of Caribbean reefs. *Coral Reefs*, **19**, 25–35.

Copper, P. (1994) Ancient reef ecosystem expansion and collapse. *Coral Reefs*, **33**, 3–11.

Davies, R., **Cartwright, J.**, **Pike, J.** and **Line, C.** (2001) Early Oligocene initiation of North Atlantic Deep Water formation. *Nature*, **410**, 917–920.

DeConto, R.M. and **Pollard, D.** (2003) Rapid Cenozoic glaciation of Antarctica induced by declining atmospheric CO_2. *Nature*, **421**, 245–249.

Feary, D.A. and **James, N.P.** (1995) Cenozoic biogenic mounds and buried Miocene (?) barrier reef on a predominantly cool-water carbonate continental margin - Eucla basin, western Great Australian Bight. *Geology*, **23**, 427–430.

Fischer, A.G. (1984) The two Phanerozoic supercycles. In: *Catastrophes and Earth History* (Eds W.A. Berggren, and J.A. Van Couvering), pp. 129–150. *Princeton University Press*, Princeton, NJ.

Flügel, E. and **Flügel-Kahler, E.** (1992) Phanerozoic reef evolution: Basic questions and data base. *Facies*, **26**, 167–278.

Ford, D. and **Golonka, J.** (2003) Phanerozoic paleogeography, paleoenvironment and lithofacies maps of the circum-Atlantic margins. *Mar. Petrol. Geol.*, **20**, 249–285.

Frakes, L.A., **Francis, J.E.** and **Syktus, J.I.** (1992) *Climate Modes of the Phanerozoic: the History of the Earth's Climate over the Past 600 Million Years.* Cambridge University Press, Cambridge, 274 pp.

Golonka, J. (2002) Plate-tectonic maps of the Phanerozoic. In: *Phanerozoic Reef Patterns* (Eds W. Kiessling, E. Flügel and J. Golonka). *SEPM Spec. Publ.*, **72**, 21–75.

Golonka, J. and **Kiessling, W.** (2002) Phanerozoic time scale and definition of time slices. In: *Phanerozoic Reef Patterns* (Eds W. Kiessling, E. Flügel and J. Golonka). *SEPM Spec. Publ.*, **72**, 11–20.

Gradstein, F.M., **Ogg, J.G.**, **Smith, A.G.**, **Agterberg, F.P.**, **Bleeker, W.**, **Cooper, R.A.**, **Davydov, V.**, **Gibbard, P.**, **Hinnov, L.A.**, **House, M.R.**, **Lourens, L.**, **Luterbacher, H. P.**, **McArthur, J.**, **Melchin, M.J.**, **Robb, L.J.**, **Shergold, J.**, **Villeneuve, M.**, **Wardlaw, B.R.**, **Ali, J.**, **Brinkhuis, H.**, **Hilgen, F.J.**, **Hooker, J.**, **Howarth, R.J.**, **Knoll, A.H.**, **Laskar, J.**, **Monechi, S.**, **Plumb, K.A.**, **Powell, J.**, **Raffi, I.**, **Röhl, U.**, **Sadler, P.**, **Sanfilippo, A.**, **Schmitz, B.**, **Shackleton, N.J.**, **Shields, G.A.**, **Strauss, H.**, **Van Dam, J.**, **van Kolfschoten, T.**, **Veizer, J.** and **Wilson, D.** (2004) *A Geologic Time Scale 2004.* Cambridge University Press, Cambridge, 589 pp.

Grigg, R.W. (1997) Paleoceanography of coral reefs in the Hawaiian-Emperor Chain: Revisited. *Coral Reefs*, **16**, S33–S38.

Hallock, P. (1988) The role of nutrient availability in bioerosion: consequences to carbonate buildups. *Palaeogeogr. Palaeoclimatol. Palaeoecol.*, **63**, 275–291.

Hallock, P. (1997) Reefs and reef limestones in Earth history. In: *Life and Death of Coral Reefs* (Ed. C. Birkeland), pp. 13–42. Chapman & Hall, London.

Hallock, P. (2005) Global change and modern coral reefs: new opportunities to understand shallow-water carbonate depositional processes. *Sed. Geol.*, **175**, 19–33.

Harriott, V.J. and **Banks, S.A.** (2002) Latitudinal variation in coral communities in eastern Australia: a qualitative biophysical model of factors regulating coral reefs. *Coral Reefs*, **21**, 83–94.

Henriet, J.P., **de Mol, B.**, **Pillen, S.**, **Vanneste, M.**, **van Rooij, D.** and **Versteeg, W.** (1998) Gas hydrate crystals may help build reefs. *Nature*, **391**, 648–649.

James, N.P. and **von der Borch, C.C.** (1991) Carbonate shelf edge off southern Australia: a prograding open-platform margin. *Geology*, **19**, 1005–1008.

Kauffman, E.G. and **Fagerstrom, J.A.** (1993) The Phanerozoic evolution of reef diversity. In: *Species Diversity in Ecological Communities* (Eds R.E. Ricklefs and D. Schluter), pp. 315–329. University of Chicago Press, Chicago.

Kemper, E. (1966) Beobachtungen an obereozänen Riffen am Nordrand des Ergene-Beckens (Türkisch-Thrazien). *Neues Jb. Geol. Paläontol. Abh.*, **125**, 540–554.

Keskin, C. (1966) Microfacies study of the Pinarhisar reef complex. *Instanbul Univ. Fakült. Mecmuasi Ser. B.*, **31**, 109–146.

Kiessling, W. (2001) Paleoclimatic significance of Phanerozoic reefs. *Geology*, **29**, 751–754.

Kiessling, W. (2002) Secular variations in the reef ecosystem. In: *Phanerozoic Reef Patterns* (Eds W. Kiessling, E. Flügel and J. Golonka). *SEPM Spec. Publ.*, **72**, 625–690.

Kiessling, W. (2003) Reefs. In: *Encyclopedia of Sediments and Sedimentary Rocks* (Ed. G. Middleton), pp. 557–560. Kluwer Academic, Dordrecht.

Kiessling, W. (2005) Long-term relationships between ecological stability and biodiversity in Phanerozoic reefs. *Nature*, **433**, 410–413.

Kiessling, W. (2006) Towards an unbiased estimate of fluctuations in reef abundance and volume during the Phanerozoic. *Biogeosciences*, **3**, 15–27.

Kiessling, W. and **Flügel, E.** (2002) Paleoreefs - a database on Phanerozoic reefs. In: *Phanerozoic Reef Patterns* (Eds W. Kiessling, E. Flügel and J. Golonka). *SEPM Spec. Publ.*, **72**, 77–92.

Kiessling, W., **Aberhan, M.** and **Villier, L.** (2008) Phanerozoic trends in skeletal mineralogy driven by mass extinctions. *Nature Geoscience*, **1**, 527–530.

Kiessling, W., **Flügel, E.** and **Golonka, J.** (2000) Fluctuations in the carbonate production of Phanerozoic reefs. In: *Carbonate Platform Systems: Components and Interactions* (Eds E. Insalaco, P.W. Skelton and T.J. Palmer). *Geol. Soc. London Spec. Publ.*, **178**, 191–215.

Kiessling, W., **Flügel, E.** and **Golonka, J.** (Eds) (2002) *Phanerozoic Reef Patterns. SEPM Spec. Publ.*, **72**, 775 pp.

Lear, C.H., **Elderfield, H.** and **Wilson, P.A.** (2000) Cenozoic deep-sea temperatures and global ice volumes from Mg/Ca in benthic foraminiferal calcite. *Science*, **287**, 269–272.

Lowry, D.C. (1970) Geology of the Western Australian part of the Eucla basin. *Geol. Surv. W. Aust. Bull.*, **122**, 1–201.

Murray, J.W. (1995) Microfossil indicators of ocean water masses, circulation and climate. In: *Marine Palaeoenvironmental Analysis from Fossils* (Eds D.W.J. Bosence and P.A. Allison). *Geol. Soc. London Spec. Publ.*, **83**, 245–264.

Parrish, J.T. (1998) *Interpreting Pre-Quaternary Climate from the Geologic Record.* Columbia University Press, New York, 338 pp.

Pearson, P.N. and **Palmer, M.R.** (2000) Atmospheric carbon dioxide concentrations over the past 60 million years. *Nature*, **406**, 695–699.

Perrin, C. (2002) Tertiary: the emergence of modern reef ecosystems. In: *Phanerozoic Reef Patterns* (Eds W. Kiessling, E. Flügel and J. Golonka). *SEPM Spec. Publ.*, **72**, 587–618.

Perrin, C., **Bosence, D.W.J.** and **Rosen, B.R.** (1995) Quantitative approaches to palaeozonation and palaeobathymetry of corals and coralline algae in Cenozoic reefs. In: *Marine Palaeoenvironmental Analysis from Fossils* (Eds D.W.J. Bosence and P.A. Allison). *Geol. Soc. London Spec. Publ.*, **83**, 181–229.

Rosen, B.R. (1988) Progress, problems and patterns in the biogeography of reef corals and other tropical marine organisms. *Helgoländer Meeresun.*, **42**, 269–301.

Roy, K., **Jablonski, D.**, **Valentine, J.W.** and **Rosenberg, G.** (1998) Marine latitudinal diversity gradients: Tests of causal hypotheses. *Proc. Nat. Acad. Sci. USA*, **95**, 3699–3702.

Royer, D.L., **Berner, R.A.**, **Montañez, I.P.**, **Tabor, N.J.** and **Beerling, D.J.** (2004) CO_2 as a primary driver of Phanerozoic climate. *GSA Today*, **14**, 4–10.

Spencer-Cervato, C., **Thierstein, H.R.**, **Lazarus, D.B.** and **Beckmann, J.P.** (1994) How synchronous are Neogene marine plankton events? *Paleoceanography*, **9**, 739–763.

Stanley, S.M. and **Hardie, L.A.** (1998) Secular oscillations in the carbonate mineralogy of reef-building and sediment-producing organisms driven by tectonically forced shifts in seawater chemistry. *Palaeogeogr. Palaeoclimatol. Palaeoecol.*, **144**, 3–19.

Tarduno, J.A., **Duncan, R.A.**, **Scholl, D.W.**, **Cottrell, R.D.**, **Steinberger, B.**, **Thordarson, T.**, **Kerr, B.C.**, **Neal, C.R.**, **Frey, F.A.**, **Torii, M.** and **Carvallo, C.** (2003) The Emperor Seamounts: Southward Motion of the Hawaiian Hotspot Plume in Earth's Mantle. *Science*, **301**, 1064–1069.

Valentine, J.W., **Roy, K.** and **Jablonski, D.** (2002) Carnivore/non-carnivore ratios in northeastern Pacific marine gastropods. *Mar. Ecol.-Prog. Ser.*, **228**, 153–163.

Veizer, J. (2004) *Isotope Data.* (http://www.science.uottawa.ca/geology/isotope_data).

Veizer, J., **Ala, D.**, **Azmy, K.**, **Bruckschen, P.**, **Buhl, D.**, **Bruhn, F.**, **Carden, G.A.F.**, **Diener, A.**, **Ebneth, S.**, **Godderis, Y.**, **Jasper, T.**, **Korte, C.**, **Pawellek, F.**, **Podlaha, O.G.** and **Strauss, H.** (1999) $^{87}Sr/^{86}Sr$, $\delta^{13}C$ and $\delta^{18}O$ evolution of Phanerozoic seawater. *Chem. Geol.*, **161**, 59–88.

Wilson, M.E.J. and **Lokier, S.W.** (2002) Siliciclastic and volcaniclastic influences on equatorial carbonates: insights from the Neogene of Indonesia. *Sedimentology*, **49**, 583–601.

Zachos, J., **Pagani, M.**, **Sloan, L.**, **Thomas, E.** and **Billups, K.** (2001) Trends, rhythms, and aberrations in global climate 65 Ma to Present. *Science*, **292**, 686–693.

Int. Assoc. Sedimentol. Spec. Publ. (2010) **42**, 35–48

Carbonate grain associations: their use and environmental significance, a brief review

PASCAL KINDLER* and MOYRA E.J. WILSON†

*Department of Geology and Palaeontology, 13, Rue des Maraichers, CH-1211, Geneva, Switzerland (E-mail: pascal.kindler@unige.ch)

†Department of Geological Sciences, Durham University, South Road, Durham, DH1 4EL, UK

ABSTRACT

A complex interplay of environmental factors may affect regionally important Cenozoic carbonate deposits, and a combination of different controls may result in comparable deposits. There is, therefore, a need for purely descriptive nomenclature to define regionally extensive groupings of carbonate deposits. This terminology is reviewed here. In addition to the well-established chlorozoan (or coralgal), chloralgal, foramol, rhodalgal, molechfor and bryomol definitions, two new terms, LB-foralgal and thermacor, are proposed. These purely descriptive groupings allow for good spatial and temporal subdivision of Cenozoic carbonate deposits. The continued use of objective terms for Cenozoic carbonate deposits does not preclude the use of other interpretative grain-association terminology, and a two-tier system might be adopted. However, it should be clearly stated when groupings based on benthic components are being used purely descriptively, and therefore in an objective sense. It is hoped that this descriptive approach to regionally important deposits will help to promote objective discussions about controls on carbonate sedimentation and environmental change during the Cenozoic.

Keywords Carbonate-grain associations, classification, nomenclature, Cenozoic.

INTRODUCTION

It has long been recognized that major variations in modern carbonate deposits across the globe reflect fundamental differences in environmental conditions which directly affect the frequency and distribution of carbonate-producing organisms (Purdy, 1963; Lees & Buller, 1972; Flügel, 2004). Over the last 30 years, many authors have attempted to group the various carbonate grains into associations of environmental significance that, in turn, could be used in the interpretation of ancient limestone successions. These geological data have then been utilized to make far-reaching inferences about the spatial and temporal nature of environmental and global changes.

Initially the main controlling parameter on grain distribution was thought to be temperature, with Lees & Buller (1972) defining a chlorozoan association, typical of warm waters, and a foramol association, often, although not exclusively, restricted to cool waters. However, since this early work, it is now clear that a wider range of parameters may exert important influences on the global variations in carbonate sediments in both space and time. These parameters include salinity (Lees, 1975), nutrients (Hallock & Schlager, 1986; Mutti & Hallock, 2003; Halfar *et al.*, 2004, 2006) and light (James, 1997; Pomar *et al.*, 2004). Other factors known to affect benthic communities such as water energy, water transparency, input of terrigenous sediments, oxygen and CO_2 concentrations, $CaCO_3$ saturation in seawater, Mg/Ca ratio, alkalinity, bathymetry and nature of the substrate are starting to be investigated as additional possible influences on grain associations (Stanley & Hardie, 1998; Pomar *et al.*, 2004).

In addition to the greater awareness that a wide range of factors may influence associations, there are now considerably more data available on carbonate deposits from previously poorly studied areas of the globe. These areas include the equatorial tropics (Carannante *et al.*, 1988; Wilson, 2002; Wilson & Vecsei, 2005), the transition from warm to cool latitudes (Betzler *et al.*, 1997; Halfar *et al.*, 2004, 2006) and cool to cold or polar latitudes (Nelson, 1988; James, 1997). Thus there is a need to re-evaluate the environmental significance of the different groupings of carbonate sediments.

Not only is this a time of major debate on the environmental significance of carbonate deposits, it is also a time of discussion on nomenclature related to the way in which the groupings are defined. Carbonate deposits have been grouped on the basis of grain associations (Lees, 1975; Carannante *et al.*, 1988; James, 1997), provinces (Betzler *et al.*, 1997), or factories (Schlager, 2000, 2003). James (1997) defined an association as "a group of sedimentary particles occurring together under similar environmental conditions and havinguniform or distinctive aspects". Grain associations have either been named based on their dominant components (Lees & Buller, 1972; Lees, 1975), mode of life of the benthic communities (James, 1997), or their latitudinal occurrence (carbonate province of Betzler *et al.*, 1997). Schlager (2000, 2003) named his carbonate factory types on a mixture of latitudinal (tropical), temperature (cool-water) and geological structure (mud mound) nomenclature, separating the three main factories based on the mode of carbonate precipitation.

In defining carbonate groupings or associations, the nomenclatures used are either descriptive or interpretative. However, interpretative environmental connotations, whether rightly or wrongly, are commonly attached to the original descriptive grain associations of Lees & Buller (1972). Some nomenclature schemes are linear, whereas others are hierarchical, the latter "associations" containing smaller subgroups or "assemblages" (Hayton *et al.*, 1995; James, 1997). In the light of the recent recognition of different parameters controlling carbonate grain distribution, it seems timely to reassess the way in which groupings of carbonate sediments are discussed.

The specific objectives of this paper are to: (1) review existing classifications of carbonate grain associations (Fig. 1); (2) discuss briefly the different environmental parameters that represent major influences on grain distributions; (3) evaluate the pros and cons, and make some recommendations, about using different terminologies and hierarchies to group carbonate sediments; and finally, if a descriptive nomenclature is to be used: (4) determine whether the current subdivisions are sufficiently distinctive to help recognize regionally or environmentally important groupings. The layout and order of the manuscript follows subsections related to these objectives.

This paper concentrates on Cenozoic deposits since this was the theme of the Potsdam "Oligo-Miocene" meeting and discussion naturally focused around this era. However, the discussion herein has implications for older time periods, although the problems of classification and interpretation are compounded because many of the older organisms do not have extant relatives, and environmental conditions are more difficult to infer.

EXISTING CLASSIFICATIONS OF CARBONATE-GRAIN ASSOCIATIONS

Nomenclature based on the nature of constituent grains

In these classifications, the names of carbonate-grain associations are usually drawn from the contracted names of their main constituents. This approach was pioneered by Purdy (1963) who defined the *coralgal* facies characterized by the abundance of fragments of hermatypic corals, calcareous green algae, molluscs, benthic foraminifera, and subordinate non-skeletal grains. Afterwards, in their seminal paper, Lees & Buller (1972) recognized two categories among carbonate skeletal grains: the *chlorozoan* association, regrouping green algae and coral fragments (named on the basis of Chlorophyta and Zoantheria), and the *foramol* association, in which foraminiferan and molluscan debris dominates. The chlorozoan association of Lees & Buller (1972) corresponds, at a larger scale, to the previously defined coralgal facies of Purdy (1963).

Lees (1975) later defined a *chloralgal* association including green algae, but devoid of corals, and three associations of non-skeletal grains bearing the name of the dominant particles (*ooid/aggregate, pellet, absent*). As noted by Carannante *et al.* (1988), the phonetic similarity between the terms coralgal (Purdy, 1963) and chloralgal (Lees, 1975) can be confusing because the defining constituents of these associations differ significantly. Carannante *et al.* (1988) further recognized two sediment types within the foramol association: the *rhodalgal* lithofacies (strictly speaking a biofacies) predominantly composed of encrusting coralline or red algae (rhodophytes) and bryozoans, and the *molechfor* lithofacies, in which echinoids and molluscs are associated with arenaceous benthic foraminifera. In the same volume, Nelson *et al.* (1988) coined the term *bryomol* to describe a third subdivision of the foramol assemblage including abundant bryozoan and mollusc fragments.

The number of skeletal groupings of the foramol association exploded with the paper of Hayton

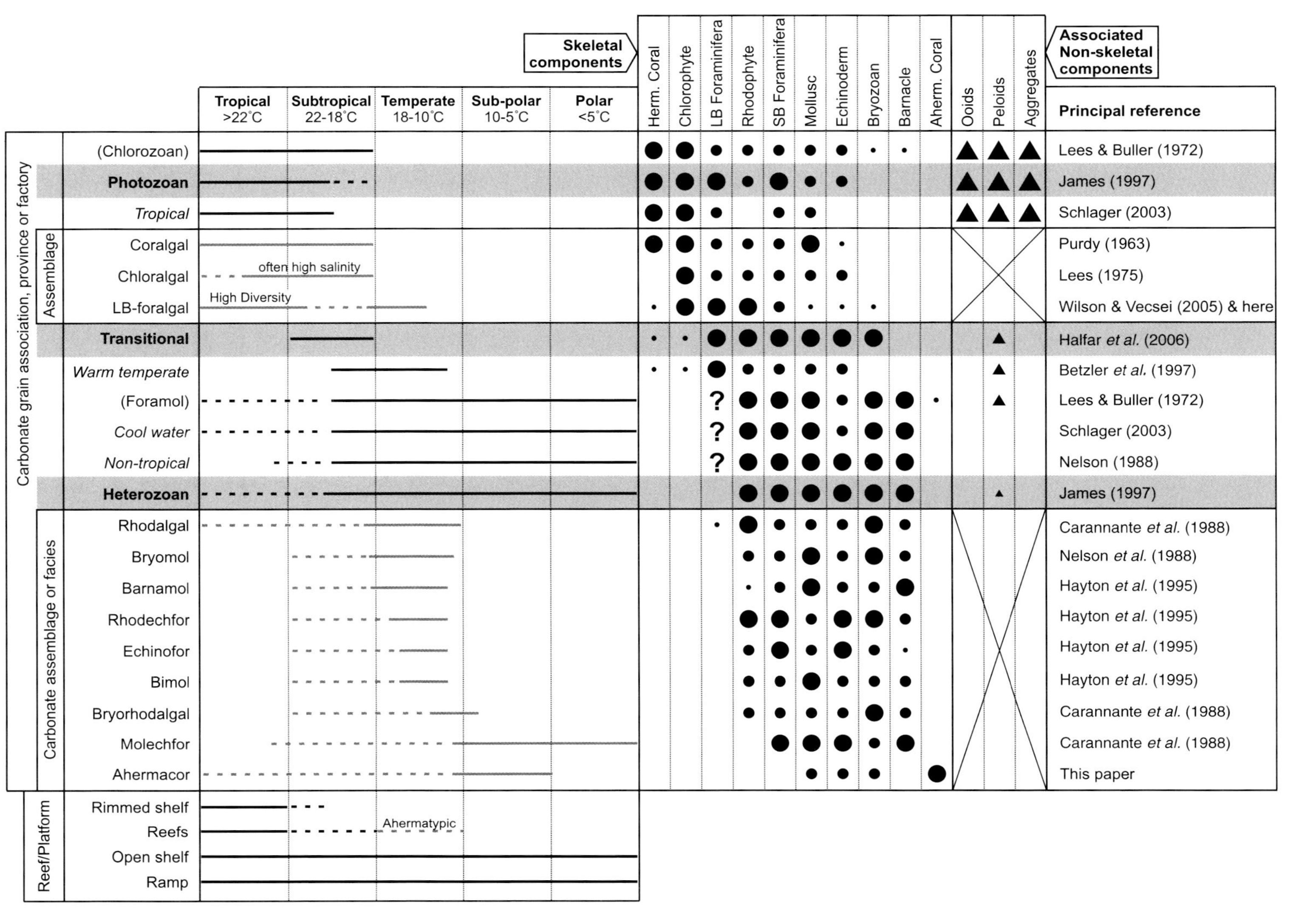

Fig. 1. Summary of regionally important Cenozoic carbonate deposit types based on benthic (skeletal and non-skeletal) components (after Wilson & Vecsei, 2005). Carbonates are grouped hierarchically with a 1st-order grouping including associations, provinces or factories, with the names based either on common components (bracketed), their light dependence, or lack thereof (bold), or environmental conditions (italics). Here the preference is to use the association terms proposed by James (1997; i.e. Photozoan and Heterozoan) complemented by an intermediate group named after Halfar *et al.* (2006; i.e. Transitional) (lines shaded grey). If a two-tier system is to be used, the associations include 2nd-order subgroups of assemblages or biofacies, named descriptively on the basis of their dominant benthic components. Chlorozoan (roughly equivalent to coralgal) and foramol terms may also be used as 2nd-order descriptive terms. Associated non-skeletal grains are shown only for the 1st-order groupings. Second-order assemblages are generally present within, but may not be restricted to, the associations they are shown with. Dominant platform types and the presence/absence of reefs are also shown for the different global provinces. Although the global distribution of the groupings are shown, for the descriptive assemblages, common distributions are shown in grey to signify that these assemblages are not necessarily restricted to particular settings. The subdivision from tropical to polar provinces is based on bottom-water temperature after James (1997). The size of circles and triangles (small-medium-large) indicates the relative abundance of specific carbonate grain types in 1st- and 2nd-order groupings.

et al. (1995). Based on petrographic and linkage cluster analyses of Cenozoic shelf limestones from New Zealand, they presented five additional groupings: the *echinofor* assemblage characterized by the co-occurrence of echinoderm fragments and benthic foraminifera; the *barnamol* assemblage (barnacles and molluscs); the *rhodechfor* assemblage (coralline algae fragments, echinoderms, benthic foraminifera and bryozoans); the *bimol* assemblage (bivalve molluscs); and the *nannofor* assemblage (nannofossils and planktonic foraminifera). However, Hayton *et al.* (1995), like Carannante *et al.* (1988), used a hierarchical approach in which they regarded their "assemblages" or "facies" as subdivisions of the broader foramol association.

In these terminologies, the names of skeletal-grain groupings are based on the dominant faunal assemblage occurring in the sediments. These names could thus be used in a purely descriptive way, but this has never truly been the case. Purdy (1963) pointed out that the coralgal facies is restricted to high-energy, rocky bottoms on the outer reaches of the Great Bahama Bank. Lees & Buller (1972) indicated that minimum and mean temperatures, but not latitude, are the main factors controlling the distribution of carbonate grains around the globe. Nonetheless, they acknowledged that, in terms of temperature, there is a considerable overlap in the ranges of their associations that cannot be explained merely by post-mortem displacement of carbonate grains. Lees (1975) emphasized the role of salinity as a further controlling factor in the repartition of skeletal and non-skeletal grains. He argued also that temperature and salinity have a "mutual compensating effect", i.e. high temperature compensates for low salinity, and vice versa. According to Lees (1975), this effect accounts for the occurrence of the foramol assemblage in warm, but slightly hyposaline, tropical waters.

Carannante *et al.* (1988) suggested that their rhodalgal lithofacies occurs in both temperate and nutrient-rich, tropical waters. They were thus among the first in noticing the importance of nutrients on carbonate-grain distribution. They remarked also that lithofacies distribution is related primarily to latitude and depth, but that water circulation, river discharge, suspended sediment load, and salinity may also play a fundamental role. Finally, they emphasized that tropical and temperate rhodalgal facies are not easily distinguishable in the fossil record. Nelson *et al.* (1988) and Hayton *et al.* (1995) finally proposed a model of non-tropical carbonate shelf where their various carbonate assemblages (barnamol, bimol, etc.) are distributed according to depth, nature of the substrate and water energy.

Nomenclature based on the mode of life of benthic organisms

The lack of evidence for a poleward extension of tropical carbonate belts during warmer episodes of the Earth's history (e.g. Cretaceous) led Ziegler *et al.* (1984) to surmise that light refraction rather than temperature is the main limiting factor controlling carbonate production. Following up on Ziegler *et al.*'s (1984) observation, James (1997) proposed a classification of carbonate sediments and rocks based on the light dependency of the main biotic constituents. This classification comprises two major groups of sedimentary particles: the *Heterozoan* and the *Photozoan* Associations. The Heterozoan Association matches the foramol assemblage of Lees & Buller (1975). It includes carbonate particles produced by light-independent benthic organisms (e.g. bryozoans, barnacles) and fragments of coralline algae. The molechfor, rhodalgal (Carannante *et al.*, 1988) and bryomol (Nelson *et al.*, 1988) lithofacies were thus grouped by James (1997) into this association.

Roughly corresponding to the chlorozoan assemblage of Lees & Buller (1975), the Photozoan Association regroups (1) all skeletal grains derived from warm-water, light-dependant benthic organisms (i.e. scleractinian corals, green algae, larger benthic foraminifera in modern seas and Cenozoic rocks, plus rudist and megalodont bivalves, fusulinid foraminifera and stromatoporoids in pre-Cenozoic limestones) and/or (2) non-skeletal particles (ooids, aggregates, and peloids), plus or minus (3) skeletons from the Heterozoan Association. The chloralgal (Lees, 1975) and coralgal (Purdy, 1963) assemblages are considered by James (1997) as simple facies of the Photozoan Association. Note that some bioclasts (small benthic foraminifera, molluscs, brachiopods and coralline algae) are found in both the Heterozoan and the Photozoan Associations.

James' (1997) classification of carbonate grains is descriptive for modern and Cenozoic sediments because the mode of life (i.e. need for light) of the producing benthic organisms can be readily observed, or may be inferred with a fair degree of certainty. It is more interpretative when applied to older deposits because the mode of life of the carbonate producers is more conjectural. James (1997) further linked the Photozoan Association with the shallow, warm-water, tropical carbonate realm, but clearly stated that Heterozoan carbonates were not necessarily cool-water sediments. Indeed, he pointed out that the interpretation of ancient limestones as cool-water deposits must rely on several criteria including: (1) the occurrence of associated glaciogenic sediments; (2) a setting in high palaeolatitudes; (3) the presence of glendonite (a hydrated calcite typical of cold waters); (4) the co-occurrence of grains from the Heterozoan Association and sediment structures suggesting shallow-water deposition; and (5) geochemical evidence (e.g. elevated $\delta^{18}O$ signatures).

Nomenclature based on latitude and/or water temperature

When considering latitude and/or water temperature, carbonate sediments have usually been partitioned into two categories: those produced in the tropical belt, and those accumulated outside of this zone. The latter have been loosely called "temperate-water" (Lees & Buller, 1972), or "non-tropical" (Nelson, 1988) carbonates. Based on their study of the Brazilian shelf, Carannante *et al.* (1988) presented a model showing trends in the distribution of the chlorozoan, chloralgal, rhodalgal and molechfor lithofacies according to latitude and depth. They further defined three zones of carbonate sedimentation: (1) a *tropical* realm ranging from the equator to 15°S; (2) a *transitional* zone between 15°S and the Tropic of Capricorn; and (3) a *cold-temperate zone* to the south of it.

Rao (1996) defined tropical, temperate and polar carbonates on the basis of their sedimentology and geochemistry. Betzler *et al.* (1997) separated a warm temperate province from the cool temperate province based on the presence of low-diversity large benthic foraminifera, calcareous green algae and zooxanthellate corals. At the same time, James (1997) identified five sediment provinces (*tropical*, *subtropical*, *temperate*, *subpolar* and *polar*; Fig. 1), each characterized by a typical range in bottom-water temperatures. From the equator to the poles, these provinces generally show a decrease in bioclasts from the Photozoan Association relative to those from the Heterozoan Association, a reduction in carbonate mud and marine cements, a shift from aragonite and high-Mg calcite to low-Mg calcite mineralogies, and a change from rimmed platforms to open shelves and ramps.

These classifications can be considered as being descriptive in modern deposits, because the latitude and water temperature of the analysed sediments can be readily determined. However, when applied to ancient limestones, they are clearly interpretative.

Nomenclature based on the mode of carbonate precipitation

Schlager (2000, 2003) proposed a classification of carbonate production systems including three factories (*tropical*, *cool-water* and *mud-mound*) identified by specific modes of carbonate precipitation. The tropical factory is dominated by biotically controlled precipitates from photo-autotrophic organisms (corals, green algae), abiotic precipitates (ooids, marine cements), and carbonate mud of mixed biotic and abiotic origin. This factory operates in warm, well-oxygenated, oligotrophic waters, between approximately 30°N and S of the equator, and usually produces flat-topped platforms with relatively steep slopes. It corresponds to the Photozoan Association of James (1997).

The cool-water factory is characterized by biotically controlled precipitates produced by both heterotrophic (molluscs, bryozoans) and photo-autotrophic organisms (coralline algae, larger benthic foraminifera), but lacks lime mud, ooids, and marine cements. This carbonate system extends poleward of the tropical factory, but also exists at depth and in upwelling areas at lower latitudes, and tends to build homoclinal and distally steepened ramps. Clearly Schlager's (2000) cool-water factory correlates with James' (1997) Heterozoan Association. Abiotic and biotically induced (e.g. microbial) modes of precipitation typify the mud-mound factory. The main products of this carbonate system are: (1) firm micrite precipitated *in situ* ("automicrite"; Wolf, 1965); and (2) abiotic marine cements precipitated in vugs within the rigid framework of automicrite.

The mud-mound factory thrives in oligophotic or aphotic, poorly oxygenated, nutrient-rich environments. Characteristic of this factory are upward-convex mounds that become flatter at shallow depths, and flat-topped platforms when sea-level is reached. Unlike other factories, the mud-mound model is essentially based on the study of pre-Cenozoic limestones and does not have a modern equivalent. It is also outside the main scope of this paper, which deals with skeletal carbonates.

Schlager's (2000) classification can be regarded as partly descriptive and partly interpretative. Identifying the deposits of the mud-mound factory from those produced by the other two carbonate systems is a rather straightforward task based on direct observation. In contrast, the distinction between the products of the tropical and cool-water factories is based on the mode of life of the producing organisms, and thus relies on interpretation when applied to ancient limestones. The mere names "tropical" and "cool-water" further give a temperature and/or a latitudinal connotation to these carbonate groupings.

INFLUENCES ON CARBONATE-GRAIN DISTRIBUTION

It is now known that a wide range of factors may influence the distribution of environmentally significant groupings of carbonate grains. These factors vary from the well-known influences of temperature, salinity and nutrients, to less studied factors such as CO_2 concentrations, Mg/Ca ratio and $CaCO_3$ saturation in seawater (Lees & Buller, 1972; Lees, 1975; Hallock & Schlager, 1986; Mutti & Hallock, 2003; Pomar *et al.*, 2004). As the controls on carbonate sediments from previously understudied regions are investigated more fully, the specific environmental context of grain associations is revealed to be more complex and the boundaries between associations are becoming more "blurred". For example, Photozoan deposits are said to be found in tropical areas where bottom-water temperatures are $>22\,^{\circ}C$, whereas Heterozoan sediments are present in subtropical ($\sim$18–22 °C) and cool temperate ($\sim$5–10 °C) areas, respectively (James, 1997).

A recent compilation of data from around the world has shown that the Photozoan and Heterozoan associations have a wide temperature range and are additionally restricted to waters with specific nutrient content, the latter being estimated by using chlorophyll-a concentrations as proxy data (Halfar *et al.*, 2004, 2006). Thus, the Photozoan Association forms at 18–27 °C in oligotrophic to slightly mesotrophic waters, the Heterozoan to Photozoan transition (1–20% photozoan components) at 15–27 °C under strong mesotrophy and eutrophy, and the Heterozoan Association ($<1\%$ photozoan components) at $<20\,^{\circ}C$ under mesotrophy and eutrophy. It is beyond the scope of this review to evaluate how all the different controls may influence the distribution of grain associations. It is also likely that, as more research is undertaken, the complex interplay of factors controlling the distribution of carbonate sediments in time and space will be better understood in the future.

At the 2005 Oligo-Miocene carbonate workshop in Potsdam, it was found that different researchers were attaching different environmental connotations to the groupings of carbonate sediments based on benthic components. These differing interpretations were not necessarily seen as being incorrect, just that because of the complex interplay of controls, a regionally significant deposit type can occur under a combination of different environmental conditions. For example, rhodalgal deposits may be found in cool-water settings within the photic zone, in warm waters where nutrients are elevated, and/or available light penetration is low, or in areas with some clastic input and unstable substrates.

In order to distinguish between different controlling influences, it is important not just to evaluate deposit type, but to integrate as many independent lines of evidence as possible. Examples of other evidence gathered included genera or species, biota morphology, non-biotic grains, nature of enclosing sediment, types and densities of encrusters and borers, geochemistry and independent regional climatic data. There is still a need to use the groupings of carbonate sediments based on benthic components as they are objective, allow for greater subdivision of the Cenozoic (see also below) and promote far ranging discussions on controlling influences on carbonate sedimentation. However, to maintain objectivity, there was a consensus that the groupings based on benthic components should be purely descriptive terms. It is hoped that through the use of this objective approach and the independent evaluation of other lines of evidence that significant steps will be possible in understanding the nature of global change during the Cenozoic.

DISCUSSION

Hierarchy of carbonate-grain associations

The various groupings of carbonate particles reviewed above apply to geographical areas that vary considerably in scale. For example, the barnamol assemblage of Hayton *et al.* (1995) appears to be limited to tide-dominated seaways in cool-water areas, whereas the foramol group (Lees & Buller, 1975) and the Heterozoan Association (James, 1997) generally extend from latitudes 30° to the poles. It follows that categories of such different status cannot be easily compared, and are thus not mutually exclusive.

Previous nomenclatures can be regrouped into two main categories: (1) those "facies" or "assemblages" representing faunal compositional changes across a carbonate platform; and (2) those of larger geographical extent. The former category embraces the descriptive benthic groupings recognized by Purdy (1963; coralgal), Lees (1975; chloralgal), Carannante *et al.* (1988; rhodalgal, molechfor), Nelson (1988; bryomol) and Hayton *et al.* (1995; echinofor, barnamol, rhodechfor and bimol). The second category includes the more interpretative nomenclature of Lees & Buller (1972; chlorozoan – foramol), of James (1997, photozoan – heterozoan), of Schlager (2000, 2003; tropical – cool water), and those classifications based on water temperature and/or latitude (Carannante *et al.*, 1988; Rao, 1996; James, 1997; Betzler *et al.*, 1997).

Use of second-order descriptive carbonate-grain groupings

Groupings named after the nature of the dominant constituent particles refer to biofacies variations across carbonate platforms caused by changes in factors such as water depth, energy level and/or nutrient concentration. The names of these categories are essentially descriptive, although many authors have attached some environmental significance. Although some acronyms are not an inconsiderable "mouthful", many names (e.g. rhodalgal, chloralgal) are well-entrenched in the literature. In addition, unstudied Cenozoic limestone units can readily be placed in one of these categories upon identification of the predominant grains. Therefore, it is proposed that carbonate-grain groups named after their main constituents should be maintained and used to describe (bio)facies or assemblage variations across carbonate platforms. It should be kept in mind that such grain assemblages occur on a variety of carbonate platforms, in diverse environmental conditions, and, in most cases, across a moderate latitudinal range. For example, the rhodalgal facies can be found at depth, on the outer portion of non-rimmed platforms in the tropics (e.g. Carannante *et al.*, 1988) and in shallower waters, on inner shelves, in cold-temperate areas (e.g. Hayton *et al.*, 1995).

It is strongly recommended that because of the range of factors that can influence individual assemblages that such descriptive groupings should *not* be used to designate carbonate provinces of latitudinal extent. Some authors who are unhappy with a two-tier approach to carbonate grain groupings may use these assemblage terms as first-order terms. However, the preference here, like that of Carannante *et al.* (1998), Hayton *et al.* (1995) and James (1997), is to regard these assemblages or facies as second-order groupings to reflect the fact that they may vary considerably across coherent regions. As discussed above, assemblage terms should be used in a descriptive sense only and additional lines of independent evidence should be sought to evaluate possible environmental interpretations.

It is further suggested that a new descriptive group should be created to include carbonate sediments dominated by diverse larger benthic foraminifera, and sometimes coralline algae and *Halimeda*, such as those described by Wilson & Vecsei (2005) in the equatorial region of SE Asia, and as a fossil example, those forming the upper unit of the Southern Marion Platform (Neogene, NE Australia; Kindler *et al.*, 2006). Wilson & Vecsei (2005) studied assemblages containing perforate larger benthic foraminifera promoted by low light levels (oligophoty). However, diverse imperforate larger benthic foraminifera together with algae are also a common assemblage in shallow-water environments, such as sea-grass meadows. Here the name *LB-foralgal* is proposed to signify the importance of diverse larger benthic foraminifera (perforate or imperforate) together with algae for this new assemblage.

Where coral-rich assemblages can be shown to be azooxanthellate, there may be some merit in introducing a new term to signify biofacies of this type, such as *thermacor*. However, there is debate as to whether a term like this should be transferred across to an assemblage term. Azooxanthellate corals rarely appear to be a dominant component in Upper Cenozoic deposits, and their presence in other groupings such as the foramol, rhodalgal, or bryomol associations/assemblages is already acknowledged.

However, recent discoveries have shown that zooxanthellate corals may build extensive "reefs", such as those forming offshore Norway in the North Atlantic today (Freiwald & Murray, 2005), as well as their fossil (i.e. Miocene–Pleistocene) counterparts off the coast of Ireland (Williams *et al.*, 2006). An "thermacor" assemblage term could be applied to these reefs and associated sediments since they form extensive deposits over a broad region, and perhaps have environmental significance. The modern examples are known to develop in cold water, perhaps promoted by nutrient upwelling along bathymetric highs or associated with methane seeps (Freiwald & Murray, 2005). In addition to abundant, low-diversity ahermatypic corals, other features that might be used to help recognize this assemblage might be particular genera of serpulids, bryozoa, or molluscs. However, ahermatypic corals may not always form reefs in the same conditions as those so far identified for the present day, although comparable examples have been observed in the Danian record (Bernecker & Weidlich, 1990).

Use of first-order carbonate-grain associations or groupings and their nomenclature

There are both advantages and disadvantages attached to using the different "high-level" groupings of carbonate sediments whether these are termed associations, provinces or factories (Table 1). These associations/level groupings are here regarded as first-order terms since they define major compositional variations of carbonate deposits across the globe, reflecting fundamental differences in environmental conditions (Lees & Buller, 1972; Ziegler *et al.*, 1984). It is further suggested that the "facies" groupings represent the building blocks of the carbonate provinces and, because of the complexity of factors controlling carbonate production and sedimentation (James, 1997; Pomar, 2001; Pomar *et al.*, 2004), they may be found in several realms. The nomenclature associated with the first-order groupings either relates to the dominant benthic components or to their mode of life (Lees & Buller, 1972; James, 1997), or to particular latitudinal, temperature or geological structure related occurrence of the deposits (Betzler *et al.*, 1987; Rao, 1996; Schlager, 2000, 2003). Out of these, it is the grain associations related to benthic communities that have received the most widespread utilization in the geological community.

There are perhaps a number of reasons why the provinces or factories approach are used much less than the associations. Use of latitudinal or temperature-related terms may imply circular reasoning since it is often such factors that the researcher is trying to infer. Unless independent evidence is available, such as from palaeomagnetic or geochemical data, it is often difficult to infer that a grain assemblage accumulated under specific temperature or latitudinal settings. Additionally the importance of other primary controls, such as light or nutrients, may be overlooked in these terminologies. Two stages of inferences need to be made to define the groupings: (1) to infer the mode of carbonate precipitation or environmental conditions under which the deposit formed; and (2) to base factory names on inferred environmental conditions or latitudinal setting (in the case of the factories for the mode of precipitation).

It is often stated, when the original terms are defined, that the name attached to a grouping is not definitive, for example development of a "cool-water" factory is not exclusive to cool-waters. In spite of this, naming a province or a factory after a latitudinal or temperature-related occurrence naturally constrains the reader to infer that a deposit type forms in those conditions. In reality, the deposit may also occur under similar or different environmental conditions at different latitudes or temperatures. Even terms such as "temperate-type" unintentionally convey an attached environmental connotation to the reader. On the basis of the discussions on the environmental significance of deposits, it is felt that terms implying a temperature or latitudinal setting are best avoided.

Carbonate grain associations, based either on main benthic components (Lees & Buller, 1972) or, more recently, their mode of life (James, 1997), are widely used in the literature. One suggestion given at the Potsdam meeting was of a non-interpretative grain association based on mineralogy but this would be unworkable due to diagenetic alteration and sometimes complete removal of traces of primary aragonitic or high-Mg calcite. As yet, it remains difficult to see how associations could be named if not based on dominant components or their mode of life, although again it could be that taphonomy or diagenesis might add a bias to the resultant associations. The associations of foramol and chlorozoan based on main benthic components are partly descriptive although a chlorozoan assignment involves recognition of zooxanthellate organisms, a difficult task in the geological record. For example, some corals may be able to switch from a dominantly photoautotrophic, to a mixotrophic or heterotrophic mode of life, involving less or no reliance on zooxanthellae, depending on the local environmental conditions (Insalaco, 1996). Many geologists use the chlorozoan term in a descriptive way, since it actually is applied to deposits rich in green algae and abundant diverse

Table 1. Advantages and disadvantages of the different grouping schemes used for carbonate associations, provinces or factories

Carbonate associations, provinces or factories	Advantages	Disadvantages
Grain associations based on, and named after, dominant benthic components (e.g. Lees & Buller, 1972; Lees, 1975; Carannante *et al.*, 1988)	• Terms are predominantly descriptive, based on dominant components. • Although original terminology relates to Cenozoic grain associations, comparable terms have arisen for Mesozoic and Palaeozoic carbonates. • Grain associations have environmental significance, but this is not implicit in their nomenclature. • The wide range of association terms that have arisen since the original chlorozoan and foramol definitions of Lees & Buller (1972), allows for greater subdivision than grain associations based on the mode of life of benthic communities.	• Resembles a lithofacies epithet (James, 1997). • Chlorozoan term of Lees & Buller (1972) relies on recognition of components that contained chlorophyll or zooxanthellae; an interpretation that is often problematic based on the fossil record. • Original terminology relates to Cenozoic grain associations (e.g. chlorozoan, foramol, rhodalgal). • Associations, whether rightly or wrongly, are often used synonymously with environmental conditions; e.g. foramol often interpreted as a cool-water association. • Use of foramol terminology does not distinguish between an abundance of larger or smaller benthic foraminifera; the first indicative of warm temperatures, the latter often dominating in cooler environments. • Lees & Buller (1972) stated that typical elements of an association, as originally defined by them, may not always be present (e.g. foraminifera or molluscs may be absent from foramol deposits). This has led to the development of a wide range of other terms for associations.
Grain associations based on, and named after, the mode of life of benthic communities (James, 1997)	• Does not resemble a lithofacies epithet • Non-taxa specific terminology designed to be applicable to modern and ancient deposits	• Photozoan term of James (1997) relies on recognition of components that depend on light and warm temperatures: an interpretation that is often problematic in the fossil record. This is particularly the case for Mesozoic and Palaeozoic deposits where many of the organisms involved are now extinct. • A defining characteristic of dominant elements in the Photozoan Association is a reliance on light. However, coralline algae, which may be a dominant element in the Heterozoan Association, are also reliant on light. • Defining associations based on mode of life does not allow for as great a subdivision as is possible for those defined on major components.

(*continued*)

Table 1. (*Continued*)

Carbonate associations, provinces or factories	Advantages	Disadvantages
Provinces named after latitudinal occurrence (Betzler *et al.*, 1997)	• Recognition that a particular suite of environmental conditions is often present at specific latitudes, and these may result in deposits with consistent characteristics.	• Definitions based on latitudinal setting rely on knowing the location at the time of deposition. For older deposits this is often difficult to infer or know precisely. • Where latitudinal setting is not known, two stages of inferences need to be made: (1) the environmental conditions the deposits formed in; and (2) based on the environmental conditions the latitude of formation. • Naming a province after a latitudinal occurrence implies that a deposit type forms in that latitude, whereas in reality these deposits may also occur under the same or different environmental conditions at different latitudes.
Factories named after latitude, temperature or geological structures and subdivided based on mode of carbonate precipitation (Schlager, 2000, 2003)	• Recognition that a particular mode of carbonate precipitation, and hence deposit type may have widespread temporal significance under specific conditions.	• The factory types are not named after one consistent feature and are instead named after a range of latitudinal (tropical) or temperature (cool-water) occurrences and geological structures (mud mounds). • Two stages of inferences need to be made to define the groupings: (1) infer the mode of carbonate precipitation; and (2) base factory names on inferred environmental conditions for the mode of precipitation. • Naming a factory after a latitudinal- or temperature-related occurrence implies that a deposit type forms in those conditions, whereas in reality the deposit may also occur under similar or different environmental conditions at different latitudes or temperatures.

corals, rather than to those known to contain zooxanthellae.

Although many workers still use these component-based association terms, there are problems because these: (1) resemble a lithofacies epithet (James, 1997); (2) create confusion if the lower-order descriptive terms are adopted; and (3) cannot be applied to periods older than the Cenozoic. In addition, foramol deposits may not contain either foraminifera or molluscs. Furthermore, the foramol term does not distinguish between small and larger benthic foraminifera (Wilson & Vecsei, 2005), the latter which are known to contain symbionts and are restricted to warm waters within the photic zone in the Cenozoic (Lee & Anderson, 1991; Hallock, 1999). For these reasons, it is here preferred not to use the grain-based terms for first-order associations.

Associations based on the dominant mode of life of benthic communities were introduced by James (1997) to avoid names resembling lithofacies epithets and are being widely used, despite reservations by some workers. Although use of this "mode of life" terminology does separate carbonate deposits into two environmentally significant associations, as with any of the interpretative nomenclature, there is the problem of invoking circular logic in better trying to understand past environmental conditions.

Particularly for extinct organisms, it may be difficult to infer whether they were reliant on light and temperature. Indeed, many "photozoan" organisms are known to be able to change their mode of life as conditions such as light change. If photozoan elements are not present, as stated by James (1997), this cannot be taken as evidence that warm conditions did not exist. One weakness of James's (1997) terminology is that grains such as coralline algae which are reliant on light, but are not temperature dependent, may be common in both the Photozoan and Heterozoan Associations. It is inferred that these components will continue to remain problematic because they may be promoted by a range of conditions, including low light (oligophoty) and elevated nutrients (eutrophy).

Although there are clearly problems to be aware of when using this terminology, it is nonetheless being widely used. One reason for this is that, notwithstanding problems of inferring light dependence, the terms can be used for all periods of the Phanerozoic. Moreover, the associations appear to have broad environmental significance and do not resemble a lithofacies epithet. Thus, there appears no reason for introducing new association terms, although, because the boundary conditions for these associations needs further refinement (e.g. Halfar *et al.*, 2006), the term "Transitional" is suggested for use in designation of an association characterized by low (1–20%) percentages of phototrophic organisms (Halfar *et al.*, 2006; Fig. 1).

As noted above, the descriptive grain-assemblage terms allow for greater subdivision of the Photozoan and Heterozoan Associations and, if used in a non-interpretative sense, they represent just one additional approach in investigating the complex interplay of controls affecting sedimentation. One possible solution to problems of association terminology is to use a two-tier approach with the groupings based on main components continuing to be used as grain-association subdivisions. The LB-foralgal assemblage would fall in the Photozoan Association (Wilson & Vecsei, 2005), whereas the thermacor facies would be assigned to the Heterozoan Association. If this approach is to be adopted, descriptive terms appropriate to the biota present during different geological periods should continue to be used. Here, only those appropriate for the Cenozoic have been discussed. Researchers familiar with the benthic component terms should maintain their use as purely descriptive and seek additional independent lines of evidence to help interpret past environmental conditions.

CONCLUSION

There is growing awareness that a wide range of environmental factors may influence regionally important carbonate deposits. In addition to the better known controls of temperature, salinity and nutrients, other factors such as water energy, CO_2 concentrations, Mg/Ca ratio, substrate, and $CaCO_3$ saturation in seawater remain poorly understood.

Following recognition that similar deposits may develop under a different combination of environmental factors, there is nonetheless benefit in applying descriptive terms to regionally important carbonate deposits. This is particularly the case when researchers are studying one major time period such as the Cenozoic, as it allows for greater spatial and temporal subdivision of deposit types than may be possible with alternative interpretative associations.

A number of descriptive names already exist for regionally important deposit types of Cenozoic age, based on dominant benthic components. In addition to coralgal (or chlorozoan), chloralgal, foramol, rhodalgal, molecfor and bryomol, two new terms, LB-foralgal and thermacor, are introduced herein to allow for fuller subdivision of Cenozoic deposits. These descriptive terms are either defined here or redefined after the original works. Although the chlorozoan term is not strictly descriptive, it is mostly used in a descriptive sense and its use is so entrenched in the literature that it may prove difficult to replace with the earlier coralgal term.

Descriptive terms may be used as standalone objective terms, although their use does not preclude a two-tier approach, perhaps in combination with an existing interpretative grain-association term. When a two-tier approach is used, additional independent data should be sought for the interpretative assignment. The proposal here is that the groupings based on benthic components should only be used in a descriptive sense.

It is hoped that, when combined with other lines of independent evidence, this objective approach to important deposit types will help promote discussion on evaluating global change and controls on carbonate sedimentation.

ACKNOWLEDGEMENTS

P.K. and M.E.J.W. had equal input into the preparation of this manuscript. We would like to acknowledge all our colleagues present at the Potsdam meeting for their input to the discussion on grain associations. In addition, we would like to thank Noel James, Maria Mutti and Lucia Simone for additional comments on the manuscript. Whilst we have tried to capture the "flavour" of the discussion at the Potsdam meeting, we as the authors are responsible for the ideas expressed in this manuscript, which will not directly correspond to the views of all present. We thank Christian Betzler, Jean Borgomano and Jochen Halfar for reviewing our initial manuscript. P.K.'s travel expenses to Potsdam were supported by the Swiss National Science Funds (grant No. 200020-107436).

REFERENCES

Bernecker, M. and **Weidlich, O.** (1990) The Danian (Paleocene) coral limestone of Fakse, Denmark: a model for ancient aphotic, azooxanthellate coral mounds. *Facies*, **22**, 103–137.

Betzler, C., **Brachert, T.C.** and **Nebelsick, J.** (1997) The warm temperate carbonate province. A review of the facies, zonations and delimitations. *Cour. Forschungs-Inst. Senckenberg*, **201**, 83–99.

Carannante, G., **Esteban, M.**, **Milliman, J.D.** and **Simone, L.** (1988) Carbonate lithofacies as palaeolatitude indicators: problems and limitations. *Sed. Geol.*, **60**, 333–346.

Flügel, E. (2004) *Microfacies of Carbonate Rocks. Springer Verlag*, Berlin, 976 pp.

Freiwald, A. and **Murray, R.J.** (2005) *Cold-Water Corals and Ecosystems. Springer Verlag*, Berlin, 1243 pp.

Halfar, J., **Godinez-Orta, L.**, **Mutti, M.**, **Valdez-Holguin, J.E.** and **Borges, J.M.** (2004) Nutrient and temperature controls on modern carbonate production: an example from the Gulf of California, Mexico. *Geology*, **32**, 213–216.

Halfar, J., **Godinez-Orta, L.**, **Mutti, M.**, **Valdez-Holguin, J.E.** and **Borges, J.M.** (2006) Carbonates calibrated against oceanographic parameters along a latitudinal transect in the Gulf of California, Mexico. *Sedimentology*, **53**, 297–320.

Hallock, P. (1999) Symbiont-bearing foraminifera. In: *Modern Foraminifera* (Ed. B. Gupta), pp. 123–139. *Kluwer Press*, Amsterdam.

Hallock, P. and **Schlager, W.** (1986) Nutrient excess and the demise of coral reefs and carbonate platforms. *Palaios*, **1**, 389–398.

Hayton, S., **Nelson, C.S.** and **Hood, S.D.** (1995) A skeletal assemblage classification system for non-tropical carbonate deposits based on New Zealand Cenozoic limestones. *Sed. Geol.*, **100**, 123–141.

Insalaco, E. (1996) Upper Jurassic microsolenid biostromes of northern and central Europe: facies and depositional environment. *Palaeogeogr. Palaeoclimatol. Palaeoecol.*, **121**, 169–194.

James, N.P. (1997) The cool-water depositional realm. In: *Cool-Water Carbonates* (Eds N.P. James and J.A.D. Clarke). *SEPM Spec. Publ.*, **56**, 1–20.

Kindler, P., **Ruchonnet, C.** and **White, T.** (2006) The Southern Marion Platform (Marion Plateau, NE Australia) during the early Pliocene: a lowstand-producing, temperate-water carbonate factory. In: *Cool-Water Carbonates: Depositional Systems and Palaeoenvironmental Control* (Eds H.M. Pedley and G. Carannante). *Geol. Soc. London Spec. Publ.*, **255**, 271–284.

Lee, J.J. and **Anderson, O.R.** (1991) Symbiosis in Foraminifera. In: *Biology of Foraminifera* (Eds J.J. Lee and O.R. Anderson), pp. 157–220. New York Academic Press, New York.

Lees, A. (1975) Possible influence of salinity and temperature on modern shelf carbonate sedimentation. *Mar. Geol.*, **19**, 159–198.

Lees, A. and **Buller, A.T.** (1972) Modern temperate-water and warm-water shelf carbonate sediments contrasted. *Mar. Geol.*, **13**, 67–73.

Mutti, M. and **Hallock, P.** (2003) Carbonate systems along nutrient and temperature gradients: some sedimentological and geochemical constraints. *Int. J. Earth Sci.*, **92**, 465–475.

Nelson, C.S. (1988) Introductory perspective on non-tropical shelf carbonates. In: *Non-tropical Shelf Carbonate: Modern and Ancient* (Ed. C.S. Nelson). *Sed. Geol.*, **60**, 3–12.

Nelson, C.S., **Keane, S.L.** and **Head, P.S.** (1988) Non-tropical carbonate deposits on the modern New Zealand shelf. In: *Non-tropical Shelf Carbonate: Modern and Ancient* (Ed. C.S. Nelson). *Sed. Geol.*, **60**, 71–94.

Pomar, L. (2001) Types of carbonate platforms: a genetic approach. *Basin Res.*, **13**, 313–334.

Pomar, L., **Brandano, M.** and **Westphal, H.** (2004) Environmental factors influencing skeletal grain sediment associations: a critical review of Miocene examples from the western Mediterranean. *Sedimentology*, **51**, 627–651.

Purdy, E.G. (1963) Recent calcium carbonate facies of the Great Bahama Bank. 2. Sedimentary facies. *J. Geol.*, **71**, 472–497.

Rao, C.P. (1996) *Modern Carbonates, Tropical, Temperate, Polar: Introduction to Sedimentology and Geochemistry. Howrath*, Tasmania, 206 pp.

Schlager, W. (2000) Sedimentation rates and growth potential of tropical, cool-water and mud-mound carbonate systems. In: *Carbonate Platform Systems: Components and Interactions* (Eds E. Insalaco, P.W. Skelton and T.J. Palmer). *Geol. Soc. London Spec. Publ.*, **178**, 217–227.

Schlager, W. (2003) Benthic carbonate factories of the Phanerozoic. *Int. J. Earth Sci.*, **92**, 445–464.

Stanley, S.M. and **Hardie, L.A.** (1998) Secular oscillations in the carbonate mineralogy of reef-building and sediment-producing organisms driven by tectonically forced shifts in seawater chemistry. *Palaeogeogr. Palaeoclimatol. Palaeoecol.*, **144**, 3–19.

Williams, T., **Kano, A.**, **Ferdelman, T.**, **Henriet, J.-P.**, **Abe, K.**, **Andres, M.S.**, **Bjerager, M.**, **Browning, E.L.**, **Cragg, B. A.**, **de Mol, B.**, **Dorschel, B.**, **Foubert, A.**, **Frank, T.D.**, **Fuwa, Y.**, **Gaillot, P.**, **Gharib, J.J.**, **Gregg, J.M.**, **Huvenne, V.A.I.**, **Léonide, P.**, **Li, X.**, **Mangelsdorf, K.**, **Tanaka, A.**, **Monteys, X.**, **Novosel, I.**, **Sakai, S.**, **Samarkin, V.A.**, **Sasaki, K.**, **Spivak, A.J.**, **Takashima, C.** and **Titschack, J.** 2006 Cold-water coral mounds revealed. *EOS Trans. Am. Geophys. Union*, **87**, 525–526.

Wilson, M.E.J. (2002) Cenozoic carbonates in Southeast Asia: implications for equatorial carbonate development. *Sed. Geol.*, **147**, 295–428.

Wilson, M.E.J. and **Vecsei, A.** (2005) The apparent paradox of abundant foramol facies in low latitudes: their environmental significance and effect on platform development. *Earth-Sci. Rev.*, **69**, 133–168.

Wolf, K.H. (1965) Gradational sedimentary products of calcareous algae. *Sedimentology*, **5**, 1–37.

Ziegler, A.M., **Hulver, M.L.**, **Lottes, A.L.** and **Schmachtenberg, W.F.** (1984) Uniformitarianism and palaeoclimates: inferences from the distribution of carbonate rocks. In: *Fossils and Climate* (Ed. P. Brenchley), pp. 3–25. John Wiley & Sons Ltd, Chichester.

Int. Assoc. Sedimentol. Spec. Publ. (2010) **42**, 49–70

Temperate and tropical carbonate-sedimentation episodes in the Neogene Betic basins (southern Spain) linked to climatic oscillations and changes in Atlantic-Mediterranean connections: constraints from isotopic data

JOSÉ M. MARTÍN*, JUAN C. BRAGA*†, ISABEL M. SÁNCHEZ-ALMAZO† and JULIO AGUIRRE*

**Departamento de Estratigrafía y Paleontología, Facultad de Ciencias, Universidad de Granada, Campus de Fuentenueva s.n., 18002 Granada, Spain (E-mail: jmmartin@ugr.es)*

†Centro Andaluz de Medio Ambiente (CEAMA), Avenida del Mediterráneo s/n, 18006 Granada, Spain

ABSTRACT

Marine, shallow-water carbonate units composed of heterozoan (temperate) and photozoan (tropical) associations alternate in the Upper Miocene–Lower Pliocene record of the Betic intermontane basins. Heterozoan carbonates appear in the Early Tortonian, latest Tortonian–earliest Messinian, and in the Zanclean. Photozoan carbonate formation took place in the earliest Tortonian, Late Tortonian, and in the Messinian. Heterozoan carbonates consist of bioclastic limestones with bryozoans, coralline algae and bivalves as major components. Hermatypic corals and calcareous green algae were the principal components in the photozoan carbonates. Fossil assemblages and stable-isotope analyses suggest that sea-surface water-temperature variations controlled the types of carbonates formed. During the Late Miocene, heterozoan carbonates accumulated on ramps in the cold stages of third-order eustatic cycles, during sea-level lowstands, while photozoan carbonates formed on shelves in warm periods, during rising and high sea levels. In the Early Pliocene, a significant change in the Atlantic-Mediterranean connections took place with the closure of the Rifian Straits and the opening of the Gibraltar Straits. A new current pattern for the western Mediterranean was implemented with temperate surface waters flowing into the Mediterranean Sea from a more northern, cooler source area. This new situation probably caused the disappearance of coral reefs in the Mediterranean Pliocene, and favoured the widespread development of heterozoan carbonates on shelves. This regional cooling is in contrast to the subtle global warming recorded in the open oceans during the Early Pliocene.

Keywords Temperate carbonates, tropical carbonates, stable isotopes, Atlantic-Mediterranean connections, Neogene basins, Betic Cordillera.

INTRODUCTION

Following James (1997), there are two main types of shelf-carbonate component associations: heterozoan and photozoan. Biogenic components in heterozoan carbonates are mainly skeletons of bryozoans, coralline red algae, bivalves and benthic foraminifera ("foramol-type" associations in the sense of Lees & Buller, 1972). Hermatypic corals and calcareous green algae are the most characteristic elements in the photozoan carbonates ("chlorozoan-type" associations in the sense of Lees & Buller, 1972). Shelf carbonates forming in marine settings with mean sea-surface temperatures of <20 °C ("temperate carbonates" of Lees & Buller, 1972; "cool-water carbonates" of Brookfield, 1988; "non-tropical carbonates" of Nelson, 1988) are composed of heterozoan associations. Subtropical and tropical carbonates accumulate in areas with mean sea-surface temperatures of >20 °C. Shallow-water carbonates in these areas are generally composed of photozoan associations. Temperature is therefore the major factor controlling the type of shallow-water carbonate association.

High nutrient levels may also favour the development of heterozoan components regardless of the climatic context (Hallock, 1981; Mutti *et al.*, 1997). However, nutrient content is often related to temperature (the higher the temperature, the lower the nutrient content) and high nutrient levels might be related to temperature-reducing phenomena such as upwelling and river discharge (Hallock & Schlager, 1986). This makes it difficult to discern the actual role of nutrients with respect to temperature in determining the component association in fossil carbonates.

Today, cool-water shelf-carbonates extend as far north as the Arctic shelf (Hosking & Nelson, 1969; Nelson & Bornhold, 1983; Freiwald & Henrich, 1994; Henrich *et al.*, 1997), although they are more widespread in the temperate climatic belt. They are particularly abundant in places such as southern Australia (James *et al.*, 1992; Boreen & James, 1993; James *et al.*, 1997; James *et al.*, 2001) and the New Zealand shelf (Nelson *et al.*, 1988), where they have been described and studied in detail in recent years.

Several units of heterozoan and photozoan shelf-carbonates occur in the Neogene Betic basins in southern Spain in the westernmost Mediterranean region. These carbonates can be traced laterally into pelagic marls containing planktonic foraminifera in which oxygen and carbon stable-isotope studies have been carried out. Some of these studies (those referring to the Upper Miocene deposits) have been reported recently (Sánchez-Almazo *et al.*, 2001, 2007) and are summarized here, while others (those concerning the Lower Pliocene deposits) are entirely new and presented here for the first time. Together with the sedimentological and palaeontological evidence, stable-isotope data confirm that sea-surface temperature was the main factor controlling the type of shelf carbonate in the case of the Betic Neogene fossil examples.

GEOLOGICAL SETTING

Betic Neogene basins: Atlantic-linked and Mediterranean-linked basins

The emergence of the Betic Cordillera, the westernmost segment of the European Alpine belt, began in the Middle Miocene. During the Late Miocene, a series of sedimentary basins (the so-called Neogene basins) surrounded the rising Betic uplands. Most of these basins opened directly to the Mediterranean Sea, although some maintained links with the Atlantic Ocean through the Guadalquivir Basin, a foreland basin that developed along the northwestern front of the cordillera (Braga *et al.*, 2003a; Fig. 1).

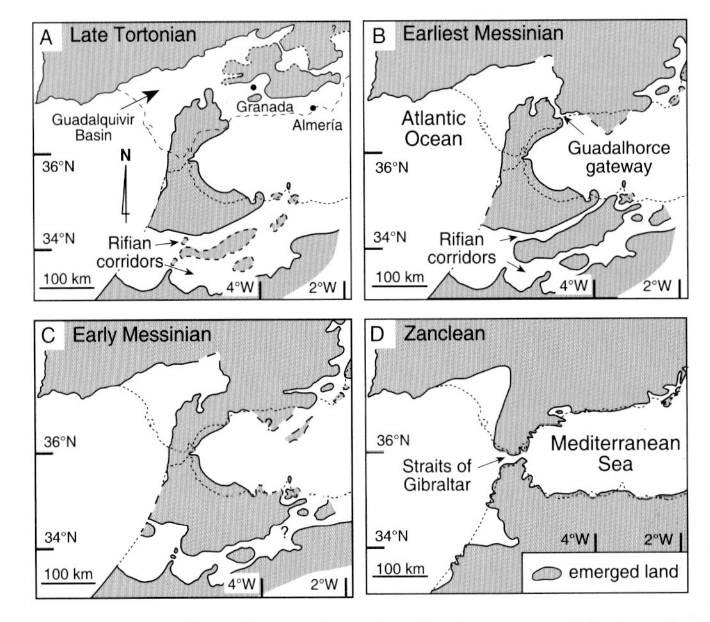

Fig. 1. Late Miocene (Tortonian–Messinian) to Pliocene (Zanclean) palaeogeographical evolution of the Betic Cordillera in the westernmost Mediterranean and changes in the Atlantic-Mediterranean connections for the same period (modified from Esteban *et al.*, 1996 and Braga *et al.*, 2003a).

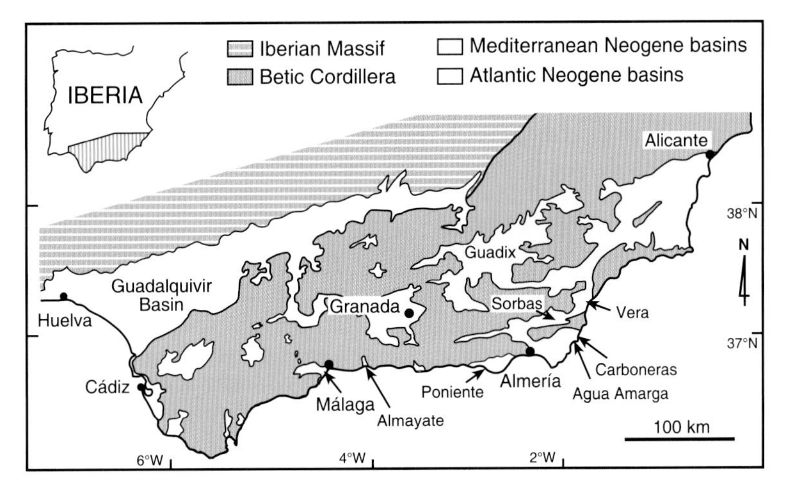

Fig. 2. Neogene sedimentary basins of the Betic Cordillera (southeastern Spain). The basins are named after local main towns in them.

Two main types of Mediterranean-linked basins (Fig. 2) can be distinguished: inner basins (such as the Granada and Guadix basins), distant from the present-day Mediterranean Sea, and outer basins (such as the Sorbas basin), near the present-day Mediterranean. During the latest Tortonian/Early Messinian, the inner basins were isolated from the Mediterranean Sea and became continental. However, the outer basins remained connected to the Mediterranean Sea for the remainder of the Miocene and, in some cases, even into the Pliocene (Fig. 1), except for a short interval in the Messinian (Riding *et al.*, 1998), during the so-called "Messinian Salinity Crisis" (Hsü *et al.*, 1973, 1977).

Atlantic-Mediterranean connections

During the Miocene, the links between the Mediterranean Sea and the world ocean were progressively reduced. After the closing of the Indopacific connections prior to the Late Miocene (Rögl, 1998; Harzhauser *et al.*, 2002), the progressive isolation of this marginal basin led to its desiccation during the Messinian (Hsü *et al.*, 1977), at around 6 Ma (Gautier *et al.*, 1994; Krijgsman *et al.*, 1999a). The Straits of Gibraltar, the current connection of the Mediterranean Sea with the Atlantic Ocean, opened in the Pliocene (Comas *et al.*, 1999). Before evaporite formation, the Mediterranean Sea and the Atlantic Ocean were connected through a series of corridors in southern Iberia (Betic corridors) and northern Africa (Rifian corridors) (Esteban *et al.*, 1996).

During the Late Tortonian, the connection to the oceans in southern Iberia was mainly through the Granada and the Guadix basins (Esteban *et al.*, 1996; Soria *et al.*, 1999; Braga *et al.*, 2003a; Betzler *et al.*, 2006) (Fig. 1). Stratigraphic and micropalaeontological data suggest that the Mediterranean to Atlantic connection through the Guadix Seaway closed at 7.8 Ma at the latest (Betzler *et al.*, 2006). Only one corridor, the Guadalhorce Strait, a narrow gateway located north of Malaga, persisted during the Early Messinian (Martín *et al.*, 2001). The flooding of the Rifian corridors took place at around 8 Ma, and maximum water depths in these passages were reached at 7.6 Ma (Krijgsman *et al.*, 1999b; Barbieri & Ori, 2000).

UPPER MIOCENE–PLIOCENE PLATFORM CARBONATES IN THE MEDITERRANEAN-LINKED BASINS

Heterozoan and photozoan units

The Mediterranean-linked, Betic intermontane basins were mainly filled by terrigenous sediments deriving from the erosion of the emergent land areas. Carbonates formed on the marine platforms around the uplands in areas of low or sporadic siliciclastic input. Shallow-water heterozoan and photozoan carbonate units alternate in the sedimentary infill of the basins. Heterozoan carbonates appear in three stratigraphic intervals: the Early Tortonian and the latest Tortonian–earliest Messinian in the Late Miocene, and the Zanclean in the Early Pliocene. Coral-reef development occurred during the earliestmost Tortonian, the Late Tortonian and the Messinian (Martín & Braga, 1994; Brachert *et al.*, 1996; Braga *et al.*, 1996a; Martín *et al.*, 1999) (Fig. 3).

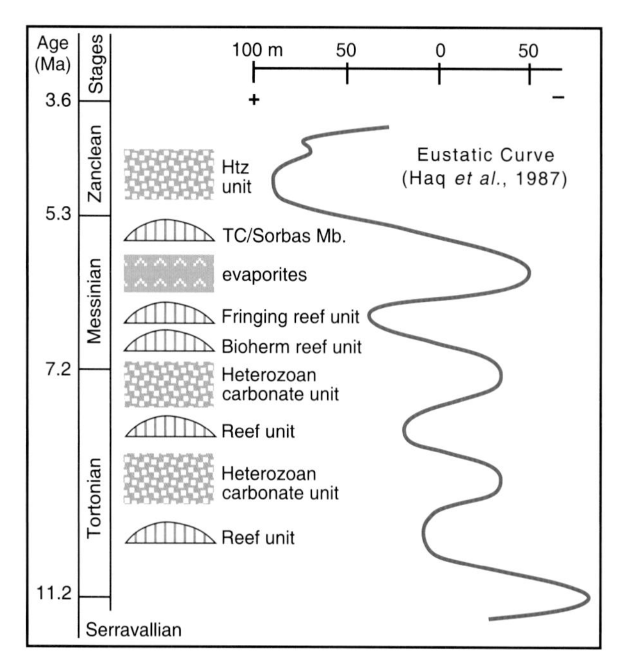

Fig. 3. Heterozoan and photozoan shelf-carbonates in the Late Miocene–Early Pliocene of the Betic Cordillera (modified from Brachert *et al.*, 1996 and Braga *et al.*, 1996a). Upper Miocene heterozoan carbonates were deposited during low sea-levels of third-order cycles, whereas Upper Miocene photozoan carbonates with coral reefs formed during high sea-levels. Lower Pliocene heterozoan carbonates show a clear departure from this trend, as they developed during a global high sea-level period. Sea-level curve is from Haq et al. (1987) with 0 m being present-day sea level.

Facies and sedimentary models

Shallow-water heterozoan carbonates

Shallow-water heterozoan carbonates in the Neogene Betic basins all comprise bioclastic limestones with varying proportions of terrigenous grains (Martín *et al.*, 1996, 2004; Betzler *et al.*, 1997a, 2000; Brachert *et al.*, 1998, 2001; Aguirre, 1998, 2000; Braga *et al.*, 2001, 2003b, 2006a). Biogenic components are mainly particles of originally calcitic skeletons of bryozoans, coralline red algae and bivalves (mainly pectinids and oysters). Benthic foraminifera (small and large), echinoids, barnacles, brachiopods and solitary corals occur as secondary constituents. Serpulids and vermetid gastropods are also locally abundant. The poor representation of originally aragonitic skeletons in the Betic Neogene examples is thought to be due to taphonomic bias caused by selective dissolution and destruction of aragonitic shells before final burial and stabilization of the sediments (Brachert *et al.*, 1998).

All the studied Neogene Betic heterozoan carbonate units show similar characteristics, regardless of age. They all formed in ramps (Fig. 4), with depositional surface profile and local hydrodynamic conditions being the major factors controlling sedimentary facies. The facies differ in relative proportions of skeletal components, internal structures and sedimentary-body geometry (Martín *et al.*, 2004; Braga *et al.*, 2006a).

Shore and near-shore sediments are well represented in the Neogene examples in southern Spain (Martín *et al.*, 1996, 2004; Betzler *et al.*, 1997a; Braga *et al.*, 2001, 2003b). Foreshore, beach deposits consist of grainstone and rudstone beds with well-developed, low-angle (10–20°), parallel lamination. They usually prograde on and laterally pass into small-scale, trough cross-bedded rudstones (shoreface deposits). Lagoonal deposits consist of horizontal silts and marls with land-plant remains, while washover-fan sediments occur as onland-dipping, parallel-laminated layers of packstones and rudstones. Grainstones with high-angle, medium- to large-scale cross-stratification represent local aeolian dunes. Spit-platform deposits consist of landward-dipping (up to 10°), packstone and rudstone beds accumulated by longshore currents and storms. Deposits that formed on low-relief rocky shores typically comprise molluscs, barnacles and coralline-algal rhodolith rudstones.

A shoal belt of medium to large-scale, trough cross-bedded packstones to rudstones occurs seawards of the shore deposits (Martín *et al.*, 1996, 2004; Betzler *et al.*, 1997a; Brachert *et al.*, 2001; Braga *et al.*, 2001, 2003b). In shoal deposits skeletal grains usually display a high degree of fragmentation and abrasion and bioturbation is a common feature, especially by irregular echinoids (*Scolicia* burrows). In some cases, washover-fan deposits appear landwards of the shoal belt, consisting of tabular cross-stratified bodies of rudstones to packstones with conspicuous *Scolicia* burrows (Betzler *et al.*, 1997a).

Carbonate production mainly took place in the areas immediately seaward of the shoals, below fairweather wave-base. Most organisms inhabited these relatively quiet environments, frequently colonized by seagrass meadows or patches. Fossil assemblages contain abundant remains of nodular and branching bryozoans, bivalves, barnacles, echinoids, red algae and other components of the heterozoan association. "Factory" facies consist of poorly-bedded, very coarse, pebble- to cobble-sized bioclastic rudstones and floatstones. The

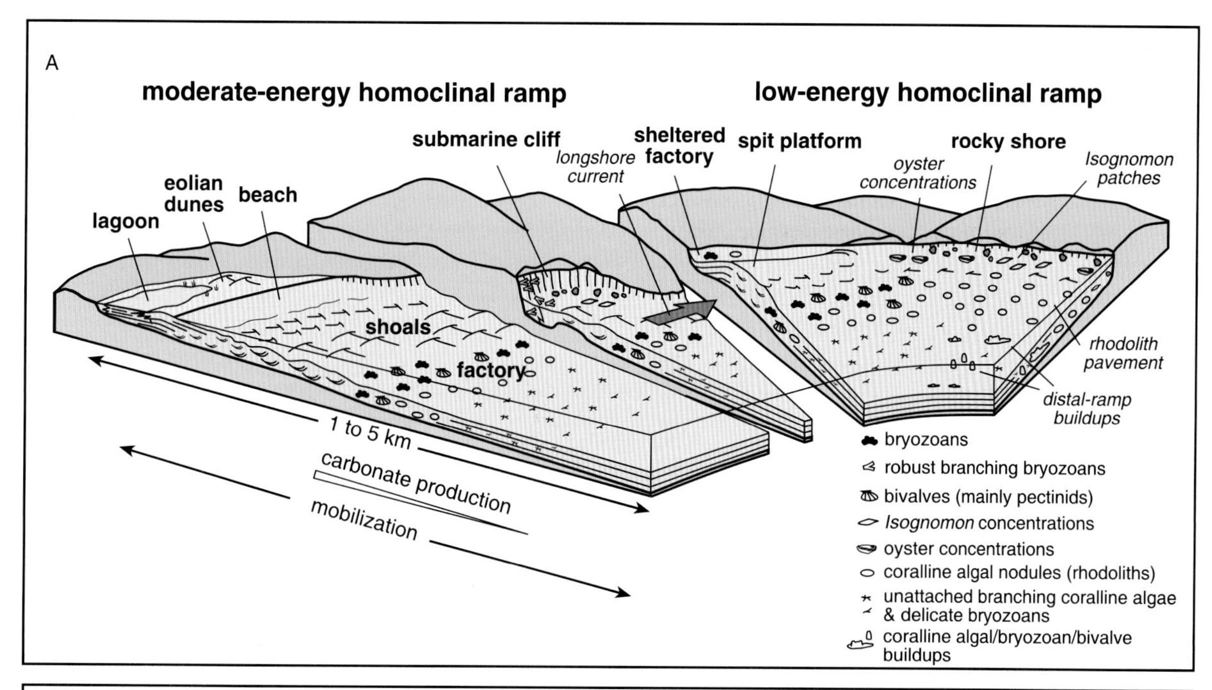

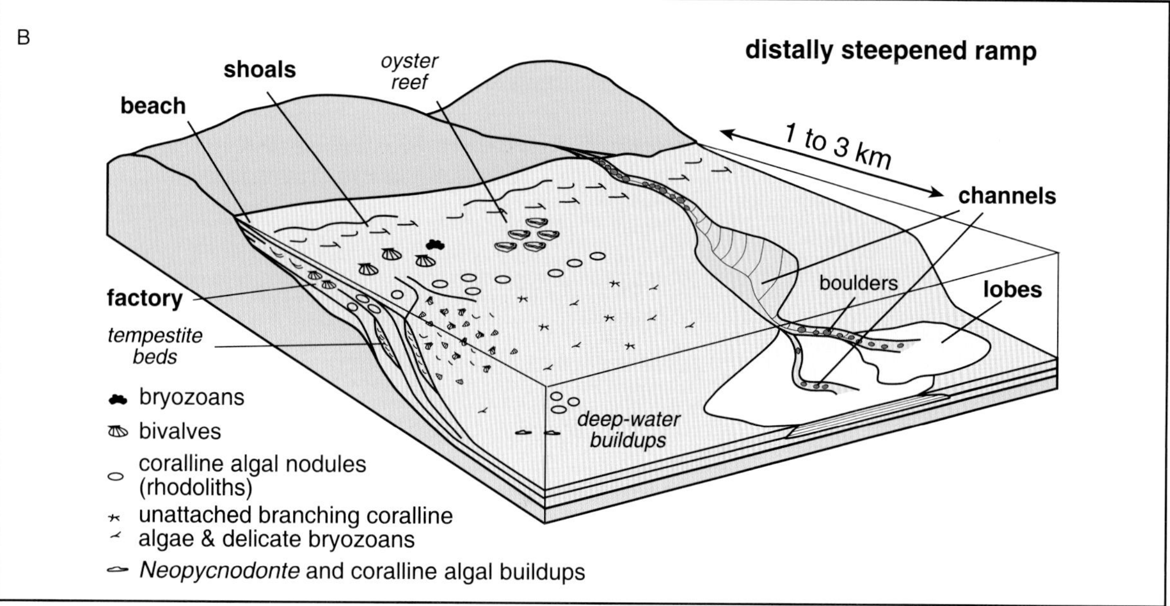

Fig. 4. Sedimentary models for heterozoan carbonates in the Neogene basins of the Betic Cordillera (modified from Braga *et al.*, 2006a). (A) Moderate- and low-energy homoclinal ramps. (B) Distally steepened ramp.

bioclasts typically show a low degree of fragmentation and abrasion and are considered as autochthonous/parautochthonous (Martín *et al.*, 1996). Bivalves (pectinids and oysters) and bryozoans are most common in the shallower settings, whereas coralline algae predominate in deeper factory areas (Braga *et al.*, 2003b, 2006a; Martín *et al.*, 2004). Production likely decreased with depth along the ramp, although in low-energy ramps, small coralline algal, bryozoan and bivalve buildups also developed in distal settings (Martín *et al.*, 2004).

Minor production of carbonate also took place at submarine cliffs. Vermetids colonised submarine cliffs and formed buildups at water depths of a few tens of metres. Robust branching bryozoans were the dominant organisms attached to other cliffs and, after detachment, their skeletons accumulated in aprons of steeply dipping rudstone beds (up to 30°) at the base of the cliffs. Barnacle patches developed locally in depressions between submarine cliffs (Betzler *et al.*, 2000).

The lack of early lithification favoured mobilization of skeletal particles during storms, resulting in landward transportation from the factory areas by waves and/or currents to accumulate in shoals, spits and beaches (Martín *et al.*, 1996). In addition, in moderate- and high-energy ramps, skeletal grains from the factory areas were also mobilized downslope along the ramp as sediment gravity flows and carbonate particles were re-deposited to accumulate in distal-ramp positions, below storm-wave base. Poorly-sorted, coarse-grained debris flows (floatstones and rudstones) grade laterally into more distal, finer-grained turbidites (rudstones) with parallel lamination (Martín *et al.*, 1996). The resulting deposits have a fan-bedded disposition, with single layers gently dipping and increasing in thickness basinwards.

Where mobilization was only sporadic, coarse-grained bioclastic layers (storm deposits) appear interbedded with the fine-grained, autochthonous/parautochthonous "background" sediment, containing abundant remains of delicate branching bryozoan and coralline algal skeletons (Martín *et al.*, 2004).

Distal ramp carbonates change laterally to silty marls and marls that accumulated in the deepest parts of the basins (Fig. 4). Redeposited carbonates also occur in submarine lobe and channel deposits within silty marls. The submarine lobes, tens of metres thick and up to 1 km wide, consist of turbiditic carbonates (rudstones and packstones rich in coralline algal, bivalve and bryozoan clasts) and mixed siliciclastics-carbonates, derived from the platform and emergent uplands. These systems were fed by channels cross-cutting platform sediments, which presumably developed as the continuation of rivers entering the sea. On the outer platform, these channels were up to several hundred metres wide and steeply cut the strata of the platform, removing significant amounts of platform sediments during their excavation (Braga *et al.*, 2001).

Heterozoan associations may form in cool, relatively deep settings, in tropical latitudes, as well as in shallow-water, non-tropical settings, with mean surface-water temperatures below 20 °C, too low for zooxanthellate coral-reef growth and significant production of carbonate by green algae (see James, 1997, among others). The occurrence of very shallow-water deposits with heterozoan associations in these Betic examples (Martín *et al.*, 1996, 2004; Betzler *et al.*, 1997a, 2000; Brachert *et al.*, 1998; Aguirre, 1998, 2000; Braga *et al.*, 2001, 2003b, 2006a) rules out the possibility of them being deep-water sediments coeval with shallow-water coral reefs. According to Pomar *et al.* (2004), this latter situation seems to be the case in some examples of heterozoan Miocene carbonates from the Balearic Islands and central Apennines.

Photozoan carbonates

Zooxanthellate coral reefs, a common feature in present-day, shallow-water, (sub)tropical areas with carbonate sedimentation, are also present in the Mediterranean-linked, Neogene basins of SE Spain (Fig. 5). These reefs appear at five distinct

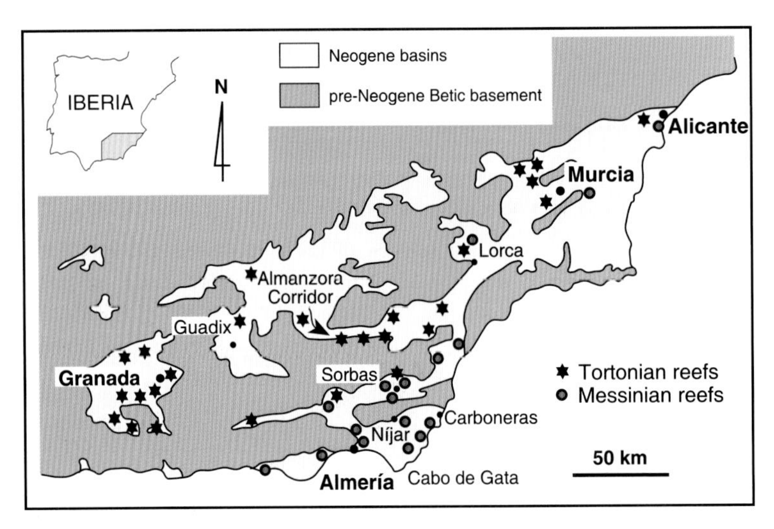

Fig. 5. Late Miocene (Tortonian–Messinian) reefs in the Neogene Betic basins (modified from Martín *et al.*, 1999).

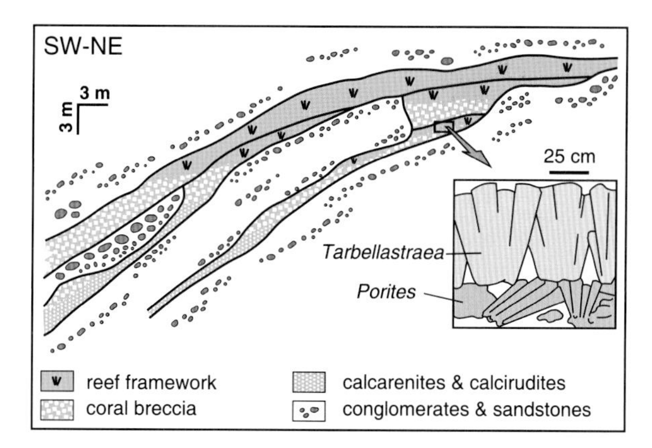

Fig. 6. Tortonian reef structure in the Granada Basin (Fig. 5; modified from Braga *et al.*, 1990). Schematic geometry of patch reefs in fan-delta deposits. Inset: simplified scheme for the development of a *Porites-Tarbellastraea* succession.

Tortonian–Messinian time-intervals (Fig. 3). Tortonian reefs usually occur within delta complexes as small patch-reefs, up to a few hundred metres in length and up to ten metres in thickness. However, in a few cases, more extensive reefs (up to a few kilometres long and some tens of metres thick), with seaward-facing forereef slopes and small backreef lagoons, fringed the palaeocoasts (Martín *et al.*, 1989; Braga *et al.*, 1990).

Tortonian reefs related to fan deltas (Fig. 6) flourished at times of reduced terrigenous sedimentation (Braga *et al.*, 1990) or in areas out of reach of coarse siliciclastic influx (Martín *et al.*, 1989). Reefs in deltas grew on abandoned, inactive lobes and channels (Santisteban & Taberner, 1988; Martín *et al.*, 1989; Braga *et al.*, 1990). *Tarbellastraea* and *Porites* were the dominant builders in Tortonian reefs (Esteban, 1979, 1996), accompanied by *Siderastrea*, *Palaeoplesiastraea*, *Platygyra* and a few other, less abundant, corals (Martín *et al.*, 1989). Upper Tortonian reefs normally exhibit an internal cyclicity consisting of *Porites-Tarbellastraea* interbeds interpreted as being a result of ecological successions (Martín *et al.*, 1989) (Fig. 6).

Messinian coral reefs appear at three successive stratigraphic levels (Fig. 3): (1) the pre-evaporitic Bioherm and (2) Fringing-reef units; and (3) the post-evaporitic Terminal Complex (Sorbas Member). The lowest Messinian reef unit (Bioherm Unit, Martín & Braga, 1994) consists of coral and algal (*Halimeda*) bioherms scattered throughout calcarenites and marls (Braga *et al.*, 1996b; Martín *et al.*, 1997). Coral bioherms, up to 10 m thick and 15 m across, grew near the shelf margin at very shallow depths (Fig. 7). *Porites* and *Tarbellastraea* are the dominant corals with rare *Siderastrea* colonies, all encrusted in places by microbial micrite. *Halimeda* bioherms developed in a more distal position, on the mid-slope, at depths between 20 m and 70 m. They are up to 40 m thick and 400 m long and consist of jumbled *Halimeda* segments, produced by the algae and stabilized and lithified close to their growth sites by micritic and peloidal microbial crusts and by isopachous marine cements (Fig. 7).

The overlying Messinian reef unit comprises prograding fringing reefs (Fringing Reef Unit, Martín & Braga, 1994) that pass basinwards to silty marls, marls, and diatomites. They are localized around basin margins and topographic highs, extending laterally for tens of kilometres and prograding seawards for more than 1 km (Dabrio *et al.*, 1981; Braga & Martín, 1996).

The reef core consists of a coral-stromatolite framework (Riding *et al.*, 1991) (Fig. 8). *Porites*, coated by thick stromatolitic crusts, is almost the only coral, accompanied solely by very rare, small *Siderastrea* colonies. The presence of echinoids, crustose corallines and *Halimeda* within the reef core, together with the scleractinians, indicates normal marine salinities throughout reef growth (Riding *et al.*, 1991). Basinwards from the core, several slope facies can be distinguished: (1) a reef-talus slope comprising framework blocks and breccias; (2) a proximal slope composed of calcirudites and calcarenites (packstones to rudstones) rich in *Halimeda*, serpulids and coralline algae; and (3) a distal slope made up of calcarenites (packstones), calcisiltites and silty marls (Dabrio *et al.*, 1981) (Fig. 8). Lagoonal facies, consisting of calcarenites to calcirudites (packstones to floatstones), occur locally in relation to aggrading phases of reef growth (Braga & Martín, 1996).

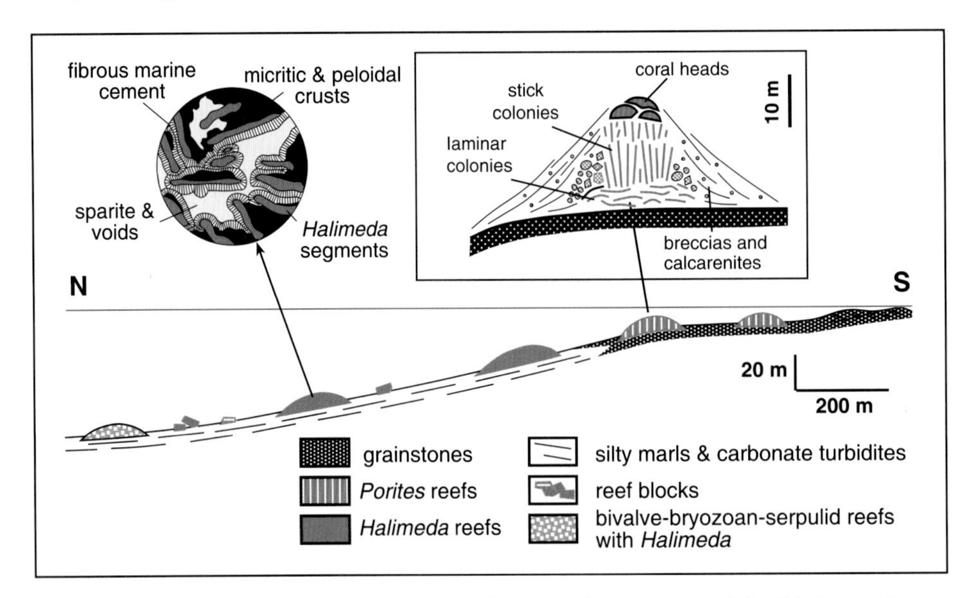

Fig. 7. Characteristics of the Lower Messinian bioherm reef unit. Sedimentary model with internal structure of a coral bioherm (rectangle) and fabric of *Halimeda* bioherms (circle). Modified from Esteban *et al.* (1996), Braga *et al.* (1996b) and Martín *et al.* (1997).

In the patch reefs of the Messinian, post-evaporitic Terminal Complex (Sorbas Member), the only coral to be found is *Porites*, encrusted by stromatolitic crusts. They are small bioconstructions, up to 3 m high and up to 10 m across, commonly associated with oolitic bars and large microbial carbonate mounds (Esteban & Giner, 1977; Riding *et al.*, 1991; Martín *et al.*, 1993; Braga *et al.*, 1995; Calvet *et al.*, 1996; Braga & Martín, 2000).

The rapid decline in coral diversity throughout the Late Miocene in the Betic basins is part of a general pattern in the Miocene Mediterranean Sea (Chevalier, 1961; Rosen, 1999) and supports the idea of cooling leading to a marginal biogeographical setting for the Messinian coral reefs in the western Mediterranean (Martín & Braga, 1994; Rosen, 1999). In modern reefs, species richness decreases with mean seawater temperatures

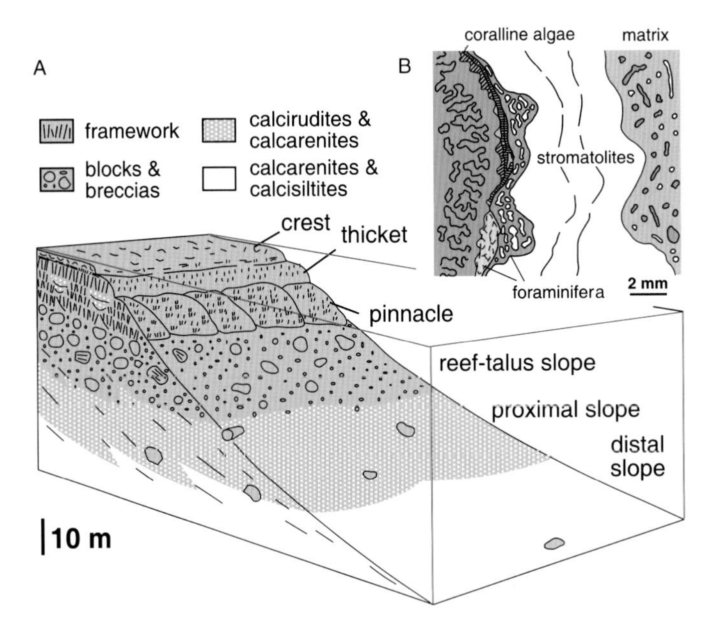

Fig. 8. Messinian fringing reefs. (A) Facies model. (B) Framework fabric. Modified from Riding *et al.* (1991).

towards the outer limits of the reef belt (Wells, 1955; Veron, 1974; Fraser & Currie, 1996).

ALTERNATING UPPER MIOCENE HETEROZOAN–PHOTOZOAN CARBONATE-SEDIMENTATION EPISODES

Carbonate lithofacies of the successive Upper Miocene stratigraphic units in southern Spain reflect several alternating non-tropical and tropical climatic episodes in the region. Heterozoan carbonates accumulated on ramps in temperate climatic conditions during sea-level lowstands, while photozoan lithofacies formed on shelves and rimmed shelves in (sub)tropical contexts during rising and high sea-levels (Martín & Braga, 1994; Brachert *et al.*, 1996). These relative sea-level fluctuations have been correlated (Martín & Braga, 1994; Brachert *et al.*, 1996; Braga *et al.*, 1996a; Esteban *et al.*, 1996; Martín *et al.*, 1999; Fig. 3) with the third-order eustatic oscillations described by Haq *et al.* (1987) and Hardenbol *et al.* (1998). In the cold stages (lowstand periods) of these low-order cycles, the western Mediterranean area was within the temperate climatic belt. In contrast, during warm periods (highstands) tropical conditions prevailed and the Mediterranean area fell well within the subtropical belt (Martín & Braga, 1994). Cycles of $\delta^{18}O$ values of benthic foraminifera in the North Atlantic interpreted as the result of ice volume variations (Miller *et al.*, 1991; Abreu *et al.*, 1998) show a frequency similar to the low-order relative sea-level changes in the Neogene basins in southern Spain.

Stable-isotope data

Stable-isotope data confirm that water temperature is indeed the major factor controlling carbonate lithofacies in upper Miocene carbonate units in southern Spain.

Oxygen isotopes

Lateral facies changes from Upper Tortonian/Messinian marginal carbonates into pelagic marls can be physically traced in well-exposed sections in the Sorbas Basin (SE Spain). The $\delta^{18}O$ values of both planktonic (*Orbulina universa*) and benthic (*Cibicidoides dutemplei* and *Uvigerina peregrina*) foraminifera from the pelagic marls coeval with heterozoan carbonate units are higher than those from marls coeval with reefs (by 1.91‰ $\delta^{18}O$ in the planktonic and 1.67‰ in the benthic records) (Fig. 9). The average $\delta^{18}O$ values of *O. universa* (0.85‰) indicate surface-water temperatures during the deposition of the uppermost Tortonian/lowermost Messinian heterozoan carbonates close to the ones in the present-day western Mediterranean (Sánchez-Almazo *et al.*, 2001) (Fig. 9).

The presence of large benthic foraminifera, mainly *Amphistegina* and *Heterostegina* in the Miocene examples, indicates minimum water temperatures in the order of 16–17 °C (Betzler *et al.*, 1995, 1997b; Langer & Hottinger, 2000; Hohenegger *et al.*, 2000). The $\delta^{18}O$ values, together with planktonic-foraminiferal assemblages, suggest a rise in sea-surface temperatures in the early Messinian as the change from heterozoan to zooxanthellate-reef carbonate deposition took place (Sánchez-Almazo *et al.*, 2001). The average temperatures were highest in the Bioherm Unit interval and then decreased during accumulation of the Fringing Reef Unit (Martín *et al.*, 1999). This temperature decrease is also reflected in the reduction of coral species diversity from the former unit to the latter (Martín & Braga, 1994).

A cooling of Mediterranean waters at the Tortonian–Messinian boundary has also been detected with the $\delta^{18}O$ record of planktonic and benthic foraminifera in the Monte Casino section in Italy (Kouwenhoven *et al.*, 1999) and in the Salé Briquetiere section in the Rifian Corridor in Morocco (Hodell *et al.*, 1994). Increase in global ice volume has been invoked as a possible causal mechanism to explain the high $\delta^{18}O$ values near the Tortonian/Messinian boundary in the Moroccan corridor (Hodell *et al.*, 1994), but this increase in $\delta^{18}O$ benthic values has also been explained by cooling due to rearrangement of the Rifian corridor currents. Increased evaporation would have a similar effect, but there is no record of widespread evaporite formation in the Mediterranean area at this time.

Ice cap expansion has been reported from the southern hemisphere at about 7 Ma, near the Tortonian–Messinian boundary (Warnke *et al.*, 1992). An expansion of polar ice caps is also suggested for the northern hemisphere during that period (Jansen & Sjøholm, 1991; Larsen *et al.*, 1994; Thiede *et al.*, 1998). The $\delta^{18}O$ values decrease approximately at the beginning of Chron C3Ar in the Monte Casino section (Kouwenhoven *et al.*, 1999). This decrease is coeval with the

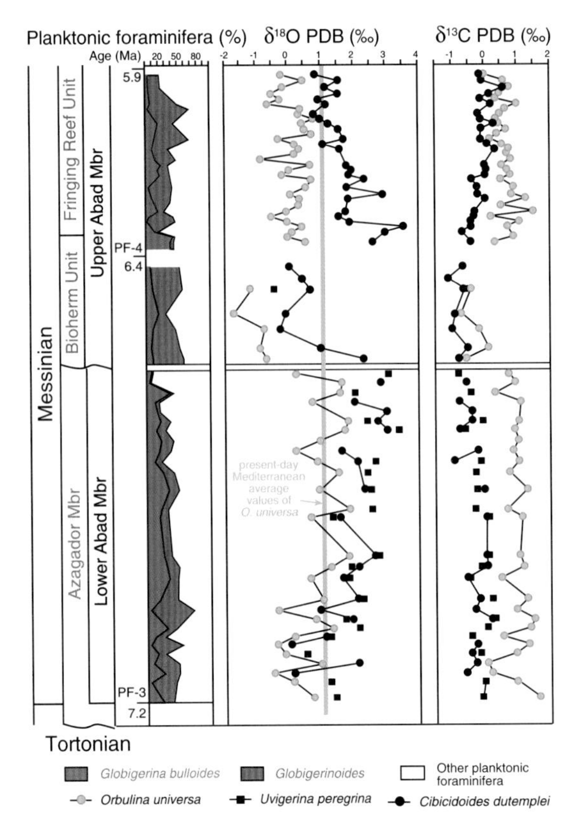

Fig. 9. Percentage of significant foraminifera groups and oxygen- and carbon-isotope values in the Upper Miocene of the Sorbas Basin plotted against estimated absolute age (from Sánchez-Almazo *et al.*, 2001).

decrease in $\delta^{18}O$ values observed in the marls laterally equivalent to the Messinian reefs in the Sorbas Basin and it points to a regional increase in seawater temperature in the Mediterranean Sea (Sánchez-Almazo *et al.*, 2001).

Cyclic relative sea-level oscillations of several tens of metres, linked to temperature changes related to precession cycles, have been recorded in the Cariatiz Messinian Fringing Reef (Sánchez-Almazo *et al.*, 2007) (Fig. 10). Correlation of the pelagic signal with cycles in the shallow-water carbonates at the basin margin indicate that *Porites* frameworks and reef-slope deposits with *Halimeda* gravels formed during surface-water temperature rises and thermal maxima (with high proportions of warm-water foraminifera and low $\delta^{18}O$ values) within the precession cycles (Fig. 11). Bioclastic packstones to rudstones with no coral growth ("lowstand inverted wedges" of Braga & Martín, 1996) that accumulated at lower relative sea levels were coeval with the lower temperature intervals recorded by the planktonic $\delta^{18}O$ signal and planktonic foraminifera assemblages (Sánchez-Almazo *et al.*, 2007) (Fig. 11).

Carbon isotopes

Carbon-isotope values of the foraminifera tests collected from pelagic marls can also yield palaeoenvironmental information about the laterally equivalent shallow-water carbonates. Upper Miocene $\delta^{13}C$ values range from +1.64‰ to −0.13‰ and show a slight general decrease from the Upper Tortonian (Fig. 9). This tendency is probably the local record of the worldwide Late Miocene Carbon Shift (Sánchez-Almazo *et al.*, 2001), which resulted in a decrease of $\delta^{13}C$ values on a global scale (Keigwin, 1979; Haq *et al.*, 1980; Vincent *et al.*, 1980; Hodell *et al.*, 1994). This trend is interrupted in the pre-evaporitic Messinian reef units by excursions to higher values (Sánchez-Almazo *et al.*, 2007).

A cyclic variation in $\delta^{13}C$ values related to precession cycles can also be observed in the smoothed planktonic signal in the Cariatiz section (Sánchez-Almazo *et al.*, 2007) (Fig. 10). This signal undergoes low-amplitude fluctuations throughout the section, which parallel changes in lithology, planktonic foraminifera assemblages, and $\delta^{18}O$ values, reflecting precession-forced cyclicity. Minimal $\delta^{13}C$

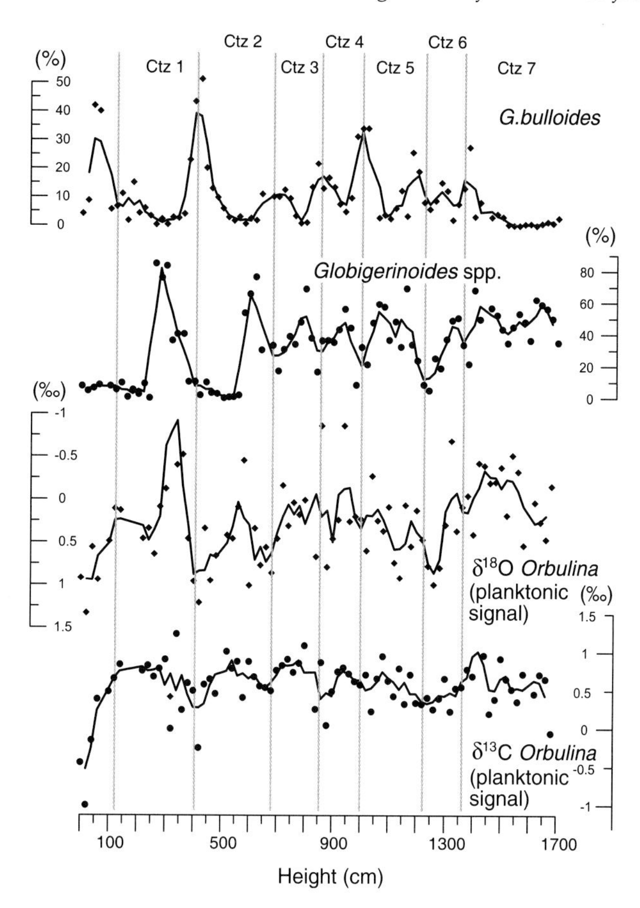

Fig. 10. Smoothed curves (three-point running averages) of variations in oxygen and carbon-isotope values and proportions of significant planktonic foraminifera taxa through the Cariatiz section. Ctz 1–7 are Cariatiz depositional cycles. From Sánchez-Almazo *et al.* (2007).

values generally correspond to maximal $\delta^{18}O$ values and a minimal abundance of warm-water planktonic foraminifera. Peaks in planktonic $\delta^{13}C$ values are coincident with ^{18}O-depleted oxygen values (Sánchez-Almazo *et al.*, 2007) (Fig. 10). Diatomitic marls formed in the basin during the coolest intervals of the precession-related cycles. The occurrence of the diatom *Thalassionema nitzschiodes* in a few diatomitic beds indicates high productivity that might be related to coastal upwelling (Saint-Martin *et al.*, 2001).

Although there are parallel variations in $\delta^{13}C$ values and temperature changes during the precessional cycles (Fig. 11), temperature itself cannot explain the fluctuations observed in carbon isotopes (Romanek *et al.*, 1992, Bemis *et al.*, 2002), nor can changes in productivity account for the $\delta^{13}C$ cyclicity in the Cariatiz section. The fluctuations in $\delta^{13}C$ values of the planktonic foraminifera track changes in $\delta^{13}C$ values in total dissolved inorganic carbon (DIC). In the photic zone the latter changes depend on primary productivity as organic matter differentially uptakes ^{12}C during photosynthesis and the DIC in the waters is consequently enriched in $\delta^{13}C$. However, in several samples of the Cariatiz record, maximal proportions of *G. bulloides*, indicating nutrient-rich, cooler waters, coincide with minimal $\delta^{13}C$ values of planktonic foraminifera (Fig. 11). This suggests that productivity is not a key factor in determining the carbon-isotope signal. Input of ^{12}C greater than the ^{12}C uptake by photosynthesis must be the reason for the minimal $\delta^{13}C$ values during the highest-productivity conditions. Input of ^{12}C from remineralization of organic matter in deeper water probably increased due to more efficient mixing in the water column during the lower sea levels, related to the cooler periods within the precessional cycles (Sánchez-Almazo *et al.*, 2007). The $\delta^{13}C$ in the Cariatiz section probably reflects the balance between input of ^{12}C to the sea-surface water and ^{12}C uptake by photosynthesis. This

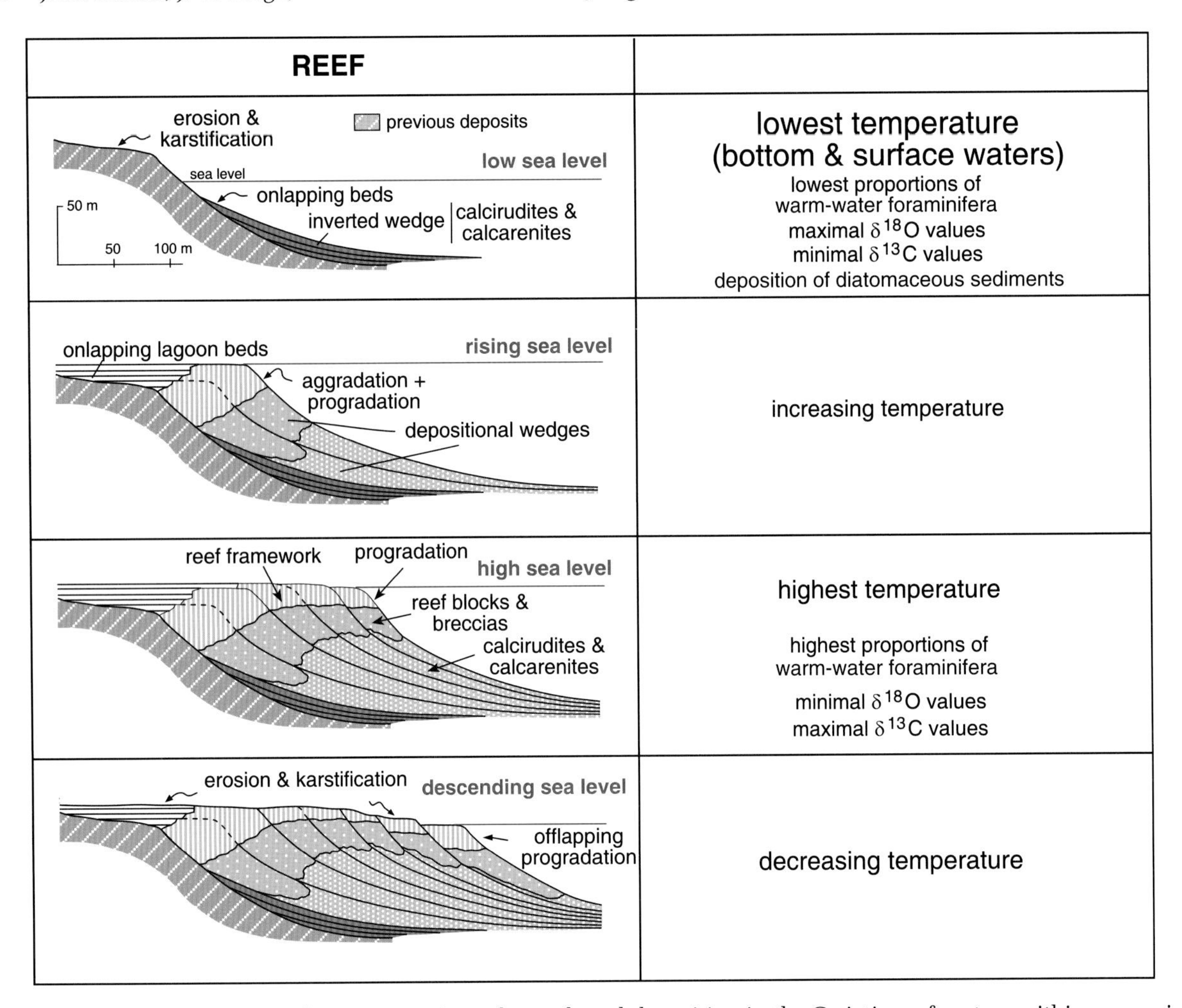

Fig. 11. Messinian fringing reefs. Variations in reef growth and deposition in the Cariatiz reef system within precession cycles and coeval changes in the forereef area (from Sánchez-Almazo *et al.*, 2007).

balance changes with sea-level and temperature oscillations within the precessional cycles (Sánchez-Almazo *et al.*, 2007).

LOWER PLIOCENE HETEROZOAN CARBONATE-SEDIMENTATION EPISODE

Photozoan carbonate lithofacies, typical of (sub) tropical carbonate platforms, occur in the Mediterranean topmost Messinian and include coral (*Porites*) reefs (Riding *et al.*, 1991; Cunningham *et al.*, 1994; Cornée *et al.*, 1996; Esteban *et al.*, 1996). In contrast, Lower Pliocene shallow-water carbonate production in the Mediterranean is dominated by heterozoan associations characteristic of carbonate sediments in temperate and/or cool-water seas (Nelson, 1988; James, 1997).

A striking paradox: Early Pliocene Mediterranean cooling

Global temperatures during the Early Pliocene were similar to or even higher than the ones prevailing in the latest Messinian. Palaeotemperature indicators measured in open oceans reveal that seawater temperatures did not change across the Messinian (latest Miocene)–Zanclean (Early Pliocene) boundary (Hodell & Kennett, 1986; McKenzie *et al.*, 1999; Hodell *et al.*, 2001; Billups, 2002) (Fig. 12). The Early Pliocene was marked by a subtle warming trend lasting until ~3.2 Ma (Zachos *et al.*, 2001). Despite this global pattern, in the Mediterranean area, zooxanthellate coral reefs paradoxically disappeared at the beginning of the Pliocene after having been extensively recorded throughout a long stratigraphic range in Miocene deposits.

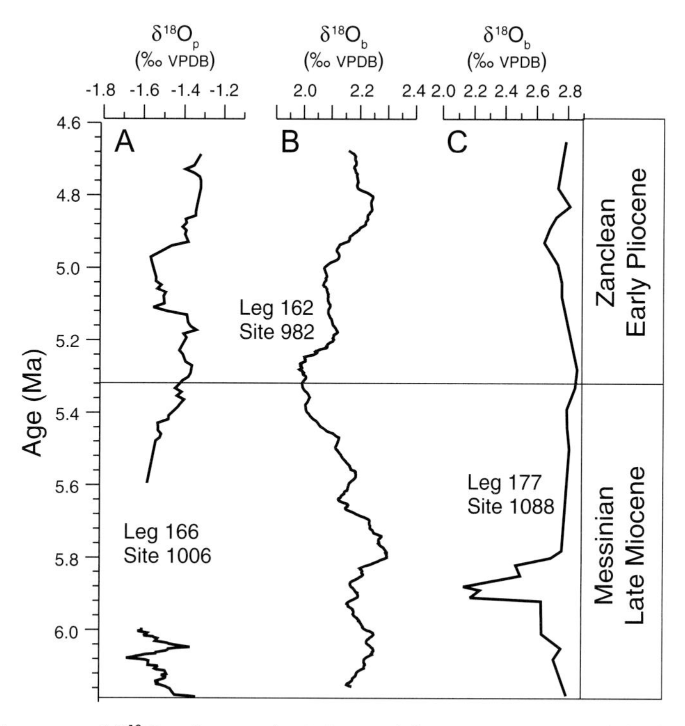

Fig. 12. Strongly smoothed curves of $\delta^{18}O$ values at the Miocene/Pliocene transition. (A) Planktonic foraminifera in the Bahamas (McKenzie *et al.*, 1999). (B) Benthic foraminifera in the North Atlantic (Hodell *et al.*, 2001). (C) Benthic foraminifera in the Southern Ocean (Billups, 2002). $\delta^{18}O$ values of Lower Pliocene foraminifera from open-ocean settings are similar to or lighter than the ones in the Messinian record.

Oxygen stable-isotope analyses carried out on planktonic foraminifera from a core (Hole 976B) in the western Mediterranean (ODP Leg 161, Site 976), reveal a temperature decrease of Mediterranean surface waters during the Early Pliocene (MPl 2 Zone), at about 4.9 Ma, which probably caused the disappearance of zooxanthellate coral reefs in the Mediterranean Pliocene and favoured the widespread development of heterozoan carbonates on shelves.

Comparative data from ODP Site 976

ODP Site 976 was drilled by Leg 161 in the Alborán Sea, off the coast of Malaga (position: 36°12.313′N, 4° 18.763′W; water depth: 1108 m), close to the present-day Gibraltar Straits (Fig. 13).

Lithology and biostratigraphy

Hole 976B exhibits the best-preserved uppermost Messinian–Lower Pliocene deep-water sequence

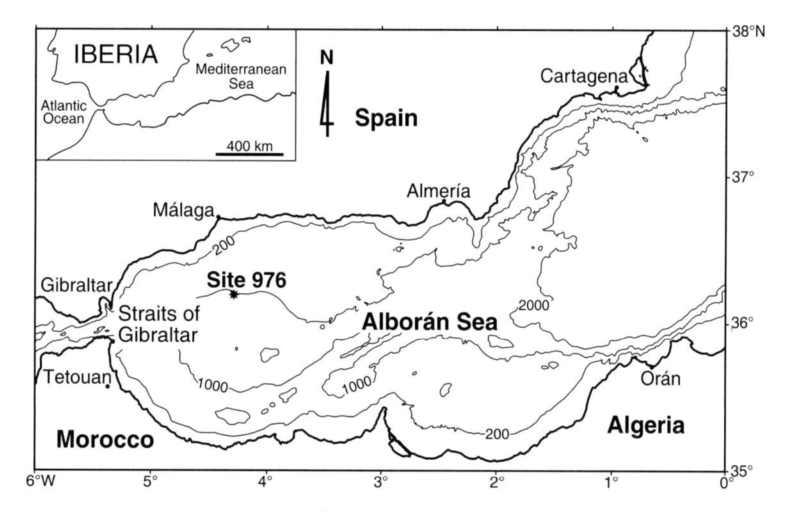

Fig. 13. Location of Site 976 (ODP Leg 161) in the Alborán Sea (western Mediterranean).

drilled in the western Mediterranean to present. These basinal deposits contain well-preserved planktonic foraminifera appropriate for isotopic studies of the Miocene–Pliocene transition. The sampled section is a monotonous succession of greyish nannofossil-rich clay and silty clay (Comas *et al.*, 1996). The Miocene–Pliocene boundary is about 574 mbsf according to the nannofossil assemblages (Siesser & de Kaenel, 1999). From 574 mbsf upwards, the common occurrence of *Globorotalia margaritae* (Comas *et al.*, 1996; Serrano *et al.*, 1999) indicates the Mediterranean planktonic foraminifera zone MPl 2 (Cita, 1975; Sprovieri, 1993). The occurrence peak of *Sphaeroidinellopsis* in the absence of *G. margaritae*, characteristic of the MPl 1 Zone (Cita *et al.*, 1978; Sprovieri, 1993), is not recorded at Site 976. The top of the sampled interval still corresponds to the MPl 2 zone (Comas *et al.*, 1996, Serrano *et al.*, 1999) due to the occurrence of *G. margaritae* with no record of *Globorotalia puncticulata*, indicative of the MPl 3 zone (Cita, 1975; Sprovieri, 1993). The presence of *Globigerina nephentes* together with *G. margaritae* indicates a maximum age interval of 5.10 to 4.18 Ma for the sampled Pliocene interval.

Stable-isotope data

Methods. A total of 136 samples were analysed, with a sampling interval of 50 cm except for the section between 573.5 and 583.1 mbsf, in which only the core catcher was recovered. Fifteen to twenty specimens of the planktonic foraminifera *Orbulina universa* were picked from the 250–350 μm fraction in each sample. Analyses were performed using a VG SIRA mass spectrometer at The Godwin Laboratory, University of Cambridge. The isotope data are reported to the VPDB international standard. The precision of the results was better than ±0.06‰ for $^{12}C/^{13}C$ and ±0.08‰ for $^{16}O/^{18}O$.

Results. Three distinct intervals of $\delta^{18}O$ values of *Orbulina universa* tests can be distinguished in the analysed samples (Fig. 14). In the lower interval (588.9–573.8 mbsf), $\delta^{18}O$ values show large excursions ranging from −1.68 to 1.33‰ within an increasing trend. The average $\delta^{18}O$ value is −0.40‰. The highest value occurs at the top of the interval, which coincides with the top of the Messinian. In the middle interval (573.8–556.9 mbsf), $\delta^{18}O$ values also vary considerably within a general increasing trend in a narrower range (from −2.60 to 0.45‰) and the average value is −1.12‰. From 556.9 mbsf to the top of the sampled interval (518.6 mbsf), $\delta^{18}O$ values change, ranging from −0.33 to 1.14‰ (average of 0.31‰). The shift to the higher $\delta^{18}O$ values recorded at 556.9 mbsf took place several hundred thousand years after the beginning of the Pliocene, dated at 5.3 Ma. The base of the recorded Pliocene in the core (573.8 mbsf) is at least at 5.10 Ma (first common occurrence datum of *G. margaritae* in the Mediterranean). The change in $\delta^{18}O$ values took place at a minimum age of 4.9 Ma, based on interpolation of the age of the base and the minimum age of the top of the sampled Pliocene interval.

Interpretation. The change to heavier $\delta^{18}O$ values in the planktonic record from Site 976 in the Early Pliocene at 556.9 mbsf (Fig. 14) can be attributed to a sudden decrease in sea surface temperature in the western Mediterranean. Other causes such as salinity or ice volume can be discounted. The Lower Pliocene sedimentary and fossil records do not support the possibility of high salinities of the western Mediterranean water that would also have led to heavier $\delta^{18}O$ values. In addition, there is no evidence of an increase in polar ice caps during the Early Pliocene that might account for the reported heavier $\delta^{18}O$ values. After this cooling, surface temperatures remained relatively constant, with only minor fluctuations during the remainder of the analyzed time interval.

The temperature decrease observed in the western Mediterranean is not recorded in the open oceans. In the southwest Pacific and South Atlantic, the benthic signal suggests a long period of warm conditions between 5.0 and 4.1 Ma (Hodell & Kennett, 1986) and the average benthic $\delta^{18}O$ values in the Southern Ocean are similar to the ones from Messinian deposits (Billups, 2002). After a period of lower values corresponding to the Messinian–Pliocene boundary, the benthic signal in the North Atlantic between 5.0 and 4.6 Ma has mean values comparable to those of the period 6.5–5.9 Ma (Hodell *et al.*, 2001), during which reefs flourished in the Mediterranean Basin. A similar pattern is shown by the planktonic signal in the central Atlantic (Bahamas): the $\delta^{18}O$ values have similar ranges during 5.5–4.6 Ma and 6.0–6.2 Ma (McKenzie *et al.*, 1999) (Fig. 12).

The oxygen stable-isotope record of planktonic foraminifera at the Messinian–Pliocene transition in the western Mediterranean demonstrates that the change in carbonate production and the disappearance of zooxanthellate coral reefs in the Mediterranean Pliocene were related to an actual

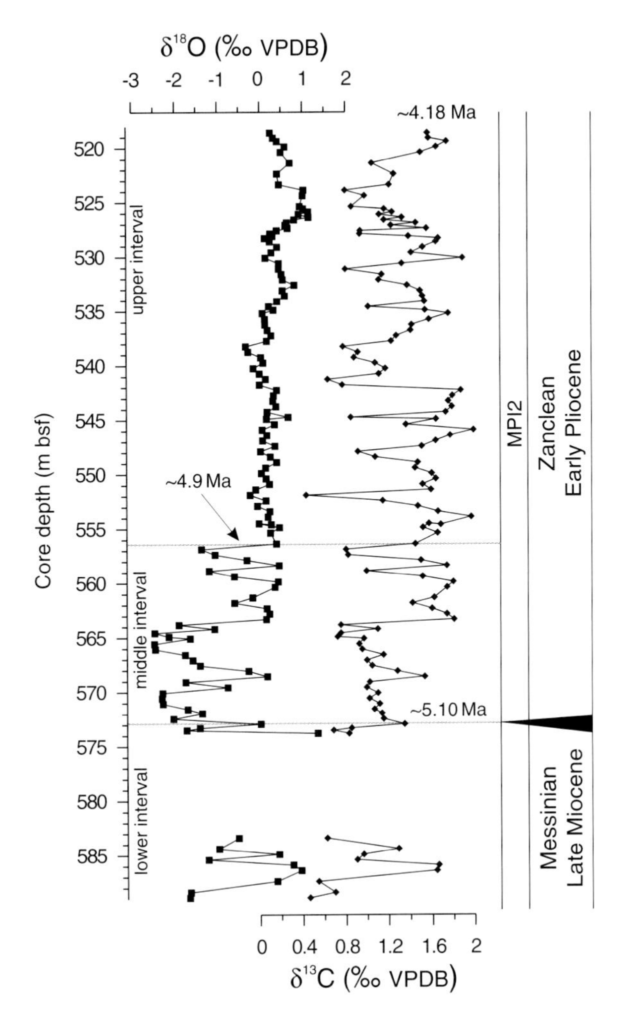

Fig. 14. $\delta^{18}O$ and $\delta^{13}C$ values of the planktonic foraminifera *Orbulina universa* from the Miocene–Pliocene transition in Site 976. Note three distinct intervals of $\delta^{18}O$ values. The shift to heavier $\delta^{18}O$ values at 556.9 mbsf (upper interval) can be attributed to a sudden decrease in sea-surface temperature. This shift took place several hundred thousand years after the Miocene–Pliocene boundary (at about 4.9 Ma).

decrease in sea-surface water temperatures. This cooling of the Mediterranean waters cannot be reconciled with global temperature variation and as such must be related to changes in regional patterns of surface water inflow into the semi-enclosed basin.

Closing of the Rifian corridors and the development of the Gibraltar Straits: consequences

The early Pliocene cooling of the Mediterranean surface waters was demonstrably a regional phenomenon related to the particular oceanography of this semi-isolated basin. It is suggested that the temperature drop was caused by the rearrangement of the gateways connecting the Mediterranean to the Atlantic, which caused the Atlantic waters to enter the Mediterranean via the Straits of Gibraltar from a more northern and cooler source area. Other contributing causes are not discarded but cannot be specified at present. As described above, at the beginning of the Late Miocene (Tortonian), the Mediterranean was connected to the Atlantic through the Rifian corridors in northern Morocco, and the Betic corridors in southern Spain (Fig. 1). The previous connections with the Indian Ocean over the Middle East and the Paratethys closed prior to the Late Miocene (Harzhauser *et al.*, 2002, Meulenkamp & Sissingh, 2003). Continued uplift of the Betic Cordillera caused the closure of several Betic corridors during the Tortonian (Esteban *et al.*, 1996; Soria *et al.*, 1999; Martín *et al.*, 2001; Betzler *et al.*, 2006). The last gateway at the Betic margin was the Guadalhorce corridor. The closure of the Guadalhorce gateway took place in the Early Messinian and Atlantic-Mediterranean communication was subsequently limited to the Rifian Straits in northern Morocco (Martín *et al.*, 2001; Fig. 1). The later closure of the Rifian Straits led to the complete isolation of the Mediterranean Sea, which resulted in the precipitation of extensive evaporites (Hsü *et al.*, 1973) at around 6 Ma (Krijgsman *et al.*, 1999a).

The post-evaporitic reflooding of the western Mediterranean Basin by normal marine water took place before the end of the Messinian, as indicated by the sedimentary record in the deep Western Mediterranean and in the emerged marginal basins of southern Spain (Riding *et al.*, 1998). At DSDP Site 372 and ODP Site 975, the evaporites are overlain by Messinian marls with planktonic foraminifera (Hsü *et al.*, 1977, 1978; Cita *et al.*, 1978, Comas *et al.*, 1996), whereas Messinian post-evaporitic deposits in southeastern Spain comprise coral reefs and other marine fossil assemblages (Martín *et al.*, 1993; Riding *et al.*, 1998; Saint-Martin *et al.*, 2000; Goubert *et al.*, 2001; Aguirre & Sánchez Almazo, 2004; Braga *et al.*, 2006b). The occurrence of zooxanthellate coral reefs together with the oxygen stable-isotope data from Site 976 indicate that the marine reflooding of the Mediterranean during the latest Messinian did not imply a major change in the Mediterranean surface-water temperature in relation to the pre-evaporitic conditions of the basin. Additional data from invertebrates (irregular echinoids) also indicate that the Messinian crisis did not mark the diversity change from warmer to colder-water conditions (Néraudeau *et al.*, 2001).

The major temperature drop took place long after the beginning of the Pliocene. This suggests that the Messinian reflooding of the Mediterranean occurred via the Rifian corridors and only several hundred thousand years later did the Straits of Gibraltar begin to act as the single connector of the Mediterranean to the Atlantic. The highly fluctuating planktonic signal is probably the result of the relative restriction of the Mediterranean Basin and the long process of final closure of the Rifian corridors and the definitive opening of the Straits of Gibraltar. The restriction of water circulation enhanced the variation of the oxygen-isotope signal, which is more sensitive to environmental changes in small, semi-isolated basins than in open ocean waters (Glaçon *et al.*, 1990).

The recorded temperature decrease of the Mediterranean surface waters can be explained, at least

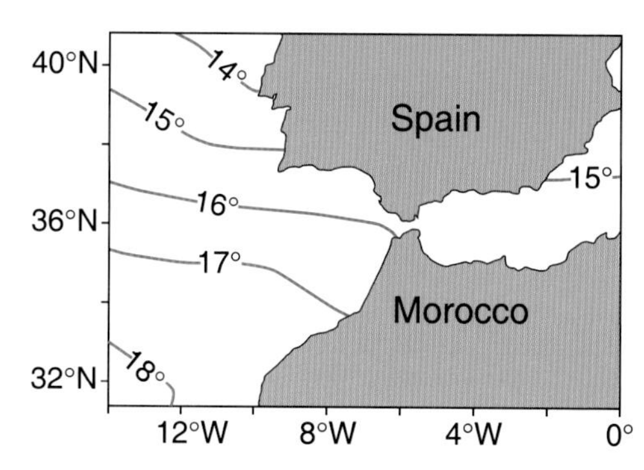

Fig. 15. Minimum winter (February) surface temperature contours on the Atlantic side of the Straits of Gibraltar (from http://iridl.ldeo.columbia.edu/expert/SOURCES/.NOAA/.NCEP/.EMC/.CMB/.GLOBAL/.Reyn_SmithOIv2/). Present-day winter surface-water temperatures on the Atlantic-side position of the Rifian corridors (see Fig. 1) are about 1.5 °C higher than on the western side of the Straits of Gibraltar.

partly, by the change of the source area of the inflowing Atlantic surface waters. Present-day Atlantic winter surface water temperatures at 34° northern latitude (at the former location of the Atlantic side of the Rifian corridors) are about 1.5 °C higher than on the western side of the Straits of Gibraltar (Fig. 15). Although this temperature decrease is small it would be sufficient to make zooxanthellate coral reefs surviving at the very edge of their tolerance limits, as it was the case in the latest Messinian (see above), to disappear.

In the Early Pliocene, the Atlantic surface water inflow through the Straits of Gibraltar caused a concomittant decrease in the temperature of the western Mediterranean surface water. The zooxanthellate coral reefs and tropical/subtropical carbonate lithofacies, which developed in the Mediterranean during the Late Miocene at unusually high latitudes, were not able to thrive in the new cooler conditions and disappeared from the basin. The maintenance of the Straits of Gibraltar as the only connection to the Atlantic ever since, has meant that major palaeoceanographic variations in the Mediterranean since the Early Pliocene have mainly been controlled by global changes.

CONCLUSIONS

Fossil assemblages in the different carbonate units and stable-isotope data for planktonic foraminifera from pelagic deposits laterally equivalent to platform carbonates, suggest that sea-surface water temperature was the major factor controlling the type of carbonate formation in the Mediterranean-linked Neogene basins in southern Spain. Isotopic values of planktonic foraminifera tests indicate surface-water temperatures similar to those of the present-day western Mediterranean (17–20 °C mean annual temperature) during episodes of heterozoan carbonate deposition. Lighter isotopic values of the planktonic foraminifera signal, as well as the occurrence of hermatypic corals, suggest higher surface-water temperatures during the formation of the photozoan, reef-bearing units.

In the Late Miocene, heterozoan carbonates were deposited during low sea-level phases and photozoan carbonates formed at high sea-level phases. These relative sea-level variations can be correlated with the Late Neogene global sea-level curve. This suggests that change in the type of carbonate formation was driven by temperature variations of seawater related to global temperature fluctuations.

At the beginning of the Pliocene, the (sub)tropical platform carbonate lithofacies common in the Mediterranean Messinian gave way to lithofacies typical of warm-temperate seas. In the Early Pliocene, at about 4.9 Ma, a significant change in the Atlantic-Mediterranean connections took place with the opening of the Gibraltar Straits. This situation favoured a new current-circulation pattern for the western Mediterranean, with surface waters flowing into the Mediterranean Sea from a more northern, cooler source area, which no doubt contributed to the persistence of temperate-water conditions at the surface ever since.

ACKNOWLEDGMENTS

Samples from ODP Site 976 were provided by the Ocean Drilling Program. We thank The Godwin Laboratory (University of Cambridge) for use of the stable-isotope analytical facilities. This work was funded by "Ministerio de Educación y Ciencia (Spain)" Project CGL2004-04342/BTE. Constructive comments by C. Dullo, J. Nebelsick and W. Piller are greatly appreciated. Christine Laurin is thanked for correcting the English text.

REFERENCES

Abreu, V.S., **Hardenbol, J.**, **Haddad, G.A.**, **Baum, G.R.**, **Droxler, A.W.** and **Vail, P.R.** (1998) Oxygen isotope synthesis: a Cretaceous ice-house. In: *Mesozoic and Cenozoic Sequence Stratigraphy of European Basins* (Eds P.C. Graciansky, J. Hardenbol, T. Jacquin and P.R. Vail). *SEPM Spec. Publ.*, **60**, 3–13.

Aguirre, J. (1998) El Plioceno del SE de la Península Ibérica (provincia de Almería). Síntesis estratigráfica, sedimentaria, bioestratigráfica y paleogeográfica. *Rev. Soc. Geol. Esp.*, **11**, 297–315.

Aguirre, J. (2000) Evolución paleoambiental y análisis secuencial de los depósitos pliocenos de Almayate (Málaga, sur de España). *Rev. Soc. Geol. Esp.*, **13**, 431–443.

Aguirre, J. and **Sánchez-Almazo, I.M.** (2004) The Messinian post-evaporitic deposits of the Gafares area (Almería-Níjar basin, SE Spain). A new view of the "Lago-Mare" facies. *Sed. Geol.*, **168**, 71–95.

Barbieri, R. and **Ori, G.G.** (2000) Neogene palaeoenvironmental evolution in the Atlantic side of the Rifian Corridor (Morocco). *Palaeogeogr. Palaeoclimatol. Palaeoecol.*, **163**, 1–31.

Bemis, B.E., **Spero, H.J.**, **Lea, D.W.** and **Bijma, J.** (2002) Temperature influence on the carbon isotopic composition of *Globigerina bulloides* and *Orbulina universa* (planktonic foraminifera). *Mar. Micropaleontol.*, **38**, 213–228.

Betzler, C., **Brachert, T.C.** and **Kroon, D.** (1995) Role of climate for partial drowning of the Queensland Plateau carbonate platform (northeastern Australia). *Mar. Geol.*, **123**, 11–32.

Betzler, C., **Brachert, T.C.**, **Braga, J.C.** and **Martín, J.M.** (1997a) Nearshore, temperate, carbonate depositional systems (lower Tortonian, Agua Amarga basin, southern Spain): implications for carbonate sequence stratigraphy. *Sed. Geol.*, **113**, 27–53.

Betzler, C., **Brachert, T.C.** and **Nebelsick, J.** (1997b) The warm temperate carbonate province. A review of facies, zonations and delimitations. *Vogel Festsch. Cour. Forsch.-Inst. Senckenb.*, **201**, 83–99.

Betzler, C., **Martín, J.M.** and **Braga, J.C.** (2000) Non-tropical carbonates related to rocky submarine cliffs (Miocene, Almería, southern Spain). *Sed. Geol.*, **131**, 51–65.

Betzler, C., **Braga, J.C.**, **Martín, J.M.**, **Sánchez-Almazo, I.M.** and **Lindhorst, S.** (2006) Closure of a seaway: stratigraphic record and facies (Guadix basin, southern Spain). *Int. J. Earth Sci. (Geol. Rundsch.)*, **95**, 903–910.

Billups, K. (2002) Late Miocene through early Pliocene deep water circulation and climate change viewed from the sub-Antarctic South Atlantic. *Palaeogeogr. Palaeoclimatol. Palaeoecol.*, **185**, 287–307.

Boreen, T.D. and **James, N.P.** (1993) Holocene sediment dynamics on a cool-water carbonate shelf: Otway, southeastern Australia. *J. Sed. Petrol.*, **63**, 574–588.

Brachert, T.C., **Betzler, C.**, **Braga, J.C.** and **Martín, J.M.** (1996) Record of climatic change in neritic carbonates: turnover in biogenic associations and depositional modes (Late Miocene, southern Spain). *Geol. Rundsch.*, **85**, 327–337.

Brachert, T.C., **Betzler, C.**, **Braga, J.C.** and **Martín, J.M.** (1998) Microtaphofacies of a warm-temperate carbonate ramp (uppermost Tortonian/lowermost Messinian, southern Spain). *Palaios*, **13**, 459–475.

Brachert, T.C., **Hultzsch, N.**, **Knoerich, A.C.**, **Krautworst, U.M.R.** and **Stückard, O.M.** (2001) Climatic signatures in shallow-water carbonates: high-resolution stratigraphic markers in structurally controlled carbonate buildups (Late Miocene, southern Spain). *Palaeogeogr. Palaeoclimatol. Palaeoecol.*, **175**, 211–237.

Braga, J.C. and **Martín, J.M.** (1996) Geometries of reef advance in response to relative sea-level changes in a Messinian (uppermost Miocene) fringing reef (Cariatiz reef, Sorbas Basin, SE Spain). *Sed. Geol.*, **107**, 61–81.

Braga, J.C. and **Martín, J.M.** (2006) Subaqueous siliciclastic stromatolites. A case history from late Miocene beach deposits in the Sorbas Basin of SE Spain. In: *Microbial Sediments* (Eds R.E. Riding and S.W. Awramik), pp. 226–232. *Springer-Verlag*, Berlin-Heidelberg.

Braga, J.C., **Martín, J.M.** and **Alcalá, B.** (1990) Coral reefs in coarse-terrigenous sedimentary environments (Upper Tortonian, Granada Basin, southern Spain). *Sed. Geol.*, **66**, 135–150.

Braga, J.C., **Martín, J.M.** and **Riding, R.** (1995) Controls on microbial dome fabric development along a carbonate-siliclastic shelf-basin transect, Miocene, SE Spain. *Palaios*, **10**, 347–361.

Braga, J.C., **Martín, J.M.**, **Betzler, C.** and **Brachert, T.C.** (1996a) Miocene temperate carbonates in Agua Amarga basin (Almería, SE Spain). *Rev. Soc. Geol. Esp.*, **9**, 285–296.

Braga, J.C., **Martín, J.M.** and **Riding, R.** (1996b) Internal structure of segment reefs: *Halimeda* algal mounds in the Mediterranean Miocene. *Geology*, **24**, 35–38.

Braga, J.C., **Martín, J.M.** and **Wood, J.L.** (2001) Submarine lobes and feeder channels of redeposited, temperate carbonate and mixed siliciclastic-carbonate platform deposits (Vera basin, Almería, southern Spain). *Sedimentology*, **48**, 99–116.

Braga, J.C., **Martín, J.M.** and **Quesada, C.** (2003a) Patterns and average rates of late Neogene-Recent uplift of the Betic Cordillera, SE Spain. *Geomorphology*, **50**, 3–26.

Braga, J.C., **Betzler, C.**, **Martín, J.M.** and **Aguirre, J.** (2003b) Spit-platform temperate carbonates: the origin of landward downlapping beds along a basin margin (Lower Pliocene, Carboneras basin, SE Spain). *Sedimentology*, **50**, 553–563.

Braga, J.C., **Martín, J.M.**, **Betzler, C.** and **Aguirre, J.** (2006a) Models of temperate carbonate deposition in Neogene basins in SE Spain: a synthesis. *Geol. Soc. London Spec. Publ.*, **255**, 123–137.

Braga, J.C., **Martín, J.M.**, **Riding, R.**, **Aguirre, J.**, **Sánchez-Almazo, I.M.** and **Dinarès-Turell, J.** (2006b) Testing models for the Messinian Salinity Crisis: the Messinian record in Almería, SE Spain. *Sed. Geol.*, **188–189**, 131–154.

Brookfield, M. E. (1988) A mid-Ordovician temperate carbonate shelf- The Black River and Trenton Limestone Group of southern Ontario, Canada. *Sed. Geol.*, **60**, 137–153.

Calvet, F., **Zamarreño, I.** and **Vallès, D.** (1996) Late Miocene reefs of the Alicante-Elche basin, southeast Spain. In: *Models for Carbonate Stratigraphy: From Miocene Reef Complexes of Mediterranean Regions* (Eds E.K. Franseen, M. Esteban, W.C. Ward and J.M. Rouchy), *SEPM Concepts Sedimentol Paleontol.*, **5**, 177–190.

Chevalier, J.P. (1961) Recherches sur les madréporaires et les formations récifales miocènes de la Méditerranée occidental. *Mém. Soc. Géol. Fr.*, **40**, 1–562.

Cita, M.B. (1975) Studi sul Pliocene e gli strati di passaggoi dal Miocene al Pliocene, VII. Planktonic foraminiferal biozonation of the Mediterranean Pliocene deep-sea record: a revision. *Riv. Ital. Paleontol. Stratigr.*, **81**, 527–544.

Cita, M.B., **Ryan, W.B.F.** and **Kidd, R.B.** (1978) Sedimentation rates in Neogene deep-sea sediments from the Mediterranean and geodynamic implications of their changes. *Init. Rep. Deep Sea Drilling Proj.*, **42A**, 991–1002.

Comas, M.C., **Zahn, R. Klaus, A.** and **Shipboard Scientists**, (1996) *Proc. ODP Prog., Init. Rep.*, **161**, 1023 pp. College Station, TX.

Comas, M.C., **Platt, J.P.**, **Soto, J.I.** and **Watts, A.B.** (1999) The origin and tectonic history of the Alboran Basin; insights from Leg 161 results. *Proc. ODP Sci. Results*, **161**, 555–580.

Cornée, J.J., **Saint Martin, J.P.**, **Conesa, G.**, **Andre, J.P.**, **Muller, J.** and **Benmoussa, A.** (1996) Anatomie de quelques plate-formes carbonatées progradantes messiniennes de Méditerranée occidentale. *Bull. Soc. Géol. Fr.*, **167**, 495–507.

Cunningham, K.J., **Farr, M.R.** and **Rakic-El Bied, K.** (1994) Magnetostratigraphic dating of an Upper Miocene shallow-marine and continental sedimentary succession in northeastern Morocco. *Earth Planet. Sci. Lett.*, **127**, 77–93.

Dabrio, C.J., **Esteban, M.** and **Martín, J.M.** (1981) The coral reef of Níjar, Messinian (uppermost Miocene), Almería province, S. E. Spain. *J. Sed. Petrol.*, **51**, 521–539.

Esteban, M. (1979) Significance of the upper Miocene coral reefs of the western Mediterranean. *Palaeogeogr. Palaeoclimatol. Palaeoecol.*, **26**, 169–188.

Esteban, M. (1996) An overview of Miocene reefs from Mediterranean areas: general trends and facies models. In: *Models for Carbonate Stratigraphy: from Miocene Reef Complexes of Mediterranean Regions* (Eds E.K. Franseen, M. Esteban, W.C. Ward and J.M. Rouchy). *SEPM Concepts Sedimentol Paleontol.*, **5**, 3–53.

Esteban, M. and **Giner, J.** (1977) El arrecife de Santa Pola. In: *Primer Seminario Práctico de Asociaciones Arrecifales y Evaporíticas* (Ed. R. Salas), *Universidad de Barcelona*, Barcelona, 11–51.

Esteban, M., **Braga, J.C.**, **Martín, J.M.** and **Santisteban, C.** (1996) Western Mediterranean reef complexes. In: *Models for Carbonate Stratigraphy: from Miocene Reef Complexes of Mediterranean Regions* (Eds E.K. Franseen, M. Esteban, W.C. Ward and J.M., Rouchy) SEPM *Concepts Sedimentol Paleontol.*, **5**, 55–72.

Fraser, R.H. and **Currie, D.J.** (1996) The species richness-energy hypothesis in a system where historical factors are thought to prevail: coral reefs. *Am. Nat.*, **148**, 138–159.

Freiwald, A. and **Henrich, R.** (1994) Reefal coralline-algal build-ups within the Arctic Circle: morphology and sedimentary dynamics under extreme environmental seasonality. *Sedimentology*, **41**, 963–984.

Gautier, F., **Clauzon, G.**, **Suc, J.P.**, **Cravatte, J.** and **Violanti, D.** (1994) Age et durée de la crise de salinité messinienne. *CR Acad. Sci. Paris*, **318**, 1103–1109.

Glaçon, G., **Grazzini, C.V.**, **Iaccarino, S.**, **Rehault, J.P.**, **Randrianasolo, A.**, **Sierro, J.F.**, **Weaver, P.**, **Channell, J.**, **Torii, M.** and **Hawthorne, T.** (1990) Planktonic foraminiferal events and stable-isotope records in the Upper Miocene, Site 654. *Proc. ODP Sci. Results*, **107**, 415–427.

Goubert, E., **Neraudeau, D.**, **Rouchy, J.M.** and **Lacour, D.** (2001) Foraminiferal record of environmental changes:

Messinian of the Los Yesos area (Sorbas Basin, SE Spain). *Palaeogeogr. Palaeoclimatol. Palaeoecol.*, **175**, 61–78.

Hallock, P. (1981) Production of carbonate sediment by selected large foraminifera on two Pacific coral reefs. *J. Sed. Petrol.*, **51**, 467–474.

Hallock, P. and **Schlager, W.** (1986) Nutrient excess and the demise of coral reefs and carbonate platforms. *Palaios*, **1**, 389–398.

Haq, B.U., Worsley, T.R., Burckle, L.H., Douglas, R.G., Keigwin, L.D., Opdyke, N.D., Savin, S.M., Sommer II, M.A., Vincent, E. and **Woodruff, F.** (1980) Late Miocene marine carbon-isotopic shift and synchronity of some phytoplanktonic biostratigraphic events. *Geology*, **8**, 427–431.

Haq, B.U., Hardenbol, J. and **Vail, P.R.** (1987) Chronology of fluctuating sea levels since the Triassic. *Science*, **235**, 1156–1167.

Hardenbol, J., Thierry, J., Farley, M.B., Jacquin, T., Graciansky, P.C. and **Vail, P.R.** (1998) Mesozoic and Cenozoic sequence chronostratigraphic framework of European basins. In: *Mesozoic and Cenozoic Sequence Stratigraphy of European Basins* (Eds P.C. Graciansky, J. Hardenbol, T. Jacquin and P.R., Vail), *SEPM Spec. Publ.*, **60**, 3–13.

Harzhauser, M., Piller, W.E. and **Steininger, F.F.** (2002) Circum-Mediterranean Oligo-Miocene biogeographic evolution - the gastropods' point of view. *Palaeogeogr. Palaeoclimatol. Palaeoecol.*, **183**, 103–133.

Henrich, R., Freiwald, A., Bickert, T. and **Schäfer, P.** (1997) Evolution of an Arctic open-shelf carbonate platform, Spitbergen Bank (Barents Sea). In: *Cool-water Carbonates*, (Eds N.P. James and J.A.D. Clarke). *SEPM Spec. Publ.*, **56**, 163–181.

Hodell, D.A. and **Kennett, J.P.** (1986) Late Miocene-early Pliocene stratigraphy and paleoceanography of the South Atlantic and southwest Pacific oceans: A synthesis. *Paleoceanography*, **1**, 285–311.

Hodell, D.A., Benson, R.H., Kent, D.V., Boersma, A. and **Rakic-El Bied, K.** (1994) Magnetostratigraphic, biostratigraphic, and stable-isotope stratigraphy of an Upper Miocene drill core from the Salé Briqueterie (northwestern Morocco): a high-resolution chronology for Messinian stage. *Paleoceanography*, **9**, 835–855.

Hodell, D.A., Curtis, J.H., Sierro, F.J. and **Raymo, M.E.** (2001) Correlation of late Miocene to early Pliocene sequences between the Mediterranean and North Atlantic. *Paleoceanography*, **16**, 164–178.

Hohenegger, J., Yordanova, E. and **Hatta, A.** (2000) Remarks on West Pacific Nummulitidae (foraminifera). *J. Foramin. Res.*, **30**, 3–28.

Hosking, C.M. and **Nelson, Jr., R.V.** (1969) Modern marine carbonate sediments, Alexander Archipelago. *Alaska. J. Sed. Petrol.*, **39**, 581–590.

Hsü, K.J., Ryan, W.B.F. and **Cita, M.B.** (1973) Late Miocene desiccation of the Mediterranean. *Nature*, **242**, 240–244.

Hsü, K.J., Montadert, L., Bernoulli, D., Cita, M.B., Erickson, A., Garrison, R.E., Kidd, R.B., Mélières, F., Müller, C. and **Wright, R.** (1977) History of the Mediterranean salinity crisis. *Nature*, **267**, 1053–1078.

Hsü, K.J., Montadert, L., Bernoulli, D., Cita, M.B., Erickson, A., Garrison, R.E., Kidd, R.B., Mélières, F., Müller, C. and **Wright, R.** (1978) *Init. Rep. Deep Sea Drilling Proj.*, **42A**, 1249 pp.

James, N.P. (1997) The cool-water carbonate depositional realm. In: *Cool-water Carbonates* (Eds N.P. James and J.A.D. Clarke). *SEPM Spec. Publ.*, **56**, pp. 1–20.

James, N.P., Bone, Y., von der Borch, C.C. and **Gostin, V.A.** (1992) Modern carbonate and terrigenous clastic sediments on a cool-water, high-energy, mid-latitude shelf: Lacepede, southern Australia. *Sedimentology*, **39**, 877–903.

James, N.P., Bone, Y., Hageman, S.J., Feary, D.A. and **Gostin, V.A.** (1997) Cool-water carbonate sedimentation during the Terminal Quaternary sea-level cycle: Lincoln Shelf, southern Australia. In: *Cool-water Carbonates*, (Eds N.P. James and J.A.D. Clarke). *SEPM Spec. Publ.*, **56**, 53–75.

James, N.P., Bone, Y., Collins, L.B. and **Kyser, T.K.** (2001) Surficial sediments of the Great Australian Bight: facies dynamics and oceanography on a vast cool-water carbonate shelf. *J. Sed. Res.*, **71B**, 549–567.

Jansen, E. and **Sjøholm, J.** (1991) Reconstruction of glaciation over the past 6 Myr from ice borne deposits in the Norwegian Sea. *Nature*, **349**, 600–603.

Keigwin, L.D. (1979) Late Cenozoic stable-isotope stratigraphy and paleoceanography of DSDP sites in the equatorial and central Pacific Ocean. *Earth Planet. Sci. Lett.*, **45**, 361–382.

Kouwenhoven, T.J., Seidenkrantz, M.S. and **van der Zwaan, G.J.** (1999) Deep-water: the near-synchronous disapperance of a group of benthic foraminifera from the Late Miocene Mediterranean. *Palaeogeogr. Palaeoclimatol. Palaeoecol.*, **152**, 259–281.

Krijgsman, W., Hilgen, F.J., Raffi, I., Sierro, F.J. and **Wilson, D.S.** (1999a) Chronology, causes and progression of the Messinian salinity crisis. *Nature*, **400**, 652–655.

Krijgsman, W., Langereis, C.G., Zachariasse, W.J., Boccaletti, M., Moratti, G., Gelati, R., Iaccarino, S., Papani, G. and **Villa, G.** (1999b) Late Neogene evolution of the Taza-Guercif Basin (Rifian Corridor, Morocco) and implications for the Messinian salinity crisis. *Mar. Geol.*, **153**, 147–160.

Langer, M.R. and **Hottinger, L.** (2000) Biogeography of selected "larger" foraminifera. *Micropaleontology*, **46**, 105–126.

Larsen, H.C., Saunders, A.D., Clift, P.D., Beget, J., Wei, W., Spezzaferri, S. and **ODP Leg 152 Scientific Party** (1994) Seven million years of glaciation in Greenland. *Science*, **264**, 952–955.

Lees, A. and **Buller, A.T.** (1972) Modern temperate-water and warm-water shelf carbonate sediments contrasted. *Mar. Geol.*, **13**, 67–73.

McKenzie, J.A., Spezzaferri, S. and **Isern, A.** (1999) The Miocene-Pliocene Boundary in the Mediterranean Sea and Bahamas: Implications for Global Flooding Event in the Earliest Pliocene. *Mem. Soc. Geol. It.*, **54**, 93–108.

Martín, J.M. and **Braga, J.C.** (1994) Messinian events in the Sorbas Basin in southeastern Spain and their implications in the recent history of the Mediterranean. *Sed. Geol.*, **90**, 257–268.

Martín, J.M., Braga, J.C. and **Rivas, P.** (1989) Coral successions in Upper Tortonian reefs in SE Spain. *Lethaia*, **22**, 271–286.

Martín, J.M., Braga, J.C. and **Riding, R.** (1993) Siliciclastic stromatolites and thrombolites, Late Miocene, S.E. Spain. *J. Sed. Petrol.*, **63**, 131–139.

Martín, J.M., **Braga, J.C.**, **Betzler, C.** and **Brachert, T.C.** (1996) Sedimentary model and high-frequency cyclicity in a Mediterranean, shallow shelf, temperate-carbonate environment (uppermost Miocene, Agua Amarga basin, southern Spain). *Sedimentology*, **43**, 263–277.

Martín, J.M., **Braga, J.C.** and **Riding, R.** (1997) Late Miocene *Halimeda* alga-microbial segment reefs in the marginal Mediterranean Sorbas Basin, Spain. *Sedimentology*, **44**, 441–456.

Martín, J.M., **Braga, J.C.** and **Sánchez-Almazo, I.M.** (1999) The Messinian record of the outcropping marginal Alboran Basin deposits: significance and implications. *Proc. ODP Sci. Results*, **161**, 543–551.

Martín, J.M., **Braga, J.C.** and **Betzler, C.** (2001) The Messinian Guadalhorce corridor: the last northern, Atlantic-Mediterranean gateway. *Terra Nova*, **13**, 418–424.

Martín, J.M., **Braga, J.C.**, **Aguirre, J.** and **Betzler, C.** (2004) Contrasting models of temperate carbonate sedimentation in a small Mediterranean embayment: the Pliocene Carboneras basin. *SE Spain. J. Geol. Soc. London*, **161**, 387–399.

Meulenkamp, J.E. and **Sissingh, W.** (2003) Tertiary palaeogeography and tectonostratigraphic evolution of the Northern and Southern Peri-Tethys platforms and the intermediate domains of the Africa-Eurasian convergent plate boundary zone. *Palaeogeogr. Palaeoclimatol. Palaeoecol.*, **196**, 209–228.

Miller, K.G., **Feigeson, M.D.**, **Wright, J.D.** and **Clement, B.M.** (1991) Miocene isotope reference section, Deep Sea Drilling Project Site 608: an evaluation of isotope and biostratigraphic resolution. *Paleoceanography*, **6**, 33–52.

Mutti, M., **Bernoulli, D.** and **Stille, P.** (1997) Temperate carbonate platform drowning linked to Miocene oceanographic events: Maiella platform margin, Italy. *Terra Nova*, **9**, 122–125.

Nelson, C.S. (1988) Non-tropical shelf carbonates – modern and ancient. *Sed. Geol.*, **60**, 71–94.

Nelson, C.S. and **Bornhold, B.D.** (1983) Temperate skeletal carbonate sediments on Scott shelf, northwestern Vancouver Island, Canada. *Mar. Geol.*, **52**, 241–266.

Nelson, C.S., **Keane, S.L.** and **Head, P.S.** (1988) Non-tropical carbonate deposits on the modern New Zealand shelf. *Sed. Geol.*, **60**, 71–94.

Néraudeau, D., **Goubert, E.**, **Lacour, D.** and **Rouchy, J.M.** (2001) Changing biodiversity of Mediterranean irregular echinoids from the Messinian to the Present-Day. *Palaeogeogr. Palaeoclimatol. Palaeoecol.*, **175**, 43–60.

Pomar, L., **Brandano, M.** and **Westphal, H.** (2004) Environmental factors influencing skeletal grain sediment associations: a critical review of Miocene examples from the western Mediterranean. *Sedimentology*, **51**, 627–651.

Riding, R., **Martín, J.M.** and **Braga, J.C.** (1991) Coral-stromatolite reef framework, Upper Miocene, Almería, Spain. *Sedimentology*, **38**, 799–818.

Riding, R., **Martín, J.M.**, **Braga, J.C.** and **Sánchez-Almazo, I.M.** (1998) Mediterranean Messinian Salinity Crisis: constraints from a coeval marginal basin, Sorbas, southeastern Spain. *Mar. Geol.*, **146**, 1–20.

Rögl, F. (1998) Palaeogeographic considerations for Mediterranean and Paratethys seaways (Oligocene to Miocene). *Ann. Naturhistor. Mus. Wien.*, **99A**, 279–310.

Romanek, C.S., **Grossman, E.L.** and **Morse, J.W.** (1992) Carbon isotopic fractionation in synthetic aragonite and calcite – Effects of temperature and precipitation rate. *Geochim. Cosmochim. Acta*, **56**, 419–430.

Rosen, B.R. (1999) Paleoclimatic implications of the energy hypothesis from Neogene corals of the Mediterranean region. In: *The Evolution of Neogene Terrestrial Ecosystems in Europe* (Eds J. Agusti, L. Rook and P., Andrews), *Cambridge University Press*, Cambridge, UK, 309–327.

Saint-Martin, J.P., **Neraudeau, D.**, **Lauriat-Rage, A.**, **Goubert, E.**, **Secretan, S.**, **Babinot, J.F.**, **Boukli-Hacene, S.**, **Pouyet, S.**, **Lacour, D.**, **Pestrea, S.** and **Conesa, G.** (2000) Interbedded faunas in the Messinian gypsum of Los Yesos (Sorbas Basin, SE Spain): consequences. *Geobios*, **33**, 637–649.

Saint-Martin, J.P., **Pestrea, S.** and **Conesa, G.** (2001) Messinian diatom assemblages of infra-gypsum diatomites in the Sorbas basin (SE Spain). *Cryptog. Algol.*, **22**, 127–149.

Sánchez-Almazo, I.M., **Spiro, B.**, **Braga, J.C.** and **Martín, J.M.** (2001) Constraints of stable isotope signatures on the depositional palaeoenvironments of upper Miocene reef and temperate carbonates in the Sorbas Basin, SE Spain. *Palaeogeogr. Palaeoclimatol. Palaeoecol.*, **175**, 153–172.

Sánchez-Almazo, I.M., **Braga, J.C.**, **Dinarès, J.**, **Martín, J.M.** and **Spiro, B.** (2007) Palaeoceanographic controls on reef deposition: the Messinian Cariatiz reef (Sorbas Basin, Almería, SE Spain). *Sedimentology*, **54**, 637–660.

Santisteban, C. and **Taberner, C.** (1988) Sedimentary models of siliciclastic deposits and coral reefs interrelation. In: *Carbonate-Clastic Transitions* (Eds L.J. Doyle and H.H. Roberts). *Dev Sedimentol.*, **42**, 35–76.

Serrano, F., **González-Donoso, J.M.** and **Linares, D.** (1999) Biostratigraphy and paleoceanography of the Pliocene at Site 975 (Menorca Rise) and 976 (Alboran Sea) from a quantitative analysis of the planktonic foraminiferal assemblages. *Proc. ODP Sci. Results*, **161**, 185–195.

Siesser, W.G. and **de Kaenel, E.P.** (1999) Neogene calcareous nannofossils: Western Mediterranean biostratigraphy and paleoclimatolgy. *Proc. ODP Sci. Results*, **161**, 223–237.

Soria, J.M., **Fernández, J.** and **Viseras, C.** (1999) Late Miocene stratigraphy and palaeogeographic evolution of the intramontane Guadix Basin (Central Betic Cordillera, Spain): implications for an Atlantic-Mediterranean connection. *Palaeogeogr. Palaeoclimatol. Palaeoecol.*, **151**, 255–266.

Sprovieri, R. (1993) Pliocene-early Pleistocene astronomically forced planktonic Foraminifera abundance fluctuations and chronology of Mediterranean calcareous plankton bioevents. *Riv. Ital. Paleont. Stratigr.*, **99**, 371–414.

Thiede, J., **Winkler, A.**, **Wolf-Welling, T.**, **Eldholm, O.**, **Myhre, A.M.**, **Baumann, K.H.**, **Henrich, R.** and **Stein, R.** (1998) Late Cenozoic history of the polar North Atlantic: results from ocean drilling. *Quatern. Sci. Rev.*, **17**, 185–208.

Veron, J.E.N. (1974) Southern geographic limits to the distribution of Great Barrier reef hermatypic corals. *Proc. 2nd International Coral Reef Symposium, Brisbane*, 465–473.

Vincent, E., **Killingley, J.S.** and **Berger, W.H.** (1980) The magnetic epoch-6 carbon shift: a change in the ocean's $^{13}C/^{12}C$ ratio, 6,2 million years ago. *Mar. Micropaleontol.*, **5**, 185–203.

Warnke, D.A., **Allen, C.P.**, **Müller, D.W.**, **Hodell, D.A.** and **Brunner, C.A.** (1992) Miocene-Pliocene Antarctic glacial evolution: a synthesis of ice-rafted debris, stable isotope, and planktonic foraminiferal indicators, ODP Leg 114. In: The Antarctic Paleoenvironment: a Perspective on Global Change (Eds J.P. Kennett and D.A. Warnke). *Am. Geophys. Union Antarct. Res. Ser.*, **56**, 311–325.

Wells, J.W. (1955) A survey of the distribution of reef coral genera in the Great Barrier Reef region. *Rep. Gt. Barrier Reef Comm.*, **6**, 21–29.

Zachos, J., **Pagani, M.**, **Sloan, L.**, **Thomas, E.** and **Buillups, K.** (2001) Trends, rhythms, and aberrations in global climate 65 Ma to present. *Science*, **292**, 686–693.

Int. Assoc. Sedimentol. Spec. Publ. (2010) **42**, 71–88

Facies models and geometries of the Ragusa Platform (SE Sicily, Italy) near the Serravallian–Tortonian boundary

CYRIL RUCHONNET[1] and PASCAL KINDLER

Section des Sciences de la Terre et de l'environement, 13 rue des Maraichers, 1205, Geneva, Switzerland (E-mail: Cyril.Ruchonnet@ihs.com)

ABSTRACT

Detailed sequence and facies analyses of ten stratigraphic sections, positioned on two platform to basin transects across the Ragusa ramp (Sicily, Italy), demonstrate that two models of facies zones must be used to describe the internal architecture of this carbonate edifice near the Serravallian/Tortonian boundary. Seven facies zones, based on the palaeobathymetric position of the carbonate-producing biota, have been defined and described for each of these stages. During the late Serravallian, the inner ramp was characterized mainly by small epiphytic biota of the foramol association, whereas the middle ramp was dominated by rhodoliths and minor benthic foraminifera, bryozoans, barnacles, serpulids and echinoids. In contrast, during the early Tortonian, the inner ramp was characterized by a chlorozoan biota (*Halimeda* and *Porites*) and the middle ramp by branching Corallinacea and rhodoliths. This seemingly minor modification in the type and locus of carbonate production, and the associated changes in hydrodynamics, had profound consequences on the overall geometry of the platform, which evolved from a distally steepened ramp in the late Serravallian, to a geometry approaching that of a flat-topped platform in the early Tortonian.

Keywords Miocene, Serravallian, Tortonian, Sicily Ragusa platform, carbonate ramp, lithofacies, facies zones, Monti Climiti Formation, Carlentini Formation, Palazzolo Formation.

INTRODUCTION

Carbonate platforms are potential reservoirs for hydrocarbons and sensitive recorders of sea-level and climatic changes. Therefore, for both economic and academic reasons, it is important to refine our understanding of their depositional architecture and of the factors that control their geometry. Schlager (2003) suggested that an important factor determining the internal facies pattern and the overall shape of a carbonate factory was the depth window of sediment production. In tropical systems, a narrow and shallow depth window creates flat-topped or rimmed platforms with relatively steep slopes and possibly deep lagoons (e.g. Schlager, 1981). In contrast, in the cool-water realm, a wider and deeper depth window puts no significant constraints on the geometry of the accumulations, which may form homoclinal or distally steepened ramps (James, 1997).

Recently, Pomar (2001) suggested that the type of produced sediment (i.e. grain size), the locus of production and the hydraulic energy collectively influence the variability of depositional profiles among carbonate platforms. The Middle and Upper Miocene neritic limestones from the Mediterranean region show a variety of facies patterns and depositional architectures that fluctuate both in space and time (Vecsei & Sanders, 1999; Brandano & Corda, 2002; Pomar *et al.*, 2004; Bassant *et al.*, 2005) in response to ecological changes (e.g. Demarcq, 1987; Esteban, 1996; Turco *et al.*, 2001; Dall'Antonia, 2002). Therefore, they represent a particularly suitable target to improve our knowledge on the dynamics of carbonate platforms.

This paper focuses on the Ragusa ramp, a well-exposed, tectonically undeformed, carbonate build-up located in southeastern Sicily. The evolution of the facies patterns and depositional profiles of this platform is documented from the late

[1]Present address: IHS Energy, 24 Chemin de la Mairie, PO Box 152, 1258 Perly-Geneva, Switzerland

Serravallian to the early Tortonian. From a regional viewpoint, this study shows that one single facies model (Buxton & Pedley, 1989; Pedley, 1998) is not sufficient to account for all the facies variability on the Ragusa ramp. From a more general standpoint, it demonstrates that subtle changes in carbonate sediment producers can profoundly affect the geometry of a carbonate system.

GEOLOGICAL SETTING

Located in SE Sicily (Fig. 1), the Ragusa platform corresponds to the outcropping portion of the Hyblean Plateau. The latter rests on the northernmost part of the African plate (Grasso & Lentini, 1982) and comprises a sedimentary succession of Triassic to Pleistocene age. The Ragusa ramp

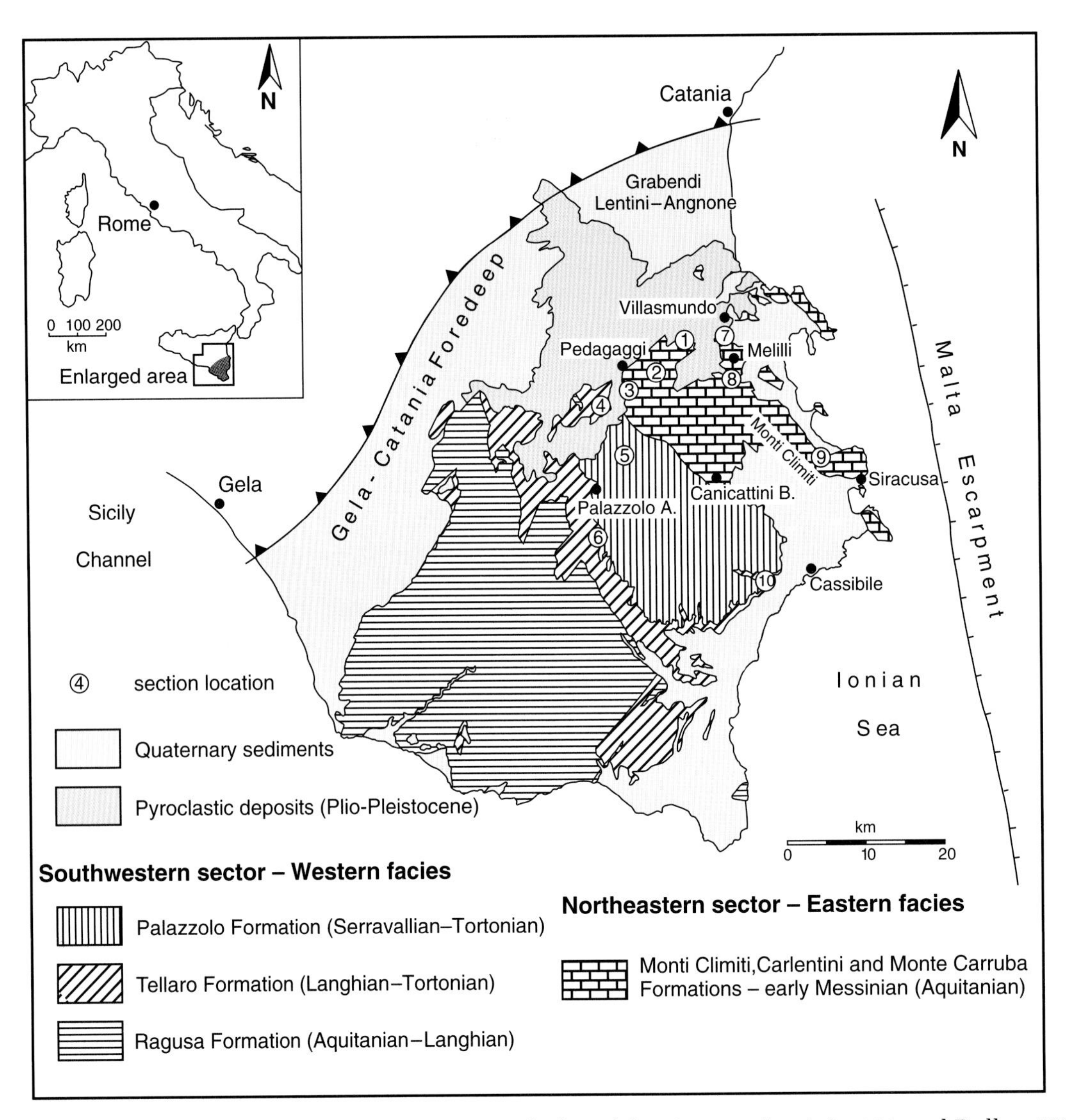

Fig. 1. Simplified geological map of the Ragusa carbonate platform (after Grasso & Lentini, 1982, and Pedley, 1981) with location of the studied sections. These are: **1.** Monte Carruba (base: WB024184, 315 m, top: WB021191, 530 m, Carta d'Italia, 1:50000, Augusta); **2.** Sortino (base: VB995126, 490 m, top: VB997128, 551 m, Carta d'Italia, 1:50000, Vizzini); **3.** Monte Venere 1, NE flank (base: VB959130, 510 m, top: VB957130, 555 m, Carta d'Italia, 1:50000, Vizzini); **4.** Monte Venere 2, SW flank (base: VB935123, 670 m, top: VB941124, 774 m, Carta d'Italia, 1:50000, Vizzini); **5.** Cassaro (base: VB 959059, 445 m, top: VB959054, 515 m, Carta d'Italia, 1:50000, Vizzini); **6.** Contrada Ciurca (base: VB926958, 535 m, top: VB927955, 610 m, Carta d'Italia, 1:50000, Vizzini); **7.** Contrada Salafia (base: WB088186, 204 m, top: WB089185, 250 m, Carta d'Italia, 1:50000, Augusta); **8.** Palombara (base: WB126142, 175 m, top: WB125141, 226 m, Carta d'Italia, 1:50000, Siracusa); **9.** Monti Climiti (base: WB152075, 230 m, top: WB149077, 340 m, Carta d'Italia, 1:50000, Siracusa); **10.** Monti d'Oro (base: WA129889, 265 m, top: WA128894, 394 m, Carta d'Italia, 1:50000, Noto).

displays gently dipping, Middle to Upper Miocene sediments preserving a nearly undeformed platform to basin transition approximately 50 km long and 40 km wide. These sediments were deposited at the front of the northwestern Sicilian-Maghrebian foreland. During the Middle and early Late Miocene, the whole platform experienced slow and homogenous tectonic subsidence provoked by the gentle thrusting of the Sicilian-Maghrebian terranes towards the southeast (Grasso & Lentini, 1982; Catalano *et al.*, 1995; Yellin-Dror *et al.*, 1997). Near the Tortonian/Messinian boundary, subsidence stopped and the sedimentary pile underwent uplift caused by flexure of the African Plate, which led to the partial subaerial exposure of the platform from the late Messinian to the present (Barrier, 1992).

STRATIGRAPHIC FRAMEWORK

The Middle/Upper Miocene sediments of the Ragusa ramp encompass two palaeogeographic domains (Figs 1 and 2): a northeastern sector broadly composed of shallow-water carbonates dominated by algal-bryozoan facies; and a southwestern sector consisting of hemipelagic and pelagic sediments. Each palaeogeographic domain includes several lithostratigraphic units defined by Rigo & Barbieri (1959) and Pedley (1981). In the northeastern sector, platform deposits include successively the Monti Climiti, Carlentini and Monte Carruba Formations (Pedley, 1981). In the southwestern sector, the basinal sediments comprise the Tellaro and Palazzolo Formations that overlie the Lower Miocene Ragusa Formation (Grasso & Lentini, 1982).

The Monti Climiti Formation of the northeastern sector is further subdivided into two members: the Melilli Member (at the base) and the Siracusa Member (at the top; Pedley, 1981; Fig. 2). The presence of *Paragloborotalia partimlabiata* in the marly intervals of the Melilli Member suggests a middle Serravallian age for the lower part of the Monti Climiti Formation (Grasso *et al.*, 1982; MMi7 Biozone of Sprovieri *et al.*, 2002). By superposition, a late Serravallian age has always been assigned to the Siracusa Member (Grasso *et al.*, 1982; Fig. 2). The thickness of this unit varies locally, but usually ranges between 30 m and 100 m.

The uppermost beds of the Monti Climiti Formation are delimited by a pronounced discontinuity surface showing evidence of subaerial exposure at many different locations (e.g. in the Monte Carruba area). Above this surface, the limestone-rich Carlentini Formation is attributed to the Tortonian by analogy of fauna with the Terravecchia Formation of Central Sicily (Chevalier, 1961; Grasso *et al.*, 1982), and because the Tortonian/Messinian boundary has been recognized in the lowermost strata of the overlying Monte Carruba Formation (Grasso *et al.*, 1982). The Carlentini Formation has a variable thickness, ranging from 80 m in the innermost, restricted part of the platform to a few metres basinwards.

In the southwestern palaeogeographic domain, the Palazzolo Formation encloses planktonic foraminifera associations indicative of a late

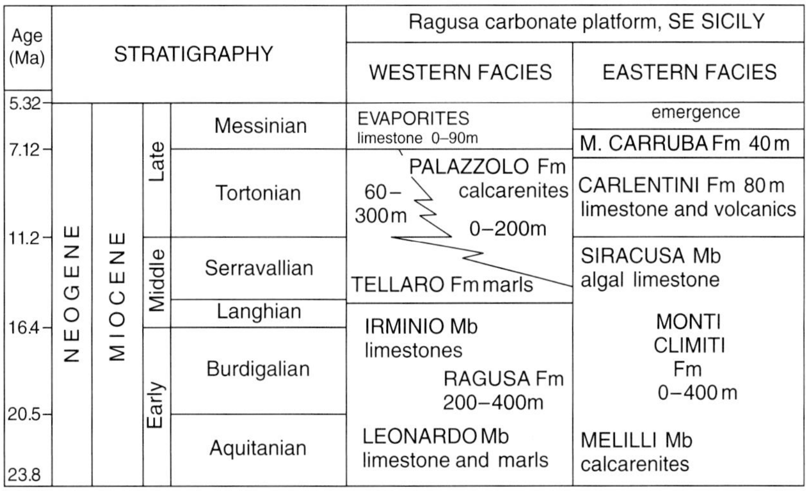

Age (Ma)	STRATIGRAPHY				Ragusa carbonate platform, SE SICILY	
					WESTERN FACIES	EASTERN FACIES
5.32	NEOGENE	MIOCENE	Late	Messinian	EVAPORITES limestone 0–90m	emergence; M. CARRUBA Fm 40 m
7.12				Tortonian	PALAZZOLO Fm calcarenites 60–300m 0–200m	CARLENTINI Fm 80 m limestone and volcanics
11.2			Middle	Serravallian	TELLARO Fm marls	SIRACUSA Mb algal limestone
				Langhian	IRMINIO Mb limestones	MONTI CLIMITI Fm 0–400 m
16.4			Early	Burdigalian	RAGUSA Fm 200–400m	
20.5				Aquitanian	LEONARDO Mb limestone and marls	MELILLI Mb calcarenites
23.8						

Fig. 2. Stratigraphy of the Ragusa carbonate platform. Modified from Pedley (1981) and Grasso & Lentini (1982); chronostratigraphy after Hardenbol *et al.* (1998).

Langhian–middle Tortonian age. The stacking pattern is dominated by decimetre- to metre-thick beds of marly limestone containing a few marl layers over a thickness of 60–300 m (Pedley, 1981). Basinwards, these sediments grade into the Tellaro Formation represented by hemipelagic marls and marly limestone of late Langhian to middle Tortonian age.

METHODS

Sediment nomenclature

The importance of latitude and temperature as controlling factors of carbonate production has long been recognized, and has led to a number of classifications and terminologies (Kindler & Wilson, this volume, pp. 35–48). Lees & Buller (1972) and Lees (1975) classified the distinctive biota associations along temperature gradients as "chlorozoan" (dominated by zooxanthellate corals and calcareous algae), "chloralgal" (dominated by calcareous green algae) and "foramol" (dominated by arenaceous foraminifera and molluscs). Carannante *et al.* (1988) further expanded and detailed this descriptive nomenclature, adding the "rhodalgal" (dominated by Rhodophyta) and the "molechfor" (characterized by benthic foraminifera, molluscs, echinoids, bryozoans and barnacles) lithofacies.

As presumed by Lees & Buller (1972), but occasionally forgotten by subsequent authors, non-tropical (i.e. foramol) assemblages are not necessarily indicative of cool-water settings. In addition to latitude and temperature, a number of factors related to oceanographic settings such as nutrient supply, water turbidity and depth can trigger the deposition of rhodalgal and foramol assemblages even in the tropics (Carannante *et al.*, 1988; Halfar *et al.*, 2004; Pomar *et al.*, 2004; Wilson & Vecsei, 2005).

To replace temperature-biased terminologies, James (1997) proposed to classify carbonate sediments according to the light dependency of their major biotic constituents. He defined a "heterozoan" association predominantly composed of light-independent organisms, ± calcareous red algae, and a "photozoan" association including the skeletons of light-dependent organisms, ± non-skeletal particles (ooids, peloids), and ± elements from the heterozoan association. This terminology is appropriate when dealing with tropical and/or cold-water carbonates, in which the respective dominance or absence of light-dependant organisms is clear. However, differences between photozoan and heterozoan sediments become less well defined in the temperate-water realm, where water temperatures are too cold to support calcareous green algae and zooxanthellate reef growth, but where other light-dependent organisms occur in abundance. Therefore, the terminologies of Lees & Buller (1972) and Carannante *et al.* (1988) are adopted in the present study (Fig. 3), since transitional terms were not defined between the photozoan and heterozoan binary end-members of James (1997).

Trophic resource and temperature influence

Organism diversity and abundance can be related to the level of trophic resource at the time of deposition. Actualistic studies show that changes in the trophic resource continuum from nutrient-rich, relatively eutrophic states to nutrient-poor, relatively oligotrophic states affect the character of the substrate, by influencing the nature of microfaunal assemblages and the turbidity of the water column (Hallock & Schlager, 1986; Hallock, 1987, 1988). However, nutrients and temperature are two negatively correlated controls. High nutrients or low temperatures can result in the production of comparable sediments. Thus, it can be problematic to distinguish carbonates formed in warm water under high nutrient conditions from those formed in cold water (Samankassou, 2002). In modern environments, it is possible to address this issue by quantifying the combined influence of nutrients and temperature on the different types of sediment produced (Halfar *et al.*, 2004). Although a similar approach cannot be performed in the fossil record, the presence and abundance of eurytherm eutrophic organisms (e.g. bryozoans) and/or stenotherm oligotrophic organisms (e.g. zooxanthellate corals, Vermetidae) in the sediments still provide valuable information on the palaeoecological conditions, and are used in the present paper.

Water-depth estimation

Accurate water-depth estimation has to rely on the presence of *in situ* biota of limited reworking potential. Unfortunately, the determination of water depth from rhodalgal-dominated sediments is hampered by the presence of biota that cover a large environmental spectrum of the neritic province (James, 1997). In these sedimentary systems, depth estimation cannot rely on shallow-water indicators, such as zooxanthellate coral reefs for

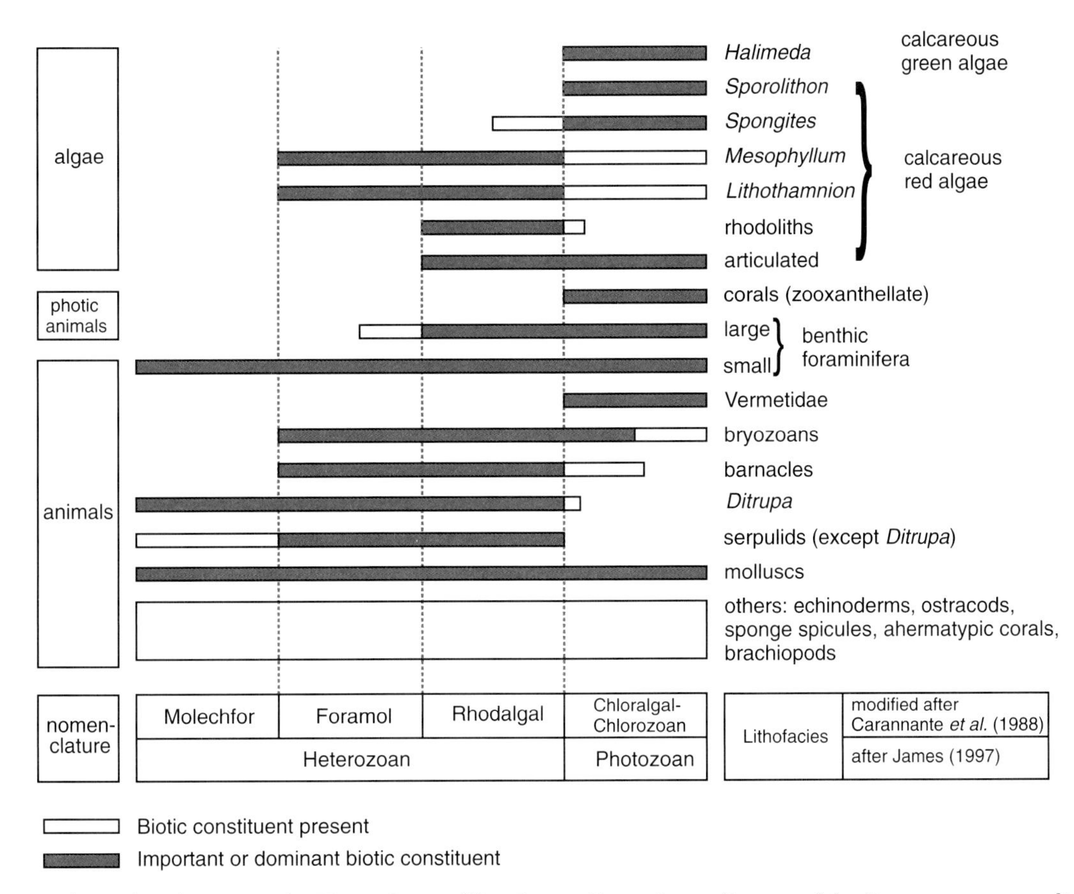

Fig. 3. Main biota distribution in the Upper Serravallian–Lower Tortonian sediments of the Ragusa ramp according to the lithofacies classifications of Carannante *et al.* (1988) and James (1997).

example. This difficulty is partly avoided in large systems with good lateral continuity among outcrops, where taphonomic distribution from inner platform to basin sediments can be observed, and discernable facies zones recognized.

In the present study, water depth is estimated based on the modern distribution of the encountered biota (Perès and Picard, 1964; Larsen, 1976; Hottinger, 1977, 1983, 1997; Adey, 1979; Adey *et al.*, 1982; Drooger, 1983; Bosence, 1985; Hallock & Glenn, 1986; Hohenegger, 1995). Additionally, models for ancient deposits, such as those of Esteban (1979), Pedley (1979, 1996), Buxton & Pedley (1989), Perrin *et al.* (1995), Bosellini *et al.* (2001, 2002), Brandano & Corda (2002) and Pomar *et al.* (2004) were also widely consulted. In rhodalgal sediments, information about depth is provided by red algae associations (Melobesoideae, Mastophoroideae), the abundance and test shape of larger benthic foraminifera (*Amphistegina*, *Heterostegina* and *Operculina*) and the abundance of planktonic foraminifera. In chlorozoan sediments, further clues are given by the occurrence of Sporolithaceae, Vermetidae, *Halimeda*, *Clypeaster* and *Porites*. Finally, carbonate-producing biota have been subdivided in three groups according to their light dependency: euphotic in shallow areas; oligophotic in deeper areas; and aphotic in all deeper ranges (Pomar, 2001).

Stratigraphic analysis

The bulk of the data used to reconstruct facies models comes from field observation and the description of ten sections positioned along two "platform to basin" transects across the ramp (Fig. 1; Ruchonnet, 2006). Each section was logged at decimetre-scale and sampled at metre-scale resolution for microfacies examination (about 650 thin sections). The sections are located in Fig. 1 and their coordinates are given in the corresponding legend. Chronostratigraphic correlation was based on sequence stratigraphic analysis integrated with planktonic foraminiferal biostratigraphy (Cita, 1959; Rigo & Barbieri, 1959; Foresi *et al.*, 1998; Di Stefano *et al.*, 2002).

Sequence stratigraphic interpretation was carried out by considering the relative position of facies zones and sequence boundaries. In the absence of evidence of subaerial exposure, the latter correspond to sharp lateral shifts of facies zones (shallower facies over deeper facies). These shifts can be physically traced and correlated among sections, giving a consistent sequence stratigraphic framework, reliable at km-scale. The criterion used to define the lateral shifts was the palaeobathymetric position of the carbonate-producing biota. Three sectors were thus defined along the ramp, based on the distribution of the producing organisms. Broadly, the inner ramp, characterized by euphotic communities, extends from 0 to about 30 m depth. The middle ramp, dominated by mesophotic-oligophotic communities, reaches down to 90 m, and the outer ramp, corresponding to the aphotic sector, extends to 200 m depth.

Sequence correlation and depositional profile definition

The Ragusa ramp was affected by a major sea-level fall at the Serravallian/Tortonian boundary which was identified in all studied sections from the inner to outer ramp settings. In the proximal part of the platform (NE sector), this event matches an exposure surface at the top of the upper Serravallian deposits (Monti Climiti Formation). In the distal, southwestern area, it corresponds to an abrupt basinward shift of the facies zones that delineates a lowstand wedge in the outer-ramp sedimentary record. The recognition of this major sequence boundary (Ser4/Tor1; Haq *et al.*, 1987) in all studied sections (Fig. 1) provides a time line for sequence correlation. In addition, the distance among the sections constrains the geometrical attributes of the sequences, as well as the depositional profile of the ramp through time.

FACIES-ZONE MODELS AND PLATFORM GEOMETRY

Microfacies analysis of the Middle/Upper Miocene sediments of the Ragusa ramp reveals the existence of two different modes of carbonate production during this time interval. In the inner and middle sectors of the ramp, the upper Serravallian sediments of the Monti Climiti Formation (Siracusa Member) belong to the foramol-rhodalgal association, whereas the lower Tortonian deposits of the Carlentini Formation enclose key-bioclasts of the chlorozoan association (Fig. 3). Subsequently, two facies models have been defined for the studied time interval. To avoid any confusion, these models are described separately (Figs 4 and 5). Typical facies characterizing both models are represented in Fig. 6.

The Ragusa ramp during the late Serravallian

Inner ramp (foramol assemblage)

Sediments from the inner ramp are poorly represented and only crop out in the northeastern sector between Melilli, the Contrada Salafia and the Monte Carruba sections (Fig. 1). The majority of these sediments has been eroded during partial exposure of the ramp linked to the major sea-level drop at the Serravallian/Tortonian boundary (Ser4/Tor1; Haq *et al.*, 1987). The most internal deposits preserved are represented by crudely stratified, subtidal, foramol facies, from which wave-related structures are systematically missing. Two facies zones have been recognized among these sediments: an inner and an outer shallow-water zone (Fig. 4). The inner shallow-water zone (FZ-1) regroups a low-diversity assemblage of small benthic foraminifera (small rotaliids, miliolids, textularids), serpulids (including *Ditrupa*), and barnacles (Fig. 6A). Red-algae fragments are scarce, well sorted, and usually smaller than 500 μm. Rock textures are dominated by wackestones.

The outer shallow-water zone (FZ-2) is characterized by the systematic occurrence of red-algae fragments with sizes ranging between 500 μm and a few mm. These wackestones to packstones comprise a more diverse fauna marked by the first appearance of Cibicididae, Anomalinidae, barnacles and rare larger foraminifera (*Heterostegina*, and inflated, lenticular *Amphistegina*). Small rotaliids, textularids and *Ditrupa* are common to abundant. Pedley (1998) reported sea-grass communities in these sediments. Arguments for palaeobathymetric estimates are limited in this part of the ramp because clear euphotic fauna are constantly missing. However, shallow-water conditions can be inferred by comparison with similar faunas from inner-ramp, shallow-water, sea-grass meadows in Australia (James 1997; Lukasik *et al.*, 2000), and by the systematic absence of deeper mesophotic-oligophotic taxa (e.g. larger

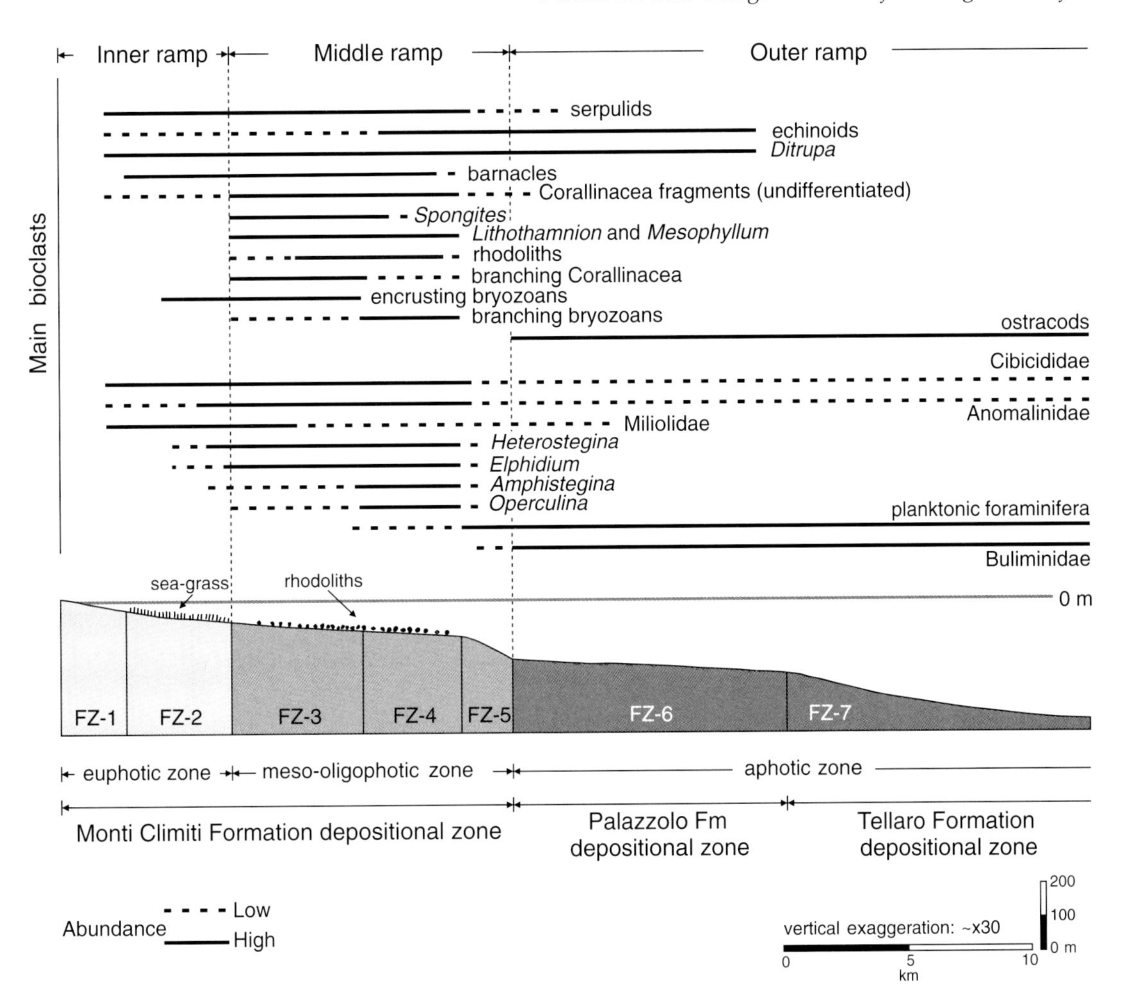

Fig. 4. Bioclast distribution, sector and facies-zone (FZ) extent on the Ragusa ramp during the Late Serravallian.

benthic foraminifera, *Lithothamnion* and *Mesophyllum* rhodoliths).

Middle ramp (rhodalgal assemblage)

The middle ramp sector extends between the Monte Venere 1 section, Cassibile, Siracusa and Melilli (Fig. 1). Sediments are easily recognized by the profusion of spherical and laminar rhodoliths composed of *Lithothamnion, Mesophyllum* and *Spongites*. They form a wide pavement extending over 8 km basinwards. Three facies zones have been defined in this sector: an inner and an outer algal pavement and a steeper slope (Fig. 4). The inner algal pavement (FZ-3) is characterized by a basinward increase (20 to 90%) in the red-algae content of the sediments (Fig. 6B), which grade laterally from algal-rich wackestones to rhodolith-dominated rudstones, locally forming metre-scale buildups. These buildups are organized "en echelon", rarely exceed 1 m of vertical development, and a few tens of metre of horizontal dimensions. Collectively, they provide a separation line with the outer algal pavement. In a synthetic model of the Oligo-Miocene ramps from the Central Mediterranean Basin, Pedley (1998) considered these buildups as an individual facies zone.

The red-algal assemblage also changes basinwards, showing an increase in the abundance of *Lithothamnion* and *Mesophyllum* relative to the more shallow-water dwelling genus *Spongites*. In the field, the gradual enrichment in red algae is emphasized by a more distinct bedding and a thickening of the strata. Associated taxa include encrusting bryozoans, serpulids and barnacles. *Amphistegina* and *Heterostegina* may be abundant (up to 15% of the biota) in several facies, while small rotaliids, Textulariidae and Sphaerogypsinidae dominate the shallowest associations. Scattered planktonic foraminifera are locally

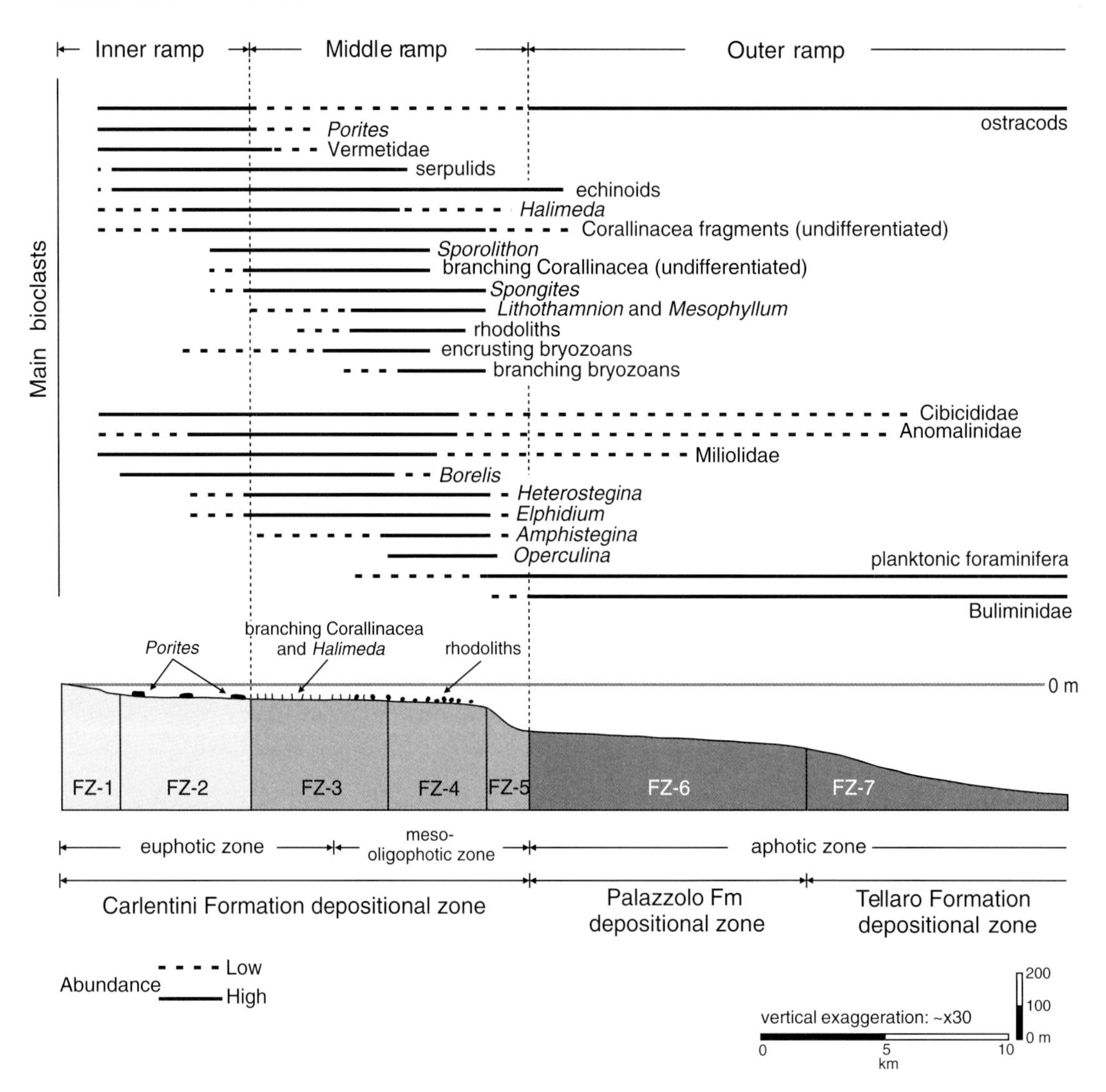

Fig. 5. Bioclast distribution, sector and facies-zone (FZ) extent on the Ragusa ramp during the Early Tortonian.

present. In the rhodolith rudstones, encrusting foraminifera such as Placopsilinidae and Homotrematidae are commonly fixed on the cm-sized rhodoliths. Evidence of early marine diagenesis occurs as sparry calcite, micritic infills, isopachous bladed cements surrounding various bioclasts, and micritic envelopes around dissolved mollusc shells.

The outer algal pavement (FZ-4) is represented by Corallinacea and foraminifera-rich wackestones. Compared to the inner algal pavement, planktonic foraminifera, branching bryozoans, flat *Amphistegina*, and echinoid fragments are more common, whereas miliolids and textularids become rare. In the innermost part of this facies zone, rhodolith accumulations (floatstones with wackestone to packstone matrix) up to 8 m thick can be observed. Basinward, rhodoliths are progressively replaced by branching shapes and reworked fragments. The abundance of larger benthic foraminifera, branching bryozoans, echinoids and planktonic foraminifera increases concurrently (Fig. 6C). In the field, these faunal changes lead to a thinning of the strata (down to about 3 m).

The steeper slope (FZ-5) marks the transition to the outer ramp. It consists of foraminiferal wackestones and packstones containing red algae and bryozoan debris, and larger benthic foraminifera in the upper part of the zone. The relative abundance of planktonic foraminifera, Cibicididae, Anomalinidae, Buliminidae, *Ditrupa* and echinoids increases basinward, whereas larger benthic foraminifera and red algae fragments disappear. The steeper slope represents the transition from the massively bedded, coarse bioclastic

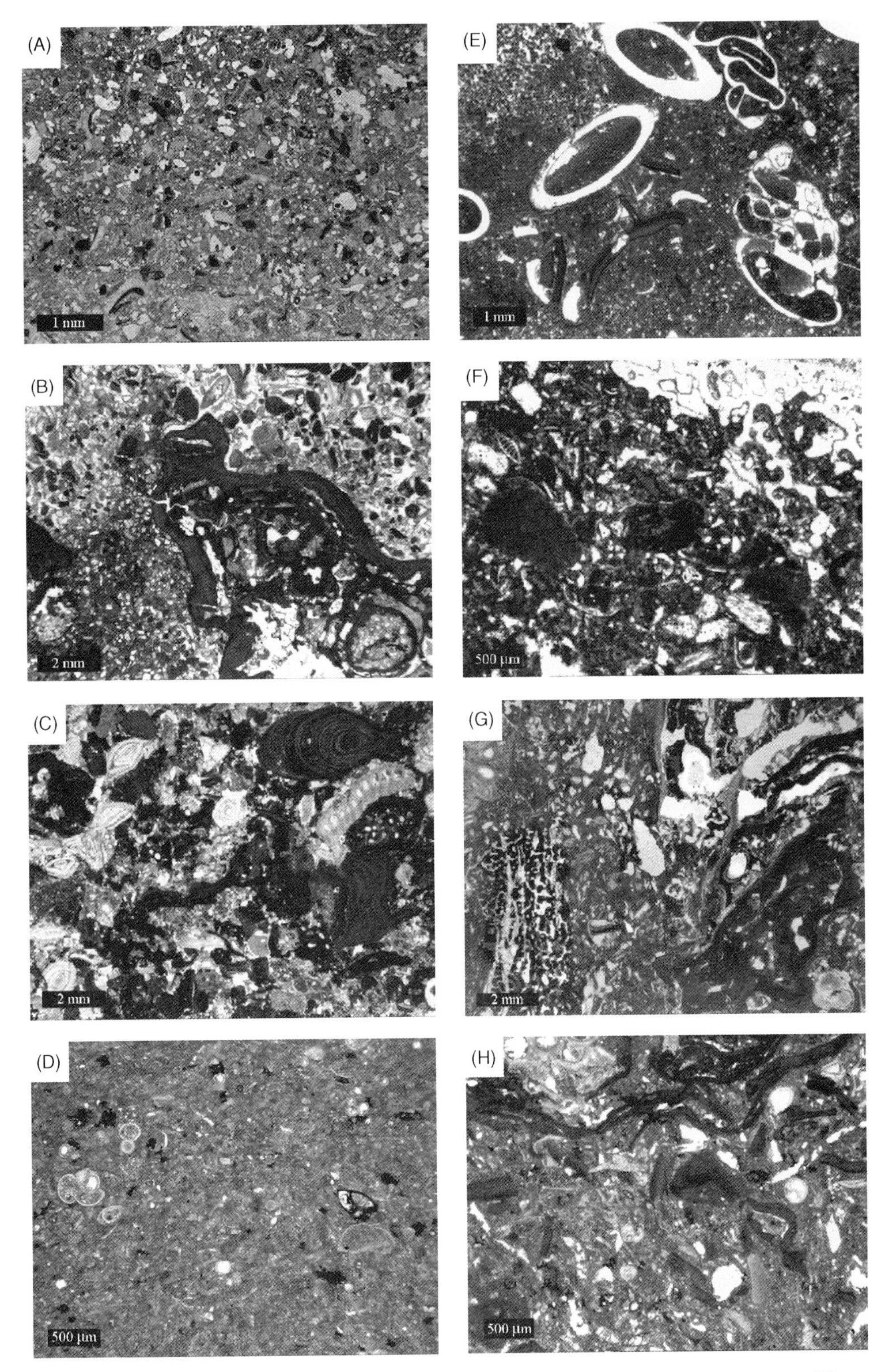

Fig. 6. Thin section photographs illustrating some facies-zones of the Ragusa ramp. (A) Upper Serravallian *outer shallow-water zone* (small rotaliids-*Ditrupa* packstone). (B) Upper Serravallian *inner algal pavement* (rhodolith floatstone with wackestone-grainstone matrix). (C) Upper Serravallian *outer algal pavement* (red algae-larger foraminiferal floatstone). (D) Upper Serravallian *outer plain* (planktonic foraminiferal mudstone). (E) Lower Tortonian *inner shallow-water zone* (Vermetidae mudstone). (F) Lower Tortonian *outer shallow-water zone* (*Porites*-red algae floatstone). (G) Lower Tortonian *inner algal pavement* (rhodolith-*Porites* floatstone with mudstone matrix). (H) Lower Tortonian *outer algal pavement* (rhodolith-planktonic foraminiferal floatstone with mudstone matrix).

accumulations of the Siracusa Member to the fine-grained, marly limestones of the Palazzolo Formation. It corresponds to the "Fine-grained Resediment Zone" of Grasso & Lentini (1982).

The modern distribution of red algae (Pérès & Piccard, 1964; Adey *et al.*, 1982) suggests that the sediments of the inner- and outer algal pavement facies zones were deposited under mesophotic-oligophotic conditions in water depths between 30 and 60 m. *Lithothamnion* and *Mesophyllum* are most abundant in water depths between 50 and 80 m, whereas *Spongites* remains usually settled at shallower depths (between 30 and 50 m; Adey *et al.*, 1982). The abundance of nodular rhodoliths and grainy facies supports the existence of bottom currents similar to those observed on the present-day Mediterranean shelves ("faciès à prâlines" of Pérès & Picard, 1964; Carannante *et al.*, 1988). In the inner algal pavement, current sweeping is verified by frequent grainstone textures, which are more common than in the deeper, outer algal pavement.

The profusion of small rotaliids and encrusting bryozoans suggests that trophic resources were locally too high for larger benthic foraminifera to thrive. Micrite-filled cavities in the rhodoliths probably reflect episodes of reduced water energy, indicating that red-algae accumulations provided a protective, armored pavement capable of trapping fine-grained sediments. The greater amount of micrite along the outer algal pavement suggests a progressive reduction in water energy. This trend is also attested by a relative increase in *Operculina* specimens, which live primarily on soft substrates (Hottinger, 1983), and by the continuous decline of spherical rhodoliths. The co-occurrence of branching bryozoans, flat *Amphistegina*, *Heterostegina*, *Operculina* and echinoids is emblematic of deep, oligophotic associations under eutrophic conditions (Corda & Brandano, 2003).

Outer ramp (molechfor assemblage)

Sediments from the outer ramp belong to the Palazzolo and Tellaro Formations. They represent transitional deposits between the neritic and the pelagic realms, and are characterized by an absence of photosynthetic organisms. Geographically, the outer ramp extends southwestward from a line linking Pedagaggi to Cassibile (Fig. 1). The innermost part of this sector comprises the marly limestones of the Palazzolo Formation which grade laterally into marls and marly limestone alternations of the Tellaro Formation. Each formation corresponds to one facies zone (Fig 4). Sediments of the Palazzolo Formation were deposited on a low-angle (<1°) outer plain (FZ-6), where uninterrupted sedimentation has produced strata up to 15 m thick. The faunal assemblage includes mostly planktonic foraminifera, and less important Cibicididae, Anomalinidae, Buliminidae, ostracods, *Ditrupa* and echinoids (Fig. 6D). The ichnofauna includes *Thalassinoides* burrows, which are particularly abundant close to discontinuity surfaces.

The strata thin basinwards, grading laterally into the marls and marly limestone alternations of the hemipelagic zone (FZ-7; Tellaro Formation). These alternations were deposited on a slightly steeper surface than the Palazzolo Formation, as demonstrated by frequent pinch and swell features present on the outcrops, and large-scale slide features visible on the northeastern side of the Tellaro Valley (Pedley *et al.*, 1992). The invariable absence of larger benthic foraminifera (*Amphistegina* and *Heterostegina*) and of autochthonous, deep, photosynthetic biota such as red algae, suggests that these sediments were deposited in the aphotic zone.

The Ragusa ramp during the early Tortonian

Inner ramp (chlorozoan assemblage)

Sediments from the Tortonian inner ramp are poorly represented, and only crop out in the north-eastern sector between Monte Carruba, Melilli and Sortino. The vertical transition from the Serravallian rhodalgal to the Tortonian chlorozoan lithofacies is closely related to the third-order sea-level fall recorded at the Serravallian/Tortonian boundary. On the inner ramp, the uppermost Serravallian packstones are truncated by a sharp, undulating, erosional surface and overlain by smooth, well-cemented, coral-rich, mudstone/wackestone beds of Tortonian age. The innermost mudstone/wackestone facies of the inner shallow-water zone (FZ-1; Fig. 5) comprises Vermetidae gastropods (Fig. 6E) associated with small Corallinacea fragments, ostracods and *Halimeda*. At rare locations (e.g. Monte Carruba), soft-sediment dwelling *Porites* corals also occur. Metre-thick strata are characteristically bounded by pronounced discontinuity surfaces.

Shelfward, coral colonies (*Porites*) (Fig. 6F) form locally decametre-thick patches of metric lateral extent associated with small rotaliids, miliolids, serpulids, and bivalves, Alveolinidae (*Borelis*) and *Clypeaster*. Collectively, these appear in the wackestones/packstones of the outer shallow-water zone (FZ-2). Vermetidae occur also in small colonies with fragments of *Halimeda* and small rotaliids and miliolids association. The muddy sediments of the most restricted part of the inner ramp reflect low-energy, euphotic conditions. Trophic resources were low enough for *Porites* to grow, suggesting oligo-mesotrophic conditions. Despite a relative high content in Miliolidae, evidence of hypersalinity is not conspicuous because most of the biota is typically stenohaline. Basinwards, the occurrence of packstones in the outer shallow-water zone supports a relative increase in water energy. The presence of *in-situ* Vermetidae constrains the water depth to less than 15 m (Bosellini *et al.*, 2001).

Middle ramp (chlorozoan-rhodalgal assemblage)

The Tortonian middle ramp forms a polygon broadly delimited by Melilli, Siracusa, Canicattini Bagni, Pedagaggi, and Sortino. Sediments predominantly consist of Corallinacea (branching shapes and spherical rhodoliths) that are associated with chlorozoan biota (*Porites* and *Halimeda*) in the shallowest part of the sector. Three facies zones have been defined (Fig. 5). The inner algal pavement (FZ-3) is dominated by Corallinacea-rich packstones/floatstones, comprising fragments of branching *Spongites* and *Lithothamnion* as well as encrusting *Sporolithon*. Subordinate biota include decametre-sized *Porites* colonies, bryozoans, serpulids, Vermetidae, and small benthic foraminifera. The stacking pattern is characterized by flat, horizontal beds, rarely exceeding 2 m in thickness.

Basinward, the transition to rhodalgal floatstones (FZ-4) is marked by a sharp increase in branching Corallinacea and *Halimeda* fragments (Fig. 6G), and by the progressive disappearance of Vermetidae and *Porites*. Large fragments of articulated Corallinacea are enclosed in a wackestone matrix, associated with bryozoans, serpulids, miliolids, and larger benthic foraminifera (including *Borelis*). In the deeper outer algal pavement (FZ-4), fragments of *Halimeda* are progressively replaced by small spherical rhodoliths (*Spongites*, *Sporolithon*, *Lithothamnion*). Flat *Amphistegina*, *Heterostegina* and *Operculina* specimens, as well as echinoids, and planktonic foraminifera complement the biota.

The more distal slope facies (FZ-5) is identical to its Serravallian counterpart, except for the notable addition of reworked fragments of *Halimeda*. Sediments of the middle ramp were likely deposited in the euphotic-mesophotic zone. The deepest associations of *Porites*, Vermetidae and *Borelis* suggest euphotic water depths (Bosellini *et al.*, 2001). Basinwards, the near synchronous disappearance of Vermetidae and *Porites* indicates a deepening into mesophotic water depths, where *Halimeda* and *Borelis* still thrived. The deep accumulations of *Halimeda* (associated with branching Corallinacea) are probably allochthonous, as documented for several Upper Miocene analogues (Braga & Martìn, 1996; Martìn *et al.*, 1997; Bosellini *et al.*, 2001). Mesophotic conditions are assumed to have extended down to the deeper rhodolith facies (FZ-4) (Fig. 6H).

Outer ramp (molechfor assemblage)

The outer ramp extends southwestward from a line linking Pedagaggi to Cassibile (Fig. 1). Except for the intercalation of shallower facies over 1–2 metres and related to the sea-level fall, the dominating facies consists of planktonic foraminiferal mudstones and wackestones lacking light-dependent biota.

DISCUSSION

The facies analysis presented in the previous sections revealed a major change of the carbonate production on the Ragusa platform between the late Serravallian and the early Tortonian. The former interval is characterized by sediments typical of the foramol/rhodalgal associations, whereas chlorozoan/rhodalgal assemblages predominate in the latter. This biotic change further modified the internal architecture and the depositional profile of the ramp.

Processes of carbonate production and distribution

During the late Serravallian, foramol sediments were produced in the shallow-water, wave-agitated, euphotic zone, in the inner part of the ramp (Fig. 7). Assemblages mostly consisted of epiphytic biota living on sea-grasses. Despite a significant potential for ecological accommodation (Pomar, 2001), sediment accumulations remained small because sea-grass meadows are presumed to

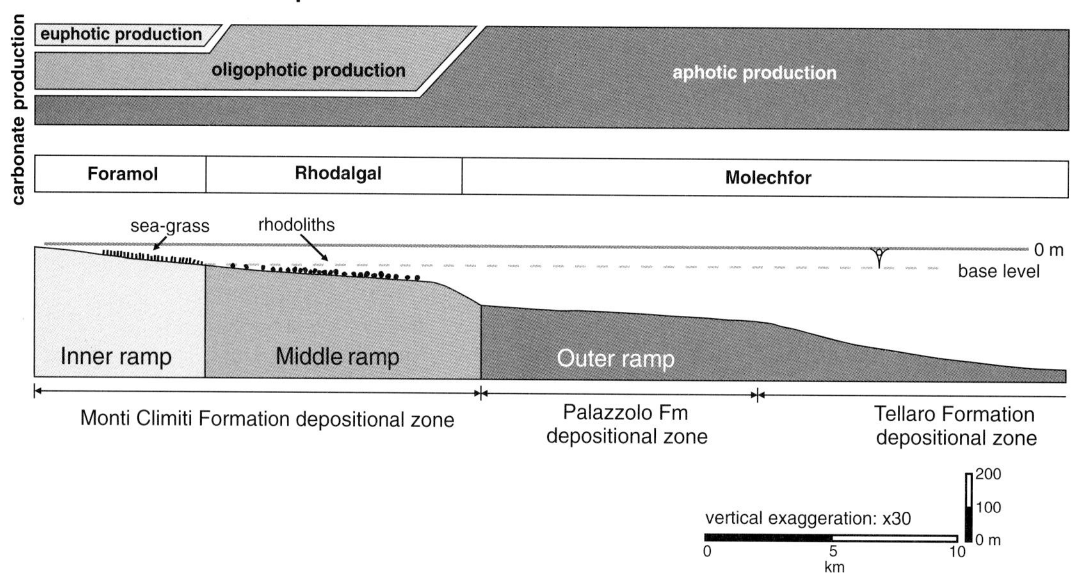

Fig. 7. Depositional profile and carbonate-type production on the Ragusa ramp during the Late Serravallian.

have colonized most of the shallow-water habitats, limiting the development of large sediment contributors (e.g. molluscs) due to space reduction. In addition, shedding towards deeper areas occurred in very limited proportions because the produced sediments were mostly baffled and trapped in the meadows. Significant accumulations of rhodalgal sediments were formed in the oligophotic zone, in the middle portion of the ramp. Because most of rhodoliths thrived below the shelf equilibrium profile (Pomar, 2001), the ambient fluid power was not strong enough to move sediments towards the basin. Shedding thus occurred only during exceptional storms.

During the early Tortonian, carbonate production in the shallow, euphotic, inner ramp, was dominated by a chlorozoan biota with low building-up capacity (Fig. 8). Small colonies of *Porites*,

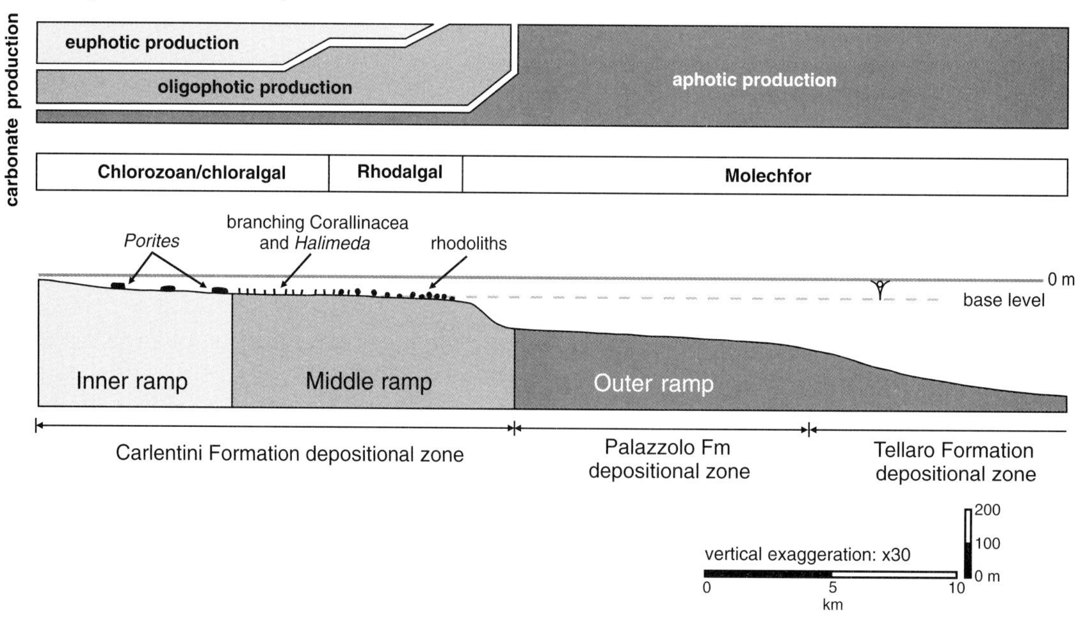

Fig. 8. Depositional profile and carbonate-type production on the Ragusa ramp during the Early Tortonian.

and encrusting Vermetidae developed only metre-scale patch reefs, and were thus unable to constitute an armoured sea-floor. Nonetheless, the occurrence of this chlorozoan assemblage suggests that rates of sediment production were somehow higher in the innermost reaches of the inner ramp (FZ-1) than during the late Serravallian. In deeper settings (FZ-2), the low building-up capacity of producing organisms, combined with a substantial hydrodynamics, favoured the downward shedding of the smaller and lighter grains. On the shallowest part of the middle ramp (FZ-3), the trapping of transported particles by branched Corallinacea contributed to keep significant rates of accumulation. Further offshore (FZ-4), the rhodolith pavement thrived in the same conditions as on the oligophotic part of the Upper Serravallian ramp.

Internal architecture and depositional profile evolution

During the late Serravallian, minor amounts of carbonate sediments were produced on both the shallow, euphotic, inner part and the deeper, aphotic, outer part of the Ragusa ramp. In contrast, significant production of coarse carbonates (rhodolith-rich rudstones and floatstones) occurred on the oligophotic middle portion of the platform. This resulted in a typical, low-angle, distally steepened ramp geometry (Fig. 9), similar to that described from other Miocene platforms (e.g. Pomar *et al.*, 2004). At the Serravallian/Tortonian boundary, carbonate production increased significantly on the inner and inner-middle ramp sectors because of the shift from a foramol-dominated to a chlorozoan-dominated biota. However, as mentioned above, limited accommodation impacted on the internal architecture and the depositional profile of the lower Tortonian ramp. Higher rates of sediment export towards the middle ramp flattened the ramp topography, and led to a basinward extension of the zone of euphotic carbonate production at the expense of the oligophotic, rhodalgal-dominated zone (Fig. 8). In the field, this progradation can be evaluated at about 3 km along the profile. The rapid progradation of euphotic-oligophotic associations in the uppermost layers of the Palazzolo Formation and the coeval appearance of coralgal associations in the upper layers of the Carlentini Formation (Grasso *et al.*, 1982) contributed to preserve this geometry through the Tortonian.

Potential factors responsible for changes in the ecological parameters

The change of carbonate production observed across the Ser4/Tor1 sequence boundary on the Ragusa ramp is attributed to modifications of the ecological parameters controlling the faunal diversity. Foramol and rhodalgal associations are known to characterize nutrient-rich and/or cool temperate environments, from shallow euphotic to deep oligophotic water depths (Carannante *et al.*, 1988; Nelson, 1988; James, 1997). Although they dominate most of the shallow-water high-latitude habitats, they are also present in warm inter-tropical regions where the nutrient and/or the terrigenous clastic content is too high for hermatypic fauna to thrive (e.g. Wilson & Vecsei, 2005). In contrast, chloralgal-chlorozoan associations characterize euphotic-mesophotic environments exposed to oligotrophic, warm conditions (Mutti & Hallock, 2003; Halfar *et al.*, 2004). These are most abundant in the inter-tropical belt, between 30°N and 30°S (Lees & Buller, 1972; James, 1997).

High salinity values represent a limiting factor for chlorozoan and chloralgal associations. The chlorozan association has tolerance limits in the range of 25–40‰ (Lees, 1975; Wilson & Vecsei, 2005), whereas chloralgal associations are tolerant of a wider range of salinity (Lees 1975). In addition, the larger benthic foraminifera (rotaliids) are typical stenohaline biota with tolerance limits in the range of 30–45‰ (Hallock & Glenn, 1986). Taking these ranges into account, reduced salinity cannot be the threshold factor limiting the growth of hermatypic corals during the Late Serravallian. The common occurrence of larger foraminifera, which have a similar tolerance with respect to salinity, rules out this hypothesis.

As a limiting factor, the chemical properties of seawater also play a major role in carbonate production. Sandberg (1983) recognized alternating intervals of "aragonitic" and "calcitic seas" during the Phanerozoic, based on the mineralogy of ooids and early marine cements. More recently, Stanley & Hardie (1998) have shown a parallel oscillation in the dominant mineralogy of both non-skeletal and skeletal components and that of marine evaporites. Those authors suggested that secular changes in the Mg/Ca ratio of seawater, driven by changes in the spreading rates along mid-ocean ridges, have controlled the long-term Phanerozoic oscillations in carbonate mineralogy and influenced

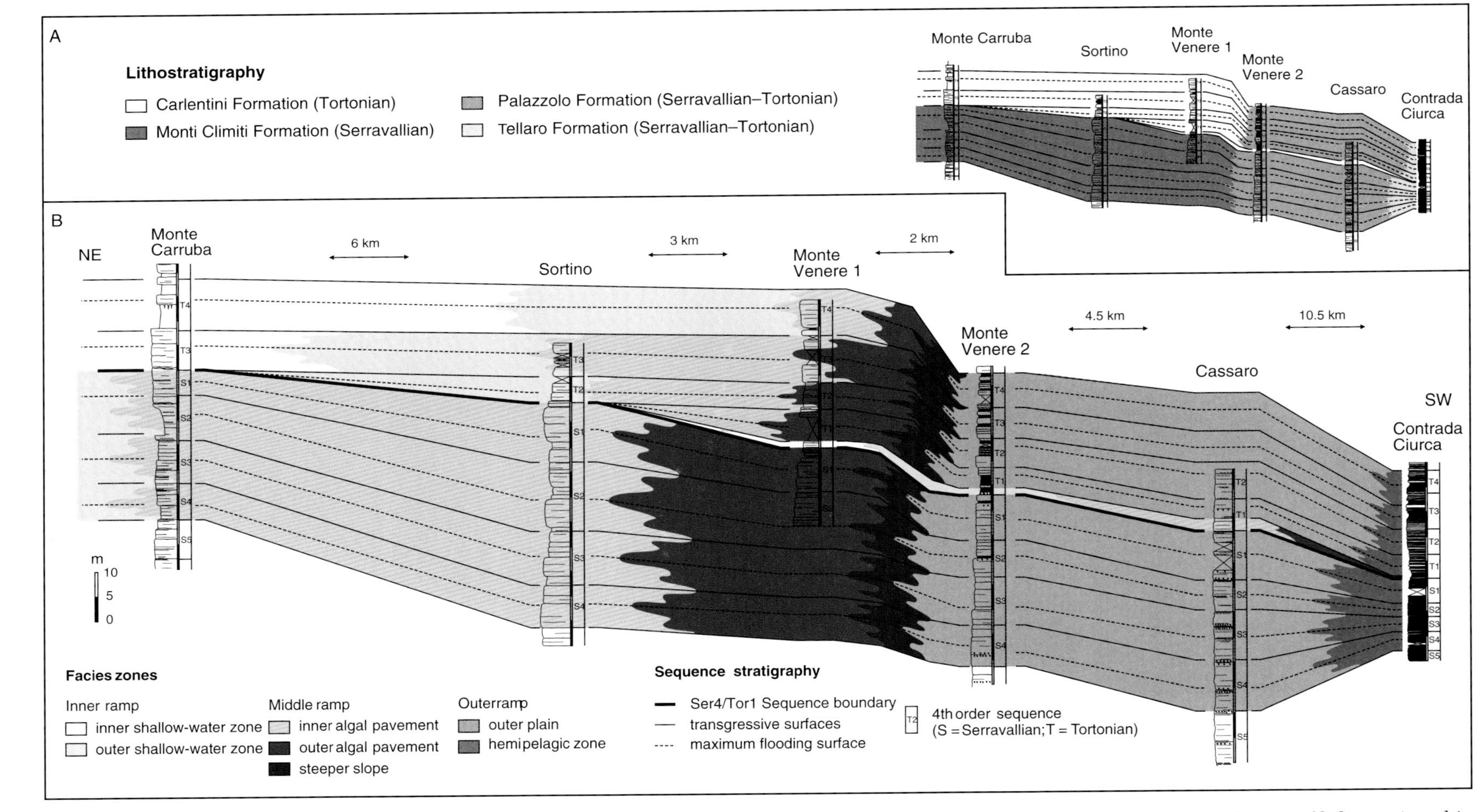

Fig. 9. Correlation of the Serravallian–Tortonian boundary interval for the westernmost profile studied across the Ragusa ramp. (A) Distribution of lithostratigraphic units. (B) Correlation of depositional sequences.

the extent of skeletal organisms. The current effects of increasing atmospheric CO_2 on hermatypic faunas are well documented (Hallock, 1996). Higher rates of volcanism at ocean ridges favour higher atmospheric CO_2 concentrations as well as higher Ca^{2+} concentrations and a lower Mg/Ca ratio in the oceans. Calcite is favoured over aragonite at higher concentrations of atmospheric CO_2 because carbonate saturation in seawater is lowered and aragonite is energetically "expensive" to precipitate at a reduced saturation state (Hallock, 2001). However, a scenario inferring high CO_2 concentration during the late Serravallian which would have prevented the growth of hermatypic corals is contradictory with the cooling trend observed at a global scale (Zachos *et al.*, 2001). Moreover, no exceptional mid-ocean ridge activity capable of lowering the seawater Mg/Ca ratio is reported from late Serravallian times.

Based on the previous discussion, nutrient delivery and water temperature are seen as the most probable parameters responsible for the transition from a rhodalgal-dominated carbonate production, during the late Serravallian, to a chlorozoan production during the early Tortonian. The amplitude of the sea-level fall responsible for the Ser4/Tor1 sequence boundary may have interrupted the upwelling inferred during the Middle Miocene in the Mediterranean Basin (Mutti *et al.*, 1999; Brandano & Corda, 2002). In parallel, a reduction of the connection with the cooler Atlantic waters may have favoured a local warming of Mediterranean surface waters.

CONCLUSIONS

This study shows that two facies models must be applied to the Ragusa Platform during the short time interval (2–3 Myr) between the late Serravallian and the early Tortonian. For each stage, seven facies zones ranging from the shallow-water, euphotic, inner ramp down to deep, aphotic, outer ramp, over a distance of about 30 km, were defined and described. Importantly, the facies zones clearly differ between the two time intervals, especially in the proximal part of the edifice.

During the late Serravallian, the euphotic inner ramp was characterized by small epiphytic biota of the foramol association, mostly thriving in seagrass meadows, whereas the oligophotic middle ramp was dominated by rhodoliths and subordinate larger benthic foraminifera, bryozoans, barnacles, serpulids and echinoids. By contrast, during the early Tortonian, the inner ramp was characterized by chlorozoan biota (*Porites*, *Halimeda*), and the middle ramp by branching Corallinacea and rhodoliths.

The seemingly minor changes in the type of carbonate production had profound consequences for the overall geometry of the platform. During the late Serravallian, the low rates of sediment production on the inner ramp, the limited amount of shedding related to the wide sea-grass meadows and the elevated rhodalgal production in deeper waters gave the Ragusa platform a typical distally steepened geometry. During the early Tortonian, the increase in carbonate production, linked to the appearance of a chlorozoan biota in the inner ramp and higher rates of basinward export of sediments resulted in the progradation of the euphotic carbonate factory and in a flattening of the proximal ramp sectors. The Ragusa ramp thus tended towards a flat-topped geometry.

The response of the Ragusa Platform to the relatively minor faunal change at the Serravallian–Tortonian boundary illustrates the great sensitivity of carbonate systems to regional and global environmental change.

ACKNOWLEDGMENTS

We thank Prof. R. Wernli (Université de Genève) for controlling the palaeontological determinations, F. Gischig and P. Desjacques (Université de Genève) for thin-section processing. We are thankful to T. Bechstaedt, M. Pedley, and W. Piller for their reviews of this paper. This study was supported by grants from the Swiss National Science Foundation (No. 21-67139.01 and 200020-107436/1).

REFERENCES

Adey, W.H. (1979) Crustose coralline algae as microenvironmental indicators in the Tertiary. In: *Historical Biogeography, Plate Tectonics and the Changing Environments* (Eds J. Gray and A.J. Boucot), pp. 459–464. *Oregon State University Press*, OR.

Adey, W.H., **Townsend, R.A.** and **Boykins, W.T.** (1982) The crustose coralline algae (Rhodophyta: Corallinaceae) of the Hawaiian Islands. *Smithson. Contrib. Mar. Sci.*, **15**, 1–74.

Barrier, E. (1992) Tectonic of a flexed foreland: the Ragusa Platform. *Tectonophysics*, **206**, 91–111.

Bassant, P., **van Buchem, F.H.P.**, **Strasser, A.** and **Gorur, N.** (2005) The stratigraphic architecture and evolution of the Burdigalian carbonate-siliciclastic sedimentary systems of the Mut Basin, Turkey. *Sed. Geol.*, **173**, 187–232.

Bosellini, F.R., **Russo, A.** and **Vescogni, A.** (2001) Messinian reef-building assemblages of the Salento Peninsula (southern Italy): paleobathymetric and paleoclimatic significance. *Palaeogeogr. Palaeoclimatol. Palaeoecol.*, **175**, 7–26.

Bosellini, F.R., **Russo, A.** and **Vescogni, A.** (2002) The Messinian reef complex of the Salento Peninsula (Southern Italy): stratigraphy, facies and paleoenvironmental interpretation. *Facies*, **47**, 91–112.

Bosence, D.W.J. (1985) Preservation of coralline algal frameworks. *Proc. 5th Int. Coral Reef Congress, Tahiti*, **2**, 39–45.

Braga, J.C. and **Martín, J.M.** (1996) Geometries of reef advance in response to relative sea-level changes in a Messinian (uppermost Miocene) fringing reef (Cariatiz reef, Sorbas Basin, SE Spain). *Sed. Geol.*, **107**, 61–81.

Brandano, M. and **Corda, L.** (2002) Nutrient, sea-level and tectonics constrains for the facies architecture of Miocene carbonate ramp in Central Italy. *Terra Nova*, **14**, 257–262.

Buxton, M.W.N. and **Pedley, H.M.** (1989) A standardized model for Tethyan Tertiary carbonate ramps. *J. Geol. Soc. London*, **146**, 746–748.

Carannante, G., **Esteban, M.**, **Milliman, J.D.** and **Simone, L.** (1988) Carbonate lithofacies as paleolatitude indicators: problems and limitations. *Sed. Geol.*, **60**, 333–346.

Catalano, R., **Infuso, S.** and **Sulli, A.** (1995) Tectonic history of the submerged Maghrebian chain from the southern Tyrrhenian sea to the Pelagian foreland. *Terra Nova*, **7**, 179–188.

Chevalier, J.P. (1961) Recherches sur les Madréporaires et les formations récifales miocènes de la Méditerranée occidentale. *Soc. Géol. Fr. Mém.*, **93**, 1–562.

Cita, M.B. (1959) Stratigrafia micropaleontologica del Miocene Siracusano. *Boll. Soc. Geol. Ital.*, **77**, 71–165.

Corda, L. and **Brandano, M.** (2003) Aphotic zone carbonate production on a Miocene ramp, Central Apennines, Italy. *Sed. Geol.*, **161**, 55–77.

Dall'Antonia, B. (2002) Short paleoecological notes on the Middle Serravallian-Basal Tortonian ostracods from the Tremiti Islands. In: *Integrated Stratigraphy and Paleoceanography of the Mediterranean Middle Miocene* (Ed. S.M. Iaccarino). *Riv. Ital. Paleontol. Stratigr.*, **108**, 298–296.

Demarcq, G. (1987) Paleothermic evolution during the Neogene in Mediterranea through the marine megafauna. Budapest Regional Committee on Mediterranean Stratigraphy, 7e Congrès Néogène Méditerranéen. Abstract. Budapest.

Di Stefano, E., **Bonomo, S.**, **Caruso, A.**, **Dinares-Turell, J.**, **Foresi, L.M.**, **Salvatorini, G.** and **Sprovieri, R.** (2002) Calcareous plankton bio-events in the Miocene Case Pelancani Section (Southeastern Sicily, Italy). In: *Integrated Stratigraphy and Paleoceanography of the Mediterranean Middle Miocene* (Ed. S.M. Iaccarino). *Riv. Ital. Paleontol. Stratigr.*, **108**, 307–323.

Drooger, C.W. (1983) Environmental gradients and evolutionary events in some larger foraminifera. *Utrecht Micropaleontol. Bull.*, **30**, 255–271.

Esteban, M. (1979) Significance of upper Miocene coral reefs of the western Mediterranean. *Palaeogeogr. Palaeoclimatol. Palaeoecol.*, **29**, 169–188.

Esteban, M. (1996) An overview of Miocene reefs from the Mediterranean areas: general trends and facies models. In: *Models for Carbonate Stratigraphy from Miocene Reef Complexes of Mediterranean Regions* (Eds E.K. Franseen, M. Esteban, W. Ward and J. Rouchy). *SEPM Concepts Sedimentol. Paleontol.*, **5**, 3–53.

Foresi, L.M., **Iaccarino, S.**, **Mazzei, R.** and **Salvatorini, G.** (1998) New data on middle to late Miocene calcareous plankton biostratigraphy in the Mediterranean area. *Riv. Ital. Paleontol. Stratigr.*, **104**, 95–114.

Grasso, M. and **Lentini, F.** (1982) Sedimentary and tectonic evolution of the eastern Hyblean plateau (southeastern Sicily) during Late Cretaceous to Quaternary time. *Palaeogeogr. Palaeoclimatol. Palaeoecol.*, **39**, 261–280.

Grasso, M., **Lentini, F.** and **Pedley, M.** (1982) Late Tortonian-Lower Messinian (Miocene) palaeogeography of SE Sicily: information from two new Formations of the Sortino Group. *Sed. Geol.*, **32**, 279–300.

Halfar, J., **Godinez-Orta, L.**, **Mutti, M.**, **Valdez-Holguìn, J.E.** and **Borges, J.M.** (2004) Nutrient and temperature controls on modern carbonate production: An example from the Gulf of California, Mexico. *Geology*, **32**, 213–216.

Hallock, P. (1987) Fluctuations in the trophic resource continuum: a factor in global diversity cycle? *Paleoceanography*, **2**, 457–471.

Hallock, P. (1988) The role of nutrient availability in bioerosion: consequence to carbonate build-ups. *Palaeogeogr. Palaeoclimatol. Palaeoecol.*, **63**, 275–291.

Hallock, P. (1996) Reef and reef limestones in Earth History. In: *Life and Death of Coral Reefs* (Ed. C. Birkeland), pp. 13–42. *Chapman & Hall*, New York.

Hallock, P. (2001) Coral reefs in the 21st century: is the past key to future? In: *Proceedings of the 10th symposium on the Geology of the Bahamas and Other Carbonate Regions* (Eds B.J. Greenstein and C.K. Carney), pp. 8–13. *Gerace Res. Center.*, San Salvador, Bahamas.

Hallock, P. and **Glenn, C.** (1986) Larger foraminifera: a tool for paleoenvironmental analysis of Cenozoic carbonate depositional facies. *Palaios*, **1**, 55–64.

Hallock, P. and **Schlager, W.** (1986) Nutrient excess and the demise of coral reefs and carbonate platforms. *Palaios*, **1**, 389–398.

Haq, B.U., **Hardenbol, J.** and **Vail, P.R.** (1987) Chronology of fluctuating sea levels since the Triassic. *Science*, **235**, 1156–1166.

Hardenbol, J., **Thierry, J.**, **Farley, M.B.**, **Jacquin, T.**, **De Graciansky, P.-C.** and **Vail, P.R.** (1998) Mesozoic-Cenozoic sequence chronostratigraphic chart. In: *Sequence Stratigraphy of European Basins* (Eds P.-C. De Graciansky, J. Hardenbol, T. Jacquin, P.R. Vail and M.B. Farley). *SEPM Spec. Publ.*, **60**, charts.

Hohenegger, J. (1995) Depth estimation by proportion of living larger foraminifera. *Mar. Micropaleontol.*, **26**, 31–47.

Hottinger, L. (1977) Distribution of larger Peneroplidae, *Borelis* and Nummulitidae in the Gulf of Elat, Red Sea. *Utrecht Micropaleontol. Bull.*, **15**, 35–109.

Hottinger, L. (1983) Processes determining the distribution of larger foraminifera in space and time. *Utrecht Micropaleontol. Bull.*, **30**, 239–253.

Hottinger, L. (1997) Shallow benthic foraminiferal assemblages as signals for depth of their deposition and their limitation. *Bull. Soc. Géol. Fr.*, **168**, 491–505.

James, N.P. (1997) The cool-water carbonates depositional realm. In: *Cool-water Carbonates* (Eds N.P. James and J.A.D. Clarke). *SEPM Spec. Publ.*, **56**, 1–20.

Larsen, A.R. (1976) Studies of Recent *Amphistegina*: taxonomy and some ecological aspects. *Israel J. Earth Sci.*, **25**, 1–26.

Lees, A. (1975) Possible influence of salinity and temperature on modern shelf carbonate sedimentation. *Mar. Geol.*, **19**, 158–198.

Lees, A. and **Buller, A.T.** (1972) Modern temperate-water and warm-water shelf carbonate sedimentation. *Mar. Geol.*, **13**, M67–M73.

Lukasik, J., **James, N.P.**, **McGowran, B.** and **Bone, Y.** (2000) An epeiric ramp: low-energy, cool-water carbonate facies in a Tertiary inland sea, Murray Basin, South Australia. *Sedimentology*, **47**, 851–881.

Martín, J.M., **Braga, J.C.** and **Riding, R.** (1997) Late Miocene *Halimeda* algal-microbial segment reef in the marginal Mediterranean Sorbas Basin, Spain. *Sedimentology*, **44**, 441–456.

Mutti, M. and **Hallock, P.** (2003) Carbonate systems along nutrient and temperature gradients: some sedimentological and geochemical constrains. *Int. J. Earth Sc.*, **92**, 465–475.

Mutti, M., **Bernoulli, D.**, **Spezzaferri, S.** and **Stille, P.** (1999) Lower and Middle Miocene carbonate facies in the Central Mediterranean: the impact of paleoceanography on sequence stratigraphy. In: *Advances in Carbonate Sequence Stratigraphy: Application to Reservoirs, Outcrops and Models* (Eds P.M. Harris, A.H. Saller and J.A. Simo). *SEPM Spec. Publ.*, **63**, 371–384.

Nelson, C.S. (1988) An introductory perspective on non-tropical shelf carbonates. In: *Non-tropical Shelf Carbonates: Modern and Ancient* (Ed. C.S. Nelson). *Sed. Geol.*, **60**, 3–12.

Pedley, M. (1979) Miocene bioherms and associated structures in the Upper Coralline Limestone of the Maltese Islands: their lithification and paleoenvironment. *Sedimentology*, **26**, 577–591.

Pedley, M. (1981) Sedimentology and palaeoenvironment of the southeast Sicilian Tertiary platform carbonates. *Sed. Geol.*, **28**, 273–291.

Pedley, M. (1996) Models for carbonate stratigraphy from Miocene reef complexes of Mediterranean regions. In: *Models for Carbonate Stratigraphy from Miocene Reef Complexes of Mediterranean Regions* (Eds E.K. Franseen, M. Esteban, W. Ward and J. Rouchy). *SEPM Concepts Sedimentol. Paleontol.*, **5**, 247–259.

Pedley, M. (1998) A review of sediment distribution and processes in Oligo-Miocene ramps of southern Italy and Malta (Mediterranean divide). In: *Carbonate Ramps* (Eds V.P. Wright and T.P. Burchette). *Geol. Soc. London Spec. Publ.*, **149**, 163–179.

Pedley, M., **Cugno, C.** and **Grasso, M.** (1992) Gravity slide and resedimentation processes in a Miocene carbonate ramp, Hyblean Plateau, southeastern Sicily. *Sed. Geol.*, **79**, 189–202.

Pérès, J.M. and **Picard, J.** (1964) Nouveau manuel de bionomie bentique de la Mer Méditerrannée. *Rev. Trav. Stn. Mar. Endoume-Marseille*, **31**, 1–137.

Perrin, C., **Bosence, D.** and **Rosen, B.** (1995) Quantitative approaches to paleozonation and paleobathymetry of corals and coralline algae in Cenozoic reefs. In: *Marine Palaeoenvironmental Analysis from Fossils* (Eds D.W.J. Bosence and P.A. Allison). *Geol. Soc. London Spec. Publ.*, **83**, 181–229.

Pomar, L. (2001) Types of carbonate platforms: a genetic approach. *Basin Res.*, **13**, 313–334.

Pomar, L., **Brandano, M.** and **Westphal, H.** (2004) Environmental factors influencing skeletal grain sediment associations: a critical review of Miocene examples from the western Mediterranean. *Sedimentology*, **51**, 627–651.

Rigo, M. and **Barbieri, F.** (1959) Stratigrafia pratica applicata in Sicilia. *Boll. Ser. Geol. Ital.*, **80**, 351–441.

Ruchonnet, C. (2006) Climatic and oceanographic evolution of the Mediterranean basin during Late Serravallian/Early Tortonian (Middle/Late Miocene): the record from the Ragusa carbonate Platform (SE Sicily, Italy). PhD dissertation, Université de Genève, Terre Environ., **63**, 156 pp.

Samankassou, E. (2002) Cool-water carbonates in paleo-equatorial shallow-water environment: The paradox of the Auering cyclic sediments (Upper Pennsylvanian, Carnic Alps, Austria-Italy) and its implications. *Geology*, **30**, 655–658.

Sandberg, P.A. (1983) An oscillating trend in Phanerozoic non-skeletal carbonate mineralogy. *Nature*, **305**, 19–22.

Schlager, W. (1981) The paradox of drowned reefs and carbonate platforms. *Geol. Soc. Am. Bull.*, **92**, 197–211.

Schlager, W. (2003) Benthic carbonate factories of the Phanerozoic. In: *New Perspectives in Carbonate Sedimentology* (Eds J.J.G. Reijmer, C. Betzler and M. Mutti). *Int. J. Earth Sci.*, **92**, 445–464.

Sprovieri, R.M., **Bonomo, S.**, **Caruso, A.**, **Di Stefano, A.**, **Di Stefano, E.**, **Foresi, L.M.**, **Iaccarino, S.M.**, **Lirer, F.**, **Mazzei, R.** and **Salvatorini, G.** (2002). An intergrated calcareous plankton biostratigraphic scheme and biochronology for the Mediterranean middle Miocene. In: *Integrated Stratigraphy and Paleoceanography of the Mediterranean Middle Miocene* (Ed. S.M. Iaccarino). *Riv. Ital. Paleontol. Stratigr.*, **108**, 2, 337–353.

Stanley, S.M. and **Hardie, L.A.** (1998) Secular oscillations in the carbonate mineralogy of reef-building and sediment-producing organisms driven by tectonically forced shifts in seawater chemistry. *Palaeogeogr. Palaeoclimatol. Palaeoecol.*, **144**, 3–19.

Turco, E., **Hilgen, F.J.**, **Lourens, L.J.**, **Shackleton, N.J.** and **Zachariasse, W.J.** (2001) Punctuated evolution of global climate cooling during the late middle to early late Miocene: High-resolution planktonic foraminiferal and oxygen isotope records from the Mediterranean. *Paleoceanography*, **16**, 405–423.

Vecsei, A.S. and **Sanders, G.K.** (1999) Facies analysis and sequence stratigraphy of a Miocene warm-temperate carbonate ramp, Montagna della Maiella, Italy. *Sed. Geol.*, **123**, 103–127.

Wilson, M.E.J. and **Vecsei, A.** (2005) The apparent paradox of abundant foramol facies in low latitudes: their environmental significance and effect on platform development. *Earth-Sci. Rev.*, **69**, 133–168.

Yellin-Dror, A., Grasso, M., Ben-Avraham, Z. and **Tibor, G.** (1997) The subsidence history of the northern Hyblean plateau margin, southeastern Sicily. *Tectonophysics*, **282**, 277–289.

Zachos, J.C., Pagani, M., Sloan, L., Thomas, E. and **Billups, K.** (2001) Trends, rhythms and aberrations in global climate 65 Ma to present. *Science*, **292**, 686–693.

Int. Assoc. Sedimentol. Spec. Publ. (2010) **42**, 89–106

The sensitivity of a tropical foramol-rhodalgal carbonate ramp to relative sea-level change: Miocene of the central Apennines, Italy

MARCO BRANDANO*, HILDEGARD WESTPHAL† and GUILLEM MATEU-VICENS‡

**Dipartimento di Scienze della Terra, Università di Roma "La Sapienza", Ple Aldo Moro, 5. I-00185 Roma, Italy (E-mail: marco.brandano@uniroma1.it)*

†Department of Geosciences, Bremen University, Loebener Straße, D-28359 Bremen, Germany

‡Departament de Ciències de la Terra, Universitat de les Illes Balears, Ctra. De Valldemossa, Km 7.5 E-07071 Palma de Mallorca, Spain

ABSTRACT

Over the past 15 years it has been demonstrated that carbonate ramps react in a distinctly different way to sea-level fluctuations than flat-topped carbonate platforms. Until now, most sequence stratigraphic studies of carbonate ramps have concentrated on temperate to cool-water carbonates and on tropical systems after a phase of forced regression. This study aims at characterizing a tropical, eutrophic carbonate ramp, the Miocene Latium-Abruzzi platform in the central Apennines. The stratigraphic architecture of the carbonate ramp deposited during a 5 Myr period reflects two 2nd-order relative sea-level cycles. In contrast, the architectural and facies patterns only reflect 3rd-order sea-level fluctuations in the inner ramp. Sea-level fluctuations that cause changes in light-penetration may produce shifts from euphotic to oligophotic or even aphotic conditions that may be recorded as facies changes. The insensitivity of middle- to outer-ramp facies belts is the result of the carbonate-producing organisms in these parts of the ramp, and of the low production of carbonate sediments on the inner ramp, limiting the export to middle- and outer-ramp settings. On this type of carbonate ramp, most voluminous carbonate sediment production takes place in the oligophotic zone by rhodalgal associations, or in the aphotic zone by bryomol or molechfor associations. Thus, carbonate sediment accumulates below the hydrodynamic base level in a water-depth interval that may be wider than the amplitude of the high-frequency sea-level fluctuation. Such ramps therefore appear not to record high-frequency sea-level fluctuations.

Keywords carbonate ramp, sequence stratigraphy, Miocene, foramol carbonates, sea-level change.

INTRODUCTION

The seminal carbonate sequence stratigraphic model that was developed in the 1980s and 1990s (e.g. Schlager *et al.*, 1994) was largely based on modern flat-topped carbonate platforms such as the Great Bahama Bank. Later models included the effect of different morphologies (such as flat-topped, steep-sided platforms versus gently inclined carbonate ramps) on carbonate productivity and sedimentary sequences (e.g. Burchette & Wright, 1992; Bachmann & Willems, 1996; Bachmann *et al.*, 1996; Pomar, 2001a). The relationship between sea-level variations, carbonate-producing biota and compositional variations as sensitive recorders is now well-established (e.g. Sarg, 1988; Pomar, 1991; Louks & Sarg, 1993; Homewood, 1996; Pomar, 2001a; Schlager, 2003).

In particular, biotic communities have been recognized to strongly influence carbonate platform morphology (Handford & Louks, 1993; Homewood, 1996; Pomar 2001b, Schlager, 2003; Pomar *et al.*, 2005). In carbonate systems, the base level for sediment accumulation depends on both physical accommodation (hydrodynamic conditions at the locus of accumulation) and ecological accommodation (building capability). Thus, platform architecture is determined to a large extent by ecological accommodation. That is, the ability of the various carbonate-producing biota to resist a different hydrodynamic energy level (Pomar, 2001b; Pomar *et al.*, 2004, 2005) by building a framework, or by producing large grain sizes (e.g. rhodoliths). Thus, accommodation depends not only on the amount, but also on the type of sediment being produced, on biological processes (binding, trapping, baffling), on the hydraulic energy at the locus of production, and also on early diagenetic processes including early cementation (Pomar, 2001a, b).

One time interval with a large number of well-studied carbonate ramps is the Miocene. Numerous studies have successfully interpreted 2nd- and 3rd-order depositional sequences of Miocene carbonate ramps (e.g. Martín *et al.*, 1996; Betzler *et al.*, 1997a; Franseen *et al.*, 1997; Vecsei & Sanders, 1999; Brandano & Corda, 2002; Cunningham & Collins, 2002; Cathro *et al.*, 2003; Lukasik & James, 2003). Several studies also identified with 4th- and 5th-order sequences in depositional environments above or close to wave base (e.g. Martín *et al.*, 1996; Betzler *et al.*, 1997a; Lukasik & James, 2003). It has been pointed out that such high-frequency fluctuations are potentially also recorded on the slope of distally steepened ramps and on distal outer ramps, where sediments shed from the inner ramp interfinger with the pelagic sediments (Franseen *et al.*, 1997; Passlow, 1997; Betzler *et al.*, 2005). However, studies of sub-wavebase deposits in middle- and outer-ramp environments revealed that they fail to record high-frequency sea-level fluctuations (Carannante & Simone, 1996; Vecsei & Sanders, 1999; Brachert *et al.*, 2001; Brandano & Corda, 2002; Cunningham & Collins, 2002).

The question concerning under what circumstances middle- to outer-ramp deposits are sensitive recorders of (relative) sea-level fluctuations has not been systematically explored to date. The present study aims at contributing to this question by examining the sensitivity of a Miocene eutrophic, rhodalgal-bryomol dominated carbonate ramp. The focus is the Late Burdigalian to Langhian succession of the Latium-Abruzzi platform in the central Apennines. The objective of this study is to test whether there is any expression of sea-level fluctuations, particularly in the subtidal parts of the ramp, that is whether the oligophotic and aphotic carbonate factories record sea-level fluctuations when located below the wave base.

GEOLOGICAL SETTING

The Apennine fold and thrust belt, representing the Neogene deformation of the passive margin of the Mediterranean Tethys, contains numerous outcrops of carbonate platforms, particularly in its central and southern sectors (e.g. Accordi & Carbone 1988, Bernoulli, 2001; Parotto & Praturlon, 2004). The Latium-Abruzzi platform is located in the central Apennines and formed on the passive side of the Apennine foredeep (Fig. 1). It consists of a thick, discontinuous succession of about 5000 m of limestone, with subordinate dolostone, of Late Triassic to Late Miocene age. A hiatus between Upper Cretaceous and Lower to Middle Miocene deposits is physically expressed by an unconformity recognizable in large parts of the Latium-Abruzzi platform.

The focus of the present study is on the Lower to Middle Miocene, represented by a carbonate platform succession known as "Calcari a Briozoi e Litotamni" Formation. This formation was deposited in a foreland basin during the early phase of the development of the Central Apennines chain. The platform records the eastward migration of the orogenic front, with moderate tectonic uplift or stasis followed by an accelerated subsidence (Carminati *et al.*, 2007). The eastward migration of the uplift is interpreted as a response to the eastward propagation of compressional deformation in the Central Apennines (Carminati *et al.*, 2007). The following accelerated tectonic subsidence is interpreted as a result from the subduction-related flexure of the Adriatic plate (Doglioni, 1991). Eventually, the platform drowned, as recorded in the Tortonian hemipelagic deposits (*Orbulina* Marls), followed by siliciclastic turbidites representing the sedimentation in the foredeep system (Fig. 2; Carannante *et al.*, 1988; Patacca *et al.* 1991). Afterwards, this succession as a whole was deformed tectonically during the Late Tortonian to Early Pliocene (Patacca *et al.*, 1991). The Aquitanian to Serravallian "Calcari a Briozoi e Litotamni" Formation of the Latium-Abruzzi platform, which forms an extensive carbonate ramp, is the focus of this paper.

METHODS

Stratigraphic and sedimentological analyses of 27 logged sections, 22 of which have been discussed previously (Brandano, 2001; Corda & Brandano, 2003), form the basis of this study (Fig. 1). The sections were examined with respect to platform architecture and macroscopic depositional facies. These observations were

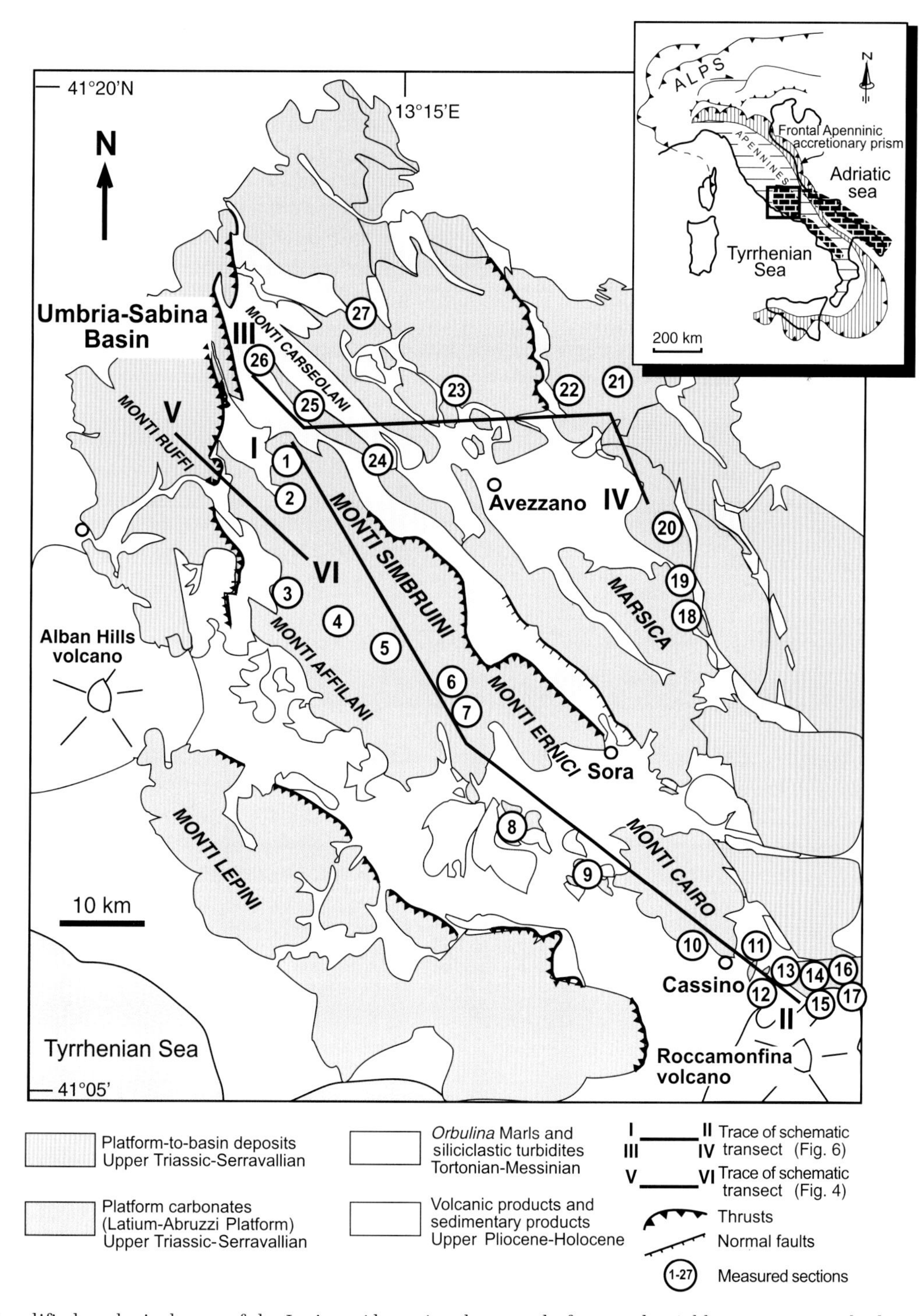

Fig. 1. Simplified geological map of the Latium-Abruzzi carbonate platform and neighbouring areas. The locations of the analyzed stratigraphic sections are indicated by numbers and are referred to in the subsequent figures. Modified from Brandano (2001).

complemented with petrographic examination of 200 thin-sections for textural characterization and identification of skeletal components. Lithostratigraphic correlation and interpretation of the platform architecture was undertaken on the basis of sedimentological criteria such as variations in skeletal-grain assemblage composition. The overall facies description is thus based on lithology,

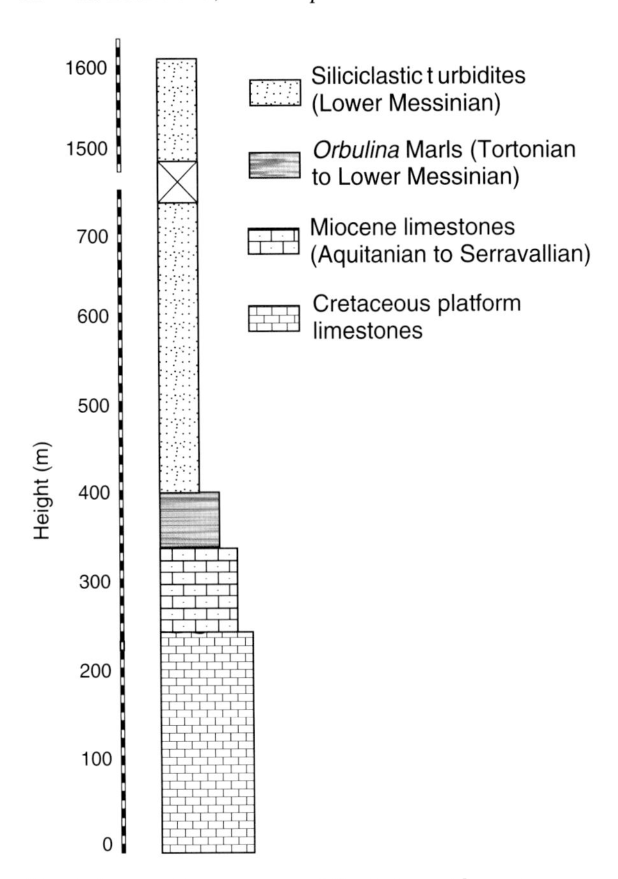

Fig. 2. Schematic summary of the Latium-Abruzzi succession (modified from Carminati *et al.*, 2007).

sediment constituents, sedimentary structures, stratification and geometric relationships.

RAMP SEDIMENTS OF THE "CALCARI A BRIOZOI E LITOTAMNI" FORMATION

Facies

The facies have been grouped into three depositional environments that correspond to inner-, middle-, and outer-ramp deposits according to the general scheme of Brandano & Corda (2002), Brandano (2003) Pomar *et al.* (2004) and Civitelli & Brandano (2005). Although the terminology of Burchette & Wright (1992) is used here to subdivide the carbonate-ramp facies, the criteria employed to characterize the various parts of the ramp follow Pomar (2001a).

Inner ramp

The inner-ramp environment spans from the littoral zone to sediment produced at the lower limit of the euphotic zone. The littoral zone is represented by slightly inclined Cretaceous substratum characterized by the occurrence of intense bioerosion, including *Gaestrochaenolites lapidicus* and *Caulostrepsis*, although *Entobia* is largely absent (Brandano, 2003).

An unequivocal palaeoshoreline is not documented, although local occurrence of *Microcodium* levels in the inner-ramp deposits imply occasional emergence (Civitelli & Brandano, 2005). The deposits of the littoral zone are composed of balanid floatstone to rudstone and lithoclastic conglomerates. The balanid floatstone to rudstone is crudely stratified with subhorizontal bedding planes. It is composed mainly of balanid macroids and balanid fragments with subordinate red algal fragments, isolated rhodoliths, extensively bioeroded bivalve shells, echinoids and bryozoans. The bioclasts are dispersed in a packstone to wackestone matrix with miliolids, rotalids and rounded micritized mollusc fragments. Lithoclastic conglomerates occur intercalated with the balanid floatstone to rudstone and are crudely stratified into beds of 20 to 50 cm thickness. Pebbles (up to 15 cm in diameter) are subangular to subspherically-shaped lithoclasts of the bedrock (Cretaceous limestone and, rarely, Palaeogene calcarenites). They show traces of bioerosion and glauconitic mineralization. The matrix consists of bioclastic packstone with micritized mollusc fragments, bryozoans and benthic foraminifera.

The above facies grade basinward into unsorted bioclastic packstone to floatstone and from rudstone to floatstone with red algal nodules and branches (Brandano, 2003). The skeletal packstone to floatstone is crudely stratified. Individual beds are 60–70 cm thick with planar bedding surfaces. Rocks are composed of well-rounded and usually micritized mollusc fragments, barnacle and echinoid fragments, bryozoans, and red algal fragments and branches. These bioclasts are dispersed in a poorly sorted bioclastic wackestone to packstone rich in epiphytic foraminifera. *In situ* mechanical breakage of large bivalve fragments, larger benthic foraminifera and barnacles is indicated by the fragments of the same test in the immediate vicinity.

The rudstones to floatstones associated with these bioclastic packstones to floatstones contain abundant red algal nodules and branches. They comprise 1.3 to 3 m thick, crudely stratified, subhorizontal beds. Subordinate components include fragments of echinoids, corals, balanids and bivalves, larger benthic foraminifera, and small benthic foraminifera (miliolids, rotalids). The bioclasts and red algae are embedded in a carbonate mud with ostracods and miliolids. The skeletal rudstone to floatstone facies, as well as the packstone to floatstone facies, are characterized by the presence of bowl-like structures up to 10 cm wide and up to 5 cm high, with tabulations interpreted by Piller & Brandano (2005) as remains of hextactinellid sponges associated with seagrass. Small ellipsoidal peloids and angular, dark, structureless grains (up to 150 μm in diameter) are also common and likely represent micritized particles of skeletal and faecal origin. The presence of seagrass is inferred from the absence of wave-related sediment structures.

The biogenic components have been subjected to fragmentation, bioerosion, and dissolution. In particular, aragonitic components are generally leached, but nevertheless in most cases they can be recognized due to previous micritization. Non-skeletal components such as ooids and grapestones are absent.

Middle ramp

The transition from the inner ramp to the middle ramp is marked by rudstone to floatstone with abundant red algae nodules and branches, interfingering with small coral build-ups that occur in the lower part of the succession (coral carpet, *sensu* Riegl & Piller, 2000). Besides the coral carpets, middle-ramp facies consist of alternating rhodolithic floatstone to rudstone, medium to coarse-grained bioclastic packstone and oyster-pectinid floatstone. No wave-related sedimentary structures are present in the middle-ramp deposits. This absence of wave-related structures is interpreted as indicating a position below wave-base.

In the coral carpets, *Porites* are associated with bivalves and are dispersed in a matrix of medium to coarse-grained packstone with echinoid fragments and thalli of red algae. Rare foraminifera include miliolids and planktonic foraminifera. Despite the presence of coral carpets, middle-ramp components belong to the rhodalgal association. The coralline algae are preserved as non-fragmented branches and rhodoliths. Generally, bryozoan colonies and acervulinid foraminifera are abundant components of the rhodoliths, interlayered between the red algal thalli. The coralline algae may also form thick laminar layers (up to 5 cm thick) encrusting double-valved neopycnodont oyster shells. A detailed description of coralline algae assemblages from the middle-ramp facies is given by Brandano & Piller (this volume, pp. 151–166). Micrite envelopes are absent, peloids are rare.

The rhodolith floatstone to rudstone is crudely stratified with subhorizontal bedding planes. The main components are rhodoliths dispersed in a medium to coarse-grained packstone with red algae, molluscs, echinoid fragments and larger benthic foraminifera.

The medium to coarse-grained bioclastic packstone is composed of 60–70 cm thick tabular beds. Components include well-rounded red algal fragments, echinoid plates and echinoid spines, benthic foraminifera and larger benthic foraminifera (*Heterostegina, Operculina, Amphistegina, Miogypsina*). Fragments of balanids, and bryozoans are subordinate. Planktonic foraminifera are rare.

The oyster-pectinid floatstone comprises tabular, massive beds (0.4 to 2 m thick) associated and interbedded with a poorly sorted bioclastic packstone to grainstone rich in bryozoan and echinoid remains, benthic and encrusting foraminifera, serpulids, and algal detritus. Oysters are locally encrusted by red algae and, as with the coralline algal crusts, are heavily bioeroded.

Outer ramp

On the basis of the biota associations and lithological characteristics the outer ramp can be subdivided into three distinct facies belts: proximal, intermediate, and distal outer ramp. Skeletal carbonate sediments of the proximal outer ramp are dominated by bryozoan colonies and bivalves (bryomol association), intermediate and distal outer-ramp sediments are characterized by fragmented bivalves, echinoids, and benthic and planktonic foraminifera (molechfor association). One feature common to all outer-ramp facies, as also for the middle-ramp facies, is the absence of wave-related sedimentary structures, interpreted as an indication that they have been deposited below wave-base.

Lithofacies of the proximal outer ramp are crudely stratified coarse-grained echinoid-foraminiferal packstone, bryozoan floatstone and bivalve floatstone (Fig. 3). The skeletal fraction of these facies is dominated by bryozoans, echinoid and bivalve fragments, and benthic and planktonic foraminifera. Rare beds of extensively bioeroded rhodoliths

Fig. 3. Calcari a Briozoi e Litotamni Formation outcrop at Tagliacozzo, Monti Carseolani. Intermediate outer-ramp deposits (A) paraconformably overlie Cretaceous limestones (K). The proximal (B) and the intermediate outer-ramp successions are often characterized by decametre-thick intervals of a single facies type.

are present (Brandano, 2002). Carbonate sediment of the mud fraction occurs in very low amounts, and compaction features are abundant. Breakage of large bivalve fragments and foraminifera, sutured grain contacts and grain interpenetration are ubiquitous.

In the intermediate outer ramp, the most abundant facies is packstone. Fine to medium-grained packstone is dominated by echinoid fragments and planktonic foraminifera; bioclastic packstone with minor wackestone beds contains benthic foraminifera, echinoid fragments and the serpulid *Ditrupa*. These facies are characterized by thick (1–2 m) layers without sedimentary structures, delimited by planar bedding surfaces. On these surfaces, *Thalassinoides* burrows are common. Pressure solution structures, sutured grain contacts, and grain interpenetration features are common. As for the proximal outer ramp, irregular stylolites crosscutting the compaction features are widespread.

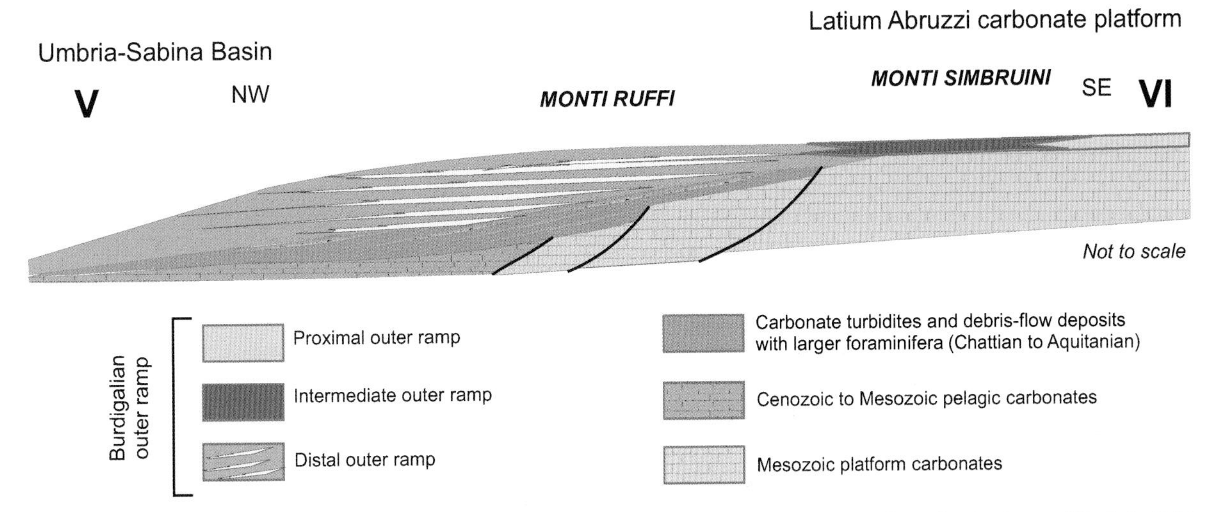

Fig. 4. Schematic cross section of the Burdigallian outer ramp and transition to the northern basinal area (Umbria Sabina Basin). Line of section is marked V–VI on Fig. 1.

Distal outer-ramp facies are represented by highly bioturbated argillaceous wackestone with planktonic foraminifera and siliceous sponges, as well as by bioclastic packstone and echinoid-benthic foraminiferal grainstone. These facies are vertically associated and form mixed carbonate-siliciclastic subtidal cycles (Fig. 4; Brandano *et al.*, 2005a). These cycles show a basal part characterized by highly bioturbated argillaceous wackestone passing upward to marly packstone made up of planktonic foraminifera and, subordinately, small fragments of echinoids, benthic foraminifera (textularids, rotalids), fragments of serpulids and subordinately siliceous sponge spicules. The fragment grain-size is extremely variable, ranging from 100–300 μm up to 600 μm. The top of the cycle is represented by bioclastic grainstone. Echinoids and benthic foraminifera (rotalids) are the main skeletal components together with rare bryozoans and planktonic foraminifera. Skeletal grain-size ranges from 500 to 1000 μm, rarely exceeding 3 mm. The top of the cycles is an omission surface expressed as an iron-stained and weakly convoluted hardground, or as a sharp flat surface overlain by argillaceous wackestones of the next cycle. The thickness of the cycles ranges between a few decimetres to up to 12 m.

The terrigenous portion of the sediments in the cycles is not derived from the coastal area of the ramp in the east of the foreland basin (i.e. the passive side of the fore-deep), but from the emergent areas of the developing Apennines ridge to the west (i.e. the active side of the fore-deep; Civitelli *et al.*, 1986). This influx of terrigenous material from the west, and neritic carbonate from the east, produced mixed siliciclastic-carbonate sedimentation on the distal outer ramp, whereas at the same time the proximal facies belts of the ramp were dominated by carbonate sedimentation with no terrigenous influx from the hinterland (Brandano, 2001; Brandano & Corda, 2002).

Photic zones of the "Calcari a Briozoi e Litotamni" formation

Based on depth of light penetration, the water column can be divided into euphotic, dysphotic and aphotic zones (e.g. Cognetti & Sarà, 1974). The euphotic zone of the benthic realm corresponding to the layer inhabitated by autotrophic plants with light intensities that permit the operation of photosynthetic processes by primary producers. In the pelagic realm, the euphotic zone is also known as the epipelagic zone. Phytoplankton ecologists have observed that at 1% of surface-light penetration depth, photosynthesis balances respiration (compensation depth). This is the lower limit of the euphotic zone (Lüning, 1990). Below the euphotic zone, the dysphotic zone corresponds to light intensities of <1%, and the aphotic zone completely lacks light. Thus, photodependent carbonate-producing biota exclusively inhabit the euphotic zone, and production corresponding to dysphotic environments may not be distinguished from that in aphotic conditions. Hence, this light zonation does not offer much information for the characterisation of the carbonate factories.

Other light zonation schemes emphasize the differences among the different organisms producing in the euphotic zone. Pomar (2001a, b), in attempting to describe the light conditions in the inner, middle and outer-ramp environments, subdivided the water column into euphotic, mesophotic, oligophotic and aphotic zones, according to qualitative criteria of light-intensity. In the euphotic zone, biota (autotrophs and mixotrophs) that thrive need relatively high light levels. In tropical settings the lower limit of this zone corresponds to the maximum depth of vigorous hermatypic coral growth (>20% of subtropical surface light intensities). The mesophotic zone contains biota that require moderate light levels (between 20 and 4%). The oligophotic zone includes biota (autotrophs and mixotrophs) requiring poor light levels (<4% of subtropical surface light intensities). Aphotic biota includes heterotrophs that do not require light, although they may live in any environment.

In the present work, Pomar's subdivisions are followed, which permit a more refined nomenclature, and avoid the use of "euphotic" as synonymous of the general and non-specific term "photic". Additionally, the photic zones of Pomar (2001a, b) relate to the zones proposed by Hottinger (1997) in that the upper photic zone presents two subzones, the first one from 0 to 40 m, which is equivalent to the euphotic zone, and a second one from 40 to 80 m, that corresponds to the mesophotic zone. The lower photic zone of Hottinger (1997), between 80 m and 120–140 m depth, is equivalent to the oligophotic zone. Similarly, James & Bourque (1992) differentiated a photic from a subphotic zone. Carannante & Simone (1996) described the light-preference of biotic communities with the terms photophile and sciaphile.

Inner ramp: euphotic zone

The lithofacies present in the inner ramp indicate euphotic conditions. The inner ramp is characterized shoreward by a high-energy environment immediately below the littoral zone as indicated by borings in the Cretaceous substratum (*Gaestrochaenolites lapidicus*, *Caulostrepsis* with *Entobia* largely being absent or insignificant; Bromley & Asgaard, 1993). The depositional substratum is colonized initially by bioeroders and then successively by balanids (balanid floatstone to rudstone). Analogues facies were described by Betzler *et al.* (2000) in the Upper Miocene of southern Spain. Basinward, these facies grade into unsorted skeletal packstone to floatstone representative of a seagrass-dominated environment (Fig. 5). The presence of a seagrass zone is supported by a typical biota association, which includes epiphytic foraminifera such as *Elphidium*, *Planorbulina*, small miliolids, discorbids and textulariids. Large benthic foraminifera such as *Amphistegina*, and *Heterostegina* may also live in such environments (e.g. Murray, 1973; Brasier, 1975; Hoffman, 1979; Sen Gupta, 1999).

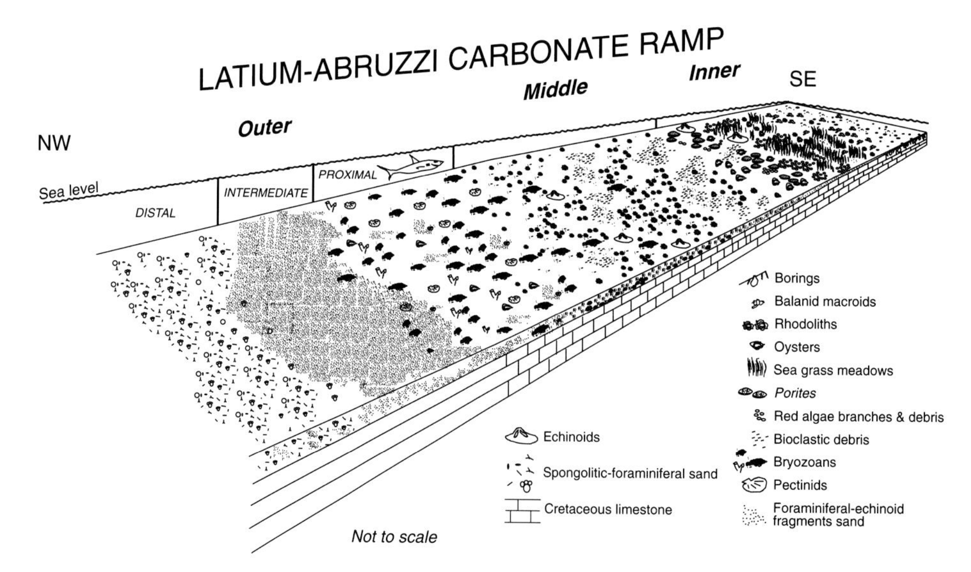

Fig. 5. Depositional model of the Latium-Abruzzi carbonate ramp during Late Aquitanian–Early Burdigalian time (modified from Brandano, 2003).

Moreover, the absence of sorting and sedimentary structures has been interpreted to be the result of sediment baffling, trapping and stabilization by seagrass (compare Pomar *et al.*, 2002).

Seagrass meadows were associated basinward with "maërl facies" (Pérés & Picard, 1964; Bosence, 1983), composed of rudstone to floatstone with red algal nodules and branches. In the present-day Mediterranean Sea, maërl deposits occupy the upper part of the circalittoral zone, just below the deepest occurrence of *Posidonia* meadows (Basso & Tomaselli, 1994; Canals & Ballesteros, 1997). In tropical settings, maërl deposits are generally confined to very shallow zones with frequent, moderate wave action (Steneck, 1986). The biota associations and ichnogenera assemblages, directly or indirectly dependent upon photosynthetic organisms such as the phanerogame plants, indicate that inner-ramp carbonate production took place in the euphotic zone.

Middle ramp: oligophotic zone

The main biota producing carbonate sediment of the middle ramp are red algae and foraminifera, with subordinate echinoids, bryozoans and bivalves. Bioeroders produce sand-size bioclastic sediment. The red algae genera associations and the larger benthic foraminifer assemblages of the middle-ramp facies are typical of the oligophotic zone below the base of wave action (Civitelli & Brandano, 2005).

Outer ramp: oligophotic to aphotic zone

On the proximal outer ramp, the dominant sediment-producing biota (bryozoans, pectinids, oysters and echinoids) and the rare rhodolith levels, together with an absence of larger benthic foraminifera, suggest a depositional environment at the boundary between the oligophotic and aphotic zones. Based on sedimentological features and biota assemblages, the intermediate to distal outer ramp is located in the aphotic zone.

Stratigraphic architecture of the "Calcari a Briozoi e Litotamni" formation

Overall architecture and 2nd-order sequences

From the relative positions of facies belts and the dependence upon light penetration of the carbonate-producing communities, it is inferred that the depositional profile during the formation of the "Calcari a Briozoi e Litotamni" was a homoclinal ramp, even though outcrops do not cover the entire system (Brandano, 2001, Pomar *et al.*, 2004; Civitelli & Brandano, 2005). This profile was the combined result of the inherited substratum (the Cretaceous carbonate platform) and it being the locus of carbonate production and accumulation. The substratum was gently inclined, dipping at $<0.01^{\circ}$ towards the NW, as a result of tectonically forced subsidence in the foreland basin when the western part of the Cretaceous platform was flooded during Aquitanian–Early Budigalian times (Brandano, 2001). The absence of zooxanthellate coral reefs (with the exception of the coral carpets at the boundary between inner and middle ramp in the lower part of the succession) resulted in a ramp morphology. Carbonate production in the shallow-water environment was generally low. The fact that the main locus of carbonate productivity and accumulation was in the deeper oligophotic (rhodalgal) and aphotic (bryomol) zones counteracted any possible tendency to form a distally steepened ramp.

The "Calcari a Briozoi e Litotamni" Formation is composed of two major sedimentary units that were interpreted as 2nd-order depositional sequences (Brandano & Corda, 2002; Civitelli & Brandano, 2005). The sequences were dated using Sr-isotope chemostratigraphy ($^{87}Sr/^{86}Sr$ ratios) of 35 bulk carbonate and skeletal samples (Fig. 3; Brandano, 2001). Ages were calculated using the $^{87}Sr/^{86}Sr$ versus age relationships of Hodell *et al.* (1991). The strontium ages for the bulk-carbonate samples correspond very closely to the strontium ages obtained from skeletal grains (echinoids and bivalves) from the same samples. The most prominent discontinuity surfaces recognized matched with major biostratigraphic events (i.e. last occurrence of *Miogypsina* cf. *globulina* at the base of the succession, first occurrence of *Orbulina universa*). Generally, the calculated strontium ages are consistent with their relative stratigraphic position and with biostratigraphic results.

The lower sequence, Aquitanian to Serravallian in age, is composed of a lowstand systems tract, a transgressive systems tract and a highstand systems tract (Brandano, 2001; Brandano & Corda, 2002; Civitelli & Brandano, 2005). The lower part of the lowstand systems tract is preserved only in the basinal area and is represented by carbonate turbidites and debris-flow deposits (Fig. 4); the larger benthic foraminiferal assemblages

(*Nephrolepidina* spp, *Miogypsinoides* spp., *Miogypsina* spp.) do not suggest deposition before the Aquitanian. All three main facies belts of the ramp (inner, middle and outer ramp) are preserved in the upper lowstand systems tract. In the transgressive and highstand systems tracts, only the middle and outer-ramp deposits are preserved.

The transgressive system tract has been determined on the basis of backstepping of the middle-ramp facies belt as a consequence of the landward (upward) migration of the photic carbonate-producing biota. During this backstepping of the middle-ramp facies belt, the proximal outer-ramp facies, dominated by heterozoan carbonate production, prograded into the basin (Fig. 6A). This means that at the same time, the ramp system shows progradation in the aphotic part and backstepping in the photic part, demonstrating that such facies shifts need to be studied carefully as part of the entire system before inferring sea-level movements. The heterozoan biota, being independent of water depth, kept producing sediment in conditions of progressive deepening. The progressive expansion of this facies belt indicates favourable conditions for bryozoans to grow and thus implies that significant amounts of suspended organic matter were present.

The upper sequence is Serravallian to Early Tortonian in age. It consists only of a lowstand systems tract and is terminated by the tectonically-driven drowning phase controlled by the eastward migration of the flexure of the Apenninic foredeep system. The top of the sequence has been dated at

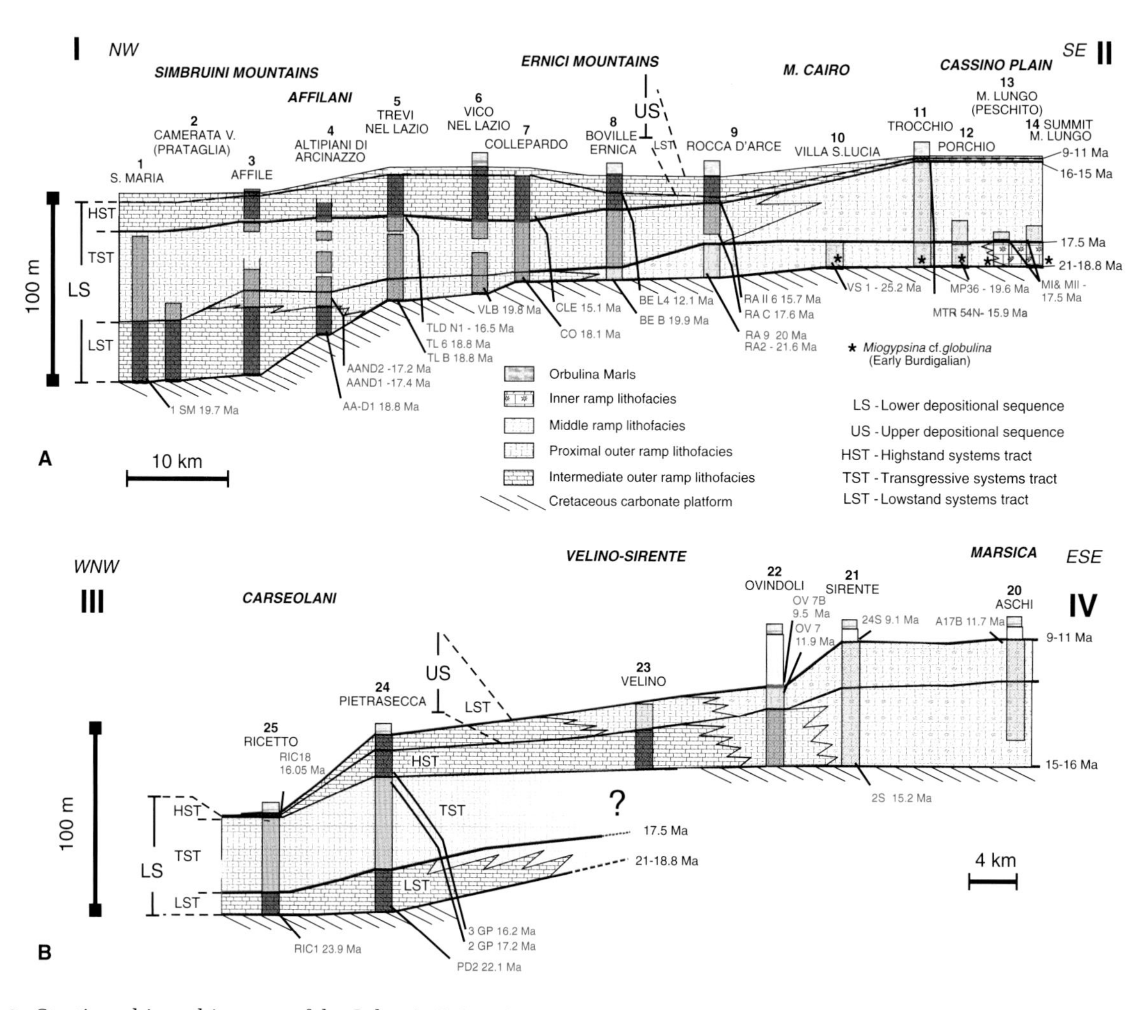

Fig. 6. Stratigraphic architecture of the Calcari a Briozoi e Litotamni Formation. (A) Section I–II in Fig. 1. On this ramp, there is no significant expression of sea-level cyclicity. Note the simultaneous backstepping and progradation of outer-ramp facies after 17.5 Ma. (B) Section III–IV in Fig. 1. Ages in blue are derived from Sr-isotope data (Brandano *et al.*, this volume, pp. 107–128). Modified from Brandano (2001) and Civitelli & Brandano (2005).

11–9 Ma, but no isotopic age is available for the base. The basal sequence boundary does not display evidence of subaerial or submarine erosion and is marked merely by a sharp facies contrast from deep to shallower facies. Considering the significant global sea-level lowstand at this time, the subtle expression of the basal unconformity is likely the result of a superimposed tectonically increased subsidence preceding the involvement in the foredeep system. The termination of this upper sequence is characterized by hemipelagic sediments followed by siliciclastic turbidites (Brandano, 2001; Brandano & Corda, 2002).

The facies architecture of the lower sequence, incorporating Late Burdigalian to Serravallian sediments, has been interpreted by Brandano & Corda (2002) to record an ongoing change in environmental parameters that then persisted during the deposition of the upper sequence. At the base of the lower sequence, scarce *Porites* colonies are present, but they disappear upwards and are replaced by an increasing proportion of suspension feeders (bryozoans, bivalves) together with abundant red algae.

High-frequency cycles

Higher frequency sea-level changes are recorded in the inner-ramp succession during the lowstand phase of the lower sequence. In the inner ramp above the Cretaceous substratum, the facies are arranged in complex vertical and lateral associations and display a general deepening-upward trend (Fig. 7). In particular, the sedimentary record consists of stacked beds that are separated by discontinuity surfaces. Lateral and vertical facies changes inside these beds monitor the variability of juxtaposed depositional environments and describe the evolution of these environments through time. Discontinuity surfaces formed when sea level dropped. The beds between these discontinuity surfaces may be considered as a sedimentary cycle. A single cycle is characterized by colonization by bioeroders and balanids during the first transgression phase, and by the spreading of seagrass maërl facies during the late transgressive and highstand phases (Fig. 7). The formation of lithoclastic conglomerates is related to successive sea-level fall. The high intensity of bioerosion in the pre-Miocene substrate substratum promoted the formation of pebbles that are reworked in the conglomeratic levels.

The stacking patterns of the cycles describe the evolution of the inner-ramp succession. The first ten metres of the inner-ramp succession record a transgressive phase represented by the backstepping of high-energy facies (balanid floatstone and lithoclastic conglomerates) laterally associated with seagrass meadows and branching red algal facies. A regressive trend is recorded by the expansion of the littoral facies (balanid floatstone) on a discontinuity surface locally characterized by *Microcodium,* indicating an episode of emergence. The next transgression phase is represented by the maërl facies passing into coral carpets and into middle-ramp deposits (Fig. 7).

Estimating the period of the high-frequency sea-level cycles is difficult due to the low resolution of biostratigraphic data. The lowstand phase of the lower sequence comprises five high-frequency

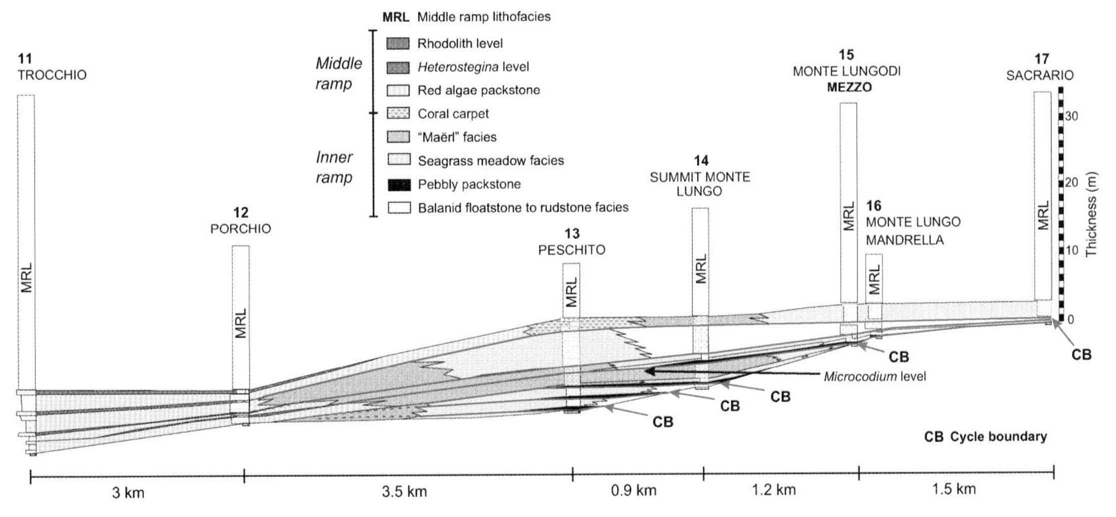

Fig. 7. Inner-middle ramp facies architecture in Latium-Abruzzi platform carbonates.

units (Fig. 7). The base of the lower sequence is dated at 20–21 Ma by Sr-isotope stratigraphy. This age is confirmed by biostratigraphic data (presence of *Miogypsina* cf. *globulina*). The upper limit of the lowstand is a discontinuity surface (transgressive surface) dated at 17.5 Ma by Sr isotopes (Brandano, 2001). Consequently, the inner-ramp succession covers a maximum time span of 3.5 Myr, and the periodicity of the five cycles within the inner-ramp deposits is roughly on the order of 0.5 Myr.

The middle ramp displays a less organized succession of rhodolithic floatstone to rudstone, medium to coarse-grained bioclastic packstone and oyster-pectinid floatstone. The coralline algal assemblages in the rhodolithic intervals may show a subtle cyclicity marked by a change in the abundance of red algae taxa (Brandano & Piller, this volume, pp. 151–166). Taxonomic analyses of rhodolithic intervals proves the presence of cycles identified by a drop in species richness upward with a dominance of sciaphile genera (*Mesophyllum* and *Lithothamnion*). These can be related to the distinct change in substrate availability, sea-level changes and/or to water turbidity. However, it is not possible to assign bathymetric ranges to these facies as they could represent lateral facies changes in a given environment at the same bathymetric depth.

The proximal and intermediate outer-ramp successions display few facies changes. In the intermediate ramp the most common facies is packstone rich in planktonic foraminifera, echinoid plates, *Ditrupa* and subordinate benthic foraminifera (*Lenticulina*). Generally the only changes observable over decameter intervals are variations in grain-size (from fine to coarse packstone) and colour (from dark grey to light brown; Civitelli & Brandano, 2005). The proximal outer-ramp successions are up to 90 m thick and are characterized by four facies. The most common facies is bryozoan floatstone that may constitute thick decametre intervals (up to 50 m) without any significant textural and compositional change. The bivalve floatstones constitute intervals up to 3 m thick with a lateral extent of several hundreds of metres. The other facies (rhodolith floatstone, light brown calcarenites) constitute only isolated horizons up to 1 m thick (Brandano, 2001; Civitelli & Brandano, 2005). Bioturbation is largely absent.

Cycle formation in the distal outer-ramp deposits, composed of alternating dominantly platform-derived carbonate sediment and dominantly terrigenous material originating from the western side of the basin, is interpreted as being the result of cyclic climate change (Brandano *et al.*, 2005a). The increased terrigenous supply is linked to humid phases, with intensified precipitation, that cyclically induced increased rates of weathering on emergent areas of the Apennine chain (Brandano *et al.*, 2005a). Hence, the mixed siliciclastic-carbonate cycles of the distal outer ramp do not represent a record of sea-level fluctuations, but are primarily a response to cyclic climate change that controlled terrigenous supply and the rate of carbonate sediment production in this environment.

DISCUSSION

The morphology of carbonate platforms (ramp versus rimmed) is one of the major controls on the response of such platforms to sea-level changes (e.g. Handford & Loucks, 1993; Pomar, 2001b; Schlager, 2003). In many cases, facies partitioning occurs where lowstand units are composed of a different set of facies than highstand units (Homewood, 1996). In particular, higher-order sea-level fluctuations influence sediment production on flat-topped platforms where vast areas are exposed during sea-level lowstands (e.g. Schlager & Ginsburg, 1981; Sarg, 1988; Burchette & Wright, 1992; Pomar, 2001b). Such "Bahama-type" platforms show significant differences in sediment composition during a relative sea-level cycle (e.g. Haak & Schlager, 1989; Reijmer, 1991; Everts, 1994; Westphal *et al.*, 1999). By contrast, on distally steepened ramps such as the Great Bahama Bank during the Early Pliocene, when a drastic sea-level fall forced regression in a fully tropical carbonate factory system, areas available for carbonate production are reduced less drastically by higher-order sea-level fluctuations, thus allowing for fairly continuous sediment production. Sea-level fluctuations in such systems result mainly in shifts of the facies belts without switching entire facies belts on and off (Eberli & Ginsburg, 1988; Westphal, 1998).

The problem with interpreting carbonate ramps in a sequence stratigraphic sense lies in their large lateral extent, shallow inclination, and comparably low relief (Burchette & Wright, 1992). The classic features used for defining sequences may not be present. For example, sequence boundaries may not be marked by erosional or emergent features. This is the case for the carbonate ramp discussed here. In such examples, the study of facies successions is important for sequence analysis and for recognizing sequence architecture (Burchette & Wright, 1992). Careful study of facies can help to recognize upward and downward shifts of the facies belts and thus the underlying sea-level fluctuations (Bachmann *et al.*, 1996; Bachmann & Willems, 1996).

The carbonate ramp studied here has been found to clearly record long-term (2nd-order) sequences and tectonically forced relative sea-level change. During the 5 Myr of deposition of the "Calcari a Briozoi e Litotamni" significant eustatic sea-level fluctuation has been proven elsewhere (three 3rd-order cycles of Haq *et al.*, 1987; four 3rd-order cycles of Hardenbol *et al.*, 1998). However, these pronounced high-frequency sea-level fluctuations of the Aquitanian to Serravallian only show a clear signature in the inner-ramp setting, but not in the sub-wavebase, meso- to oligophotic middle and outer-ramp settings. The frequency recorded in the inner-ramp (roughly 0.5 Myr) corresponds to the time-span of 3rd-order frequencies, demonstrating that higher frequencies left no signature even in the inner ramp. Moreover, the ramp shows a long-term behaviour of simultaneous backstepping and progradation after 17.5 Ma that resulted from simultaneous sea-level rise and an increase in trophic levels (Brandano & Corda 2002; Pomar *et al.*, 2004). This demonstrates that the different parts of such a carbonate depositional system may react to different environmental factors in complex ways.

Different rates of sediment production in the different depositional environments control the depositional profile as well as the facies architecture (Pomar *et al.*, 2004). Volumetrically minor inner-ramp facies, dominated by loose fine-grained sediments and the absence of rigid framework, strongly dampen the impact of high-frequency sea-level fluctuations. Volumetrically important middle- and outer-ramp facies accumulate below the hydrodynamic base level and consequently, only major sea-level fluctuations of the 2nd order impinge shifts of facies belts persistent enough to be reflected in the sediment record (Fig. 6A). Only in the inner ramp, where facies accumulated above the hydrodynamic base level, can a moderate expression of 3rd-order sea-level cyclicity be seen expressed as the landward/basinward migration of facies belts, resulting in rudstone with free-living branching red-algal and seagrass-meadow deposits alternating with conglomeratic layers and balanid floatstone (Fig. 7). Basic accretional units corresponding to parasequences (*sensu* Van Wagoner *et al.*, 1990) or simple sequences (*sensu* Vail *et al.*, 1991) are not recognized in the "Calcari a Briozoi e Litotamni" Formation.

The few attempts in literature to interpret heterozoan carbonate ramps in a sequence stratigraphic context based on stacking patterns show similar results. Only above wave-base are such ramps sensitive to high-frequency sea-level fluctuations. For the Oligocene to Upper Miocene ramp succession from the Montagna Maiella (Italy), Vecsei & Sanders (1999) recognized 3rd-order sequences by depositional geometries and by facies changes, but no proof for subaerial exposure was found. Within the transgressive systems tracts of these sequences, laterally extensive sediment bodies are present in the middle ramp, separated by flooding surfaces that have been interpreted as parasequences, whereas no parasequences are recognized in the highstand systems tract. Similarly, the Tortonian ramp of the Apostoli Basin (Crete) is composed of a transgressive cycle, which culminates in the alternation of Rethymnon bioclastic limestones with marls, documenting the Tortonian marine transgression (Pomoni-Papaioannou *et al.*, 2002). For these rhodalgal to echinofor sediments, no depositional higher-frequency sequences have been distinguished.

In contrast, in the Melilla Basin, northeastern Morocco, an Upper Miocene carbonate ramp exhibits a 3rd-order depositional sequence where the transition from the transgressive to the highstand systems tract corresponds to a shift from a heterozoan carbonate factory to a photozoan carbonate factory (Cunningham *et al.*, 1997; Cunningham & Collins, 2002). In the transgressive systems tract, the authors distinguished only one parasequence, although for the high-stand systems tract, they distinguished three parasequences.

For the uppermost Tortonian to Lower Messinian bryomol ramp of the Agua Amarga Basin (Spain), a lowstand systems tract of a 4th-order sequence has been subdivided into three sedimentary cycles (Martín *et al.*, 1996). From the carbonate factory, located seaward of some shoals, skeletal components were washed landwards and at the same time accumulated in fans downslope. The factory-area and fan-bedded sediments show five well-defined, calcarenites/fine-grained calirudites cycles interpreted as Milankovitch cycles of the precession band.

Clearly, on heterozoan carbonate platforms the absence of a rigid framework and the volumetrically importance of production below wave base (in the oligophotic zone) reduce the impact of high-frequency sea-level fluctuations in creating facies architectural heterogeneities (Pomar, 2001a; Pomar *et al.*, 2004). On a global scale, rhodalgal facies reached peak abundances from the Burdigalian to Tortonian and commonly replaced coral-reef environments (Halfar & Mutti, 2005). The rhodalgal facies characterizes depositional environments placed in many cases below the storm wave-base in a wide depth range controlled by light penetration (e.g. Adey & Macintyre, 1973; Bosence, 1983; Pomar, 2001a; Brandano *et al.*, 2005b; Wilson & Vecsei, 2005). For example in the modern Ryukyu Islands, the interval dominated by oligophotic red algae is between 50 and 135 m depth (Iryu *et al.*, 1995). In the modern Mediterranean the infralittoral zone extends down to 50 m (lower seagrass meadows limit), the oligophotic zone corresponds to the upper circalitoral zone, and living red algae are known from down to 98 m (Basso, 1998) in a facies assemblage corresponding to the rhodalgal association (Carannante *et al.*, 1988; Betzler *et al.*, 1997b). Hence a sea-level excursion of 50 m may not produce significant facies change if the sea floor remains in the subwavebase oligophotic zone.

As a consequence, the stratigraphic architecture of a carbonate platform may not reflect high frequency sea-level cyclicity (4th order and higher) when the important carbonate sediment production (carbonate factory) takes place in the oligophotic zone (by rhodalgal associations) or in the aphotic zone (by bryomol or molechfor associations). This is because the carbonate sediments produced accumulate below the hydrodynamic base level and the associated depositional facies are present in a bathymetric interval that may be wider than the high-frequency sea-level amplitude. Only sea-level fluctuations large enough to produce shifts from the euphotic to the oligophotic or the aphotic zone or significant changes in light-penetration (e.g. due to oceanographic changes) may be recorded as facies shifts.

CONCLUSION

The Latium-Abruzzi Platform gives insights into the reasons for the different sensitivity of such ramp depositional systems to low- versus high-frequency sea-level changes. The carbonate factory of the Latium-Abruzzi carbonate ramp was characterized by rhodalgal facies in the oligophotic zone, and bryomol and molechfor associations in the aphotic zone below wave base. Limited sediment production in the inner-ramp environment and the absence of a rigid framework made the sedimentary system less sensitive to high-frequency sea-level fluctuations. The carbonate sediment produced accumulated below the hydrodynamic base level, while the associated depositional facies were present in a bathymetric interval that may be wider than the high-frequency sea-level amplitude. As a consequence, despite being deposited in the context of increasing accommodation space induced by a slow tectonic subsidence, and despite the fact that sea-level cyclicity was significant during the 5 Myr depositional period, only major sea-level fluctuations of the 2nd order induced shifts of facies belts sufficient to create platform heterogeneities. Sea-level fluctuations able to produce shifts from the euphotic to the oligophotic or the aphotic zone may be recorded as facies changes as well as changes in light-penetration that, in addition, may be related to changes in oceanographic conditions.

ACKNOWLEDGMENTS

Luis Pomar and Giacomo Civitelli are gratefully acknowledged for their suggestions and useful comments during the study. The reviews by Christian Betzler, Cedric John and Jeff Lukasik helped to improve the manuscript and are gratefully acknowledged. The authors are grateful to Jason A. Bradley for the English revision. This research has been funded by MIUR (Progetti della Facoltà di Scienze M.F.N., Università di Roma "La Sapienza") and by the Spanish Ministerio de Educación y Ciencia (Project MEC-DGI CGL2005-00537). Guillem Mateu-Vicens has a PhD grant from the Balearic Islands Government (Conselleria d'Economia, Hisenda i Innovació).

REFERENCES

Accordi, G. and **Carbone, F.** (1988) Sequenze carbonatiche mesocenozoiche. In: *Note Illustrative alla Carta delle Litofacies del Lazio-Abruzzo ed Aree Limitrofe.* (Eds G. Accordi, F. Carbone, G. Civitelli, L. Corda, D. de Rita, D. Esu, R. Funiciello, T. Kotsakis, G. Mariotti and A. Sposato). *CNR Quad. Ric. Sci.*, **114**, 11–92.

Adey, W.H. and **Macintyre, I.G.** (1973) Crustose coralline algae: A reevaluation in the geological sciences. *Geol. Soc. Am. Bull.*, **84**, 883–904.

Bachmann, M. and **Willems, H.** (1996) High-frequency cycles in the Upper Aptian carbonates of the Organyà basin, NNE Spain. *Geol. Rundsch*, **85**, 586–605.

Bachmann, M., **Bandel, K.**, **Kuss, J.** and **Willems, H.** (1996) Sedimentary processes and intertethyal comparisons of two Early/Late Cretaceous carbonate ramp systems (NE-Africa and Spain). In: *Global and Regional Controls on Biogenic Sedimentation. II. Cretaceous Sedimentation.* (Eds J. Reitner, F. Neuweiler and F. Gunkel). *Gött. Arb. Geol. Paläontol. Sonderband*, **3**, 151–163.

Basso, D. (1998) Deep rhodolith distribution in the Pontian Islands, Italy: a model for the paleoecology of a temperate sea. *Palaeogeogr. Palaeoclimatol. Palaeoecol.*, **137**, 173–187.

Basso, D. and **Tomaselli, V.** (1994) Paleoecological potentiality of rhodoliths: a Mediterranean case history. In: *Studies on Ecology and Paleoecology of Benthic Communities* (Eds R. Matteucci, M.G. Carboni and J.S. Pignatti). *Boll. Soc. Paleontol. Ital. Spec. Vol.*, **2**, 17–27.

Beccaluva, L., **Bianchini, G.** and **Siena, F.** (2004) Tertiary-Quaternary volcanism and tectono-magmatic evolution in Italy. In: *Geology of Italy* (Eds V. Crescenti, S. D'Offizi, S. Merlino and L. Sacchi). *Soc. Geol. Ital. Spec. Vol*, 153–160.

Bernoulli, D. (2001) Mesozoic-Tertiary carbonate platforms, slopes and basino of the external Apennines and Sicily. In: *Anatomy of an Orogen: the Apennines and Adjacent Mediterranean Basins* (Eds G.B. Vai and I.P. Martini), pp. 307–326. Kluwert Acadamic Publishers, UK.

Betzler, C., **Brachert, T.C.**, **Braga, J.C.** and **Martín, J.M.** (1997a) Nearshore, temperate, carbonate depositional systems (Lower Tortonian, Agua Amarga Basin, southern Spain): implications for carbonate sequence stratigraphy. *Sed. Geol.*, **113**, 27–53.

Betzler, C., **Brachert, T.C.** and **Nebelsick, J.** (1997b) The warm temperate carbonate province - A review of facies, zonations, and delimitations. *Cour. Forschungs-Inst. Senckenb.*, **201**, 83–99.

Betzler, C., **Martín, J.M.** and **Braga, J.C.** (2000) Nontropical carbonates related to rocky submarine cliffs (Miocene, Almeria, Southern Spain). *Sed. Geol.*, **131**, 51–65.

Betzler, C., **Saxena, S.**, **Swart, P.K.**, **Isern, A.** and **James, N.P.** (2005) Cool-water carbonate sedimentology and eustasy; Pleistocene upper slope environments, Great Australian Bight. *Sed. Geol.*, **175**, 169–188.

Bosence, D.W.J. (1983). The occurrence and ecology of recent rhodoliths. In: *Coated Grains* (Ed. T.M. Peryt), pp. 225–242. *Springer*, Berlin.

Brachert, T.C., **Hultzsch, N.**, **Knoerich, A.C.**, **Krautworst, U.M.R.** and **Stückrad, O.M.** (2001) Climatic signatures in shallow-water carbonates: high-resolution stratigraphic markers in structurally controlled carbonate buildups. *Palaeogeogr. Palaeoclimatol. Palaeoecol.*, **175**, 211–237.

Brandano, M. (2001) *Risposta fisica delle aree di piattaforma carbonatica agli eventi più significativi del Miocene nell'Appennino centrale.* Unpubl. PhD thesis, Univ. of Rome "La Sapienza", 180 pp.

Brandano, M. (2002) "La Formazione dei "Calcari a Briozoi e Litotamni" nell'area di Tagliacozzo (Appennino Centrale) e considerazioni paleoambientali sulle facies rodalgali. *Boll. Soc. Geol. Ital.*, **121**, 179–186.

Brandano, M. (2003) Tropical/Subtropical Inner Ramp Facies in Lower Miocene "Calcari a Briozoi e Litotamni" of the Monte Lungo Area (Cassino Plain, Central Apennines, Italy). *Boll. Soc. Geol. Ital.*, **122**, 85–98.

Brandano, M. and **Corda, L.** (2002) Nutrients, sea level and tectonics: constrains for the facies architecture of a Miocene carbonate ramp in central Italy. *Terra Nova*, **14**, 257–262.

Brandano, M., **Corda, L.** and **Mariotti, G.** (2005a) Orbital forcing recorded in subtidal cycles from a Lower Miocene siliciclastic-carbonate ramp system, (Central Italy). *Terra Nova*, **17**, 434–441.

Brandano, M., **Vannucci, G.**, **Pomar, L.** and **Obrador, A.** (2005b) Rhodolith assemblages from the Lower Tortonian carbonate ramp of Menorca (Spain): environmental and paleoclimatic implications. *Palaeogeogr. Palaeoclimatol. Palaeoecol.*, **226**, 307–323.

Brasier, M.D. (1975) An outline history of seagrass communities. *Palaeontology*, **18**, 681–702.

Bromley, R.G. and **Asgaard, U.** (1993) Endolithic community replacement on a Pliocene rocy coast. *Ichnos*, **2**, 93–116.

Burchette, T.P. and **Wright, V.P.** (1992) Carbonate ramp depositional systems. *Sed. Geol.*, **79**, 3–57.

Canals, M. and **Ballesteros, E.** (1997) Production of carbonate particles by phytobenthic communities on the Mallorca-Menorca shelf, northwestern Mediterranean Sea. *Deep-Sea Res. II*, **44**, 611–629.

Carannante, G. and **Simone, L.** (1996) Rhodolith facies in the central-southern Apennines mountains, Italy. In: *Models for Carbonate Stratigraphy from Miocene Reef Complexes of Mediterranean Regions* (Eds E.K. Franseen, M. Esteban, W. Ward and J. Rouchy). *SEPM Concepts Sedimentol. Paleontol.*, **5**, 261–275.

Carannante, G., **Esteban, M.**, **Milliman, J.D.** and **Simone, L.** (1988) Carbonate lithofacies as paleolatitude indicators: problems and limitations. *Sed. Geol.*, **60**, 333–346.

Carminati, E., **Corda, L.**, **Mariotti, G.** and **Brandano, M.** (2007) Tectonic control on the architecture of a Miocene carbonate ramp in the Central Apennines (Italy): insights from facies and backstripping analyses. *Sed. Geol.*, **198**, 233–254.

Cathro, D.L., **Austin, J.A.** and **Moss, G.D.** (2003) Progradation along a deeply submerged Oligocene–Miocene heterozoan carbonate shelf: How sensitive are clinoforms to sea level variations? *AAPG Bull.*, **87**, 1547–1574.

Civitelli, G. and **Brandano, M.** (2005) Atlante delle litofacies e modello deposizionale dei Calcari a Briozoi e Litotamni nella Piattaforma carbonatica laziale-abruzzese. *Boll. Soc. Geol. Ital.*, **124**, 611–643.

Civitelli, G., **Corda, L.** and **Mariotti, G.** (1986) Il bacino sabino: 3, evoluzione sedimentaria ed inquadramento regionale dall'Oligocene al Serravalliano. *Mem. Soc. Geol. Ital.*, **35**, 399–406.

Cognetti, M. and **Sarà, M.** (1974) *Biologia Marina.* Calderini, Bologna, 439 pp.

Corda, L. and **Brandano, M.** (2003) Aphotic Zone Carbonate Production on a Miocene Ramp, Central Apennines, Italy. *Sed. Geol.*, **161**, 55–70.

Cunningham, K.J., **Benson, R.H.**, **Rakic-El Bied, K.** and **McKenna, L.W.** (1997) Eustatic implications of late Miocene depositional sequences in the Melilla Basin, northeastern Morocco. *Sed. Geol.*, **107**, 147–165.

Cunningham, K.V. and **Collins, L.S.** (2002) Controls on facies and sequence stratigraphy of an upper Miocene carbonate ramp and platform, Melilla basin, NE. *Sed. Geol.*, **146**, 285–304.

Doglioni, C. (1991) A proposal for the kinematic modelling of W-dipping subductions; possible applications to the Tyrrhenian-Apennines system. *Terra Nova*, **3**, 423–434.

Eberli, G. P. and **Ginsburg, R.N.** (1988) Aggrading and prograding Cenozoic seaways, northwest Great Bahama Bank. In: *Atlas of Seismic Stratigraphy* (Ed. A.W. Bally). *AAPG Stud. Geol.*, **27**, 97–103.

Everts, A.J.W. (1994) *Carbonate Sequence Stratigraphy of the Vercors (French Alps) and its Bearing on Cretaceous Sea Level.* Unpubl. PhD Thesis, Vrije University, Amsterdam, The Netherlands, 176 pp.

Franseen, E.K., **Goldstein, R.H.** and **Farr, M.R.** (1997) Substrate-slope and temperature controls on carbonate ramps: revelations from Upper Miocene outcrops, SE Spain. In: *Cool-Water Carbonates* (Eds N.P. James and J.A.D. Clarke). *SEPM Spec. Publ.*, **56**, 271–290.

Haak, A.B. and **Schlager, W.** (1989) Compositional variations in calciturbidites due to sea-level fluctuations, Late Quaternary, Bahamas. *Geol. Rundsch.*, **78**, 477–486.

Halfar, J. and **Mutti, M.** (2005) Global dominance of coralline red-algal facies: a response to Miocene oceanographic events. *Geology*, **33**, 481–484.

Handford, C.R. and **Louks, R.G.** (1993) Carbonate depositional sequences and systems tracts response of carbonate platforms to relative sea-level changes. In: *Carbonate Sequence Stratigraphy Recent Developments and Applications* (Eds R.G. Loucks and F. Sarg). *AAPG Mem.*, **57**, 3–41.

Haq, B.U., **Hardenbol, J.** and **Vail, P.R.** (1987) Chronology of fluctuating sea level since the Triassic. *Science*, **235**, 1156–1166.

Hardenbol, J., **Thierry, J.**, **Farley, M.B.**, **Jacquin, T.**, **de Graciansky, P.C.** and **Vail, P.R.** (1998) Mesozoic and Cenozoic sequence chronostratigraphic chart. In: *Mesozoic and Cenozoic Sequence Stratigraphy of European Basins* (Eds P.C. Graciansky, J. Hardenbol, T. Jacquin and P.R. Vail). *SEPM Spec. Publ.*, **60**, 3–13.

Hodell, D.A., **Mueller, P.A.** and **Garrido, J.R.** (1991) Variations in the strontium isotopic composition of seawater during the Neogene. *Geology*, **19**, 24–27.

Hoffman, A. (1979) Indian ocean affinities of a Badenian (Middle Miocene) seagrass-associated macrobenthic community of Poland. *Ann. Géol. Pays Hellén. Tome hors sér.*, **II**, 537–541.

Homewood, P. W. (1996) The carbonate feedback system: interaction between stratigraphic accommodation, ecological succession and the carbonate factory. *Bull. Soc. Géol. Fr.*, **167** (6), 701–715.

Hottinger, L. (1997) Shallow benthic foraminiferal assemblages as signals for depth of their deposition and their limitations. *Bull. Soc. Géol. Fr.*, **168**, 491–505.

James, N.P. and **Bourque, P.A.** (1992) Reefs and Mounds. In: *Facies Models: Response to Sea Level Change* (Eds R.G. Walker and N.P. James), pp. 323–347. Geological Association of Canada, St. Johns.

Iryu, Y., **Nakamori, T.**, **Matsuda, S.** and **Abe, O.** (1995) Distribution of marine organisms and its ecological significance in the modern reef complex of the Ryukyu Islands. *Sed. Geol.*, **99**, 243–258.

Louks, R.G. and **Sarg, J.F.** (1993) Carbonate sequence stratigraphy: recent developments and applications. *AAPG Mem.*, **57**, 545 pp.

Lukasik, J.F. and **James, N.P.** (2003) Deepening-upward subtidal cycles, Murray basin, South Australia. *J. Sed. Res.*, **73**, 653–671.

Lüning, K. (1990) *Seaweeds. Their Environment, Biogeography and Ecophysiology.* John Wiley & Sons, Inc., New York, 527 pp.

Martín, J.M., **Braga, J.C.**, **Betzler, C.** and **Brachert, T.** (1996) Sedimentary model and high-frequency cyclicity in a Mediterranean, shallow-shelf, temperate-carbonate environment (uppermost Miocene, Agua Amarga Basin, Southern Spain). *Sedimentology*, **43**, 263–277.

Murray, J.W. (1973) *Distribution and Ecology of Living Benthic Foraminiferids.* Heinemann, London, 274 pp.

Parotto, M. and **Praturlon, A.** (2004) The Southern Apennine Arc. In: *Geology of Italy* (Eds V. Crescenti, S. D'Offizi, S. Merlino and L. Sacchi). *Ital. Geol. Soc. Spec. Vol.*, 33–58.

Passlow, V. (1997) Slope sedimentation and shelf to basin sediment transfer: a cool-water carbonate example from the Otway Margin, southeastern Australia. In: *Cool-water Carbonates* (Eds N.P. James and J.A.D. Clarke). *SEPM Spec. Publ.*, **56**, 107–125.

Patacca, E., **Scandone, P.**, **Bellatalla, M.**, **Perilli, N.** and **Santini, U.** (1991) La zona di giunzione tra l'arco appenninico settentrionale e l'arco appenninico meridionale nell'Abruzzo e nel Molise. In: *Studi Preliminari all'Acquisizione Dati del Profilo **CROP 11** Civitavecchia-Vasto* (Eds M. Tozzi, G.P. Cavinato and M. Parotto). *Stud. Geol. Camerti Vol. Spec.*, **1991/2**, 417–441.

Pérès, J.M. and **Picard, J.** (1964) Nouveau manual de bionomie benthique de la Mer Méditerranée. *Trav. Stat. Mar. Endoume-Marseille*, **47**, 5.

Piller, W.E. and **Brandano, M.** (2005) Record of non-skeletal biota from the inner ramp of a Burdigalian carbonate platform. *12th Congress RCMNS, 6–11 September 2005, Vienna*, Abstract book, p. 185.

Pomar, L. (1991) Reef geometries, erosion surfaces and high-frequency sea-level changes, upper Miocene Reef Complex, Mallorca, Spain. *Sedimentology*, **38**, 243–269.

Pomar, L. (2001a) Ecological control of sedimentary accommodation: evolution from carbonate ramp to rimmed shelf, Upper Miocene, Balearic Islands. *Palaeogeogr. Palaeoclimatol. Palaeoecol.*, **175**, 249–272.

Pomar, L. (2001b) Types of carbonate platforms: A genetic approach. *Basin Res.*, **13**, 313–334.

Pomar, L., **Brandano, M.** and **Westphal, H.** (2004) Environmental factors influencing skeletal-grain sediment associations: A critical review of Miocene examples from the Western Mediterranean. *Sedimentology*, **51**, 627–651.

Pomar, L., **Gili, E.**, **Obrador, A.** and **Ward, W.C.** (2005) Facies architecture and high-resolution sequence stratigraphy of an Upper Cretaceous platform margin succession, southern central Pyrenees, Spain. *Sed. Geol.*, **175**, 339–365.

Pomar, L., **Obrador, A.** and **Westphal, H.** (2002) Sub-wavebase cross-bedded grainstones on a distally steepened carbonate ramp, upper Miocene, Menorca, Spain. *Sedimentology*, **49**, 139–169.

Pomoni-Papaioannou, F., **Drinia, H.** and **Dermitzakis, M.D.** (2002) Neogene non-tropical carbonate sedimentation in a warm temperate biogeographic province (Rethymnon Formation, Eastern Crete, Greece). *Sed. Geol.*, **154**, 147–157.

Reijmer, J.J.G. (1991) *Sea Level and Sedimentation on the Flanks of Carbonate Platforms.* Unpubl. PhD Thesis, Vrije Universiteit, Amsterdam, The Netherlands, 162 pp.

Riegl, B. and **Piller, W.E.** (2000) Biostromal coral facies: a Miocene example from the Leitha Limestone (Austria) and its actualistic interpretation. *Palaios*, **15**, 399–413.

Sarg, J.F. (1988) Carbonate sequence stratigraphy. In: *Sea-Level Changes: an Integrated Approach* (Eds C.K. Wilgus, B.S. Hasting, C.G.St.C. Kendall, H.W. Posamentier, C.A. Ross and J.C. van Wagoner). *SEPM Spec. Publ.*, **42**, 155–182.

Schlager, W. (2003) Benthic carbonate factories of the Phanerozoic. *Int. J. Earth. Sci.*, **92**, 445–464.

Schlager, W. and **Ginsburg, R.N.** (1981) Bahama carbonate platforms: the deep and the past. *Mar. Geol.*, **44**, 1–24.

Schlager, W., **Reijmer, J.J.G.** and **Droxler, A.W.** (1994) Highstand shedding of carbonate platforms. *J. Sed. Petrol.*, **64**, 270–281.

Sen Gupta, B.K. (1999) Foraminifera in Marginal Marine Environments. In: *Modern Foraminifera* (Ed. B.K. Sen Gupta), pp. 141–160. Kluwer Academic Publishers, Great Britain.

Steneck, R.S. (1986) The ecology of coralline algal crusts: convergent patterns and adaptive strategies. *Ann. Rev. Ecol. Syst*, **17**, 273–303.

Vail, P.R., **Audemard, F.**, **Bowman, S.A.**, **Eisner, P.N.** and **Perez-Cruz, C.** (1991). The stratigraphic signatures of tectonics, eustasy and sedimentology: an overview. In: *Cycles and Events in Stratigraphy* (Eds. G. Einsele, W. Ricken and A. Seilacher), pp. 617–659. Springer, Berlin.

van Wagoner, J.C., **Mitchum, R.M.J.**, **Campion, K.M.** and **Rahmanian, V.D.** (1990) Siliciclastic sequence stratigraphy in well logs, cores and outcrops. *AAPG Methods Explor. Ser.*, **7**, 1–55.

Vecsei, A. and **Sanders, D.G.K.** (1999) Facies analysis and sequence stratigraphy of a Miocene warm-temperate carbonate ramp, Montagna della Maiella, Italy. *Sed. Geol.*, **123**, 103–127.

Westphal, H. (1998) Carbonate Platform Slopes - A Record of Changing Conditions. The Pliocene of the Bahamas. *Lect. Notes Earth Sci.*, **75**, 197 pp.

Westphal, H., **Reijmer, J.J.G.** and **Head, M.J.** (1999) Input and Diagenesis on a Carbonate Slope (Bahamas): Response to Morphology Evolution and Sea-Level Fluctuations. In: *Advances in Carbonate Sequence Stratigraphy: Applications to Reservoirs, Outcrops, and Models* (Eds P.M. Harris, A. Saller and J.A. Simo) *SEPM Spec. Publ.*, **63**, 247–274.

Wilson, M.E.J. and **Vecsei, A.** (2005) The apparent paradox of abundant foramol facies in low latitudes: their environmental significance and effect on platform development. *Earth-Sci. Rev.*, **69**, 133–168.

Int. Assoc. Sedimentol. Spec. Publ. (2010) **42**, 107–128

Facies and sequence architecture of a tropical foramol-rhodalgal carbonate ramp: Miocene of the central Apennines (Italy)

MARCO BRANDANO*†, LAURA CORDA* and FRANCESCA CASTORINA*†

**Dipartimento di Scienze della Terra, Università di Roma "La Sapienza", P. Aldo Moro 5, Roma, 00185, Italy (E-mail: laura.corda@uniroma1.it)*

†Istituto di Geologia Ambientale e Geoingegneria-CNR, Sezione "La Sapienza", Roma, 00185, Italy

ABSTRACT

Integrated data from sedimentological and microfacies analyses have been used to reconstruct the stratigraphic and architectural evolution of a carbonate ramp exposed in the central Apennines. The platform was a wide homoclinal ramp characterized by rhodalgal, bryomol and foramol skeletal-grain associations that developed under tropical-subtropical conditions. Based on facies distribution and palaeoecology, the ramp can be subdivided into: (1) an inner zone whose main components include balanids, molluscs and branching red algae; (2) a middle zone characterized by rhodoliths and larger foraminifera; and (3) an outer ramp consisting of: (a) a proximal zone dominated by bryozoan colonies, bivalves and echinoids: (b) an intermediate zone where the main components are benthic and planktonic foraminifera, worm tubes, fragmented echinoids and bivalves; and (c) a distal zone where marls and calcarenites contain silica-sponge spicules, planktonic foraminifera and mollusc-bryozoan-echinoid debris.

Strontium-isotope dating has been essential in generating a detailed chronostratigraphy. The reconstructed architecture reflects two second-order depositional sequences. The lower sequence (uppermost Aquitanian to Serravallian) is composed of a lowstand systems tract, a well-developed transgressive systems tract and a thin highstand systems tract. The architectural pattern of the lower sequence first shows aggradation, followed by retrogradation and finally a pronounced contemporaneous progradation. Changes in trophic resources and tectonic phases have been interpreted as the main factors controlling the sequence architecture. The upper sequence, starting in the Serravallian, was strongly controlled by tectonic subsidence and its diachronous termination was a consequence of the downflexure of the foreland related to the eastward migration of the Apennine foredeep system.

Keywords Tropical carbonate ramp, Miocene, sequence stratigraphy, Mediterranean.

INTRODUCTION

Miocene carbonates in the Mediterranean region have received significant interest during the last few decades, both from a palaeoecological and a stratigraphic–architectural point of view (Carannante *et al.* 1988; Pomar & Ward, 1995; Esteban, 1996; Esteban *et al.*, 1996; Brachert *et al.*, 1996, 2001; Mutti *et al.*, 1997, 1999; Braga *et al.*, 2001; Pomar, 2001a; Brandano & Corda, 2002; Pomar *et al.*, 2002; Corda & Brandano, 2003). During Miocene time, rhodalgal (cf. Carannante *et al.*, 1988), bryomol (cf. Nelson *et al.*, 1988) and foramol facies (cf. Lees and Buller, 1972) were predominant, while chlorozoan facies (Esteban, 1996) containing *Porites* colonies were well developed during the early Messinian. The dominance of rhodalgal and bryomol facies associations has been interpreted in some cases (Nebelsick, 1989; Vecsei & Sanders, 1999) as the response of carbonate systems to a cooler phase (e.g. during the Early Miocene) connected with global glacio-eustatic oscillations. In other examples it has been related to changes of oceanic circulation and environmental conditions in tropical waters (Pedley, 1996; Mutti *et al.*, 1997, 1999; Pomar, 2001a; Brandano & Corda, 2002; Cunningham & Collins, 2002; Mutti and Bernoulli, 2003).

In the central and southern Apennines, Miocene platform carbonates are predominantly characterized by rhodalgal, bryomol and molechfor associations. In particular, the Miocene succession of the

Latium-Abruzzi platform (central Apennines) is represented by the "Calcari a Briozoi e Litotamni" Formation. These carbonates form a 40–100 m-thick succession containing red algae, bryozoans, molluscs, echinoids and foraminifera. To date, the chronostratigraphical position of this formation has not been precisely defined, as biostratigraphical data are extremely poor.

The major aims of this paper are to: (1) describe the general facies patterns and sequence stratigraphy of the Latium-Abruzzi Miocene carbonate ramp; (2) develop a high-resolution stratigraphy using Sr isotope analyses; (3) present a case-study that tests the relationships between facies belt shifting and sea-level changes; (4) show how the interactions between relative sea-level changes, regional tectonics and oceanographic conditions can control facies associations and sequence stratigraphic packaging; and (5) use new data to reconstruct the palaeoenvironmental conditions that existed during the Early-Middle Miocene in the central Mediterranean region.

GEOLOGICAL SETTING

The Apennine fold-and-thrust belt is mainly a Neogene mountain chain derived from the deformation of the western Adria passive margin. The backbone of the central Apennine chain is represented by at least 3000 m of shallow-water carbonates deposited on a long-lived isolated platform that persisted from the Late Triassic to the Miocene. Carbonate deposition took place on the passive margin of a foreland basin and was terminated by the input of terrigenous sediments (hemipelagites followed by siliciclastic turbidites) during the Tortonian to the Tortonian/Messinian, when the platform became involved in the eastward migrating thrust-belt/foredeep system (Patacca & Scandone, 1989; Patacca *et al.*, 1991; Cipollari & Cosentino, 1995). The NW–SE trending thrust tectonics affected the Mesozoic to Cenozoic sedimentary cover of the Adriatic passive margin and the overlying Cenozoic synorogenic sediments (Patacca & Scandone, 1989).

In the central Apennines the carbonate platform domain is known as the Latium-Abruzzi platform (Fig. 1), where Cretaceous beds of different ages are unconformably overlaid by Miocene carbonates of the "Calcari a Briozoi e Litotamni" Formation (Accordi & Carbone, 1988; Damiani *et al.*, 1991). In a limited number of cases the contact is marked by pebble deposits and/or locally by the interposition of a few metres of Palaeogene deposits.

The age of the Miocene transgression ranges from Aquitanian–Early Burdigalian in the western sector of the Latium-Abruzzi platform, to Langhian in the east (Brandano & Corda, 2002; Corda & Brandano, 2003). Equivalent Miocene limestones also widely crop out in other sectors of the Apennines, to the east in the Maiella Mountains (Mutti *et al.*, 1997, 1999; Vescsei & Sanders, 1999) and to the south in the Campanian Apennines (Barbera *et al.*, 1978; Simone & Carannante, 1985; Carannante *et al.*, 1988; Carannante & Simone, 1996). They typically consist of rhodalgal and bryozoan associations.

The Miocene limestones of the Latium-Abruzzi platform, whose thickness ranges from a few tens (40 m) to one hundred metres, are composed of five main lithofacies associations that reflect different environments. The relative position of the facies belts, biota components, and their relative dependence upon light penetration and bedding patterns suggest that these limestones were deposited on a homoclinal ramp under tropical-subtropical conditions (Brandano & Corda, 2002).

The measured sections used in this study are presently located in three different thrust bodies, but it has been possible to make correlations between the recorded events through the use of biostratigraphy and isotope stratigraphy. The ramp is oriented roughly SE–NW and dips towards the NW, with the shallower facies located in the SE sector (Fig. 3). The Miocene transgression age is Aquitanian–Burdigalian in the western part (Sections 1–13 and 20–22, Fig. 1) and Langhian in the eastern part (Sections 14–19, Fig. 1) of the study area. The downflexure of the Adriatic foreland, linked to the migration of the Apenninic thrust belt–foredeep system (Patacca *et al.*, 1991; Cipollari & Cosentino, 1995) and the consequent drowning of the Latium-Abruzzi platform, is younger from SW (Tortonian; Cosentino *et al.*, 1997) to NE (Tortonian/Messinian; Patacca *et al.*, 1991).

METHODS

The sequence stratigraphic interpretation was performed by considering the relative position of facies belts and sequence boundaries, which can be physically traced and chronostratigraphically (in some cases biostratigraphically) correlated. Changes in accommodation were evaluated by interpreting the

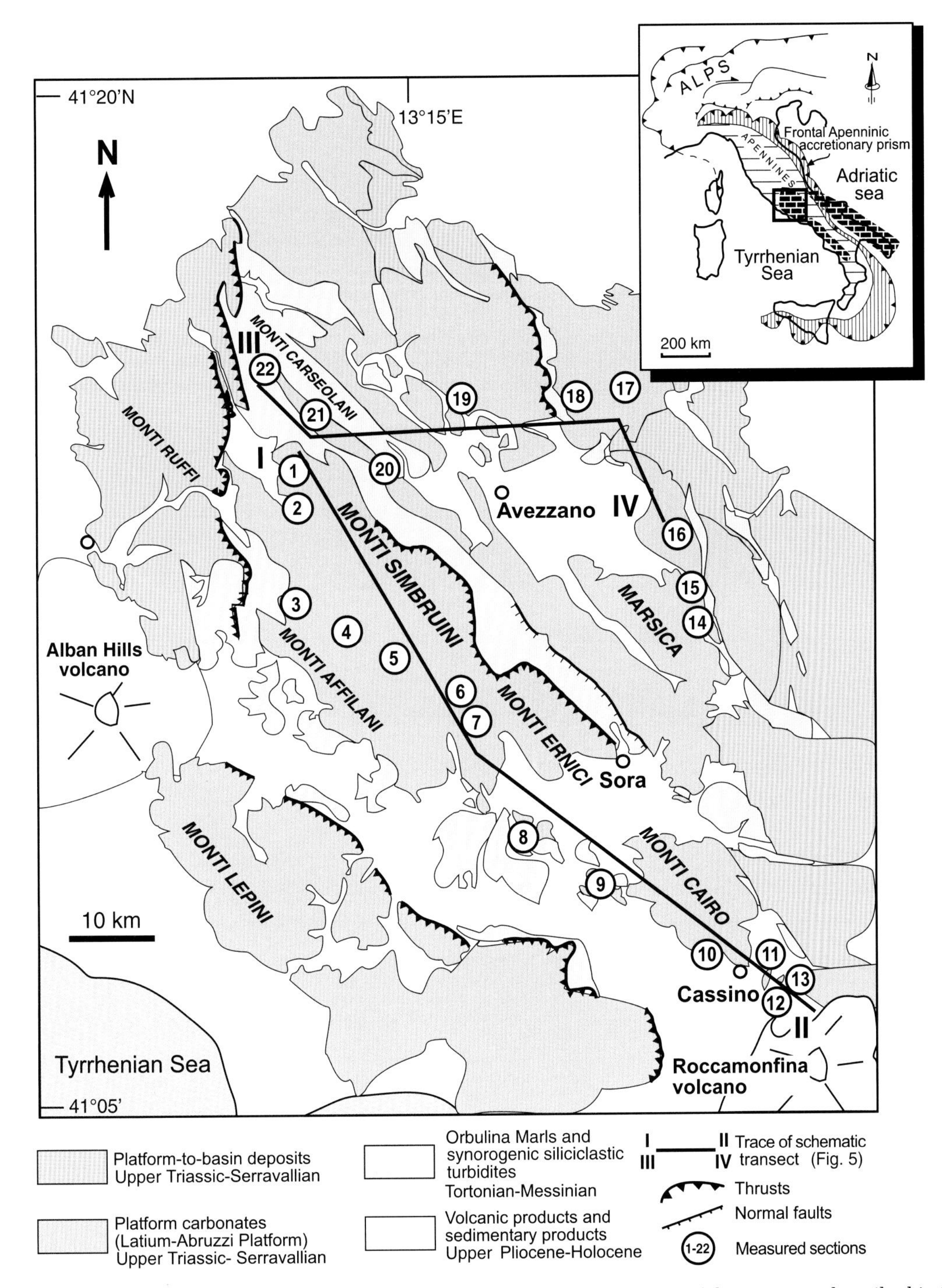

Fig. 1. Schematic geological map of the Latium-Abruzzi platform area and location of the transects described in Fig. 5. The numbers indicate the location of the measured stratigraphic sections of Fig. 2. Main Apennine platform domains are indicated in the inset.

palaeobathymetry of the biota; sedimentological and stratigraphic analyses were integrated with taxonomic and palaeoecological data.

Strontium-isotope stratigraphy was used to date the sections and facies changes. A total of 33 Sr analyses were carried out on fragments of hand-picked echinoids, bivalves, rhodoliths, and bulk carbonate samples from 22 sections crossing the entire platform. In the case of bulk carbonate analyses, the samples were first broken into chips and clean fragments were selected for crushing in a stainless steel mortar. About 30 mg of sample was

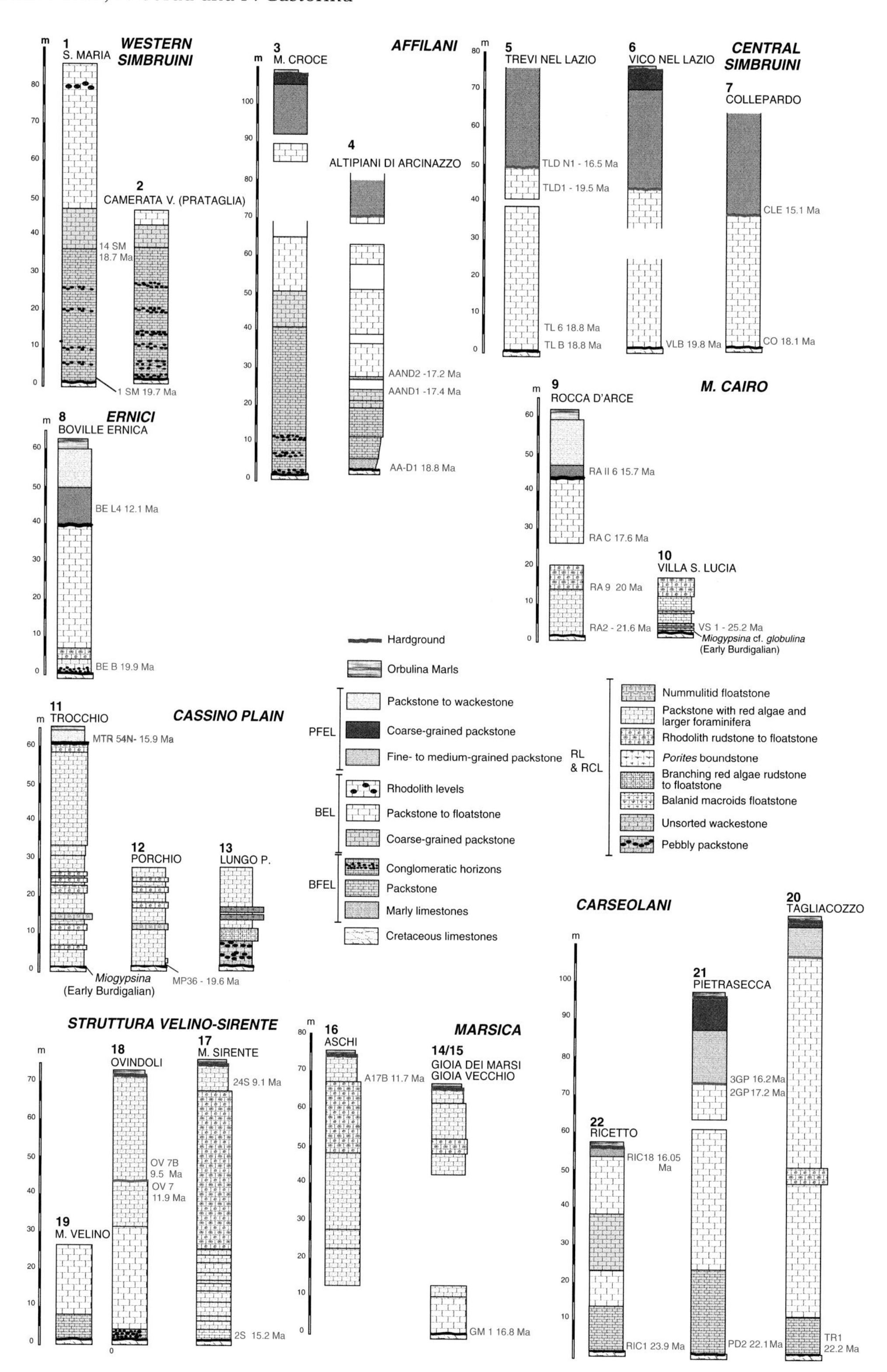

Fig. 2. Measured stratigraphic sections showing microfacies associations. Section numbers are shown in Fig. 1. Letters with numbers indicate samples taken for Sr-isotope analyses and their assigned ages. PFEL = planktonic foraminifera-echinoid limestones; BEL = bryozoan-echinoid limestones; BFEL = benthic foraminifera-echinoid limestones; RL = rhodolithic limestones; RCL = red-algae limestones and seagrass meadows.

rapidly dissolved using 2.0 M ultra-pure HCl. The insoluble residue was separated by centrifugation and then the remaining solution was loaded onto a standard Bio-Rad AG50-X8 cation exchange resin. To avoid the leaching of radiogenic ^{87}Sr and Rb from the non-carbonate constituents of the samples, the dissolution and Sr separation of the samples from the matrix were carried out in rapid succession; Sr was collected in 2.0 M HCl and evaporated to dryness. Isotopic analyses were performed at IGAG-CNR (Dipartimento di Scienze della Terra, University of Rome "La Sapienza") using a FINNIGAN MAT 262RPQ multicollector mass spectrometer. All samples were loaded on a double Re filament as nitrate and analyzed in static mode. Sr-isotope analyses were normalized to a $^{86}Sr/^{88}Sr$ ratio of 0.1194.

The Sr analytical blank was 1 ng. Internal precision ("within-run" precision) of a single analytical result is given as two standard errors of the mean (2se) and was obtained from more than 200 ratios collected on each sample using a stable beam of 2.5 V. These errors generally underestimate the actual reproducibility of the analyses (Hodell *et al.*, 1990). In this work, sample reproducibility was estimated by: (1) repeated analysis of a SRM 987 standard solution; (2) replicate analysis of selected samples subjected to the complete chemical procedure; and (3) repeated measurement of the Sr isotope ratio in samples of modern Mediterranean seawater which had been passed through the complete chemical procedure.

In the period during which mass-spectrometric analyses were performed, 30 measurements of SRM 987 gave a mean of 0.710235 ± 10. Repeated measurements of the Sr isotope ratio of modern Mediterranean water gave a ratio of 0.710170 ± 12 ($2\ \sigma$, $n = 5$).

Petrographic and geochemical criteria were previously applied to eliminate diagenetically altered samples and to minimize contamination from non-carbonate minerals. Visible residue, which is considered one of the major indicators for diagenetic alteration of Sr isotope ratios, was negligible in all analysed samples. An additional indicator of diagenetic alteration may be represented by a consistent offset between biostratigraphy and Sr isotope stratigraphy, or by the presence of scattered ages along the considered stratigraphic sections. Table 1 lists the samples analyzed for Sr isotope composition from each section, the measured $^{87}Sr/^{86}Sr$ ratio and the ages calculated using the $^{87}Sr/^{86}Sr$ age relationships from Hodell *et al.* (1991). Sr ages calculated using bulk carbonate samples correspond very closely to those obtained from the skeletal grains within the same sample. On the whole, the calculated ages are consistent with their relative stratigraphic position, suggesting that late diagenetic alteration of the original isotope composition is negligible.

The occurrence of *Miogypsina*, cf. *globulina*, at the base of the Miocene carbonates in the SE sector of the ramp and the first occurrence of *Orbulina universa* confirm the strontium data from the basal and upper parts of the Miocene, allowing the generation of an acceptable stratigraphy. Only a few samples yield Sr ages that are significantly older than their stratigraphic position, such as VS-1 from Villa S. Lucia (Section 10, Fig. 1) and RIC 1 (Section 22, Fig. 1) from the Ricetto section; these have been interpreted to suffer from the presence of reworked fragments from the underlying Cretaceous substratum. A few other Sr-isotope ratio anomalies observed in bivalve fragment samples (e.g. from Trevi nel, Section 6, Fig. 1) may be related to recrystallization processes in fractured limestones. Altogether the small analytical error, the stratigraphic consistency of the $^{87}Sr/^{86}Sr$ values from different levels, and the control given by some biostratigraphic data provide confidence in the good quality of the strontium isotope analyses for these Miocene limestones.

Twenty-two sections across the platform were measured (Figs. 1 and 2) and the identified 24 microfacies types were arranged in 5 lithofacies associations. The most important facies characteristics are listed in Table 2 (for further details see Brandano, 2001; Corda & Brandano, 2003; Civitelli & Brandano, 2005). The recognized lithofacies associations (Fig. 3) are: red-algae limestones and seagrass meadows (RCL), rhodolithic limestones (RL), bryozoan-echinoid limestones (BEL), benthic foraminifera-echinoid limestones (BFEL), and planktonic foraminifera-echinoid limestones (PFEL). Towards the basin the "Calcari a Briozoi e Litotamni" Formation grades into the Guadagnolo Formation, which consists of marls and calcarenites with silica-sponge spicules, planktonic foraminifera and mollusc-bryozoan-echinoid debris. This facies association is indicated as SEPU in Fig. 3.

DEPOSITIONAL ENVIRONMENT

The depositional profile of the Latium-Abruzzi carbonates is interpreted as being a homoclinal ramp,

Table 1. Sr-isotope results and calibrated ages for 19 sections in the Latium-Abruzzi carbonate ramp

Section	Sample	$^{87}Sr/^{86}Sr \pm 2se^{*}$	Age (Ma)**	Uncertainty
M. Trocchio	MTR 54 N	0.708730 ± 5	15.97	0.08
	MTR 11	0.708302 ± 7	23.04	0.11
M. Porchio	MP 3	0.708519 ± 8	19.46	0.13
	MP 36	0.708510 ± 7	19.60	0.11
Rocca d'Arce	RA II 6	0.708744 ± 6	15.74	0.10
	RA C	0.708063 ± 8	17.60	0.13
	RA 9	0.708482 ± 5	20.07	0.08
	RA 2	0.708389 ± 10	21.60	0.16
Villa S. Lucia	VS 1	0.708166 ± 17	25.29	0.28
Boville Ernica	BL 4	0.708810 ± 8	14.65	0.13
	BE B	0.708493 ± 5	19.89	0.08
	BE 1	0.708549 ± 6	18.96	0.10
Collepardo	CL E	0.708740 ± 8	15.80	0.13
	CO 1	0.708602 ± 6	18.08	0.10
	CL B	0.708128 ± 14	25.92	0.23
S. Maria	SM 14	0.708564 ± 7	18.71	0.11
	SM 1	0.708505 ± 6	19.69	0.10
Vico nel Lazio	VL B	0.708495 ± 4	19.85	0.07
Trevi nel Lazio	TLD N 1	0.708694 ± 9	16.56	0.15
	TL B	0.708554 ± 7	18.88	0.11
	TL 6	0.708557 ± 7	18.83	0.11
	TL D1	0.708512 ± 7	19.57	0.11
Altipianidi Arcinazzo	AA ND1	0.708642 ± 4	17.42	0.07
	AA ND2	0.708652 ± 6	17.26	0.10
	AA D1	0.708556 ± 1	18.84	0.18
Ricetto	RIC 18	0.708725 ± 6	16.05	0.10
	RIC 1	0.708248 ± 6	23.93	0.10
Pietrasecca	GP 1	0.708728 ± 7	16.00	0.11
	GP 3	0.708714 ± 8	16.23	0.13
	GP 2	0.708651 ± 7	17.28	0.11
	P 36	0.708511 ± 7	19.59	0.11
	P 1	0.708437 ± 13	20.81	0.21
	P 2	0.708465 ± 8	20.35	0.13
	PD 2	0.708359 ± 5	22.10	0.08
Tagliacozzo	TA 70	0.708610 ± 9	17.95	0.15
	TA 51 (GUS)	0.708520 ± 4	19.44	0.07
	TA 68	0.708523 ± 7	19.39	0.11
	TA 51 (SED)	0.708433 ± 6	20.88	0.10
	TR 1	0.708351 ± 6	22.23	0.10
Valpara	VP 9	0.708822 ± 5	14.45	0.08
	VP 1	0.708563 ± 10	18.73	0.16
	VP 7	0.708519 ± 6	19.46	0.10
M. Moro	MM 2	0.708553 ± 1	18.89	0.18
Ovindoli	OV 7B	0.708872 ± 8	9.50	0.13
	OV 7	0.708816 ± 6	11.90	0.10
	OV 1	0.708548 ± 6	19.98	0.10
Sirente	24 S	0.708881 ± 6	9.10	0.10
	2 S	0.708737 ± 20	15.20	0.33
Aschi	A 17B	0.708819 ± 5	11.70	0.08
Gioia dei Marsi	GM 1	0.708678 ± 9	16.83	0.15

*2 standard error × 10^5
**Ages calculated according to Hodell *et al.* (1991)

based on the relative position of facies belts and the dependence of some skeletal components on light penetration (Fig. 3). Although the terminology proposed by Burchette & Wright (1992) is used here to subdivide the carbonate-ramp facies, the criteria used to characterize the various parts of the ramp follow Pomar (2001a). The ramp can be divided into three main parts: inner; middle; and outer.

Table 2. Microfacies associations and interpreted depositional environments

Lithofacies associations	Microfacies	Texture	Grain size	Components	Environment
RCL	MF 20	Pebbly floatstone	>2 mm	Rounded Cretaceous lithoclasts and coarse skeletal fragments in packstone matrix	Inner ramp
	MF 21	Floatstone	>2 mm	Larger foraminifera (*Heterostegina, Operculina, Amphistegina*), benthic foraminifera (*Elphidium*, miliolids), bivalves, red algae (branching) embedded in fine-grained bioclastic wackestone	
	MF 22	Rudstone	>2 mm	Rhodoliths (*Mesophyllum, Spongites, Lithothamnium*), larger foraminifera (*Heterostegina, Operculina, Amphistegina*), benthic foraminifera (*Elphidium*), bivalves	
RL	MF 24	Framestone	>2 mm	Corals (*Porites*), balanids, larger foraminifera (*Heterostegina, Operculina, Amphistegina*), benthic foraminifera (*Elphidium*, miliolids), bivalves, red algae (branching and rhodoliths)	Middle ramp
	MF 6	Packstone	80–200 μm	Benthic foraminifera (rotalids, textulariids), encrusting foraminifera (acervulinds, homotrematids), fragments of: larger foraminifera, balanids, echinoids, bivalves, and bryozoans	
	MF 11	Packstone	50 μm–2 mm	Larger foraminifera (*Operculina, Amphistegina*), benthic foraminifera (*Elphidium*), serpulids, fragments of: balanids, echinoids, bivalves, bryozoans, red-algae, and subordinate planktonic foraminifera	
	MF 11a	Packstone/ wackestone	50 μm–2 mm	Larger foraminifera (*Operculina*), planktonic foraminifera, echinoids, bivalves, bryozoans and subordinate red-algae	
	MF 15	Grainstone	400 μm–2 mm	Larger foraminifera (*Miogypsina, Amphistegina*), benthic foraminifera (*Elphidium*, textulariids, rotalids), serpulids, fragments of: balanids, echinoids, red algae, bivalves, bryozoans, and rare Cretaceous lithoclasts	
	MF 17	Floatstone	>2 mm	Larger foraminifera (*Operculina, Heterostegina*) embedded in well-sorted packstone with fine grained fragments of: red algae, bivalves, echinoids as well as benthic and planktonic foraminifera	
	MF 22	Rudstone	>2 mm	Rhodoliths (*Mesophyllum, Spongites, Lithothamnium, Sporolithon*) and subordinate bryozoans and larger foraminifera; matrix is represented by bioclastic packstone with serpulids, fragments of: balanids, echinoids, and bivalves	

Table 2 (*Continued*)

Lithofacies associations	Microfacies	Texture	Grain size	Components	Environment
BEL	MF 10	Packstone	500 μm–2 mm	Bryozoans, echinoids, bivalves, benthic foraminifera (rotalids, textulariids, buliminids), rare larger foraminifera (*Amphistegina*)	Outer ramp proximal
	MF 12	Packstone	200–500 μm	Benthic foraminifera (textulariids, rotalids), planktonic foraminifera, incrusting foraminifera (acervulinds, homotrematids), fragments of: bryozoans, echinoids, and bivalves	
	MF 13	Packstone	500–800 μm	Fragments of: molluscs, echinoids, serpulids, balanids, bryozoans, red algae, benthic foraminifera (*Elphidium*, textulariids and rotalids), and rare well-rounded lithoclasts	
	MF 16	Packstone to rudstone	500 μm–5 mm	Echinoids, bivalves, serpulids and subordinate bryozoans, red algae, as well as benthic and planktonic foraminifera	
	MF 18	Floatstone	>2 mm	Bryozoan colonies embedded in bioclastic packstone-wackestone with benthic and planktonic foraminifera, as well as phosphatic and glauconitic grains	
	MF 19	Floatstone	>2 mm	Bryozoan colonies embedded in bioclastic packstone-grainstone with unsorted fragments of: bivalves, echinoids, benthic and planktonic foraminifera	
	MF 23	Floatstone	>2 mm	Rhodoliths (*Mesophyllum, Lithothamnium, Sporolithon*) embedded in poorly sorted packstone-wackestone with fragments of: bivalves, bryozoans, as well as benthic and planktonic foraminifera	
BFEL	MF 1	Wackestone	80–500 μm	Echinoid and bivalve fragments, and subordinate planktonic foraminifera (globigerinids), benthic foraminifera (rotalids, textulariids, buliminids), and serpulids (*Ditrupa*)	Outer ramp intermediate
	MF 5	Packstone	80–200 μm	Fragments of echinoids and bivalves, planktonic foraminifera (globigerinids), benthic foraminifera (rotalids, textulariids, buliminids), and subordinate serpulids (*Ditrupa*)	
	MF 8	Packstone	100–500 μm	Echinoid fragments, planktonic foraminifera (globigerinids) and benthic foraminifera. Dolomization is frequent	
	MF 9	Packstone	200 μm–2 mm	Echinoid fragments and planktonic foraminifera (globigerinids); rare larger foraminifera (*Amphistegina*), serpulids, and bryozoan fragments	

		MF 14	Grainstone	80–500 μm	Echinoid fragments, planktonic foraminifera (globigerinids) and benthic foraminifera (rotalids, textulariids, buliminids)	
PFEL	Upper unit	MF 4	Packstone to wackestone	50–200 μm	Planktonic foraminifera (globigerinids) and subordinate benthic foraminifera (rotalids, textulariids); fragments of echinoids and red algae	Outer ramp intermediate
		MF 3	Wackestone	80 μm–2 mm	Fragments of: molluscs, echinoids, serpulids, balanids, bryozoans, red algae, planktonic (globigerinids) and benthic foraminifera (textulariids and rotalids)	
	Lower unit	MF 2	Wackestone	50 μm–1 mm	Planktonic foraminifera (globigerinids), red-algae fragments and subordinate benthic foraminifera (buliminids)	
		MF 7	Packstone	200–400 μm	Planktonic foraminifera, fragments of echinoids and subordinatlely benthic foraminifera	
SEPU		MFB	Argillaceous wackestone	≤50 μm	Planktonic foraminifera, siliceous sponge spicules and subordinate small echinoid fragments. Traces of bioturbation	Outer ramp distal
		MFC	Wackestone-Packstone	100–300 μm up to 600 μm	Planktonic foraminifera and small fragments of echinoids and serpulids, benthic foraminifera (rotalids, textulariids) and rare siliceous sponge spicules. Traces of bioturbation	
		MFD	Packstone to grainstone	500 μm–1 mm rarely exceeds 3 mm	Benthic foraminifera (rotalids), fragments of echinoids, bryozoans, red algae and larger foraminifera (*Amphistegina*); rare planktonic foraminifera. Glauconitic grains	
		MFG	Packstone	100–300 μm	Planktonic foraminifera and subordinate benthic foraminifera as well as echinoids fragments. Moderate sorting	
		MFF	Packstone-grainstone	200–300 μm	Benthic foraminifera and fragments of red algae, echinoids, and bryozoans. Good sorting	
		MFE	Grainstone	1.5–3 mm	Bryozoans, echinoids, red algae and subordinate, fragments of larger foraminifera (*Amphistegina*) and molluscs, benthic foraminifera and rare planktonic foraminifera.	

Inner ramp

The inner-ramp lithofacies (RCL in Fig. 3) crop out in the SE sector of the platform (Cassino plain, Figs. 1 and 2) and only within the late Aquitanian–Early Burdigalian interval. The lithofacies included in this association (Table 2, Fig. 4F) are: balanid macroids floatstone; pebbly packstone; bioclastic packstone to floatstone; and branching red-algae rudstone to floatstone.

Balanid macroid floatstone and pebbly packstone constitute the basal part of the Miocene succession, directly overlying the highly bioeroded Cretaceous substratum. The 20 to 50 cm thick conglomerate layers are composed of rounded to subangular clasts that are mainly derived from the underlying Cretaceous limestones and a few from Eocene limestones. The clasts are often bioeroded and are embedded in bioclastic packstone to floatstone made up of unsorted skeletal debris that has abundant silt and is rich in benthic foraminifera (rotalids, miliolids, *Elphidium*). Micritized skeletal fragments are common. The absence of sedimentary structures together with the skeletal components suggests that this facies represents a seagrass meadow environment (see Brandano *et al.*, this volume, pp. 89–106). The branching red-algae rudstone lithofacies consists of branching red algae embedded in a carbonate mud, with ostracods and miliolids organized in 30–40 cm thick homogeneous beds. Scattered balanid macroids, bryozoan colonies, echinoids and highly bioeroded oyster shells are common. This facies can be compared with the Maërl facies (*sensu* Pérès and Picard, 1964).

The biota associations indicate that inner-ramp carbonate production took place in the euphotic zone (*sensu* Pomar, 2001b). The proposed model for this sector of the ramp (Fig. 3) consists of a high-energy environment immediately below the littoral zone (balanid macroids floatstone and pebbly packstone) that grades into sea grass meadows, associated Maërl facies, and, farther seaward (north-westward, Fig. 3), into the middle-ramp deposits described below. Coral carpets characterize the passage between the inner and middle zones. The coral colonies, which are up to a few centimetres high, are often encrusted with red algae; traces of bioeroders are common on bivalves and on corallinacean algal branches. The coral-bearing matrix is composed of bioclastic wackestone–packstone with micritized fragments of molluscs, echinoids, red algae, and encrusting foraminifera. The coral intervals can be compared with the coral carpets reported by Riegl & Piller (1999, 2000).

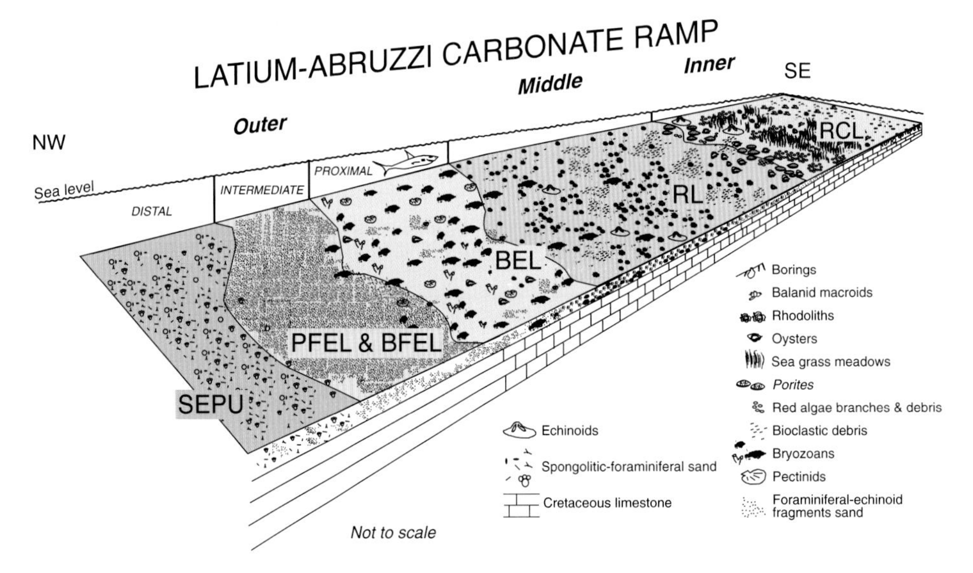

Fig. 3. Latium-Abruzzi ramp model showing the distribution of lithostratigraphic units and main biota associations. RCL (inner zone), rhodolithic limestones with lithoclasts and seagrass meadows; RL (middle zone), rhodolithic limestones with larger foraminifera; BEL (proximal outer zone), bryozoan-echinoid limestones; BFEL-PFEL (intermediate outer ramp), benthic-foraminifera-echinoid limestones and planktonic-foraminifera-echinoid limestones; SEPU (distal outer zone), marls and limestones with silica sponge spicules, planktonic foraminifera and skeletal fragments of bryozoans, echinoids and red algae (Guadagnolo Formation). See Fig. 2 for explanation of acronyms.

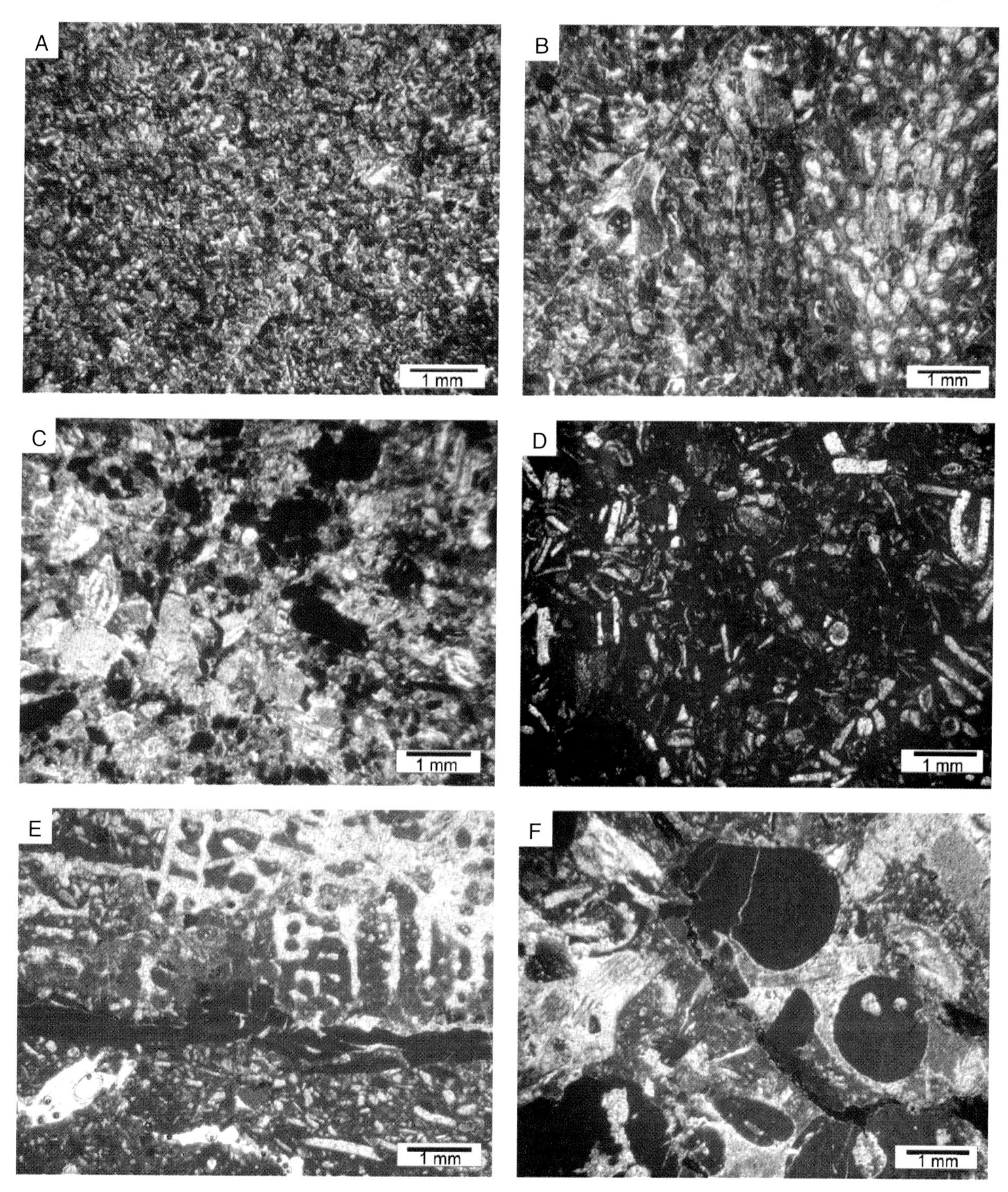

Fig. 4. Thin-section photomicrographs of representative lithofacies. (A) Packstone with echinoid fragments, planktonic foraminifera and benthic foraminifera (MF 8, intermediate outer ramp). (B) Floatstone with bryozoan colonies embedded in bioclastic packstone-grainstone (MF 19, proximal outer ramp). (C) Packstone with larger foraminifera (*Amphistegina*,), benthic foraminifera, fragments of echinoids and of red algae (MF 11, middle ramp). (D) Packstone to wackestone with larger foraminifera fragments, planktonic foraminifera, plates and echinoid spines and subordinate red algae fragments. In the limited intergranular cavities a calcisiltite matrix is present (MF 11a, middle ramp). (E) Coral boundstone encrusted by red algae (MF 24, middle ramp); (F) rudstone with red algae nodules and branches (MF 22, inner ramp). All images are plain polarized light.

Middle ramp

The middle-ramp facies, cropping out in the SE and eastern sector of the Latium Abruzzi platform (Piana di Cassino, Monti Ernici, Marsica, Monte Velino-Sirente, Figs. 1 and 2) are represented by the rhodolithic limestones (RL in Fig. 3). These units are poorly stratified with sub-horizontal bedding planes and are made up of different microfacies (Table 2, Fig. 4C–E). Medium-coarse bioclastic packstone to grainstone alternate with rhodolith rudstone to floatstone with a bioclastic packstone matrix. The rhodolithic facies form tabular bodies that vary in thickness from 20 to 200 cm. A detailed description of the rhodolithic facies is given in Brandano & Piller (this volume, pp. 151–166). A few oyster-pectinid-bearing levels (20–40 cm thick) and one key bed (1.5 m-thick), which is particularly rich in *Heterostegina* and *Operculina*, characterize the lower part of the RL unit. The latter occurs in Sections 11 and 12 (Figs 1 and 2).

Bioclastic packstone-grainstone consists of fragments of red algae, bryozoans, bioeroded bivalves, balanids, serpulids, echinoids, benthic foraminifera (rotalids, textularids, buliminids), and larger foraminifera (*Heterostegina, Operculina, Elphidium* and *Amphistegina)*. In the upper part of Sections 16, 17 and 18 (Fig. 5) this facies (RLa) exhibits a grain-size decrease and an increased percentage of planktonic foraminifera.

Sedimentological characteristics, the recognized genera in the red-algae associations, and the *Amphistegina* test shape (Mateu-Vicens *et al.*, 2005) place the depositional environment in the oligophotic zone.

Outer ramp

Outer-ramp facies are widely represented throughout the entire Latium-Abruzzi platform, particularly in the western and central sectors. The outer ramp can be sub-divided into a proximal, an intermediate and a distal outer ramp (Fig. 3). The bryozoan-echinoid limestones (BEL) represent the facies of the proximal outer ramp. They consist of poorly stratified coarse-grained bryozoan-pectinid floatstone and echinoid-foraminiferal packstone (Table 2, Fig. 4A and B). Scattered rhodolithic levels are also present. The main sediment-producing biota are primarily represented by bryozoan colonies and, to a lesser extent, by bivalves and echinoids; benthic foraminifera (rotalids, textularids, buliminids) are common. The faunal association and the absence of larger foraminifera such as *Heterostegina* and *Operculina* suggest a depositional environment located at the transition between the oligophotic and aphotic zone (Fig. 3).

Two lithofacies associations are interpreted to be representative of the intermediate outer-ramp zone (Fig. 3): the benthic foraminifera-echinoid limestones (BFEL) and the planktonic foraminifera-echinoid limestones (PFEL). These two associations are located at the base and the top, respectively, of the analyzed succession. The lower, BFEL lithofacies, consist of thick (1–2 m) structureless bodies, delimited by planar master-bedding surfaces along which burrows can be seen. Main microfacies are fine to medium packstone to wackestone with benthic and planktonic foraminifera, echinoid fragments, and serpulids (Table 2, Fig. 4A). In the NW sector of the Latium-Abruzzi platform (Monti Simbruini and Monti Affilani, Figs. 1 and 2), episodes of fine-grained conglomerates with rounded, reddish clasts are present in 10 to 50 cm thick beds; the 1–3 cm wide clasts consist primarily of Cretaceous limestones derived from the underlying substratum.

The upper lithofacies (PFEL) can be divided into two sub-units that are separated by a discontinuity marked by a sharp textural change. The lower sub-unit consists of medium to fine wackestone-packstone with planktonic and benthic foraminifera, echinoid fragments, and, subordinately, fragments of red algae and molluscs (Table 2). In the second sub-unit, consisting of medium- to coarse-grained packstone–wackestone, an increasing percentage and grain size of skeletal components (i.e. fragmented red algae, serpulids, and molluscs) and benthic foraminifera can be observed (Table 2).

The distal outer-ramp lithofacies (sponge echinoid planktonic unit, SEPU; Fig. 3, Table 2) is represented by the Guadagnolo Formation, which consists of alternating marlstones, fine bioclastic marly limestones and medium to coarse bioclastic limestones that are organized in subtidal cycles (Mariotti *et al.*, 2002; Brandano *et al.*, 2005). The main microfacies are: wackestone to packstone with planktonic foraminifera and silica sponge spicules; medium- to fine-grained packstone with benthic and planktonic foraminifera, as well as fragmented bryozoans and echinoids; and medium to coarse-grained packstone-grainstone with fragments of bryozoans, echinoids, red algae, and benthic foraminifera (for more detail see Barbieri *et al.*, 2003/2004).

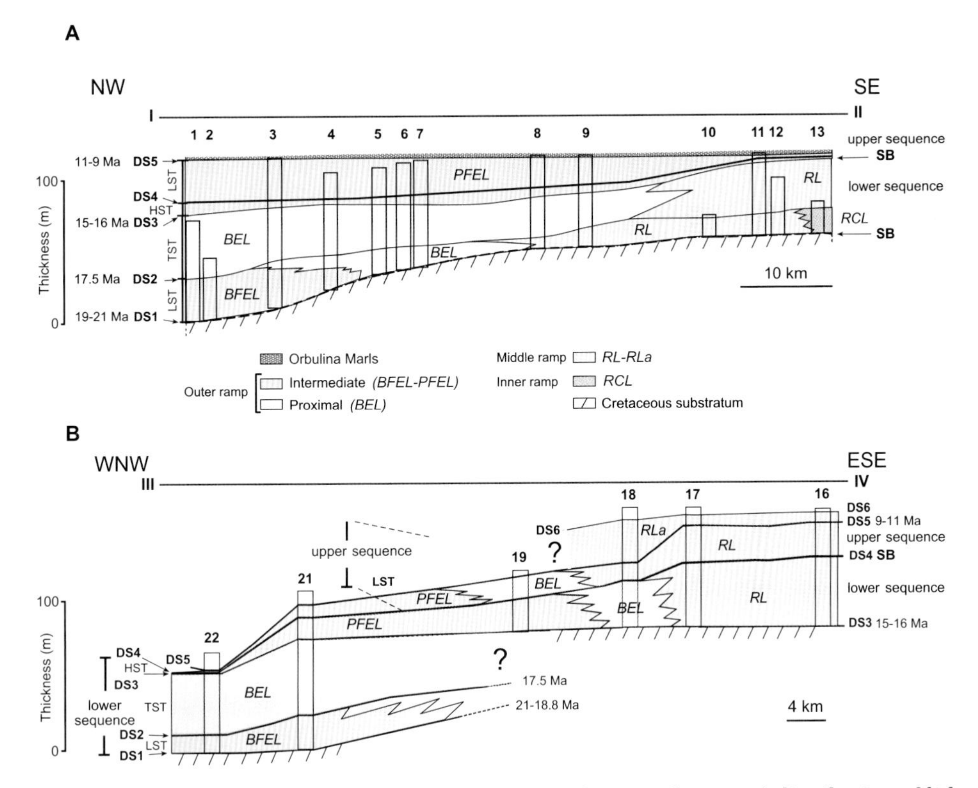

Fig. 5. Cross-section profiles of the Latium-Abruzzi carbonate ramp showing the spatial distribution of lithostratigraphic units during the Early-Middle Miocene. (A) Transect I–II, roughly oriented NW–SE. (B) Transect III–IV, roughly oriented E–W. See Fig. 1 for transect and section locations. DS = discontinuity surface; other acronyms as defined in Fig. 2.

Based on sedimentological features and biota assemblages, the outer ramp, from the intermediate to the distal zone, is placed in the aphotic zone.

MAIN DISCONTINUITY SURFACES

Six main discontinuity surfaces (*sensu* Clari *et al.*, 1995) have been recognized in the area, DS1 to DS6 (Fig. 5). They are marked by sharp facies changes (textural and compositional) and in most cases, are conformable at the outcrop scale. Almost all the surfaces are correlated using Sr-isotope stratigraphy and, in some cases, biostratigraphy.

DS1 (Fig. 5) marks the contact between the Upper Cretaceous limestones and the overlying Lower Miocene limestones. Passing from SE to NW (down-ramp direction) along the discontinuity it is possible to note a gradual change of the Miocene lithofacies that rest unconformably on a bioeroded Cretaceous substratum: RCL, RL, BEL and BFEL (Fig. 5A). Most of the time a styolitic suture marks DS1, and no evidence of subaerial exposure is present. However, erosional features are locally distinguishable, particularly in the SE sector (Sections 7, 12 and 13 in Fig. 2), where DS1 is represented by an undulating irregular surface with scattered conglomerates. The first Miocene deposits, directly overlying the Cretaceous substratum, yield Sr ages between 21.6 and 18.8 Ma. The presence of *Miogypsina* cf. *globulina* at the base of the Miocene succession in Sections 10, 11 and 12 (Fig. 2) confirms this age.

DS2 (Fig. 5) has been dated at 17.5 Ma. Bulk-rock Sr-isotope values from immediately below and above this surface provide ages between 17.6 and 17.26 (Table 1). DS2 is marked by a sharp superposition of different facies that are variable along the discontinuity in the down-ramp direction. In the southernmost and shallowest sector of the Latium-Abruzzi platform (Section 13, Fig. 5A) it corresponds to a contact between coral framestone facies (transition between inner and middle-ramp zone) and overlying rhodolith floatstone and bioclastic packstone (middle ramp). Moving along dip (north-westward) DS2 occurs first inside the middle-ramp facies associations (RL; Sections 11 and 12, Fig. 5A), where it is marked by the occurrence of the previously mentioned *Heterostegina-Operculina*-rich bed, and then corresponds to a sharp change between the underlying middle-ramp facies (RL) and the overlying proximal outer-ramp facies (BEL; Sections 8 and 9, Fig. 5A). More basinward (NNW), in the deepest sector of the ramp, DS2 marks the sharp contact between the underlying intermediate outer-ramp deposits (BFEL) and the overlying proximal outer-ramp lithofacies (BEL; Sections 1, 2, 20, 21 and 22, Fig. 5). The facies changes across DS2 suggest increased water depth.

DS3 has been isotopically dated between 15.2 and 16.8 Ma (Table 1); 16 Ma was used as a mean value of the age range. Biostratigraphic data (i.e. first occurrence of *Orbulina universa* in the overlying sediments of Sections 21 and 22, Fig. 2) confirm this age. DS3 is a flat surface which is well exposed throughout the entire platform. In the easternmost sector of the Latium-Abruzzi platform (Sections 14/15, 17, 18 and 19, Figs 1, 2 and 5) DS3 marks the beginning of Miocene sedimentation (RL facies) on the Cretaceous limestones. The age of this Miocene transgression gets progressively younger from west (DS1, 21–19 Ma) to east (16 Ma). DS3 is a styolitic surface that does not show any evidence of subaerial exposure, but which represents a significant time gap.

Dipping along the depositional profile (Section 11, Fig. 5A; SE sector of the platform), DS3 is expressed as a hardground that marks the sharp passage between the middle-ramp facies (RL) and the overlying intermediate outer-ramp facies (PFEL), without the interposition of the BEL. Phosphatized and glauconitized sediments fill macroborings, bryozoan and red-algae intracavities, and planktonic foraminifera. Farther down the depositional dip (Sections 3, 4, 5, 6, 7, 8, 9, 21 and 22, Fig. 5) DS3 is marked by glauconitic grains and by a sharp vertical change from the underlying proximal outer-ramp (BEL) to the intermediate outer-ramp facies associations (PFEL). The presence of DS3 suggests low sedimentation rates and is interpreted to represent a general increasing water-depth.

DS4 has not been dated isotopically; however, it is presumed to be of Serravallian age, based on its stratigraphic position (i.e. located between dated discontinuities, DS3 and DS5). It is marked by a sharp facies contrast and does not display evidence of subaerial or submarine erosion. The most dramatic facies contrast is observed in the eastern sector of the investigated area, where middle-ramp facies (RL) overlie proximal outer-ramp facies (BEL) (Section 18, Fig. 5B) and proximal outer-ramp facies (BEL) overlie intermediate outer-ramp facies (PFEL; Section 19, Fig. 5B). In the western sector (Sections 3–9, 21, Fig. 5) DS4 is recognizable as a sharp textural and compositional change within the intermediate outer-ramp facies associations (PFEL). Here medium- to fine-grained packstone with planktonic foraminifera and echinoid fragments are overlain by coarse packstone to wackestone consisting of mollusc, echinoid and balanid fragments, serpulids, benthic and planktonic foraminifera, and a fine red-algae debris (Table 2, lower and upper unit respectively). Based on the stacking of the facies belt, textural changes and main biota associations, DS4 is interpreted as being the result of a moderate decreased water depth.

DS5 shows Sr-isotope ages from 9.1 and 11.9 Ma, based on samples from the eastern Latium-Abruzzi platform (Table 1); biostratigraphic data from hemipelagic marls above DS5, in the NW sector, indicate the CN8 nannofossil zone (approximately 9.5 Ma; Pampaloni *et al.*, 1994). This surface has been labelled with the same abbreviation throughout the platform because of its age, even if it is characterized by two different features that can be linked to different processes.

In the NW and SE sectors of the platform (Sections 3, 8, 9, 11, 21 and 22, Figs 1 and 5) DS5 represents the drowning of the platform as the result of the rapid flexural stage of the foreland basin. These sectors, that have an internal position with respect to the Apennine orogen, were the first sectors of the foreland to be involved in the NE-migrating thrust-belt-foredeep system (Cipollari & Cosentino, 1995). DS5 is not very evident in the easternmost sector of the Latium-Abruzzi platform and is marked by a textural and compositional change within the middle-ramp facies association. Across the surface it is only possible to observe a grain-size decrease and an increased percentage of planktonic foraminifera (RLa; Sections 16–18, Fig. 5B). This microfacies change may suggest a moderate increase in water depth.

DS6 defines the end of Miocene carbonate sedimentation in the easternmost sector of the platform. It corresponds to a mineralized hardground that marks the passage to the overlying Orbulina Marls (Sections 14/15, Fig. 2; Sections 16, 17 and 18, Fig. 5B). Sr-isotope data from the uppermost platform limestones are not available, nevertheless biostratigraphic analyses from the overlying hemipelagic marls (Patacca *et al.* 1991) point to the

Tortonian/Messinian as the start time for the definitive involvement of the eastern platform in the flexural downwarp of the foreland basin.

SEQUENCE STRATIGRAPHY

Large-scale sediment geometries and bedding patterns are poorly exposed in the Latium-Abruzzi carbonate ramp successions. Consequently, the sequence architecture of the "Calcari a briozoi e litotamni" Formation has been reconstructed by considering migration and superposition of facies belts, together with features of the main recognized discontinuities and their correlations.

The reconstructed architecture reflects two second-order depositional sequences (Fig. 5). The lower sequence (uppermost Aquitanian to Serravallian) was influenced by the pre-existing topography; it is approximately wedge-shaped in a roughly NW–SE-trending cross-section. It is well developed in the western and southern sector of the Latium-Abruzzi platform, whereas it is incomplete in the easternmost sector where only the upper part of the sequence was deposited. The upper sequence started in the Serravallian and was terminated by a tectonically-driven drowning phase, when the sector was influenced by the approaching Apennine foredeep system.

Lower sequence

The lower boundary unconformity of the lower sequence is formed by the contact between Miocene and Cretaceous limestones and coincides with the previously described DS1. The upper bounding surface, DS4, is identified as a sharp textural change that is easily recognizable across the exposed platform. This wedge-shaped lower sequence reaches its maximum thickness (about 80 metres) in the NW sector of the platform (Sections 3 and 21, Fig. 5) and thins to 50 m towards the SE (Section 11, Fig. 5A). It is composed of a lowstand systems tract, a well-developed transgressive systems tract and a thin highstand systems tract.

Lowstand systems tract (LST)

The LST has an aggradational geometry and rests directly on the Cretaceous limestones, the lower bounding surface (DS1) representing a long hiatus. The "Palaeogene hiatus" is a characteristic feature of the central Apennine carbonate platform. The duration of the hiatus is variable, spanning from the Late Cretaceous to the Middle Miocene, although very rare outcrops of shallow-water Palaeogene successions have been described in the inner sector of the Latium-Abruzzi platform (Damiani *et al.*, 1991). Most of the sediments produced during this time interval consist of larger foraminifera-rich calcarenites and calcirudites resedimented into the pelagites at the margins of the platform (Civitelli *et al.*, 1986a, b). Some authors (e.g. Cipollari & Cosentino, 1995) have proposed that the Adria plate, on which the Latium-Abruzzi platform developed, underwent lithospheric folding and compressional deformations during the Middle Eocene, resulting from intraplate stress between the Apennine and Dinaride systems. As a consequence, the area underwent a relative sea-level drop which induced a regional erosive event. At present the debate regarding the nature of the Palaeogene hiatus (non-depositional vs. erosional) is still ongoing.

Above the DS2 surface, the facies belts moved landward, showing a clear back-stepping trend. Therefore, DS2 is interpreted to represent a transgressive surface separating the LST from the transgressive systems tract. The Late Aquitanian flooding of the Cretaceous platform, which occurred during the final stages of a global sea-level fall (Abreu & Anderson, 1998), is related to increased tectonic subsidence due to the eastward migrating Apennine foredeep system (Carminati *et al.*, 2007).

The LST is composed of the inner-, middle- and outer-ramp facies in the SE, central and NW sectors of the platform (transect I–II, Fig. 1). Lowstand deposits do not occur in the easternmost sectors of the Latium-Abruzzi platform (Fig. 5). The thickness of this systems tract changes from about 16 m to at least 40 m, resulting in a west-dipping ramp where the depositional profile was probably influenced by the inherited morphology.

The presence above DS1, inside the intermediate outer-ramp deposits, of discontinuous 20–50 cm-thick conglomerates, whose clasts are mostly made of Cretaceous limestones, suggests that the palaeomorphology of the Cretaceous substratum should be irregular and that pre-existing steps survived during the early stages of the Miocene sedimentation (see western sides of Fig. 5A and B).

Transgressive systems tract (TST)

This systems tract is a 30–40 m thick, laterally-widespread carbonate body that consists mainly of the BEL facies association, grading to the SE into the rhodolithic facies (RL, Fig. 5). Transgressive deposits are not recorded in the easternmost sectors of the platform (Fig. 5). The TST shows a rapid spreading of the bryomol lithofacies, resulting in contemporaneous retrogradation and progradation of the facies belt (landward to the SE and basinward to the NW, respectively, Fig. 5). This is particularly interesting if one considers that this facies was produced and accumulated in the aphotic zone, which is commonly considered to be less productive than the photic zone (see also Corda & Brandano, 2003).

At the top of the body, an early cemented and glauconitized surface (DS3) is well exposed, which is interpreted to be a maximum-flooding surface. It merges towards the east (landward) with the transgressive surface capping the Cretaceous limestones (DS3, Fig. 5).

Highstand systems tract (HST)

The highstand systems tract developed above DS3. During this phase, Miocene carbonates also covered the Cretaceous limestones in the easternmost sector of the Latium-Abruzzi platform (Sections 16–18, Fig. 5B) where the middle-ramp facies belt (RL) reaches its maximum thickness of about 40 metres. The rhodolithic limestones (RL) grade to the west into the outer-ramp facies belt (BEL of the proximal sector) and, more westward, into foraminiferal wackestone and packstone (PFEL of the intermediate sector; Fig. 5). The HST, which is dominantly aggradational and does not show evidence of progradation, displays a progressive decrease in thickness (up to a few metres) towards the NW (Fig. 5B) and SE (Fig. 5A). The HST is also characterized by a reduction of the proximal outer-ramp lithofacies and the lateral expansion of the middle and intermediate outer-ramp lithofacies.

The different thickness between the middle and outer-ramp lithofacies has been interpreted as being the result of two different carbonate factories having different efficiencies: high productivity by red algae and larger benthic foraminifera in the middle ramp; slow rates of production by echinoids and small benthic and planktonic foraminifera in the intermediate outer-ramp zone (NW and SE sectors), where rapid deepening was occurring due to the early stages of subducting plate downflexure. Indeed, starting from 16 Ma, the NW and SE sectors of the platform (corresponding to transect I–II in Fig. 1) underwent an accelerated tectonic subsidence (Carminati *et al.*, 2007) while approaching the Apennine subduction zone. The rapid deepening led to the spreading of relatively deeper facies (PFEL) on all these sectors (Fig. 5). For example, note the superposition of intermediate outer-ramp facies (PFEL) directly on middle-ramp facies (RL; Section 11, Fig. 5A), with a glauconitized hardground between them. At this time the ramp deepened towards the west, with the shallower facies limited to the more eastern zones (Fig. 5).

The upper bounding surface of this tract (DS4) is marked by a sharp textural change, and shallower facies overlie deeper facies.

Upper sequence

The upper sequence is exposed in almost the entire Latium-Abruzzi region, although it varies in thickness. The lower boundary (DS4) in the western and southern sectors (Sections 1–11, 21 and 22, Fig. 5) is marked by the sharp textural and compositional change within the intermediate outer-ramp facies associations, i.e. PFEL upper unit on PFEL lower unit (Table 2). This change is more significant in the central sector where middle-ramp facies rest on proximal outer ramp and proximal outer ramp on intermediate outer ramp (Sections 18 and 19 respectively, Fig. 5B). Based on its stratigraphic position (see the above paragraph on main discontinuity surfaces), DS4 has been correlated to the Middle-Serravallian sea-level fall on the averaged isotope curve (Abreu & Anderson, 1998); therefore, it is interpreted as a sequence boundary.

The sequence has its maximum thickness, about 40 metres, to the NE (Sections 16–18, Fig. 5B) and thins (a few metres) to the west. The upper bounding surface is represented by DS5 in the western and southern sectors of the Latium-Abruzzi platform, corresponding to M. Carseolani and M. Simbruini-M. Cairo of Fig. 1 (see also Sections 1–13, 21 and 22 in Fig. 5), whereas in the NE sectors it is represented by the younger DS6 (Sections 16–18, Fig. 5B). The eastward-younging of the drowning of the platform is related to the NE-directed migration of the Apennine foredeep system.

The upper sequence is tentatively interpreted in terms of sequence stratigraphy. It is composed of a lowstand systems tract in the entire platform and a transgressive systems tract developed only in the eastern sector.

Lowstand systems tract (LST)

In the NE sector of the platform (Sections 16–18, Fig. 5B) the lowstand systems tract consists of middle-ramp lithofacies; no inner facies are recorded. Basinward they grade into proximal outer-ramp lithofacies (Section 19, Fig. 5B) that, more westward, gradually pass to intermediate outer-ramp lithofacies, PFEL (Section 21, Fig. 5B). The upper bounding surface of the lowstand (DS5), in the western and SE sectors of the platform (Fig. 5), is also the drowning surface separating the PFEL unit from the overlying hemipelagic marls; it represents the tectonically-induced termination of the upper sequence.

In the NE sector (Sections 16–18, Fig. 5B) the upper bounding surface of the lowstand (DS5), separating the RL from the overlying RLa, is represented by a discontinuity surface with glauconitic grains that marks a relative sea-level rise, as shown by a relative increase in the planktonic versus the benthic assemblages (Fig. 4D and Table 2, MF 11a).

The limited basinward shift of the facies belt together with the absence of exposure and/or erosive phase suggest that the sea-level fall at the base of the sequence was nearly balanced by an accelerated tectonic subsidence that caused a moderate decrease in the accommodation space. Data from backstripping analyses performed on the Miocene Latium-Abruzzi platform limestones (Carminati *et al.*, 2007) account for accelerated rates of subsidence since 16 Ma, with values around 20 mm kyr^{-1}. Likewise the LST has been interpreted to be conditioned by tectonic subsidence.

At the top, the NW and SE sectors of the Latium-Abruzzi platform (the more internal with respect to the Apennine orogen) underwent a tectonic-driven drowning as a consequence of Apennine tectonics. This conclusion is supported by the occurrence of a few cm-thick glauconitized and bioeroded hardground, capping the shallow-water carbonates, followed by the hemipelagic Orbulina Marls (Sections 3, 6, 8, 11, 21 and 22, Fig. 2).

Transgressive systems tract (TST)

The transgressive systems tract, which is only recorded in the north-easternmost part of the platform, rests on the rhodolith floatstone of the RL unit (Sections 16–18, Fig. 5B). It consists of a few metres up to 25 m of packstone with fragments of red algae, larger foraminifera, echinoids, and benthic and planktonic foraminifera (RLa in Fig. 5B). Based on these components, this facies association is interpreted as being deposited on the middle sector of the ramp, as the RL depositional environment. Nevertheless, the increased percentage of planktonic foraminifera (Fig. 4 and Table 2) suggests relatively deeper conditions compared with the RL depositional environment. As cited above, a significant occurrence of detrital glauconite grains marks the boundary between the LST and TST (DS5). At this time, the NW and SE sectors of the ramp were rapidly subsiding as a result of the subduction-related downflexure of the Adria plate (Patacca *et al.*, 1991). Hardground and hemipelagic marls covering the shallow-water limestones, again represent the termination of this sequence also in the NE sector of the platform (DS6).

DISCUSSION AND CONCLUSION

From the Early to the mid-Late Miocene, the area, which now represents most of the Latium-Abruzzi Apennines, was a 100 km wide carbonate ramp. The stacking pattern of the facies belts indicates progradation towards the WNW; similar progradation has been observed in other Mediterranean platforms, such as the Apulia (Bosellini *et al.*, 1993) and Maiella Mountains (Mutti *et al.*, 1996; van Konijnbenburg *et al.*, 1999).

The Latium-Abruzzi ramp, like most Miocene Apennine platforms, is characterized by rhodalgal, bryomol and foramol skeletal-grain associations that have been used as indicators of non-tropical conditions in ancient platforms (Danese 1999; Vecsei & Sanders, 1999). In contrast, Brandano & Corda (2002) used a palaeoecological reconstruction based on biota (red algae and foraminifera) associated with rare *Porites* colonies, to suggest that the central Apennine shallow-water limestones were deposited under tropical conditions. Pomar *et al.* (2004) confirmed this reconstruction. However, the sequence architecture of the platform was strongly influenced by tectonics as well as by environmental parameters.

Lower sequence

During the late Aquitanian to Early Burdigalian, after a long hiatus accompanied by a significant erosional phase, Early Miocene sedimentation occurred during the final stage of a sea-level fall defined using the averaged isotope curve of Abreu

& Anderson (1998). In the Mediterranean region the Maiella platform records a Late Aquitanian erosional unconformity that bounds, at the base, a second-order sequence corresponding to the Upper Bryozoan Limestones (Bernoulli *et al.*, 1992; Vecsei & Sanders, 1999). A tectonic pulse should have enhanced the flooding of the Cretaceous Latium-Abruzzi platform.

The geometry and stacking pattern of the facies belts (Fig. 5) indicate aggradation (in the LST), followed by retrogradation and contemporaneous progradation above DS2 (during the TST), In particular, the proximal outer ramp is characterized by excess carbonate production during the TST, as testified by a rapid bloom of bryozoan-dominated facies. During the TST a significant change in the carbonate factory occurred. The demise of coral communities, the significant amount of other surviving tropical-subtropical biota among the coralline-algae and larger-foraminifera associations, and the bloom of bryozoans have been interpreted to be the result of high water fertility rather than a temperate palaeoclimate (Brandano & Corda, 2002). The TST, spanning between 17.5 and 16 Ma, is characterized by a wide expansion of bryomol-dominated lithofacies. Data on bulk-rock carbon-isotope compositions of the Miocene Latium-Abruzzi limestones show for the same time interval a positive shift in $\delta^{13}C$ values (from isotopic values of between 0–1‰, to values scattering around 2.1–2.2‰; Brandano *et al.*, 2004, 2010). Both observations (spread of suspension-feeding organisms, as well as a positive C-isotope excursion) may indicate increased surface water fertility.

Based on global deep-sea oxygen ($\delta^{18}O$) records for the Middle Miocene interval, Zachos *et al.* (2001) reported a warm phase that peaked between 17 and 15 Ma, as well as positive carbon ($\delta^{13}C$) isotope excursions suggestive of perturbations in the global carbon cycle. The increased $\delta^{13}C$ values of the Lower Miocene limestones outcropping in central and southern Italy, as well as certain features observed in Lower Miocene submarine hardgrounds of the Maiella platform, have been interpreted by Mutti *et al.* (1997, 1999) and Mutti & Bernoulli (2003) as a response to higher water fertility. The present results, coming from palaeoecological and C-isotope evidence, suggest that primary productivity in the central Mediterranean region should have increased significantly between approximately 17.5 Ma and 16 Ma.

Above DS3 (dated 16 Ma), during the subsequent highstand systems tract, Miocene sedimentation extended onto the more eastern sectors of the platform, where shallow-water sediments started to deposit on the Cretaceous limestones. This phase can be stratigraphically correlated with the Lower-Middle Miocene highstand event recognised in many sectors of the central Mediterranean region (Grasso *et al.*, 1994). The record of this event here supports the idea that it may represent an important tool for correlating the Middle Miocene deposits of the Mediterranean region in tectonically active areas. The HST shows a primary eustatic control, although this signal is enhanced in the western sectors of the platform (more internal with respect to the Apennine orogen) by an accelerated subsidence related to the early stages of downflexure of the subducting plate. The result is a thick aggradational HST in the shallower eastern platform that thins basinward in the deeper western sectors.

During the HST the significant spread of the intermediate outer-ramp facies belt throughout the platform parallels the decrease of bryozoans as the main sediment producers. This change in the carbonate factory is also accompanied by a return to lower $\delta^{13}C$ values (values ranging around 1‰; Brandano *et al.*, 2004, 2010). Both pieces of evidence may indicate decreased productivity of the surface waters. The positive $\delta^{13}C$ excursion highlighted in our samples (spanning from 17.5 and 16 Ma) is more restricted in time than the Monterey event, which occurred between approximately 18 and 13 Ma (Vincent & Berger, 1985). Our positive d13C shift has been interpreted as the effect of disturbance of several factors working together. Changes in the Mediterranean hydrological cycling and water circulation influenced by the voluminous volcanic activity and the closure of the Indo-Pacific connections, led to a dramatic increase in water fertility overlapping and partially masking the globally-recorded Monterey event (Brandano et al., 2010)

Upper sequence

A sea-level fall defined using the averaged isotope curve of Abreu & Anderson (1998) brought to an end (DS4) the first sequence recognized in the platform architecture. The lower boundary of the sequence is marked by a limited basinward shift of the facies belts across it. This observation is interpreted as being a consequence of accelerated tectonic subsidence that, superimposed on the sea-level fall, caused a moderate decrease in accommodation space. This subsidence acceleration was likely linked to the subduction-related downflexure of the Adriatic plate.

The entire geometry of the sequence is strongly influenced by tectonics. During this time interval, in the NW and SE sectors of the platform, only a LST developed before the final drowning. In contrast, the eastern sector experienced the aggradation both of the LST and the TST deposits before the drowning. Hence the platform experienced aggradation in the eastern sectors, while the western and more internal sectors underwent accelerated subsidence as a consequence of a rapid flexural stage before entering the foredeep. The sequence ends with the eastward-younging of the platform drowning, linked to the NE-directed migration of the Apennine foredeep system.

SUMMARY AND CONCLUDING REMARKS

The stratigraphic architecture of the Miocene platform within the central Apennines offers a case study that is useful for outlining the interactions between sea-level fluctuations, tectonics and trophic resources. Furthermore, the dynamic evolution of this carbonate system may contribute to the general knowledge of the environmental conditions in the Mediterranean region during the Early-Middle Miocene.

The Miocene Latium-Abruzzi platform was a WNW-dipping ramp that prograded towards the NW. The main carbonate producers were bryozoans, red algae and larger foraminifera (whose recognized genera are typical of tropical to subtropical climatic zones), plus bivalves and echinoids.

The sequence architecture of the Apennine platform, reconstructed using Sr-isotope dating of facies boundaries and, subordinately, biostratigraphy, mostly depends on the balance between changes in accommodation rates (sea-level and tectonic subsidence) and changes in trophic levels. Ages for discontinuity surfaces DS1, DS3 and DS4 and their correlations with the sea-level fluctuations on the averaged isotope curve of Abreu & Anderson (1998), suggest a primary eustatic control, although a tectonic overprint may, in some cases, be recognized.

The lower sequence, spanning from the Late Aquitanian to Serravallian, consists of a LST, a TST and a HST. The LST directly covered the Cretaceous substratum. The TST is characterized by a contemporaneous retrogradation and progradation of the proximal outer-ramp facies, testifying that the volume of sediment produced in the proximal sector of the outer ramp was particularly high. Such rates in the aphotic zone, which are quite unusual in tropical settings, were promoted by high trophic levels. The HST coincides with a new change in the main carbonate factory accompanied by a return to more depleted $\delta^{13}C$ concentrations. This change is interpreted here to be related to decreased water fertility.

Since that time, the platform was progressively involved in the eastward-migrating Apennine thrust belt-foredeep system which promoted the downflexure of the Adria plate, hence an acceleration of tectonic subsidence and then the progressive drowning of the platform (first in the western sectors and then in the eastern sectors).

The tectonic overprint is demonstrated by: (1) the HST geometry of the lower sequence, and particularly its small thickness and facies evolution observable in the western and southern sectors of the platform; (2) the poor evidence of the boundary between lower and upper sequence, because subsidence nearly balanced the eustatic fall; and (3) the incompleteness of the upper sequence (LST on the entire platform and TST only on the eastern sector) and its termination (hardground and hemipelagic marls) as a result of the progressive drowning into the foredeep.

ACKNOWLEDGEMENTS

We thank J. Reijmer and T. Simo for their critical reviews that greatly improved the manuscript. We are also grateful to G. Mariotti and C. Betzler for their constructive comments. This research was supported by MIUR (Progetti della Facoltà di Scienze M. F. N., Università "La Sapienza" di Roma e Programmi di rilevante interesse nazionale).

REFERENCES

Abreu, V.S. and **Anderson, J.B.** (1998) Glacial eustasy during the Cenozoic: sequence stratigraphic implications. *AAPG Bull.*, **82**, 1385–1400.

Accordi, G. and **Carbone, F.** (1988) Sequenze carbonatiche meso-cenozoiche. In: *Note Illustrative alla Carta delle Litofacies del Lazio-Abruzzo ed Aree Limitrofe* (Eds G. Accordi, F. Carbone, G. Civitelli, L. Corda, D. de Rita, D. Esu, R. Funiciello, T. Kotsakis, G. Mariotti and A. Sposato), *CNR Quad. Ric. Sci.*, **114**, 11–92.

Barbera, C., Simone, L. and **Carannante, G.** (1978) Depositi circalittorali di piattaforma aperta nel Miocene Campano, Analisi sedimentologica e paleoecologica. *Boll. Soc. Geol. Ital.*, **97**, 821–834.

Barbieri, M., Castorina, F., Civitelli, G., Corda, L., Mariotti, G. and **Milli, S.** (2003/2004) La sedimentazione di rampa carbonatica dei Monti Prenestini (Miocene inferiore, Appennino centrale): sedimentologia, stratigrafia sequenziale e stratigrafia degli isotopi dello stronzio. *Geol. Romana*, **37**, 79–95.

Bernoulli, D., Eberli, G.P., Pignatti, J.S., Sanders, D. and **Vecsei, A.** (1992) Sequence Stratigraphy of Montagna della Maiella. *Quinto Simposio di Ecologia e Paleoecologia delle Comunità Bentoniche. Paleobenthos: Roma, Libro-Guida delle escursioni*, pp. 85–109.

Bosellini, A., Neri, C. and **Luciani, V.** (1993) Platform margin collapses and sequence stratigraphic organization of carbonate slopes: Cretaceous-Eocene, Gargano Promontory, southern Italy. *Terra Nova*, **5**, 282–297.

Brachert, C.T., Betzler, C., Braga, J.C. and **Martin, J.M.** (1996) Record of climatic change in neritic carbonates: turnover in biogenic associations and depositional modes (Late Miocene, Southern Spain). *Geol. Rundsch.*, **85**, 327–337.

Brachert, C.T., Hultzsch, N., Knoerich, A.C., Krautworst, U. M.R. and **Stückrad, O.M.** (2001) Climatic signatures in shallow-water carbonates: high-resolution stratigraphic markers in structurally controlled carbonate buildups. *Palaeogeogr. Palaeoclimatol. Palaeoecol.*, **175**, 211–237.

Braga, J.C., Martín, J.M. and **Wood, J.L.** (2001) Submarine lobes and feeder channels of redeposited, temperate carbonate and mixed siliciclastic-carbonate platform deposits (Vera basin, Almeria, southern Spain). *Sedimentology*, **48**, 99–116.

Brandano, M. (2001) *Risposta fisica delle Aree di Piattaforma Carbonatica Agli Eventi piu Significativi del Miocene nell'Appennino Centrale.* Unpubl. PhD thesis, Univ. Rome "La Sapienza", 180 pp.

Brandano, M. and **Corda, L.** (2002) Nutrient, sea level and tectonics constrains for the facies architecture of Miocene carbonate ramp in Central Italy. *Terra Nova*, **14**, 257–262.

Brandano, M., Brilli, M., Corda, L. and **Mariotti, G.** (2004) Trophic control on the architecture of a Miocene carbonate ramp (central Italy): $\delta^{13}C$ fluctuations and biofacies assemblages. *23rd Meeting Int. Assoc. Sedimentol.*, Coimbra, Portugal, 15th–17th September 2004, p. 65.

Brandano, M., Corda, L. and **Mariotti, G.** (2005) Orbital forcing recorded in subtidal cycles from a lower Miocene siliciclastic-carbonate ramp system (central Italy). *Terra Nova*, **17**, 434–441.

Brandano, M., Brilli, M., Corda, L. and **Lustrino, M.** (2010) Miocene C-isotope signature from thecentral Apennine successions (Itali): Monterey vs. regional controlling factors. *Terra Nova*, **22**, 125–130.

Burchette, T.P. and **Wright, V.P.** (1992) Carbonate ramp depositional systems. *Sediment. Geol.*, **79**, 3–57.

Carannante, G. and **Simone, L.** (1996) Rhodolith facies in the central-southern Apennines mountains, Italy. In: *Models for Carbonate Stratigraphy from Miocene Reef Complexes of Mediterranean Regions.* (Eds E.K. Franseen, M. Esteban, W. Ward and J. Rouchy). *SEPM Concepts Sedimentol. Paleontol*, **5**, 261–275.

Carannante, G., Esteban, M., Milliam, J.D. and **Simone, L.** (1988) Carbonate lithofacies as paleolatitudeindicators: problems and limitations. *Sed. Geol.*, **60**, 333–346.

Carminati, E., Corda, L., Mariotti, G. and **Brandano, M.** (2007) Tectonic control on the architecture of a Miocene carbonate ramp in the central Apennines (Italy): insights from facies and backstripping analyses. *Sed. Geol.*, **198**, 233–253.

Cipollari, P. and **Cosentino, D.** (1995) Miocene unconformities in the Central Apennines: geodynamic significance and sedimentary basin evolution. *Tectonophysics*, **252**, 375–389.

Civitelli, G. and **Brandano, M.** (2005) Atlante delle litofacies e modello deposizionale dei Calcari a Briozoi e Litotamni nella Piattaforma carbonatica laziale-abruzzese. *Boll. Soc. Geol. Ital.*, **124**, 611–643.

Civitelli, G., Corda, L. and **Mariotti, G.** (1986a) Il bacino sabino: 2) sedimentologia della serie calcarea e marnoso spongolitica (Paleogene-Miocene). *Mem. Soc. Geol. Ital.*, **35**, 33–47.

Civitelli, G., Corda, L. and **Mariotti, G.** (1986b) Il bacino sabino: 3) evoluzione sedimentaria ed inquadramento regionale dall'Oligocene al Serravalliano. *Mem. Soc. Geol. Ital.*, **35**, 399–406.

Clari, P.A., de la Pierre, F. and **Martire, L.** (1995) Discontinuities in a carbonate succession: identification, interpretation and classification of some Italian examples. *Sed. Geol.*, **100**, 97–121.

Corda, L. and **Brandano, M.** (2003) Aphotic zone carbonate production on a Miocene ramp, central Apennines, Italy. *Sed. Geol.*, **161**, 55–70.

Cosentino, D., Carboni, M.G., Cipollari, P., Di Bella, L., Florindo, F., Laurenzi, A.M. and **Sagnotti, L.** (1997) Integrated stratigraphy of the Tortonian/Messinian boundary: the Pietrasecca composite section (central Apennines, Italy). *Eclogae Geol. Helv.*, **90**, 229–244.

Cunningham, K.J. and **Collins, L.S.** (2002) Controls on facies and sequence stratigraphy of an upper Miocene carbonate ramp and platform, Melilla Basin, NE Morocco. *Sed. Geol.*, **145**, 285–304.

Damiani, A.V., Chiocchino, M., Colacicchi, R., Mariotti, G., Parlotto, M., Passeri, L. and **Praturlon, A.** (1991) Elementi litostratigrafici per una sintesi delle facies carbonatiche meso-cenozoiche dell'Appennino centrale. *Stud. Geol. Camerti Vol. Spec.*, **1991/92**, CROP 11, 187–213.

Danese, E. (1999) Upper Miocene carbonates ramp deposits from the southernmost of Maiella Mountain (Abruzzo, central Italy). *Facies*, **41**, 41–54.

Esteban, M. (1996) An overview of Miocene reefs from Mediterranean areas: general trends and facies models. In: *Models for Carbonate Stratigraphy from Miocene Reef Complexes of Mediterranean Regions* (Eds E.K. Franseen, M. Esteban, W.C. Ward and J.M. Rouchy). *SEPM Concepts Sedimentol. Paleontol.*, **5**, 3–53.

Esteban, M., Braga, J.C., Martín, J. and **de Santisteban, C.** (1996) Western Mediterranean reef complexes. In: *Models for Carbonate Stratigraphy from Miocene Reef Complexes of Mediterranean Regions* (Eds E.K. Franseen, M. Esteban, W.C. Ward and J.M. Rouchy). *SEPM Concepts Sedimentol. Paleontol.*, **5**, 55–72.

Grasso, M., Pedley, H.M. and **Maniscalco, R.** (1994) The application of a late Burdigalian-early Langhian event in correlating complex tertiary orogenic carbonate successions within the central Mediterranean. *Géol. Méditerr.*, **21**, 69–83.

Hodell, D.A., **Mead, G.A.** and **Mueller, P.A.** (1990) Variation in the strontium isotopic composition of seawater (8 Ma to present): implications for chemical weathering rates and dissolved fluxes to the oceans. *Chem. Geol.*, **80**, 291–307.

Hodell, D.A., **Mueller, P.A.** and **Garrido, J.R.** (1991) Variations in the strontium isotopic composition of seawater during the Neogene. *Geology*, **19**, 24–27.

Lees, A. and **Buller, A.T.** (1972) Modern temperate-water and warm-water shelf carbonate sediments contrasted. *Mar. Geol.*, **13**, M67–M73.

Mariotti, G., **Corda, L.**, **Brandano, M.** and **Civitelli, G.** (2002) Cyclostratigraphy of Burdigalian deposits in the Ruffi Mountains (central Apennines). *Boll. Soc. Geol. Ital. Vol. Spec.*, **1**, 603–611.

Mateu-Vicens, G., **Hallock, P.** and **Brandano, M.** (2005) Amphistegina, Red Algae and Paleobathymetry: The Lower Tortonian Distally Steepened Ramp of Menorca, Balearic Islands (Spain). In: *Geologic Problem Solving with Microfossil* (Eds R.F. Waszczak and T.D. Demchuk). SEPM, Houston, Texas, March 6–11, 2005, p. 24.

Mutti, M. and **Bernoulli, D.** (2003) Early marine lithification and hardground development on a Miocene ramp (Maiella, Italy): key surfaces to track changes in trophic resources in nontropical carbonate settings. *J. Sed. Res.*, **73**, 296–308.

Mutti, M., **Bernoulli, D.**, **Eberli, G.P.** and **Vecsei, A.** (1996) Gepositional geometries and facies associations in an upper Cretaceous prograding carbonate platform margin (Orfento supersequence, Maiella, Italy). *J. Sed. Res.*, **66**, 749–765.

Mutti, M., **Bernoulli, D.** and **Stille, P.** (1997) Temperate carbonate platform drowning linked to Miocene oceanographic events: Maiella platforms margin, Italy. *Terra Nova*, **9**, 122–125.

Mutti, M., **Bernoulli, D.**, **Spezzaferri, S.** and **Stille, P.** (1999) Lower and Middle Miocene carbonate facies in the Central Mediterranean: the impact of paleoceanography on sequence stratigraphy. In: *Advances in Carbonate Sequence Stratigraphy: Application to Reservoirs, Outcrops and Models* (Eds P.M.M. Harris, A.H. Saller and J.A.T. Simo). *SEPM Spec. Publ.*, **63**, 371–384.

Nebelsick, J.H. (1989) Temperate water carbonate facies of the early Myocene Paratethys (Zogelsdorf Formation, lower Austria). *Facies*, **21**, 11–40.

Nelson, C.S., **Keane, S.L.** and **Head, P.S.** (1988) Non-tropical carbonate deposits on the modern New Zealand shelf. *Sed. Geol.*, **60**, 71–94.

Pampaloni, L., **Pichezzi, M.R.**, **Raffi, I.** and **Rossi, M.** (1994) Calcareous planktonic biostratigraphy of the marne a Orbulina Unit (Miocene, Italy). *Giorn. Geol.*, **56**, 139–153.

Patacca, E. and **Scandone, P.** (1989) Post-Tortonian mountains building in the Apennines. The role of the passive sinking of a relic lithospheric slab. In: *Tthe lithosphere in Italy. Advances in Earth Science Research*, (Eds. A. Boriani, M. Bonafede, G.B. Piccardo and G.B. Vai). *It. Nat. Comm. Int. Lith. Progr., Mid-term Conf. (Rome, 5–6 May 1987), Atti Conv. Lincei*, **80**, 157–176.

Patacca, E., **Scandone, P.**, **Bellatalla, M.**, **Perilli, N.** and **Santini, U.** (1991) La zona di giunzione tra l'arco appenninico settentrionale e l'arco appenninico meridionale nell'Abruzzo e nel Molise. *Stud. Geol. Camerti Vol. Spec.*, **1991/92**, CROP 11, 417–441.

Pedley, H.M. (1996) Miocene reef distribution and their associations in the central Mediterranean region: an overview. In: *Models for Carbonate Stratigraphy from Miocene Reef Complexes of Mediterranean Regions* (Eds E.K. Franseen, M. Esteban, W. Ward and J. Rouchy). *SEPM Concepts Sedimentol. Paleontol.*, **5**, 73–87.

Pérès, J.M. and **Picard, J.** (1964) Nouveau manuel de bionomie benthique de la Mer Méditerranée. *Trav. Stat. Mar. Endoume-Marseille. Fasc. Hors. Ser. Suppl.*, **31**, 5–138.

Pomar, L. (2001a) Ecological control of sedimentary accommodation: evolution from a carbonate ramp to rimmed shelf, Upper Miocene, Balearic Islands. *Palaeogeogr. Palaeoclimatol. Palaeoecol.*, **175**, 249–272.

Pomar, L. (2001b) Types of carbonate platforms: A genetic approach. *Basin Res.*, **13**, 313–334.

Pomar, L. and **Ward, W.C.** (1995) Sea-level changes, carbonate production and platform architecture: the Llucmajor Platform, Mallorca, Spain. In: *Sequence Stratigraphy and Depositional Response to Eustatic, Tectonic and Climatic Forcing* (Ed. B.U. Haq), pp. 87–112. Kluwer Academic Press, Dordrecht, The Netherlands.

Pomar, L., **Obrador, A.** and **Westphal, H.** (2002) Sub-wavebase cross-bedded grainstones on a distally steepened carbonate ramp, upper Miocene, Menorca, Spain. *Sedimentology*, **49**, 139–169.

Pomar, L., **Brandano, M.** and **Westphal, H.** (2004) Environmental factors influencing skeletal grain sediment associations: a critical review of Miocene examples from the western Mediterranean. *Sedimentology*, **51**, 627–651.

Riegl, B. and **Piller, W.E.** (1999) Framework revisited: Reefs and coral carpets of the nothern Red Sea. *Coral Reefs*, **18**, 305–316.

Riegl, B. and **Piller, W.E.** (2000) Biostromal coral facies - A Miocene example from the Leitha Limestone (Austria) and its actualistic interpretation. *Palaios*, **15**, 399–413.

Simone, L. and **Carannante, G.** (1985) Evolution of a carbonate open shelf from inception to drowing: the case of the Southern Appennines. *Rendiconti Acc. Scienze Fisiche e Matematiche, Serie IV*, Napoli, **53**, 1–43.

van Konijnenburg, J.H., **Bernoulli, D.** and **Mutti, M.** (1999) Stratigraphic architecture of a lower tertiary carbonate base-to-slope succession: Gran Sasso d'italia (central Apennines, Italy). In: *Carbonate Sequence Stratigraphy: Application to Reservoirs, Outcrops and Models* (Eds P. M.M. Harris, A.H. Saller and J.A.T. Simo). *SEPM Spec. Publ.*, **63**, 291–315.

Vecsei, V. and **Sanders, D.** (1999) Facies analysis and sequence stratigraphy of a Miocene warm-temperate carbonate ramp, Montagna della Maiella, Italy. *Sed. Geol.*, **123**, 103–127.

Vincent, E. and **Berger, W.H.** (1985) Carbon dioxide and polar cooling in the Miocene: the Monterey hypothesis. In: *The Carbon Cycle and Atmospheric CO_2: Natural Variations Archean to Present* (Eds E.T. Sundquist and W.S. Broeccker). *Am. Geophys. Union Monogr.*, **32**, 455–468.

Zachos, J., **Pagani, M.**, **Sloan, L.**, **Thomas, E.** and **Billups, K.** (2001) Trends, rhythms, and aberrations in global climate 65 Ma to present. *Science*, **292**, 686–693.

Int. Assoc. Sedimentol. Spec. Publ. (2010) **42**, 129–148

Facies and stratigraphic architecture of a Miocene warm-temperate to tropical fault-block carbonate platform, Sardinia (Central Mediterranean Sea)

MERLE-FRIEDERIKE BENISEK*[1], GABRIELA MARCANO†, CHRISTIAN BETZLER* and MARIA MUTTI†

*Geologisch-Paläontologisches Institut der Universität Hamburg, Bundesstr. 55, D-20146 Hamburg, Germany (E-mail: merben@statoil.com)
†Universität Potsdam, Institut für Geowissenschaften, Karl-Liebknecht-Strasse 24, Haus 25, D-14476 Golm, Germany

ABSTRACT

A Miocene (Burdigalian) carbonate platform in Northern Sardinia was studied in order to unravel its facies and stratigraphic architecture. The Sedini Limestone unit formed at the western margin of the Pèrfugas sub-basin on a fault-bounded topographic high. The 10–60 m sedimentary succession contains two depositional sequences separated by a major erosional unconformity, and several high-frequency sequences reflecting the occurrence of higher-order base-level fluctuations. The deposits are massive limestones, bedded limestones and marlstones, which form a carbonate platform with an extension of about 19 km^2. Sequence 1 consists of a homoclinal ramp with beaches, minor patch reefs, longshore bars, and finer-grained outer-ramp sediments. Sequence 1 deposits were formed by a warm temperate carbonate factory with abundant coralline algae, frequent larger benthic foraminifera (*Heterostegina*, *Amphistegina*, *Borelis*), barnacles, bryozoans, molluscs, and minor corals. Sequence 2 documents a steepening of the carbonate platform slope. In the lower part of this sequence, a belt of submarine dunes separated the platform-interior from deeper water bioclastic deposits. Dunes were locally stabilized by coralline algal bindstones. In the upper part of Sequence 2, the depositional system consisted of an extensive reef flat with a marked slope break formed by coralline algal bindstones and rhodolithic clinoform beds dipping at up to 27°. The carbonate factory of Sequence 2 is rich in reef-building zooxanthellate corals and therefore can be unequivocally assigned to a tropical carbonate factory. Spectacular outcrops in the Sedini Limestone unit allow a detailed observation of the facies, the bedding geometries, and the stratigraphic architecture of carbonates formed at the transition from the temperate to the tropical realm.

Keywords Miocene, carbonate platform, coralline red algae, Northern Sardinia, Mediterranean Sea.

INTRODUCTION

In neritic carbonates, palaeoceanography or palaeoclimate determine the type of carbonate factory (Carannante *et al.*, 1988; Nelson, 1988; Jones & Desrochers, 1992; James, 1997; Schlager, 2005), with ambient water temperature as the major driver defining the carbonate-producing associations (James, 1997; Schlager, 2005), but with additional factors such as nutrient input (Hallock & Schlager, 1986) also being important. The tropical carbonate factory is dominated by photo-autotrophic organisms; symbiont-free heterotrophs are common but non-diagnostic (Schlager, 2005). In the cool-water factory, heterotrophic organisms dominate, and nutrient levels are generally higher than in the tropical factory. The transition between the tropical and the cool-water factories is gradual, and was a termed warm-temperate carbonate province by Betzler *et al.* (1997).

A modern example of warm-temperate carbonate sediments was described by Collins *et al.* (1993) and James *et al.* (1999) from southwestern Australia. Seagrass and macrophytes are rooted in sediments rich in coralline algae and larger benthic

[1]Present address: Statoil ASA, Forusbeen 50, N-4035 Stavanger, Norway

foraminifera above coralline-encrusted hardgrounds. Heterotroph elements such as bryozoans, molluscs, and small foraminifera are abundant. *Halimeda* is poorly calcified and is not a significant carbonate sediment source. Coral reefs occur as small fringing reef complexes, but also as shelf-edge reefs embedded in rhodolith gravel. Another example of modern warm-temperate sedimentation are the carbonates in the Gulf of California reported by Halfar *et al.* (2004). Here, carbonate is produced in pocket bays and is dominated by rhodoliths, with frequent molluscs and coral patches.

Lower and Middle Miocene Mediterranean neritic carbonate facies have biotic associations resembling these recent warm-temperate assemblages (Esteban, 1996; Pedley, 1996; Brandano, 2003). The Neogene carbonates contain abundant heterozoan elements with intercalations of photozoan deposits (e.g. Pedley, 1996; Cherchi *et al.*, 2000; Galloni *et al.*, 2001). An intermediate, subtropical water temperature for these deposits can be deduced from the coral species and diversity. According to Rosen (1999), Mediterranean coral associations indicate minimum sea surface water temperatures of 19–21 °C for the Aquitanian and Burdigalian, and of 18–20 °C for the Langhian and Serravallian. This is the cool end of the tropical province and the transitional zone to the warm-temperate province.

Here, facies models and the sequence stratigraphy of Burdigalian warm-temperate and tropical carbonates deposited in the Sardinia rift basin (Martini *et al.*, 1992; Cherchi *et al.*, 2000; Bassi *et al.*, 2006) are described. Carbonate sedimentation is patchy and locally mixed with siliciclastic sediments. Outcrop quality is excellent. The carbonate factory mainly consists of coralline algae, bryozoans, molluscs, larger benthic foraminifera, and corals (Cherchi *et al.*, 2000; Simone *et al.*, 2001; Vigorito *et al.*, 2006). Coral diversity is low, with *Porites*, *Tarbellastrea*, *Montastrea*, *Favites*, and *Thegioastrea* (Cherchi *et al.*, 2000; Galloni *et al.*, 2001). The carbonates were interpreted as temperate and subtropical in origin by Murru *et al.* (2001), Vigorito *et al.* (2005, 2006) and Bassi *et al.* (2006).

The purpose of this contribution is to report on the facies of these warm-temperate carbonates, and to show their stratigraphic architecture. A turnover in platform geometry from a ramp, *sensu* Burchette & Wright (1992), to a rimmed platform with slope angles of 25° is accompanied by a change from a warm-temperate to a tropical carbonate factory.

GEOLOGICAL SETTING AND STRATIGRAPHY

The Neogene sediments were deposited in halfgrabens that had been formed in the Oligo-Miocene. Halfgraben formation was related to the counter-clockwise rotation of the Corsica-Sardinian block due to the opening of the western Mediterranean back-arc basin and the subduction of Neotethyan oceanic crust to the east of Sardinia (e.g. Cherchi & Montadert, 1982; Thomas & Gennesseaux, 1986; Faccenna *et al.*, 2002; Speranza *et al.*, 2002). In Northern Sardinia, the multiphase structural movements, accompanied by extensive volcanism, led to the formation of two N–S oriented halfgraben systems. The western branch consists of the Porto Torres Basin in the north, separated from the southern Logudoro Basin by a transfer zone (Thomas & Gennesseaux, 1986; Funedda *et al.*, 2000; Fig. 1). The eastern branch contains the Castelsardo Basin to the north and the Pèrfugas Basin to the south (Thomas & Gennesseaux, 1986; Sowerbutts, 2000). Faults in the Castelsardo Basin dip to the west, whereas they dip to the east in the Pèrfugas Basin. Both areas are separated by a range of hillocks, located between the small town of Castelsardo and the village of Sedini (Fig. 1), which may indicate buckling in an accommodation zone (Faulds & Varga, 1998). Basins in Northern Sardinia contain marine Miocene siliciclastics and carbonates (Pomesano Cherchi, 1971; Arnaud *et al.*, 1992; Martini *et al.*, 1992; Sowerbutts, 2000; Monaghan, 2001; Vigorito *et al.*, 2006).

Based on outcrop studies in the Castelsardo and Pèrfugas basins, Sowerbutts (1997, 2000) distinguished three tectonic phases. The first, Late Oligocene phase is the rift formation, followed by a Late Chattian to Early Aquitanian second phase of W–E to NW–SE oriented faulting. A third phase of N–S to NNW–SSE striking faulting occurred during the Early Burdigalian. The relief, which was flooded during the Burdigalian, was shaped by the third tectonic phase and the deposition of a series of dacitic ignimbrites between 19 and 17 Ma (Lecca *et al.*, 1997). The ignimbrites are overlain by the Pèrfugas Formation (Sowerbutts, 2000; i.e. "Lacustre" *Auct.*), which consists of volcanoclastics and lacustrine sediments. Siliciclastic

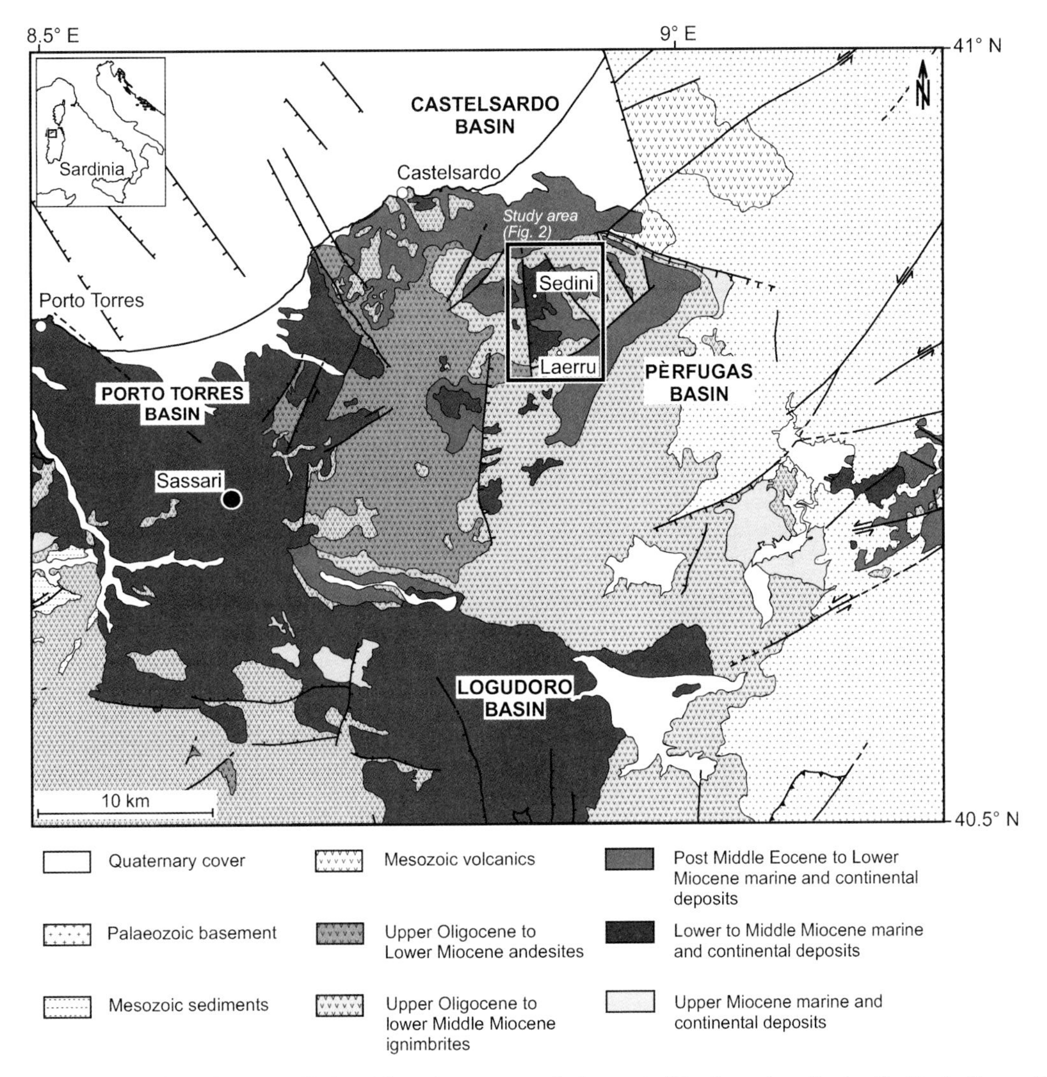

Fig. 1. Geological map of Northern Sardinia with major structural elements. The Logudoro Basin, the Porto Torres Basin, the Castelsardo Basin, and the Pèrfugas Basin are halfgrabens with Neogene sedimentary infills. Based on Thomas & Gennesseaux (1986), Funedda *et al.* (2000), Sowerbutts (2000) and Carmignani *et al.* (2001).

sediments were deposited subsequently in a fluvial to alluvial environment.

Marine carbonates, termed the Sedini Limestone (Thomas & Gennesseaux, 1986), overlie the lacustrine and alluvial deposits and the volcanics (Figs. 2 and 3). Arnaud *et al.* (1992) assigned the carbonates to the TB 2.1 and TB 2.2 cycles of Haq *et al.* (1988), i.e. latest Aquitanian and Burdigalian. However, according to Sowerbutts (2000), the Sedini Limestone is Late Burdigalian to Langhian in age (18–15 Ma). Calcareous nannoplankton associations indicate that the deposits were formed in the upper part of NN4 and the lowermost part of NN5 (C. Müller, pers. comm. 2006). The Sedini Limestone unit consists of two depositional sequences separated by an erosional unconformity. Deposits are massive limestones, bedded limestones and marlstones that form a carbonate platform with an extension of approximately 19 km^2 (Fig. 2). This study analyses the Burdigalian to Langhian limestone succession (Fig. 1) in the Pèrfugas Basin.

THE SEDINI LIMESTONE UNIT

The Sedini Limestone (Thomas & Gennesseaux, 1986) is a 10–60 m thick succession situated between Sedini and Laerru, south of Castelsardo (Fig. 1). The stratigraphic architecture of the deposits is shown in Fig. 3. Fig. 2 presents a detailed geological map. Overlying the volcanic

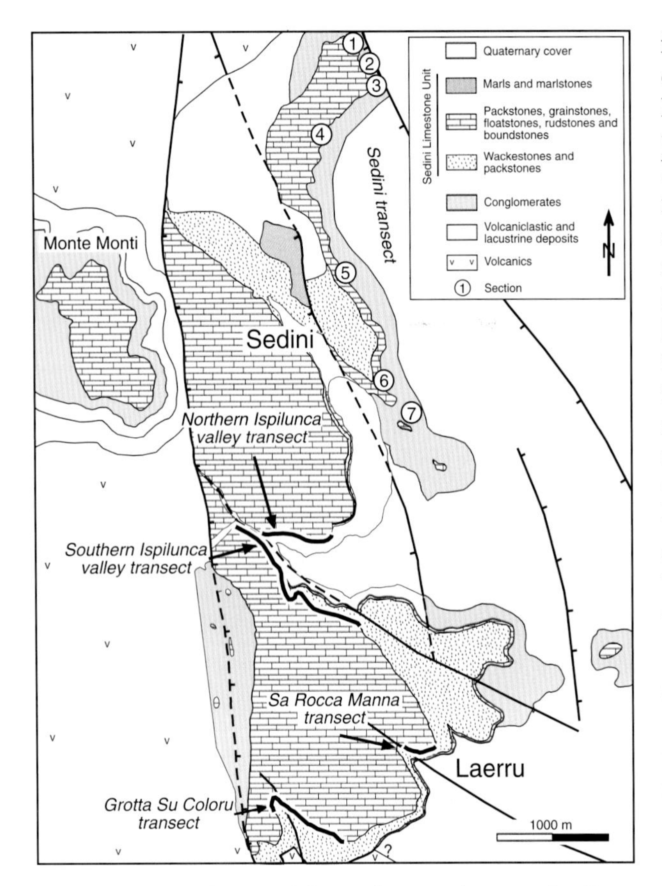

Fig. 2. Geological map of the study area showing the distribution of main lithologies, with the locations of the transects discussed in the text. Numbers refer to sections included in the Sedini transect. Based on Albertsen (2005), Gienapp (2005) and Nagel (2005).

rocks, there is a volcanoclastic and lacustrine unit, followed by cross-bedded fluvial conglomerates. Marine deposits are carbonates, which consist of packstones to rudstones with boundstones, and fine-grained wackestones and packstones. A small carbonate platform is dissected by a series of NNW–SSE oriented faults (Fig. 2). Offsets indicate postsedimentary activity of these faults. It cannot be excluded that these movements occurred as reactivations of older structures because the fluvial conglomerates, and the lowermost marine carbonates, wedge out along faults. Such a process of local fault reactivation is imaged in the seismic line published by Thomas & Gennesseaux (1986) in the Castelsardo Basin to the north. In this line, the sequence that corresponds to the Sedini Limestone unit drapes a fault block relief, but some of the faults also propagate into this sequence. Mapping shows that the youngest marine deposits in the study area are partially silicified bedded marls and marlstones with frequent sponge spicules, benthic foraminifera and fish remains located to the NNE of the village of Sedini (Fig. 2).

SELECTED TRANSECTS

Geometries, facies and stratigraphy of the carbonate platform are well exposed in a series of ravines cross-cutting the carbonate body. For this study, five transects along valleys and cliffs were selected (Fig. 2). Photographic panoramas were taken from four of the transects and, where accessible, sections were measured. In the case of the Sedini transect, no such photographic panorama could be made, owing to the wide extent of this transect and the degree of vegetation cover.

Ispilunca valley transects

The Ispilunca valley cross-cuts the Sedini Limestone unit in a NW–SE direction (Fig. 2). The

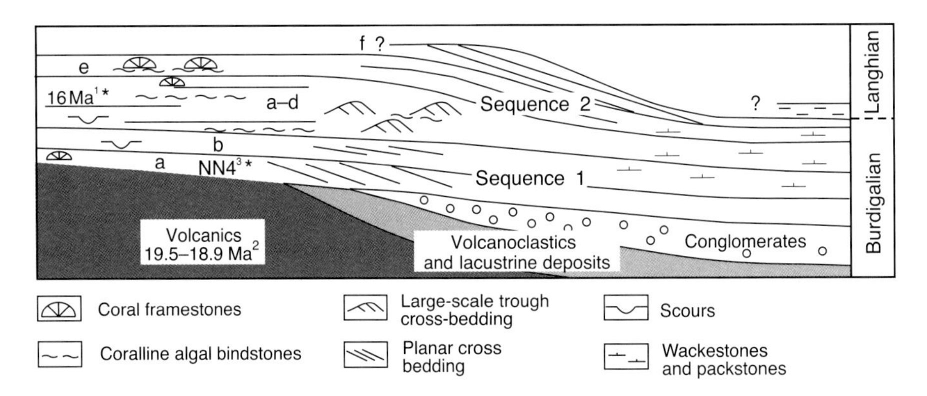

Fig. 3. Miocene stratigraphy in the northern part of the Pèrfugas Basin in the area of Sedini and Laerru. Age information compiled from: (1) Sowerbutts (1997); (2) Beccaluva *et al.* (1985); (3) calcareous nannoplankton (C. Müller, pers. comm. 2006). Discrete sediment packages are denoted by lower case letters. Not to scale.

northern and southern cliffs of this valley provide excellent insight into the internal geometries of most of the succession.

Northern Ispilunca Valley Transect

The transect (Fig. 2) is roughly oriented NNW–SSE, and is approximately 40 m high and 520 m long. It contains the lower and middle part of the Sedini Limestone unit, the basal contact between the conglomerates and the carbonates is not exposed. The cliff is partially accessible and four sections were measured. The locations of these sections (S5–S8) are shown in Fig. 4.

Section S5 is 5 m thick and comprises 4.20 m of floatstones and rudstones with large-scale straight-crested foresets arranged in simple sets (Anastas *et al.*, 1997) at its base. The dip of the cross beds is to the east. The floatstones and rudstones contain frequent branches of coralline algae, rhodoliths (3–5 cm) and *Heterostegina* and *Amphistegina* as the main components. Other constituents include bivalves, echinoids and bryozoans. The top of the cross-bedded interval is truncated (surface indicated by Roman numeral I in Fig. 4) and is overlain by a 30–40 cm thick interval with medium-scale trough cross-bedding. This interval is followed by a 40–50 cm thick floatstone with abundant rhodoliths and barnacles. A further floatstone to rudstone interval with large-scale straight-crested foresets overlies the rhodolith floatstone. One erosive surface (I), divides Section S5 into two sequences. A detail of this situation is shown in Fig. 5. To the SSE, the upper cross-bedded interval interfingers with poorly lithified wackestones and packstones.

Section S6 is 4.50 m thick and contains 1.50 m of poorly lithified bioturbated wackestones and packstones with echinoids and bivalves at its base. Above a sharp surface, indicated as "II" in Fig. 4, a 3 m thick interval of packstones and grainstones with abundant echinoid debris follows. To the SSE, these packstones and grainstones grade into wackestones and packstones. To the NNW, the packstones and grainstones onlap surface II. Here, the deposits display small-scale trough cross-bedding and cut-and-fill-structures. Farther to the NNW, surface II is directly overlain by 2 m thick coralline algal bindstones (Fig. 6), which contain encrusting forms of *Lithophyllum*, *Mesophyllum* and *Neogoniolithon*. Other components are bivalves (e.g. *Spondylus*, *Pecten* and *Isognomon*), bryozoans, larger benthic foraminifera, encrusting foraminifera, echinoid spines, and complete echinoids.

Section S7 is 18.50 m thick and starts with a 50 cm thick bioturbated wackestone and packstone with complete echinoids, rare small rhodoliths, and bivalves. Above a sharp surface (III, Fig. 4), 4 m of packstones and grainstones with coralline

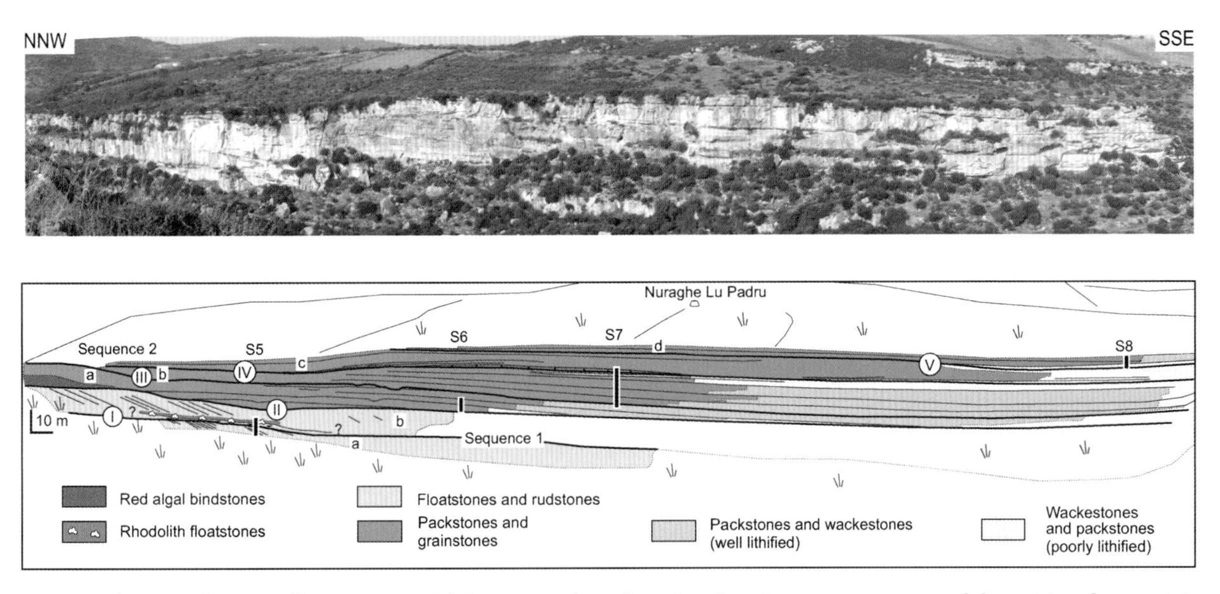

Fig. 4. Northern Ispilunca valley transect with interpretation of erosional surfaces, sequences, and depositional geometries. Bars and numbers S5–S8 indicate positions of lithological sections. Roman numerals refer to erosional surfaces. Discrete sediment packages are denoted by lower case letters.

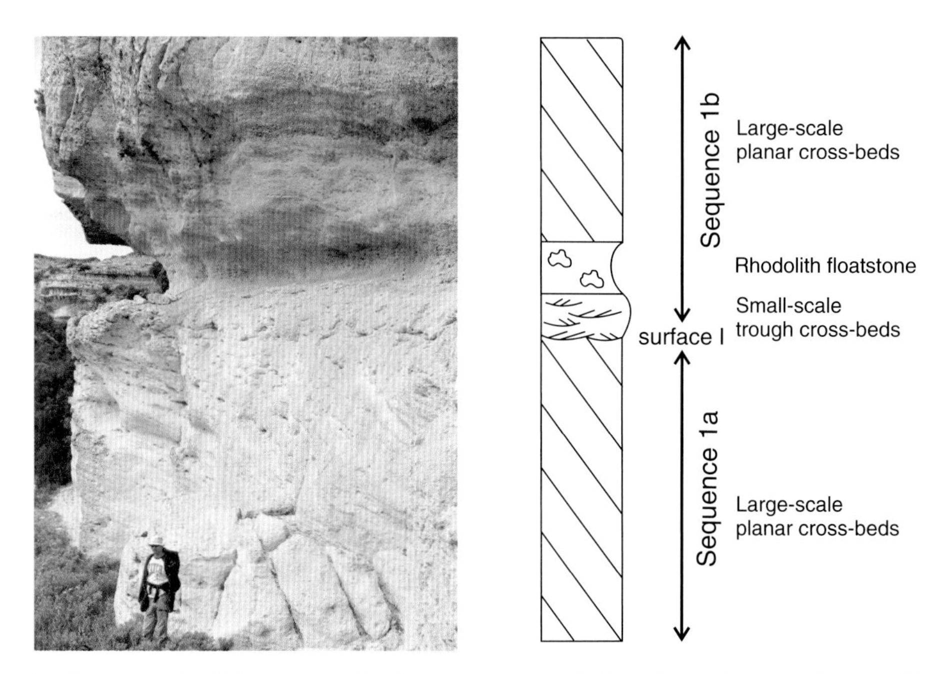

Fig. 5. Large-scale planar cross-bedding in deposits of sequences 1a and 1b in the northern Ispilunca valley transect. Note that the topmost part of the lower planar cross-beds is reworked and overlain by an interval with trough cross-bedding. The boundary between sequences 1a and 1b is placed at the base of the trough cross-bedded interval.

algal branches, echinoid and bivalve debris, bryozoans, *Heterostegina*, *Amphistegina*, and rare coral fragments follow (Fig. 7). To the NNW, surface III is characterized by channel-shaped cut-and-fill structures. Above this interval, 14 m of packstones and grainstones (with intercalated floatstones and rudstones) with rhodoliths, pectinid fragments, barnacles, bryozoans, some coral fragments, and miliolid foraminifera occur. A sharp surface (IV, Fig. 4) is overlain by a 50 cm thick and 60 m wide lense of floatstones and rudstones with rhodoliths and oysters. Overlying packstones and grainstones drape this lense. To the SSE, the deposits of S7 grade into strongly bioturbated bedded wackestones and packstones with fragmented bivalves, bryozoans, echinoids, and complete irregular echinoids, which occur mostly at the base of fining-upward cycles.

Fig. 6. Coralline algal bindstone of sequence 2a. Outcrop is located at the northern termination of the Ispilunca valley (S6 in Fig. 4). Pen is 14 cm long.

Fig. 7. Packstone with the larger benthic foraminifera *Amphistegina*, and debris of the larger benthic foraminifera *Heterostegina*, bivalve debris, coralline algal debris and other bioclasts. Sequence 2b, northern Ispilunca valley transect, Section S7 (Fig. 4).

As shown in Fig. 4, these deposits show different degrees of lithification ranging from well to poorly lithified. Poorly lithified limestones are finer grained and less sorted than the well-lithified deposits.

Section S8 is 10 m thick. The base is surface V (Fig. 4). Above this surface, 1.50 m of bioturbated wackestones and packstones occur, followed by 50 cm of a rhodolith floatstone and 8 m of packstones and grainstones with abundant echinoid fragments, coral fragments, and coralline algal debris. To the SSE, the packstones and grainstones grade into well-lithified wackestones and packstones.

Interpretation

In the northern Ispilunca valley transect, six sediment packages occur, subdivided by five surfaces. The most prominent of these is surface II, an erosional feature cutting down into the underlying deposits up to 5 m. This surface is interpreted as a sequence boundary triggered by a relative sea-level fall dividing the succession of the northern Ispilunca valley into two sequences (1 and 2). Furthermore, both sequences consist of high-frequency lower-order sequences (1a and b and 2a–d).

Large-scale cross-bedded floatstones and rudstones in sequences 1a and 1b are interpreted to have formed in submarine bars. Bars migrated to the east, the floatstones and rudstones interfinger with finer-grained, bioturbated deposits. Preservation of bioturbation and the occurrence of well-preserved complete echinoids are interpreted to represent an environment below regular wave-base.

In the upper part of sequence 1a, the well-defined contact between the straight foresets and the trough cross-bedded interval is interpreted to reflect a change of the hydrodynamic conditions. Migration of the bar system was interrupted, and the top of the bar was reworked. Trough cross-beds indicate that this reworking has been triggered by wave action. It is proposed that this is a consequence of a wave-base lowering, most probably triggered by a relative sea-level fall.

A subsequent deepening of the depositional system is witnessed by the rhodolith floatstones, which locally drape the underlying trough cross-bedded deposits. Restoration of a submarine bar system is indicated by the occurrence of the straight crested bars overlying the coralline algal interval.

Sequence 2 is subdivided into four higher frequency sequences (2a–d) by surfaces III–V. The lowermost deposits are infills of the erosional topography created by the sequence boundary (II). Coralline algal bindstones, which directly overlie the sequence boundary in the westernmost part of the outcrop, reflect a deepening after the relative sea-level lowstand, which triggered formation of the sequence boundary. According to Bosence (1991), the occurrence of the coralline algal genera *Lithophyllum*, *Mesophyllum* and *Neogoniolithon* in the coralline algal bindstones of sequence 2a points to water depths between 20 and 40 m or even shallower.

The lateral and downdip change from limestones to marlstones and marls indicates a change from hydrodynamically more agitated to quieter waters, which is interpreted as a deepening from a position around storm wave-base to below storm wave-base towards the SSE. This is indicated by the cut-and-fill-structures in the packstones and grainstones of sequences 2a and b, which are interpreted to be caused by storms. The equivalent, finer grained packstones and wackestones occurring in the SSE are well bedded and locally contain tempestite layers and are therefore thought to have been deposited in a deeper and quieter environment below storm wave-base.

Southern Ispilunca valley transect

The southern Ispilunca valley transect (Fig. 2) is oriented NW–SE, has a length of around 1500 m, and is approximately 40–50 m high. The steep cliff is only accessible at a few places and therefore mainly depositional geometries are shown in Fig. 8. The surfaces were correlated visually with the ones occurring in the northern Ispilunca valley transect at the NW termination of the valley, where it is only 50 m wide.

The basal contact of the limestones to the underlying conglomerates is not exposed in this transect. In the lowermost portion of the cliff (1a), floatstones and rudstones comprising large-scale cross-beds with straight foresets dipping to the east are exposed (Fig. 8). The floatstones and rudstones are composed of echinoid and coralline algal debris, with *Amphistegina* as the main component. Additional constituents are bivalves and bryozoans. Sediment package 1b physically corresponds to the second set of cross-beds observed in the northern Ispilunca valley transect. However, it is a massive to poorly bedded sediment package in this transect. The two units are separated by surface I, which cannot be traced farther to the SE because of the vegetation cover.

Above surface II (Fig. 8), which displays a SE-oriented downward flexure, a thick package with large-scale trough cross-bedding (2a + 2b) occurs. In the NW, this surface is directly overlain by coralline algal bindstones. In the southeastern part of this package, there is evidence for a dividing surface, which cannot be traced farther to the NW. This surface is tentatively correlated to surface III of the previously described northern Ispilunca transect. The large-scale trough cross-beds are truncated by surface IV. This surface is cut by the recent topography and cannot be traced farther to the NW. However, it displays an inclined geometry. The following unit (2c) displays no visible sedimentary structures, but drapes the moderate slope defined by the lower bounding surface IV. Surface V is the lower bounding surface of a gently inclined wedge-shaped sediment package (2d), which in its upper portion contains small coral reefs associated with oyster framestones at their base. Deposits of the package 2e overlying surface VI are clinoforms dipping with up to 25° to the ESE. Clinoforms contain abundant rhodoliths, and locally are rhodolith rudstones. Upslope, clinoform floatstones and rudstones interfinger with coralgal

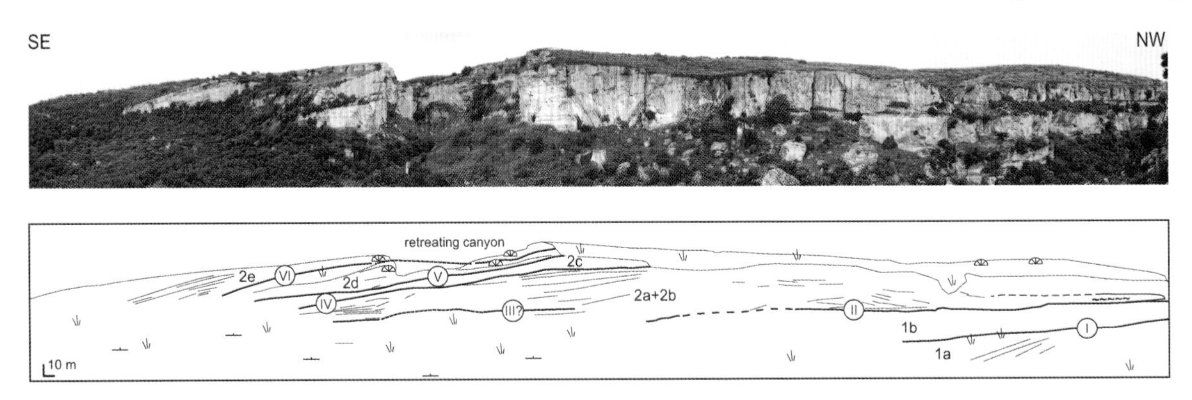

Fig. 8. Southern Ispilunca valley transect with position of the sequence boundary (II) and of the high-frequency sequence limits (roman numerals). As most of this outcrop is not accessible, no lithological interpretation is provided. The position of accessible coral reefs is indicated (coral framestones; key in Fig. 3). Discrete sediment packages are denoted by lower case letters.

Fig. 9. Coral framestone of sequence 2e at the northern end of the southern Ispilunca valley transect (Fig. 8).

boundstones up to 6 m thick. Corals form crusts alternating with crustose coralline algae. These coral deposits occupy the entire plateau at the top of the cliff, laterally extending for almost 1 km, and locally forming framestones (Fig. 9).

Interpretation

Surface II subdivides this succession into two sequences (1 and 2). The erosional character of the sequence boundary, however, is more pronounced in the northern Ispilunca valley transect. Above the sequence boundary, a gradual increase in inclination of both surfaces and depositional geometries documents a steepening of the depositional relief throughout the evolution of the carbonate platform. Steepening coincides with the inception of boundstones, for example the coralline bindstones of sequence 2a + b and the coral reefs of sequences 2d and 2e. In addition to coralline algal bindstones, sequence 2a + b consist of large-scale trough cross-bedded packstones and grainstones, which are interpreted as a system of submarine dunes.

During deposition, a platform geometry formed with submarine dunes at the platform margin locally stabilized by boundstones and framestones. At the latest stage of carbonate platform development (sequence 2e), the depositional system consisted of a reef flat with a marked slope break and fore-reef bioclastic and rhodolithic limestones. The increase of slope inclination appears much more pronounced in the southern than in the northern Ispilunca valley transect. This is the consequence of transect orientation, which in the northern transect is oblique to the platform slope.

Sa Rocca Manna transect

This transect is located along a cliff 1.8 km SSE of the Ispilunca Valley (Fig. 2). Being almost completely accessible, it was investigated in detail, and seven sections were measured. The WNW–ESE oriented cliff is 200 m wide and 20 m high (Fig. 10).

In its lower part, the cliff consists of a 17 m thick interval of packstone and grainstone beds dipping at up to 15° to the east. These deposits grade into flat-bedded biodetritic wackestones and packstones, which farther to the east grade into finer grained deposits and marlstones. Packstones contain fragments of echinoids, bivalves, as well as branches of coralline algae. Molluscs are preserved as moulds, mostly with micritic rims. Coral detritus is a minor component. The amount of coralline algal debris increases upsection. Wackestones and packstones are rich in benthic foraminifera, bryozoans, and serpulids. Marlstones contain

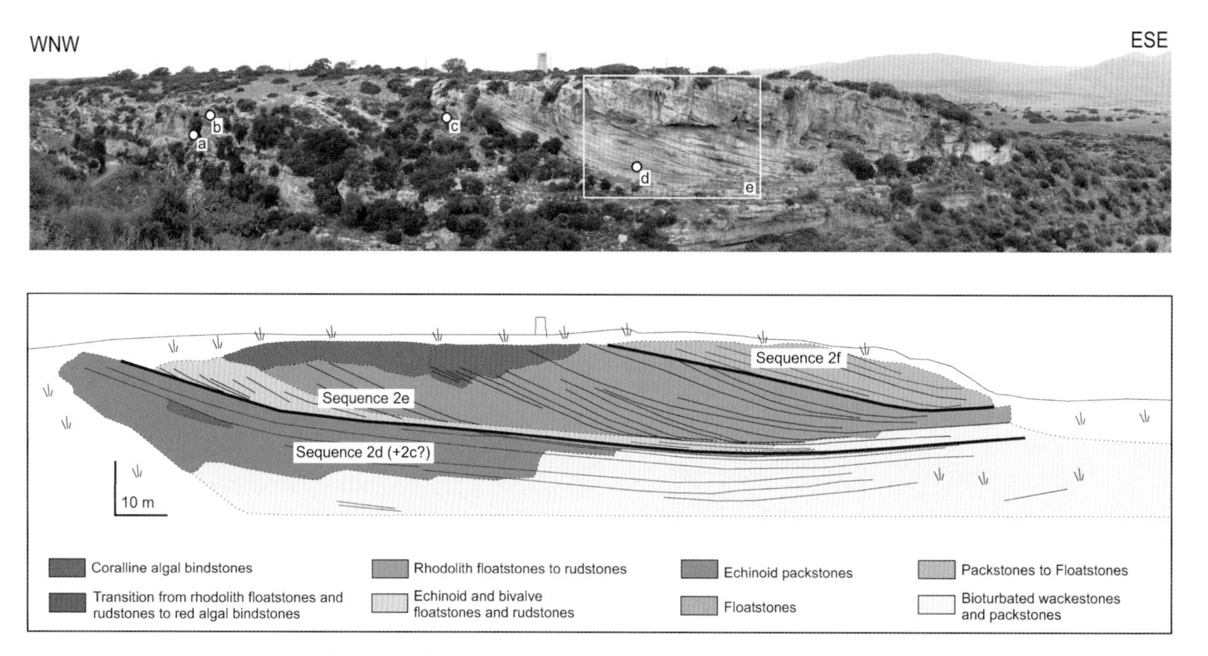

Fig. 10. Sa Rocca Manna transect (Fig. 2) with interpretation of stratigraphic architecture in high-frequency sequences 2d–f. Letters a–e in the photograph refer to detail photographs in Fig. 11.

planktonic foraminifera, echinoids, bryozoans, and bivalves, which are preserved as moulds. In the zone with inclined bedding, 2 m thick and 5–6 m wide coralline algal bindstone lenses with *Lithophyllum*, *Mesophyllum* and *Sporolithon* occur (Fig. 10). A 50 cm thick bioclastic floatstone bed with bivalves, rhodoliths, and coral fragments overlies the echinoid packstones and the flat-bedded wackestones and packstones. The bed can be traced over a distance of approximately 110 m (Fig. 10).

The top of this lower sedimentary unit is located at a surface, where a significant increase in dip of the beds up to 15–20° occurs (Figs. 10 and 11A). Above this surface, a 30 cm thick layer of fine-grained packstones is overlain by floatstones and rudstones with abundant echinoid and bivalve debris, as well as debris of corals (Fig. 11B). The fine-grained packstones are the bottomset deposits of the steeply dipping coarse-grained limestones.

Floatstones and rudstones are overlain by a 2 m thick coralline algal bindstone in the NW part of the outcrop. To the SE, topset bindstones are enriched in rhodoliths (Fig. 10). Downslope, this bindstone grades into a rhodolith rudstone arranged in 20 m high clinoforms dipping at up to 27°. A detailed view of the topsets of the clinoforms is shown in Fig. 11C. Topset beds have wavy surfaces. Rhodoliths are ellipsoidal to discoidal, and are up to 10 cm in diameter. They are closely packed and have been fused in a later stage by continued coralline algal encrustation.

Clinoform beds have complex internal geometries, and each of the clinoform beds is a composite body (Fig. 11D) with subhorizontal internal layering, similar to the rudimentary stratification described in rhodolithic deposits in New Zealand (Burgess & Anderson, 1983). Some of the clinoform beds wedge out upslope (Fig. 11E). The bottomset of the clinoforms is depicted in Fig. 11E. Each clinoform bed terminates in a 5–10 cm thick layer of bivalve rudstone grading into an echinoid floatstone away from the slope. In the bivalve rudstone, well-preserved pectinids occur. The bottomsets climb up through this sediment package.

Along the clinoforms, rhodolith size decreases downslope from approximately 10 to 5 cm. Upslope, rhodoliths predominate, while towards the bottomsets the amount of matrix between the rhodoliths increases. Rhodoliths in the upper part of the clinoforms are both sphaeroidal and ellipsoidal, the amount of ellipsoidal rhodoliths increasing downslope. In thin section, the rhodoliths are composed of alternating layers of coralline algae, encrusting foraminifera, and celleporiform bryozoans. Coralline algal growth is laminar, warty and fruticose, resulting in frequently branched rhodoliths and some laminar concentric rhodoliths (*sensu* Bosence, 1983a, and Woelkerling *et al.*, 1993). Rhodoliths with both growth patterns also occur.

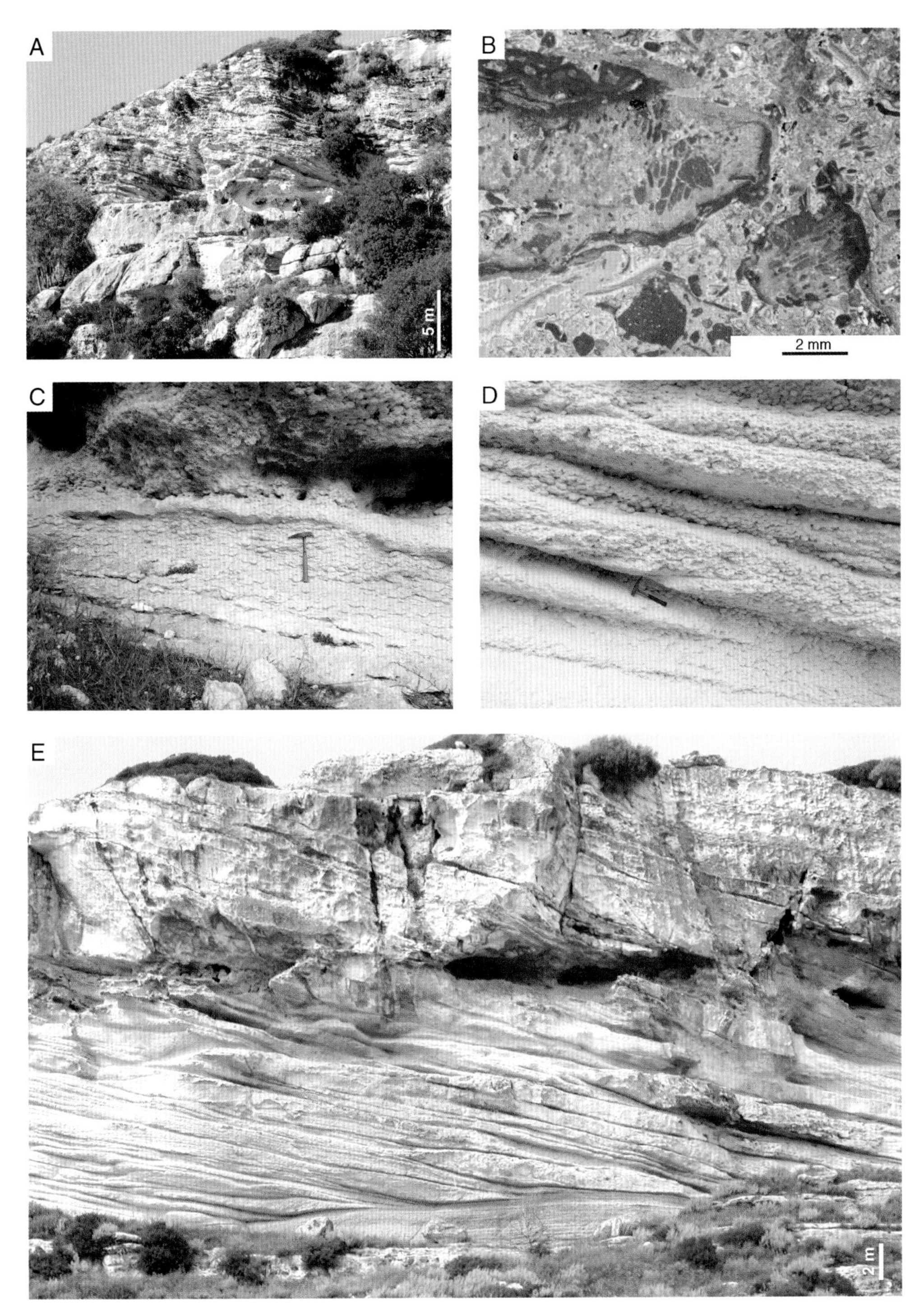

Fig. 11. Details of the Sa Rocca Manna transect. (A) Change from packstones and grainstones of sequence 2c/d to steeply dipping large-scale planar cross-bedded echinoid and bivalve floatstones and rudstones. The surface separating both units is the boundary between high-frequency sequences 2d and 2e. (B) Thin section photograph of the large-scale cross-bedded floatstones and rudstones with coralline algal encrusted coral debris. (C) Transition from a coralline algal bindstone to rhodolithic floatstones and rudstones in the topset area of the rhodolithic clinoforms. This facies forms the platform edge in sequence 2e. Hammer shaft is 30 cm long. (D) Rhodolithic clinoforms. Note subhorizontal internal layering. Hammer shaft is 30 cm long. (E) View of the clinoforms, which are approximately 20 m high. Note that each clinoform bed passes into planar floatstone bed only a few centimetres thick. There is no indication of slumping, channels or any type of redeposition in this transect. See Fig. 10 for detailed locations of photographs.

The upper limit of the sediment package is defined by a toplap configuration (Fig. 10). The youngest sedimentary unit in this transect is a bioclastic packstone to floatstone with large-scale planar cross-beds that are only preserved in a small wedge. Beds of this unit downlap onto the unit's lower bounding surface.

Interpretation

Physical tracing of sediment bodies and geometries between the two Ispilunca valley transect and the Sa Rocca Manna transect shows that this outcrop contains sequence 2 deposits. It is not currently unequivocally testable if the lower part of the cliff only contains sequence 2d, or if it also includes part of sequence 2c. Sequence 2e makes up the middle part of the succession. An additional high-frequency sequence 2f forms the youngest deposits of the cliff.

Facies and depositional geometries of sequence 2c/d are similar to the facies described in the Ispilunca valley transects, with coralline algal bindstones and large-scale trough cross-bedded submarine dune deposits. The bindstones are embedded in the large-scale trough cross-bedded deposits and are interpreted to act as sediment stabilizers. The limit between sequences 2c/d and e reflects a turnover of the carbonate factory and depositional system. The Sa Rocca Manna sequence 2e deposits are interpreted to have formed along the slope of the reef-bearing carbonate platform. Inner platform facies with coral framestones, however, are not exposed in Sa Rocca Manna, but 800 m NW of the outcrop.

Upper slope deposits are coralline algal bindstones grading into a rhodolithic slope (Figs. 10 and 11C). Subhorizontal internal layering of the composite clinoform bodies indicates that these bodies were not formed by sole off-platform shedding of rhodoliths. It is proposed that rhodoliths also grew *in situ* along the slope, which resulted in a subhorizontal internal layering. *In situ* rhodolith accumulation on slopes has been described in the large-scale rhodolithic clinobeds of Menorca (Obrador *et al.*, 1992; Pomar, 2001; Pomar *et al.*, 2002; Brandano *et al.*, 2005) and on the fore-reef slopes discussed by Bosence (1983b).

Toplaps of rhodolith clinoforms in the last progradational episodes of sequence 2e indicate post-depositional erosion before formation of sequence 2f. Deposits of sequence 2f are not discussed any further because of the poor control of this unit due to missing outcrops.

Grotta Su Coloru transect

This 1 km long cliff along the eastern flank of the valley with the Grotta Su Coloru (Figs. 2, 12A and D) has a NW–SE orientation and is 40–50 m high. Most of the cliff, which consists of the lower part of the Sedini Limestone unit, is not accessible. Therefore no sections were measured.

The base of the Sedini Limestone succession is not exposed in this cliff. At its northwestern termination, the lower part of the cliff consists of bedded packstones to wackestones (Fig. 12A–C). Above a sharp surface, these are overlain by low-angle planar cross-beds dipping toward the SE that interfinger with medium-scale trough cross-bedded grainstones and packstones (Fig. 12B). These grade into planar-bedded wackestones and fine-grained packstones toward the SE (Fig. 12A).

This succession is followed by a 3 m thick interval of finer grained bioturbated, bioclastic packstone (Fig. 12B). At the southeastern end of the Grotta Su Coloru transect, this interval thickens to 6.5 m of bioturbated wackestones and packstones with *Thalassinoides* burrows at the top (Fig. 12A). Fine-grained packstones are truncated by erosional incisions (Fig. 12B and C) that are 4–5 m broad, cutting down to 1.5 m. Infills of the erosive depressions consist of pectinid packstones to floatstones arranged in medium-scale trough cross-beds (Fig. 12B). Overlying the pectinid packstones to floatstones, there is a 3 m thick interval of large-scale low-angle planar cross-bedded packstones and grainstones dipping toward the NE (Fig. 12C).

Interpretation

Most of the Grotta Su Coloru transect consists of sequence 1 limestones, with the lower part of sequence 2 at the cliff top. High-frequency sequence 1a is separated from high-frequency sequence 1b by a sharp surface (Fig. 12B and C). Lateral facies variations in sequence 1b are typical for a beach depositional system (Fig. 12B). Low-angle planar cross-bedding indicates sedimentation in the foreshore area, trough cross-bedding is representative of the shoreface area. Planar bedded wackestones and fine-grained packstones with *Thalassinoides* burrows (Fig. 12A) are interpreted to have formed in a quieter hydrodynamic environment, below fairweather wave-base.

As the contact between the sequence 1a wackestones (Fig. 12A–C) and the overlying interval of sequence 1b beach deposits (Fig. 12B) is a sharp

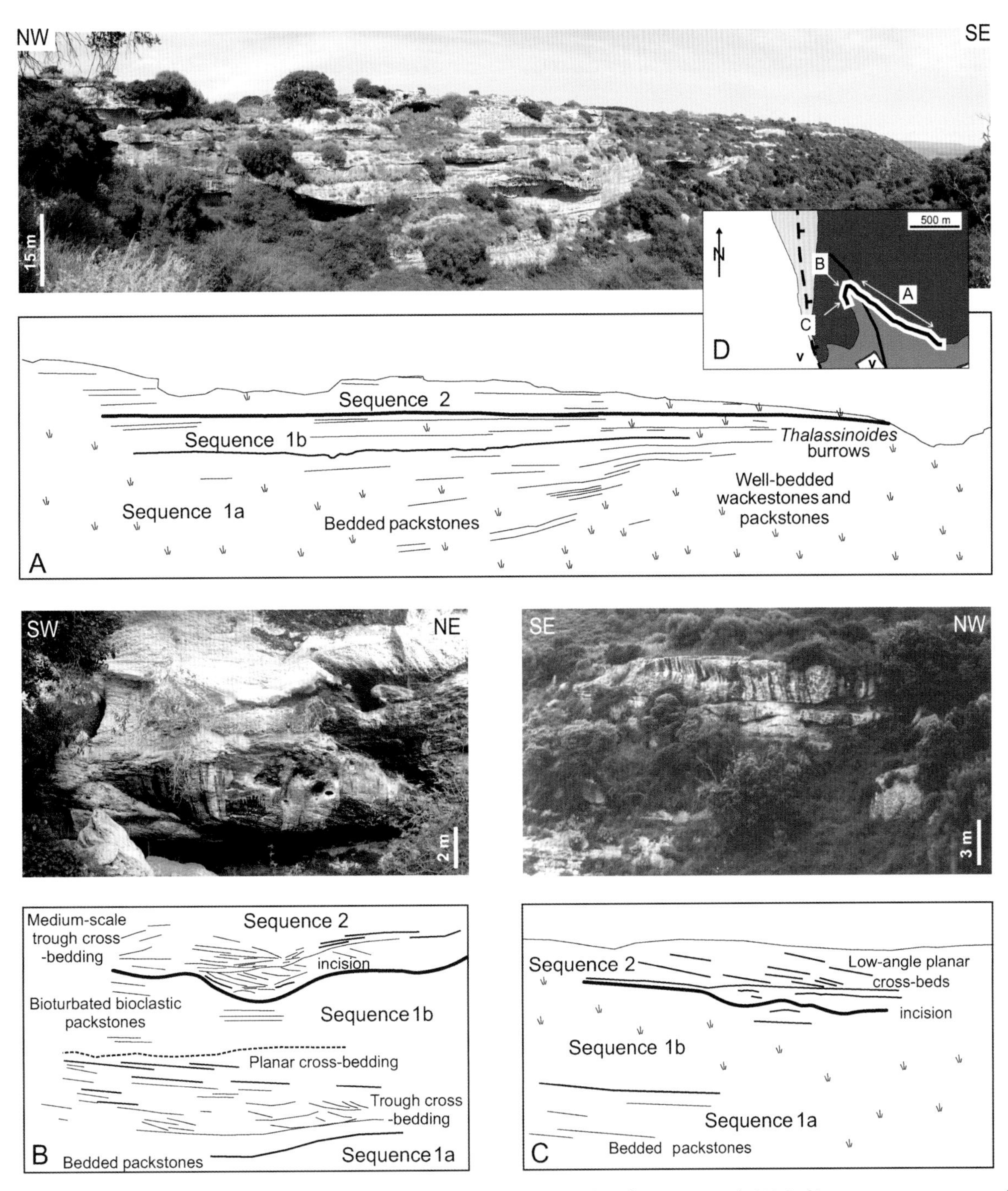

Fig. 12. Grotta Su Coloru transect. (A) Eastern valley cliff (approximate height is 40–50 m). (B) Sedimentary structures at the entrance of the Grotta Su Coloru. The height of the wall is approximately 11 m. (C) Channel incision and low-angle planar cross-bedding in packstones and grainstones in the upper part of the western valley cliff. Sediment package with cross beds is approximately 3 m thick. (D) Geological map of the Grotta Su Coloru valley (detail of Fig. 2) with location of photographs A–C.

surface, it is suggested that it reflects a forced regression. An increase of the accommodation space during deposition of sequence 1b is reflected by the turnover from beach sediments (Fig. 12B) to the bioturbated finer-grained wackestones to packstones above the trough cross-bedded deposits (Fig. 12B and C).

A renewed base-level lowering is indicated by the incision of channels (Fig. 12B and C) at the top of the bioturbated limestones. This limit is

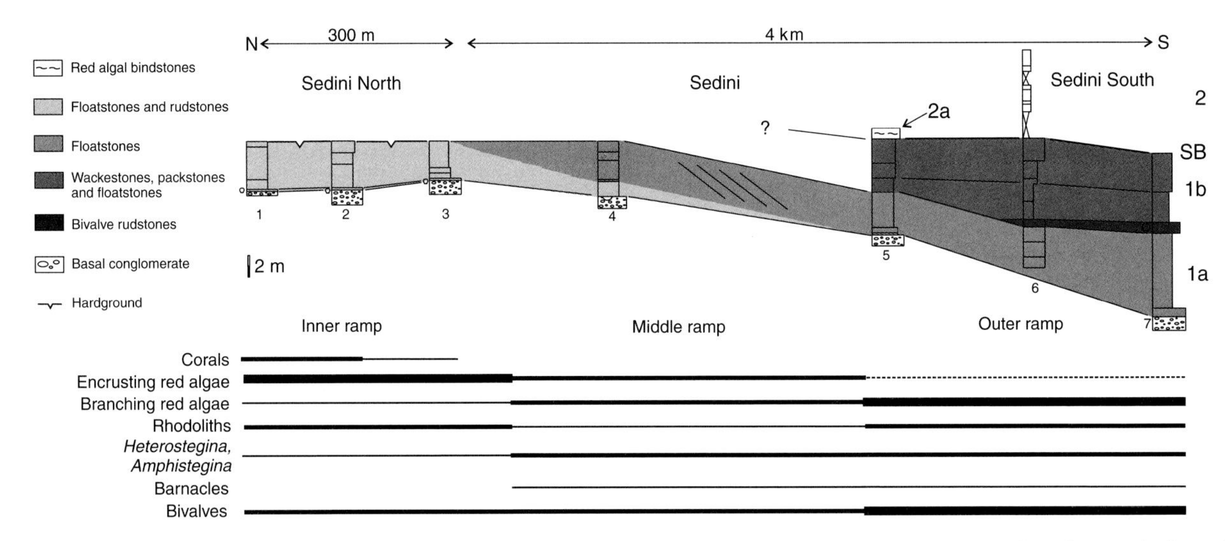

Fig. 13. Sedini transect. Location of lithological sections is shown in Fig. 2. Bars indicate the relative abundance of selected biota, ranging from rare to very abundant. Note the occurrence of a hardground at the top of the three northern sections. The coralline algal bindstone at the top of section 5 is interpreted as the lowermost part of sequence 2a.

interpreted to represent the limit between sequences 1 and 2. The later infill of these channels with the medium-scale trough cross-bedded pectinid packstones and floatstones show that the top of the succession at the Grotta valley transect records a base-level rise. Subsequent infill resulting in a reduction of accommodation space is indicated by low-angle planar cross-bedded packstones and grainstones (Fig. 12C). The sedimentary structures identify these deposits as the foreshore area of a beach depositional system.

Sedini transect

The Sedini transect is 4 km long, and runs in an approximate N–S direction along the cliff below the village of Sedini (Fig. 2). It is 6 m thick in the north and thickens to 17 m in the south. It was reconstructed using seven sections (Figs. 2 and 13). The transect of its length and vegetation cover is not documented by photographs.

Most of the sections record the transition between the Sedini Limestone unit and the underlying fluvial conglomerates. This transition consists of a conglomerate grading upsection into limestones with minor amounts of quartz, feldspar and biotite. The lower part of the limestones contains abundant reworked siliciclastics, with granite boulders up to 15 cm in size. In sections 1–3 (Fig. 13), a layer with thick-shelled oysters occurs in the upper part of the conglomerates. Overlying limestones are floatstones to rudstones containing coralline algal branches, rhodoliths, bivalves, echinoids, bryozoans, and barnacles. In section 1, deposits also contain coral fragments, and the large benthic foraminifera *Borelis* (Fig. 14A). Some

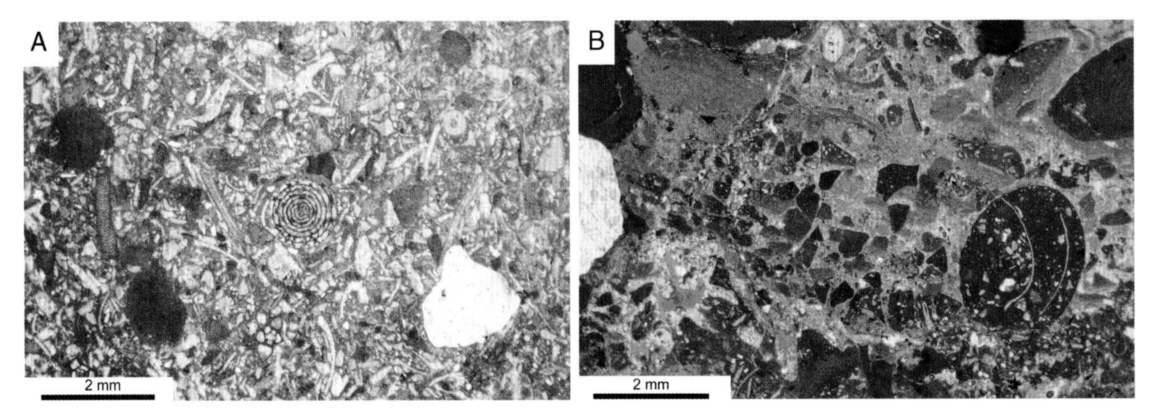

Fig. 14. Facies of sequence 1a in the northern part of the Sedini transect (Fig. 13). (A) Packstone with the larger benthic foraminifera *Borelis*, coralline algal debris and other bioclasts. Sequence 1a, section 1. (B) Floatstone with coralline algal debris, intraclasts and a *Lithophaga* boring. Sequence 1a, section 1.

horizons contain frequent intraclasts, and the top of the floatstones to rudstones is bored by *Lithophaga* (Fig. 14B).

The lower part of section 4 (Fig. 13) comprises floatstones and rudstones with bivalves, barnacles, echinoderm debris, and coralline algae. A 40 cm deep and 2 m broad incision infilled with oysters and barnacle fragments occurs in this interval. In the middle part of section 4, there is a layer with irregular echinoids. The upper part of section 4 is made of floatstones and rudstones with very abundant branches of coralline algae and encrusting coralline algae. In addition, bivalves, echinoids, and the larger benthic foraminifera *Amphistegina* and *Heterostegina* occur. To the south, the degree of fragmentation of the coralline algae increases.

Floatstones and rudstones dominated by coralline algal branches make up the lower parts of sections 5–7 (Fig. 15). Limestones in all sections are strongly bioturbated, with exception of an outcrop in the area between sections 4 and 5, where a 3.5 m thick body of large-scale cross-beds dipping at 15° to the east is exposed.

In the middle part of sections 6 and 7 (Fig. 13) there is a 1 m thick floatstone to rudstone layer with coarse bivalve detritus in section 6 and well-preserved thick-shelled oysters in section 7. In section 7, this layer also contains *in situ* barnacles and intraclasts. Barnacles and clasts are encrusted by coralline algae. Further bioclasts are branches of coralline algae, echinoids and larger benthic foraminifera (*Amphistegina* and *Heterostegina*). The floatstone to rudstone layer disappears between sections 6 and 5.

Above the rudstone to floatstone layer, in sections 6 and 7, well-sorted, bioclastic wackestones and packstones with abundant echinoid debris occur. Limestones coarsen upwards, and are rich in coralline algae in the upper part of the sections. The upper part of section 5 is made up of a coarsening-upward package of packstones and floatstones containing abundant coralline algal branches, rhodoliths and bivalves. The top of section 5 consists of a coralline algal bindstone. The upper part of section 6 contains a 7 m thick succession of wackestones (Fig. 13). The contact between the wackestones and the underlying floatstones is not exposed.

Interpretation

The Sedini transect consists of sequence 1 deposits, with a very minor part of sequence 2 in the upper part of sections 5 and 6 (Fig. 13). The limit between sequences 1a and 1b is below the floatstone to rudstone layer that contains thick-shelled oysters and *in situ* barnacles. In the northern part of the transect, this boundary is a hardground with *Lithophaga* borings. The hardground can be traced between sections 1 and 3.

The facies changes in sequence 1a are interpreted to reflect a N–S increase of the water depth along a ramp morphology. The inner ramp consists of shallow-water deposits, with coral fragments, red-algal encrustations, intraclasts and the larger

Fig. 15. Rudstone of coralline algal branches in sequence 1b in the Sedini transect, section 7. Diameter of coin 22 mm.

benthic foraminifera *Borelis*. The middle-ramp environment is characterized by the occurrence of the larger benthic foraminifera *Heterostegina* and *Amphistegina* accompanied by fragments of coralline algae. This facies resembles the *Calcarenite biancastra a macroforaminiferi* of Civitelli & Brandano (2005). This assemblage is typical for tropical and subtropical environments at water depths between 40 and 70 m (Brandano, 2003). In the middle to outer ramp, the occurrence of large-scale tabular cross-beds indicates the presence of longshore bars, similar to those in the northern transect of the Ispilunca valley. Incisions in the lower part of section 4 are interpreted as channel incisions, triggered by bottom currents.

Formation of a hardground at the top of sequence 1a attests to an interruption of sedimentation. Between sections 1 and 3, the ichnogenus *Gastrochaenolites* gives evidence for early lithification, in sections 6 and 7 *in situ* barnacles and thick-shelled oysters reflect the occurrence of a lithified sea floor. Above the hardground, fine-grained packstones and wackestones indicate sedimentation in relatively deep water. The lower sequence 1b is interpreted to reflect the consequence of a relative sea-level fall followed by a sea-level rise, leading first to the establishment of a hard-bottom community, and then to deposition of deeper-water finer-grained material. Coarsening-upwards of the overlying sediments of sequence 1b is interpreted as a shallowing of water depth.

Occurrence of the bindstone in the topmost part of section 5 shows a renewed formation of a stable substrate later colonized by encrusting coralline algae. The bindstone is interpreted to pertain to sequence 2a; the lower bounding surface of the bindstone is thought to correspond to the sequence boundary between sequences 1 and 2. Wackestones in the uppermost part of section 6 are outer platform deposits of sequence 2.

DISCUSSION

The Sedini Limestone unit is a well-exposed example of a fault-block carbonate platform (*sensu* Bosence, 2005) with a complex stratigraphic architecture. Several unconformities, the temporal changes in carbonate factory, and changes of the depositional geometries, subdivide the Sedini Limestone unit into two depositional sequences and several higher-frequency sequences (Fig. 16).

Carbonate platform facies and stratigraphy

The lower sequence 1 of the Sedini Limestone unit is a carbonate ramp depositional system that was deposited in an embayment opening to the SE (Fig. 17). The inner ramp consists of different facies associations. In the SW (Grotta Su Coloru transect), beach deposits formed, with foreshore and shoreface bedding passing into planar-bedded wackestones and packstones (Fig. 17). In the NE (Sedini transect) small patch reefs grew along the shoreline (Fig. 17). Corals are fragmented, and occur together with encrusting coralline algae, rhodoliths, and intraclasts. Reef debris deposits interfinger with floatstones and rudstones consisting of branches of coralline algae, bivalves, larger benthic foraminifera, and bryozoans. Sedimentary structures are small troughs and channels.

In the middle to outer ramp, deposits are floatstones and rudstones containing branches of coralline algae, small rhodoliths, fragmented echinoids,

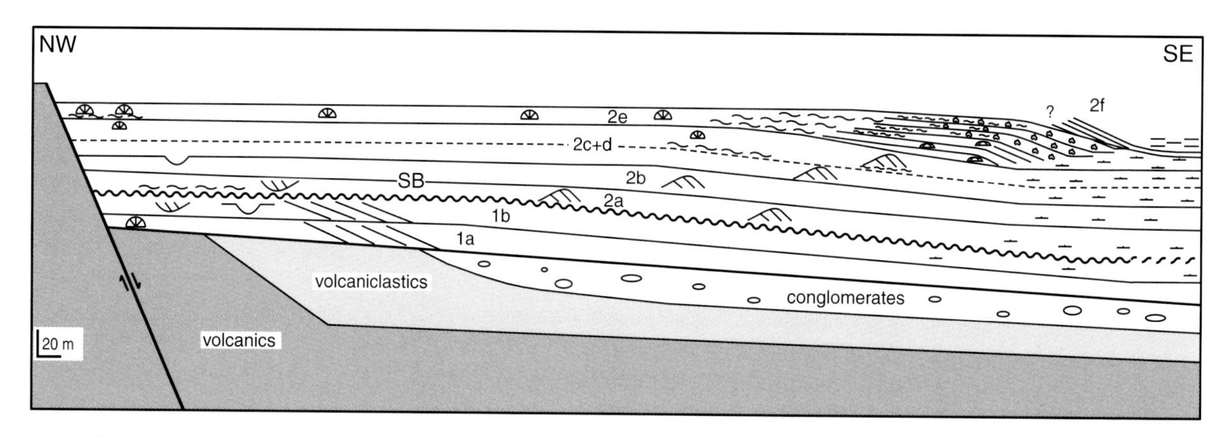

Fig. 16. Stratigraphic architecture of the Sedini Limestone unit, with a carbonate ramp in sequence 1 and a turnover from a ramp to a steep-flanked platform in sequence 2. Facies key in Fig. 3.

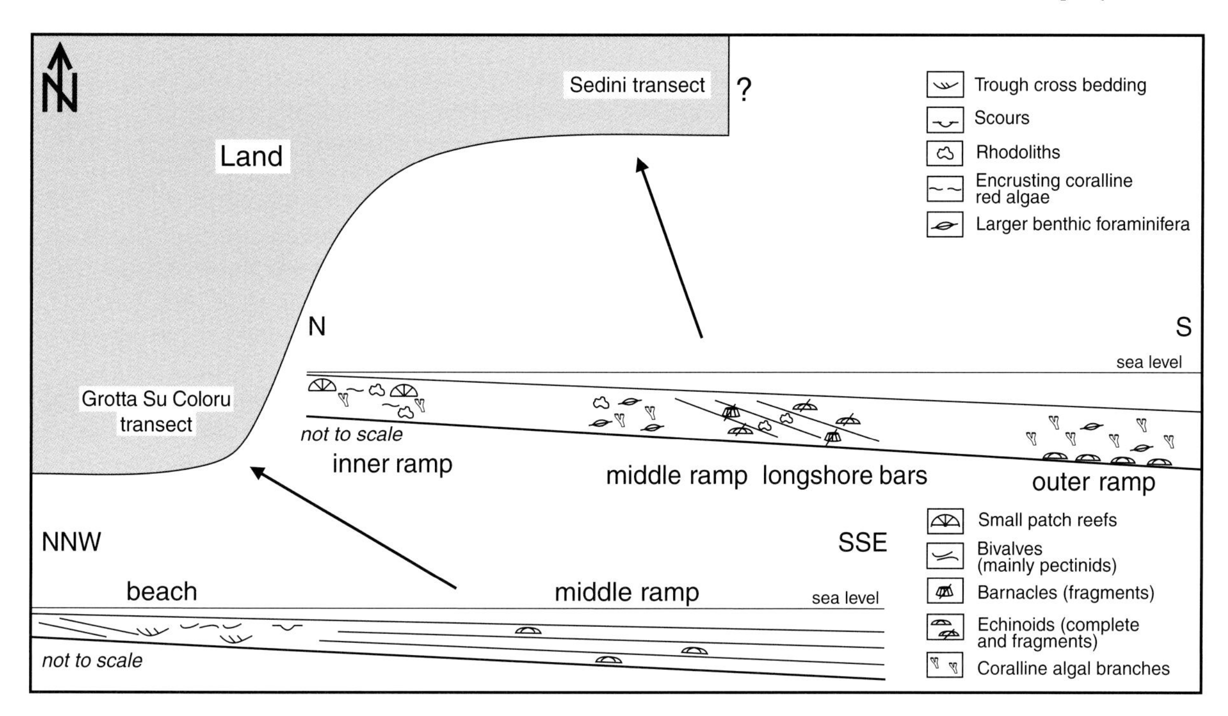

Fig. 17. Palaeoenvironmental interpretation of the study area during the formation of sequence 1. Not to scale.

and bivalves. Sedimentary structures are large-scale cross-beds with straight foresets, which are interpreted as longshore bars moved by east-directed currents. Subordinate directions are SE to NE. Small-scale trough cross-bedding in the topset area indicates partial reworking at the top of the bars by wave action. The outer ramp is represented by packstones and floatstones dominated by branches of coralline algae. Further bioclasts are echinoids, bryozoans and bivalves.

Deposits of sequence 2 document an evolution from the ramp to a platform with a defined slope break (Fig. 16). During the early stages of platform growth, in sequences 2a–2d, the inner platform facies consists of coralline algal bindstones. The lowermost coralline algal bindstone overlies the sequence boundary and forms a wedge that pinches out into a proximal direction (Fig. 8). In the mid-platform environment, submarine dunes occur. Dunes are locally stabilized by 0.4–0.5 m thick coralline algal bindstones (2a-2d). Off-platform, the dunes grade into bedded fine-grained bioclastic packstones and wackestones.

In sequences 2d and 2e, the inner platform contains coral reefs. Reef deposits are framestones, up to six metres thick (Fig. 9), and form a reef flat with a surface of approximately 4 km^2. Towards the platform edge, coralline algae are more abundant and form a bindstone at the slope break (Figs. 10 and 11C). Slope deposits at this platform stage are rhodolith rudstones arranged in a series of spectacular clinoform beds, 20 m high and dipping by up to 27° (Figs. 10, 11D and E). Although slope angles are steep, there are no breccia, slumps or debris-flow deposits, and each clinoform terminates in a floatstone and rudstone bed only a few centimetres thick (Fig. 11E). This minor off-platform shedding is interpreted as a consequence of the persistent coralline algal encrustation that stabilized the slope.

The latest stage of the platform evolution of the Sedini Limestone unit is only poorly recorded. In one of the transects (Sa Rocca Manna) it is documented that rhodolith formation at the slope ends at the upper limit of high-frequency sequence 2e. The youngest carbonates of sequence 2f are bioclastic deposits. Neritic carbonate sedimentation in the Pèrfugas Basin seems to have terminated after formation of this last platform sediment package. The last marine deposits of the area are marls with sponge spicules, which, however, crop out at only one locality overlying the fine-grained packstones and wackestones forming the basin deposits of sequence 2 (Figs. 2 and 3).

Turnover in geometry

Carbonate ramps and steep-flanked platforms are often closely related in time (e.g. Wilson, 1975;

Read, 1985; Bosence, 2005). A common succession in the geological record is the up-section change from ramps to reef-rimmed platforms at different scales ranging from small platforms of km-size structures (e.g. Martín & Braga, 1994; Brachert *et al.*, 1996) to bodies several hundreds of kilometres across such as the Great Bahama Bank (Eberli & Ginsburg, 1987, 1989; Betzler *et al.*, 1999). It has been proposed that palaeoceanographical or climate fluctuations trigger such geometrical changes (Martín & Braga, 1994; Betzler *et al.*, 1995; Brachert *et al.*, 1996), though ramp to platform turnovers may be largely a consequence of the stratigraphic architecture (Homewood, 1996). Landward stepping units, which form during base-level rises, have a residual bathymetry at the seaward end and, following seaward steps, will lead to clinoform formation. According to Schlager (2005), the turnover in geometries is a consequence of the tropical carbonate factory itself, as it has the tendency to prograde and steepen its slope. The ramp stage is understood as a transient stage due to delayed rim development during a rapid transgression.

In the Sedini Limestone, change of buildup geometry goes along with a change of the carbonate factory, and seems to be especially related to the advent of sediment binding by coralline algae in sequence 2. The lower sequence 1, in general, is dominated by coralline algae. Other representative components are larger benthic foraminifera (frequent *Heterostegina*, abundant *Amphistegina*, and rare *Borelis*), barnacles, bryozoans, and molluscs. Corals are scattered, and only recorded in the nearshore deposits of the Sedini transect. Micritic rims are rare in sequence 1 deposits (Marcano, 2008). These features assign the factory to the upper limit of the warm-temperate realm as defined by Betzler *et al.* (1997).

Similar Miocene carbonate depositional systems at the transition from the warm-temperate to the tropical realm were described from the Apennines by Brandano (2003) and Civitelli & Brandano (2005). They accumulate on carbonate ramps because the carbonate factory does not produce any wave- or current-resistant rim structure. The carbonate factory of sequence 2 is rich in reef-building zooxanthellate corals and therefore can be unequivocally assigned to a tropical carbonate factory (James, 1997). Coral reefs formed a reef flat, and extensive coralline algal encrustations stabilized the platform edge and slope. Prograding submarine dunes became locally stabilized by coralline algal encrustations and early cements (Marcano, 2008), and therefore increasingly acted as a wave-resistant rim as described in Schlager (2005), forming a rimmed platform in the late stage.

CONCLUSIONS

A sedimentological and stratigraphical model for a fault-block-carbonate platform is presented. This platform evolves from a ramp to a steep-flanked platform. The geometrical turnover goes along with a change of the carbonate factories from warm-temperate to tropical. The warm-temperate ramp contains small patch reefs, beaches, long-shore bars as well as outer-ramp bioclastic and red algal packstones to rudstones. The overlying steep-flanked platform stage contains coral reef framestones forming a reef flat, rhodolithic slope deposits and deeper-water peri-platform fine-grained carbonates. Rhodolith beds are spectacularly exposed, and facies can be traced from the topset to the bottomset of individual clinoform beds. Steepening of the depositional relief of the carbonate platform is gradual and linked to the inception of coralline algal bindstones. This study presents a further example for the close relation between the carbonate factory and the stratigraphic architecture of carbonate platforms.

ACKNOWLEDGEMENTS

Lucia Simone, Gabriele Carannante, and Marco Murru kindly introduced us to Sardinia and its Miocene carbonates. Giacomo Oggiano and Vincenzo Pascucci are thanked for their support. We especially want to thank Carla Müller for the determination of the calcareous nannoplankton. Funding by the Deutsche Forschungsgemeinschaft (grants Be 1272/12 and Mu 1680/6) is gratefully acknowledged. Reviews by Marco Brandano, Giacomo Oggiano and Werner Piller helped to improve the manuscript.

REFERENCES

Albertsen, J.C. (2005) *Kartierung des Miozäns bei Sedini und Fazies miozäner Karbonate bei Sedini/Nordsardinien.* Diplomarbeit, Universität Hamburg, 86 pp.

Anastas, A.S., Dalrymple, R.W., James, N.P. and **Nelson, C.S.** (1997) Cross-stratified calcarenites from New Zealand: subaqueous dunes in a cool-water, Oligo-Miocene seaway. *Sedimentology*, **44**, 869–891.

Arnaud, M., **Magné, J.**, **Monleau, C.**, **Négretti, B.** and **Oggiano, G.** (1992) Nouvelles données sur le Miocéne du Nord-Ouest de la Sardaigne (Italie). *CR Acad. Sci. Paris*, **315 I**, 965–970.

Bassi, D., **Carannante, G.**, **Monleau, C.**, **Negretti, B.** and **Oggiano, G.** (2006) Rhodalgal/bryomol assemblages in temperate type carbonate, channelized depositional systems: the Early Miocene of the Sarcidano area (Sardinia, Italy). In: *Cool-Water Carbonates: Depositional Systems and Palaeoenvironmental Controls* (Eds G. Carannante and H.M. Pedley). *Geol. Soc. London Spec. Publ.*, **255**, 35–52.

Beccaluva, L., **Civetta, L.**, **Macciotta, G.**, **Ottelli, L.** and **Ricci, C.A.** (1985) Geochronology in Sardinia: results and problems. *Rend. Soc. Ital. Min. Petrol.*, **40**, 57–72.

Betzler, C., **Brachert, T.C.** and **Kroon, D.** (1995) Role of climate in partial drowning of the Queensland Plateau carbonate platform (northeastern Australia). *Mar. Geol.*, **123**, 11–32.

Betzler, C., **Brachert, T.C.** and **Nebelsick, J.** (1997) The warm temperate carbonate province. A review in the facies, zonations and deliminations. *Cour. Forschungsinst. Senckenb.*, **201**, 83–99.

Betzler, C., **Reijmer, J.J.G.**, **Bernet, K.**, **Eberli, G.P.** and **Anselmetti, F.S.** (1999) Sedimentary patterns and geometries of the Bahamian outer carbonate ramp (Miocene and Lower Pliocene, Great Bahama Bank). *Sedimentology*, **46**, 1127–1146.

Bosence, D. (1983a) Description and classification of Rhodoliths (Rhodoids, Rhodolithes). In: *Coated Grains* (Ed T.M. Peryt), pp. 217–224. *Springer*, Berlin, Heidelberg.

Bosence, D. (1983b) The occurrence and ecology of Rhodoliths (Rhodoids, Rhodolithes). In: *Coated Grains* (Ed T.M. Peryt), pp. 225–242. *Springer*, Berlin, Heidelberg.

Bosence, D. (1991) Coralline algae: mineralization, taxonomy, and palaeoecology. In: *Calcarous Algae and Stromatolites* (Ed R. Riding), pp. 98–113. *Springer*, Berlin.

Bosence, D. (2005) A genetic classification of carbonate platforms based on their basinal and tectonic settings in the Cenozoic. *Sed. Geol.*, **175**, 49–72.

Brachert, T.C., **Betzler, C.**, **Braga, J.C.** and **Martín, J.M.** (1996) Record of climatic change in neritic carbonates: turnover in biogenic associations and depositional models (Late Miocene, southern Spain). *Geol. Rundsch.*, **85**, 327–337.

Brandano, M. (2003) Tropical/subtropical inner ramp facies in Lower Miocene "Calcari a Briozi e Litotamni" of the Monte Lungo area (Cassino Plain, Central Appenines, Italy). *Boll. Soc. Geol. Ital.*, **122**, 85–98.

Brandano, M., **Vannucci, G.**, **Pomar, L.** and **Obrador, A.** (2005) Rhodolith assemblages from the lower Tortonian carbonate ramp of Menorca (Spain): Environmental and palaeoclimatic implications. *Palaeogeogr. Palaeoclimatol. Palaeoecol.*, **226**, 307–323.

Burchette, T.P. and **Wright, V.P.** (1992) Carbonate ramp depositional systems. *Sed. Geol.*, **79**, 3–57.

Burgess, C.J. and **Anderson, J.M.** (1983) Rhodoids in Temperate Carbonates from the Cenozoic of New Zealand. In: *Coated Grains* (Ed T.M. Peryt), pp. 243–258. *Springer*, Berlin.

Carannante, G., **Esteban, M.**, **Milliman, J.D.** and **Simone, L.** (1988) Carbonate lithofacies as palaeolatitude indicators: problems and limitations. *Sed. Geol.*, **60**, 333–346.

Carmignani, L., **Barca, S.**, **Oggiano, G.**, **Pertusati, P.C.**, **Salvadori, I.**, **Conti, P.**, **Eltrudis, A.**, **Funedda, A.** and **Pasci, S.** (2001) Geologia della Sardegna, Note illustrativa della Carta Geologica della Sardegna in scala 1:200 000. *Mem. Descrit. Carta Geol. Ital.*, **60**, 283 pp. two maps.

Cherchi, A. and **Montadert, L.** (1982) Oligo-Miocene rift of Sardinia and the early history of the Western Mediterranean Basin. *Nature*, **298**, 736–739.

Cherchi, A., **Murru, M.** and **Simone, L.** (2000) Miocene carbonate factories in the syn-rift Sardinia graben subbasins (Italy). *Facies*, **43**, 223–240.

Civitelli, G. and **Brandano, M.** (2005) Atlante delle litofacies e modello deposizionale dei Calcari a Briozoi e Litotamni nella Piattaforma carbonatica laziale-abruzzese. *Boll. Soc. Geol. Ital.*, **124**, 611–643.

Collins, L.B., **Zhu, Z.R.**, **Wyrwoll, K.-H.**, **Hatcher, B.G.**, **Playford, P.E.**, **Eisenhauer, A.**, **Chen, J.H.** and **Bonani, G.** (1993) Holocene growth history of a reef complex on a cool-water carbonate margin: Easter Group of the Houtman Abrolhos, eastern Indian Ocean. *Mar. Geol.*, **115**, 29–46.

Eberli, G.P. and **Ginsburg, R.N.** (1987) Segmentation and coalescence of Cenozoic carbonate platforms, northwestern Great Bahama Bank. *Geology*, **15**, 75–79.

Eberli, G.P. and **Ginsburg, R.N.** (1989) Cenozoic progradation of northwestern Great Bahama Bank, a record of lateral platform growth and sea-level fluctuations. In: *Controls on Carbonate Platforms and Basin Development* (Eds P.D. Crevello, J.L. Wilson, J.F. Sarg and J.F. Read). *SEPM Spec. Publ.*, **44**, 339–351.

Esteban, M. (1996) An overview of Miocene reefs from Mediterranean areas: General trends and facies models. In: *Models for Carbonate Stratigraphy from Miocene reef complexes of Mediterranean regions* (Eds E.K. Franseen, M. Esteban, W.C. Ward and J.M. Rouchy). *SEPM Concepts Sedimentol. Palaeontol.*, **5**, 3–53.

Faccenna, C., **Speranza, F.**, **D'Ajello Caracciolo, F.**, **Mattei, M.** and **Oggiano, G.** (2002) Extensional tectonics on Sardinia (Italy): insights into the arc-back-arc transitional regime. *Tectonophysics*, **356**, 213–232.

Faulds, J.E. and **Varga, R.J.** (1998) The role of accommodation zones and transfer zones in the regional segmentation of extended terranes. In: *Accommodation Zones and Transfer Zones: The Regional Segmentation of the Basin and Range Province* (Eds J.E. Faulds and J.H. Steward). *Geol. Soc. Am. Spec. Pap.*, **323**, 1–45.

Funedda, A., **Oggiano, G.** and **Pasci, S.** (2000) The Logudoro Basin: a key area for the tertiary tectono-sedimentary evolution of North Sardinia. *Boll. Soc. Geol. Ital.*, **119**, 31–38.

Galloni, F., **Cornee, J.J.**, **Rebelle, M.** and **Ferrandini, M.** (2001) Sedimentary anatomies of early Miocene coral reefs in South Corsica (France) and South Sardinia (Italy). *Géol. Mediterr.*, **28**, 73–77.

Gienapp, H.P. (2005) *Die miozänen Karbonate im Gebiet Nulvi, Region Anglona (N-Sardinien).* Unpubl. Diplomarbeit, Universität Hamburg, 78 pp.

Halfar, J., **Godinez-Orta, L.**, **Mutti, M.**, **Valdez-Holguin, J.** and **Borges, J.** (2004) Nutrient- and temperature controls on modern carbonate production: an example from the Gulf of California. *Mexico. Geology*, **32**, 213–216.

Hallock, P. and **Schlager, W.** (1986) Nutrient excess and the demise of coral reefs and platforms. *Palaios*, **1**, 389–398.

Haq, B.U., **Hardenbol, J.** and **Vail, P.R.** (1988) Mesozoic and Cenozoic Chronostratigraphy and cycles of sea-level change. In: *Sea-Level Changes: An Integrated Approach.* (Ed B. Lidz). *SEPM Spec. Publ.*, **42**, 71–108.

Homewood, P. (1996) The carbonate feedback system: interaction between stratigraphic accommodation, ecological succession and the carbonate factory. *Bull. Soc. Géol. France*, **167**, 701–715.

James, N.P. (1997) The cool-water depositional realm. In: *Cool-water Carbonates* (Eds N.P. James and J.A.D. Clarke). *SEPM Spec. Publ.*, **56** 1–20.

James, N.P., **Collins, L.B.** and **Bone, Y.** (1999) Subtropical carbonates in a temperate realm: Modern sediments on the southwest Australian Shelf. *J. Sed. Res.*, **69**, 1297–1321.

Jones, B. and **Desrochers, A.** (1992) Shallow platform carbonates. In: *Facies Models: Response to Sea Level Changes* (Eds R.G. Walker and N.P. James), pp. 277–301. *Geol. Assoc.*, Canada.

Lecca, L., **Lonis, R.**, **Luxoro, S.**, **Melis, E.**, **Secchi, E.** and **Brotzu, P.** (1997) Oligo-Miocene volcanic sequences and rifting stages in Sardinia: a review. *Period. Min.*, **66**, 7–61.

Marcano, G. (2008) *Investigation on Sedimentology and Early Diagenesis in Shallow Water Warm-temperate to Tropical Miocen Carbonates: A Case Study from Northern Sardinia, Italy.* Unpubl. PhD Thesis, Universität Potsdam.

Martín, J.M.M. and **Braga, J.C.** (1994) Messinian events in the Sorbas Basin and their implications in the recent history of the Mediterranean. *Sed. Geol.*, **90**, 257–268.

Martini, I.P., **Oggiano, G.** and **Mazzei, R.** (1992) Siliciclastic-carbonate sequences of Miocene grabens of Northern Sardinia, western Mediterranean Sea. *Sed. Geol.*, **76**, 63–78.

Monaghan, A. (2001) Coeval extension, sedimentation and volcanism along the Cainozoic rift system of Sardinia. In: *Peri-Tethys Memoir 6: Peri-Tethyan Rift/wrench Basins and Passive Margins* (Eds P.A. Ziegler and A.H.F. Cavazza). *Mem. Mus. Natl. Hist. Nat. Paris*, **186**, 707–734.

Murru, M., **Simone, L.** and **Vigorito, M.** (2001) Carbonate channel network in the Miocene syn-rift Sardinia basins. *Géol. Mediterr.*, **28**, 133–137.

Nagel, B. (2005) *Kartierung der neogenen Schichtenfolge zwischen Sedini und Laerru, NW-Sardinien.* Unpubl. Diplomarbeit, Universität Hamburg, 54 pp.

Nelson, C.S. (1988) An introductory perspective on non-tropical shelf carbonates. *Sed. Geol.*, **60**, 3–12.

Obrador, A. Pomar and **Taberner, C.** (1992) Late Miocene breccia of Menorca (Balearic Islands): a basis for the interpretation of a Neogene ramp deposit. *Sed. Geol.*, **79**, 203–223.

Pedley, M. (1996) Miocene reef distributions and their associations in the Central Mediterranean region: an overview. In: *Models for Carbonate Stratigraphy from Miocene Reef Complexes of Mediterranean Regions* (Eds E.K. Franseen, M. Esteban, W.C. Ward and J.M. Rouchy). *SEPM Concepts Sedimentol. Palaeontol.*, **5**, 73–88.

Pomar, L. (2001) Ecological control of sedimentary accommodation: evolution from a carbonate ramp to rimmed shelf, Upper Miocene, Balearic Islands. *Palaeogeogr. Palaeoclimatol. Palaeoecol.*, **175**, 249–272.

Pomar, L., **Obrador, A.** and **Westphal, H.** (2002) Sub-wavebase cross-bedded grainstones on a distally steepened carbonate ramp, Upper Miocene, Menorca, Spain. *Sedimentology*, **49**, 139–169.

Pomesano Cherchi, A. (1971) Microfaune planctoniche di alcune serie mioceniche del Logudoro (Sardegna). In: *Proceedings of the II Planktonic Conference* (Eds A. Farinacci and R. Matteucci), pp. 1003–1016. Roma.

Read, J.F. (1985) Carbonate Platform Models. *AAPG Bull.*, **69**, 1–21.

Rosen, B.R. (1999) Palaeoclimatic implications of the energy hypothesis from Neogene corals of the Mediterranaen. In: *Hominoid Evolution and Climatic Change in Europe* (Eds J. Agusti and L. Rook), pp. 309–327. *Cambridge University Press.*

Schlager, W. (Ed.) (2005) *Carbonate Sedimentology and Sequence Stratigraphy. SEPM Concepts Sedimentol. Paleontol.*, **8**, 200 pp.

Simone, L., **Murru, M.** and **Vigorito, M.** (2001) A syn-rift temperate-type carbonate margin: morphology and depositional geometries. In: *21st Meeting Int. Assoc. Sedimentol., Davos (Abstracts).*

Sowerbutts, A.A. (1997) *Coeval Sedimentation, Extension and Arc-volcanism Along the Oligo-Miocene Sardinian Rift.* Unpubl. PhD Thesis, University of Edinburgh, 521 pp.

Sowerbutts, A. (2000) Sedimentation and volcanism linked to multiphase rifting in an Oligo-Miocene intra-arc basin, Anglona, Sardinia. *Geol. Mag.*, **137**, 395–418.

Speranza, F., **Villa, I.M.**, **Sagnotti, L.**, **Florindo, F.**, **Cosentino, D.**, **Cipollari, P.** and **Mattei, M.** (2002) Age of the Corsica-Sardinia rotation and Liguro-Provencal Basin spreading: new palaeomagnetic and Ar/Ar evidence. *Tectonophysics*, **347**, 231–251.

Thomas, B. and **Gennesseaux, M.** (1986) A two stage rifting in the basins of the Corsica-Sardinian straits. *Mar. Geol.*, **72**, 225–239.

Vigorito, M., **Murru, M.** and **Simone, L.** (2005) Anatomy of a submarine channel system and related fan in a foramol/rhodalgal carbonate sedimentary setting: a case history from the Miocene syn-rift Sardinia Basin. *Sed. Geol.*, **174**, 1–30.

Vigorito, M., **Murru, M.** and **Simone, L.** (2006) Architectural patterns in a multistorey mixed carbonate-siliciclastic submarine channel, Porto Torres Basin, Miocene, Sardinia, Italy. *Sed. Geol.*, **186**, 213–236.

Wilson, J.L. (1975) *Carbonate Facies in Geologic History. Springer*, Berlin, 471 pp.

Woelkerling, W.J., **Irvine, L.M.** and **Harvey, A.S.** (1993) Growth-forms in Non-geniculate Coralline Red Algae (Corallinales, Rhodophyta). *Aust. Syst. Bot.*, **6**, 277–293.

Int. Assoc. Sedimentol. Spec. Publ. (2010) **42**, 149–164

Coralline algae, oysters and echinoids – a liaison in rhodolith formation from the Burdigalian of the Latium-Abruzzi Platform (Italy)

MARCO BRANDANO* and WERNER E. PILLER†

*Dipartimento di Scienze della Terra, Università di Roma "La Sapienza", Ple Aldo Moro 5, I-00185, Roma, Italy (E-mail: marco.brandano@uniroma1.it)

†Institute of Earth Sciences, Geology and Palaeontology, University of Graz, Heinrichstrasse 26, A-8010, Graz, Austria

ABSTRACT

The sedimentology and palaeontology of a rhodolith interval in a middle-ramp setting of the Early Miocene Latium-Abruzzi carbonate platform in central Italy is described. This rhodolith interval (7 m thick) shows a vertical facies sequence from a coralline branch rudstone at the base, to a rhodolith floatstone, a rhodolith rudstone and a coralline algal–oyster rudstone to bind/framestone, topped by a further coralline branch rudstone. Rhodolith nucleation took place mainly on neopycnodont oyster shells or bryozoan colonies. The coralline algal flora is dominated by the genera *Lithothamnion* and *Mesophyllum*, pointing to a slightly deeper-water setting. A total of 12 taxa are systematically described. The dominant species, *Lithothamnion* cf. *macrosporangicum*, is morphologically highly variable and has the capability to grow on different substrates, but also to produce a carbonate framework on and between a neopycnodont oyster pavement. All sedimentological and biogenic characters of the rhodolith levels point to a low-energy palaeoenvironment. The movement of rhodoliths can therefore not be attributed to waves or currents, but is ascribed to biogenic activity, particularly to that of regular echinoids.

Keywords Burdigalian, coralline algae, palaeoecology, carbonate ramp.

INTRODUCTION

Coralline algal sediments are widespread in the Miocene of the Mediterranean (Franseen *et al.*, 1996), but are well documented only in the Middle and Late Miocene. Although Early Miocene coralline algae and algal carbonates have been described repeatedly (e.g. Airoldi, 1932; Conti, 1943; Ogniben, 1958; Mastrorilli, 1974; Barbera *et al.*, 1978; Vannucci *et al.*, 1993, 1996; Carannante & Simone, 1996), modern taxonomic descriptions of the algae and detailed environmental interpretations are still missing.

Coralline algae in general, and particularly algal nodules and rhodoliths, have frequently been used as palaeoecological indicators (e.g. Bosence & Pedley, 1982; Braga & Martín, 1988; Braga & Aguirre, 2001). In addition to light and temperature as predominant factors limiting the distribution of coralline algae, hydrodynamic energy is considered to be most important in affecting rhodolith shape, structure and distribution (Bosence, 1983a, b; Basso, 1998). However, in many settings bioturbation may replace or support water motion as a mechanism for rhodolith production and turn-over (Steneck, 1986; Prager & Ginsburg, 1989; Marrack, 1999; Gischler & Pisera, 1999). Rhodolith growth can be initiated by coralline algal branch break-up and their further growth as unattached fragments (non-nucleated rhodolith; Freiwald, 1994) or by repeated encrustation of a nucleus. Small rock clasts or shells or bioclasts of corals, bryozoan or even abraded rhodoliths frequently occur as nuclei. In any case, a mechanism for periodic overturning has to be present.

In this study, a rhodolith facies from the Burdigalian middle ramp of the Latium-Abruzzi carbonate platform (Brandano, 2001; Brandano & Corda, 2002) is described. In the middle ramp, rhodolith pavements alternate with bioclastic sediments dominated by red algae detritus, larger foraminifera, bivalves and echinoid remains. The lateral and vertical changing taxonomic composition of rhodoliths and the accompanying fauna provide insight into the processes of rhodolith formation and give clues about concurrent environmental conditions.

GEOLOGICAL SETTING

The investigated rhodolith interval belongs to the upper portion of the Miocene carbonate platform succession outcropping in the Monte Lungo area located on the Cassino plain (Central Appenines, central Italy; Fig. 1). The Cassino plain belongs to the wide neritic domain known as Latium-Abruzzi platform which consists of a thick discontinuous succession of about 5000 m of Upper Triassic to Upper Miocene carbonates. Since the Late Miocene (Tortonian) the Latium-Abruzzi platform has been involved in a thrust-belt/fore-deep system with a progressive drowning of the carbonate platform testified by initial hemipelagic marls (*Orbulina* Marls) followed by siliciclastic turbidites (Patacca *et al.*, 1991).

The Miocene carbonate succession of the Latium-Abruzzi platform belongs to the "Calcari a Briozoi e Litotamni Formation" (Bryozoan and *Lithothamnion* limestones) that paraconformably overlies Cretaceous limestones, locally with the interposition of a few metres of Palaeogene strata (Devoto, 1967; Accordi *et al.*, 1967; Catenacci *et al.*, 1982; Damiani, 1990; Brandano, 2001).

On the basis of its stratigraphic position (underlying the *Orbulina* Marls), many authors supported a Langhian–Serravallian age for the "Calcari a Briozoi e Litotamni Fm." (Accordi *et al.*, 1967; Parotto & Praturlon, 1975; Damiani, 1990). However, by combining strontium-isotope stratigraphy and biostratigraphy, a Late Aquitanian to Early Tortonian age was proposed for the "Calcari a Briozoi e Litotamni" by Brandano (2001).

The formation is interpreted to have been deposited on a carbonate ramp (Brandano, 2001; Brandano & Corda, 2002; Pomar *et al.*, 2004). According to Brandano & Corda (2002), the depositional profile was a homoclinal ramp characterized by three depositional environments: inner, middle and outer ramp. The inner ramp was a euphotic, high-energy environment immediately adjacent to the littoral zone passing seaward first into seagrass meadows and associated Maërl facies and, farther seaward, into coral carpets (*sensu* Riegl & Piller, 1999). Local emergence is demonstrated by the occurrence of *Microcodium*, which is well preserved in the Maërl facies (Civitelli & Brandano, 2005).

In the middle ramp, the main carbonate producers were larger foraminifera and red algae and subordinately echinoids, bryozoans and bivalves. The biota assemblages and the absence of wave-related structures place this depositional environment in the deeper oligophotic zone, below the base of wave action. The outer ramp is further sub-divided into: (1) a proximal outer ramp characterized by crudely stratified, structureless, coarse-grained echinoid-foraminiferal packstone and pectinid-bryozoan floatstone; (2) an intermediate outer ramp with fine to medium-grained echinoid-planktonic foraminiferal packstone; (3) a distal outer ramp characterized by limestone-marl alternations with siliceous sponge spicules, echinoid fragments and planktonic foraminifera. On the distal outer ramp, bioturbation is ubiquitous. The faunal associations suggest that the outer-ramp depositional environments were situated below the limit of light penetration, i.e. in the aphotic zone.

Facies architecture of the investigated deposits reflects two second-order depositional sequences (SD1 and SD2; Fig. 2). These developed over a period of approximately 6 Myr with eustatic oscillations, regional tectonic processes and trophic changes driving their evolution. The deposition of SD1 is restricted to the Late Aquitanian–Serravallian interval. Sequence SD 1 contains an upper lowstand systems tract (LST), a transgressive systems tract (TST) and a highstand systems tract (HST). Inner, middle and outer-ramp facies are present only in the LST, whereas the TST and the HST are characterized exclusively by middle and outer-ramp deposits. In the TST, the middle-ramp facies stepped back as a consequence of the landward (upward) migration of the oligophotic biota producing sediment (rhodalgal association). At the same time the proximal outer-ramp facies, dominated by aphotic biota producing sediment (bryomol association), stepped back but also prograded (Fig. 2). This biota, not being dependent on light, was able to continue sediment production even under conditions of progressive deepening. The progressive expansion of the proximal outer-ramp facies belt indicates significant production of suspended organic matter and, consequently, excellent conditions for bryozoans to grow (Brandano, 2001; Brandano & Corda, 2002).

Sequence SD2 ranges from the Serravallian to the Tortonian. The termination of this sequence is

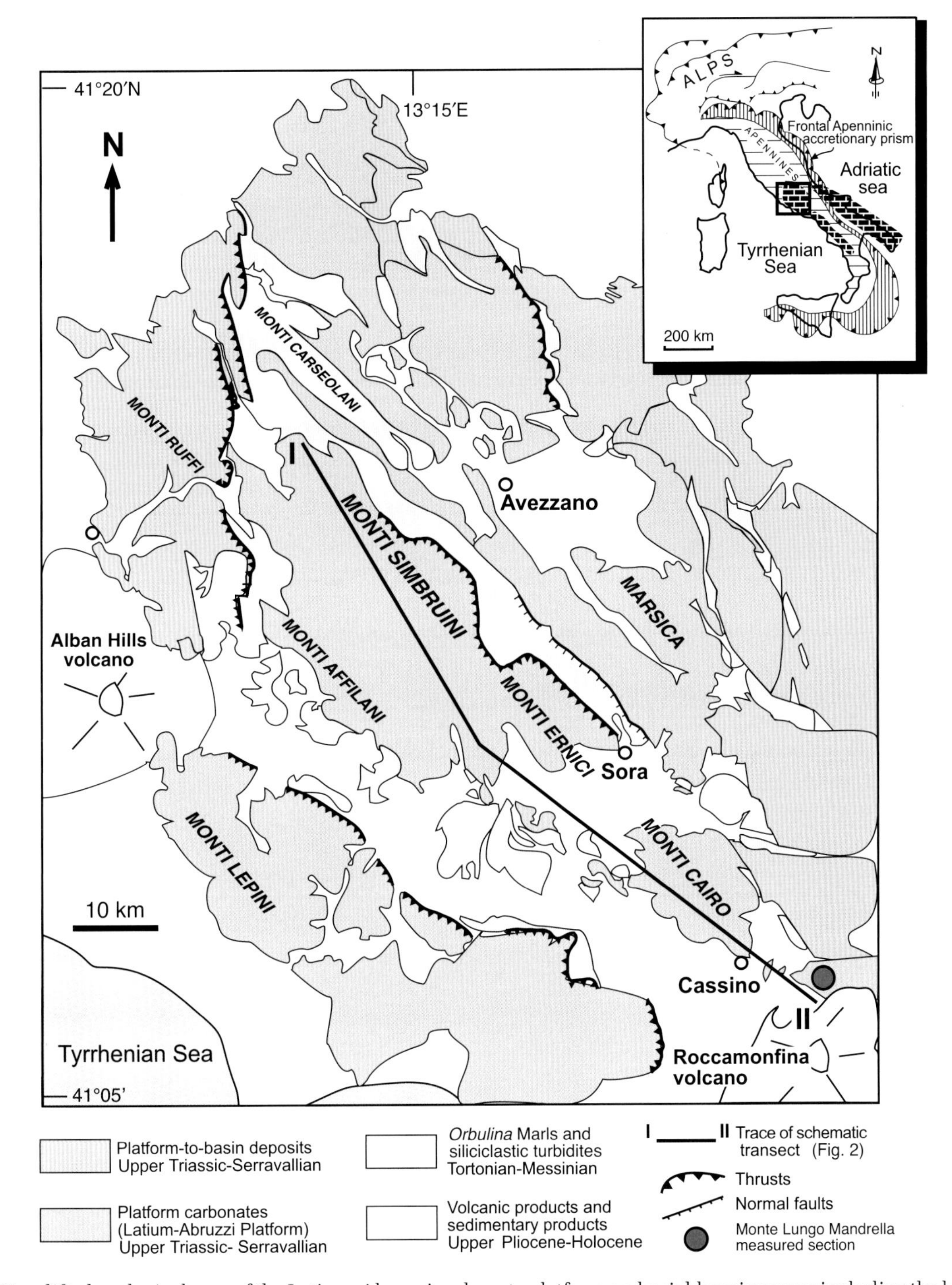

Fig. 1. Simplified geological map of the Latium-Abruzzi carbonate platform and neighbouring areas including the location of the analyzed outcrop (modified from Brandano, 2001). Cross-section I–II is illustrated in Fig. 2.

expressed by a drowning phase, which brings hemipelagic sediments first, and then siliciclastic turbidites as a response to the eastward migration of the Apenninic foredeep. The investigated rhodoliths come from the middle-ramp facies of the TST of sequence SD1 (Fig. 2).

METHODS

The rhodolith structure has been analyzed on polished hand-samples and thin-sections. Shape, structure, nucleus and branching density of rhodoliths are described according to Bosellini &

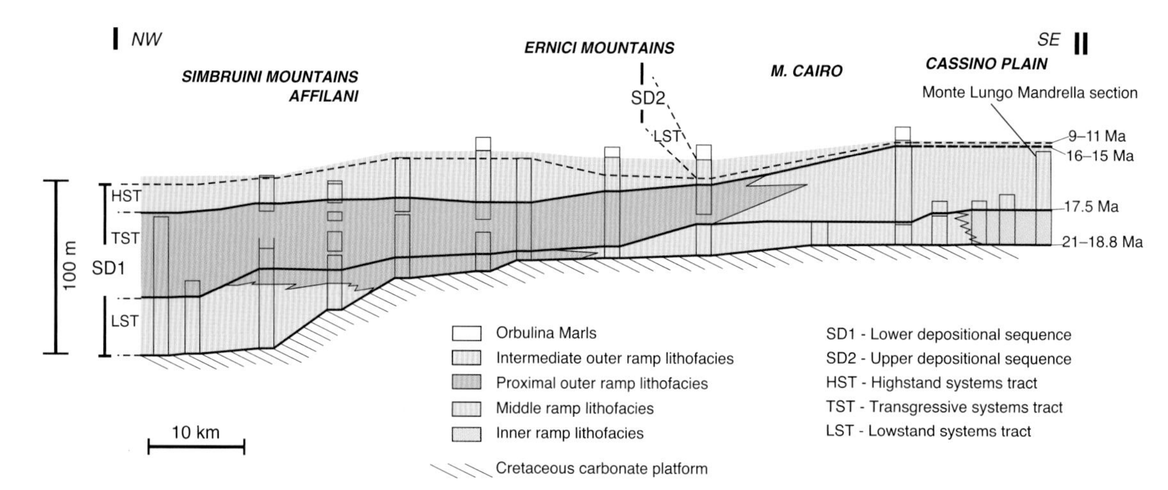

Fig. 2. Stratigraphic architecture of "Calcari a Briozoi e Litotamni Formation", section I–II in Fig. 1. Note the simultaneous backstepping and progradation of outer-ramp facies after 17.5 Ma (after Brandano, 2001 and Civitelli & Brandano, 2005).

Ginsburg (1971) and Bosence (1983a, 1991). The terminology for taxonomic features follows Woelkerling (1988) and Rasser & Piller (1999). Terminology concerning coralline algal growth form is used following that of Woelkerling *et al.* (1993).

A total of 47 thin-sections ranging from 5 × 5 cm to 10 × 6 cm have been studied to identify the red algae at the most precise taxonomic level. The relative abundance of different taxa has been estimated from thin-sections using a range scale (abundant, common, rare, very rare, absent). Measurements were performed using a stereomicroscope (Leica M420, Leica MZ 16) or polarizing microscope (Zeiss Axioplan2) connected to an image analysis and archiving system (Image Access 5.0).

RESULTS

The Monte Lungo Mandrella section

The studied material comes from the Monte Lungo Mandrella section representing a thickness of 63 m. The Miocene limestone paraconformably rests on Turonian cuneoline-miliolid wackestone (Damiani *et al.*, 1991; Fig. 3). The basal Miocene deposits are made up of 6 m of inner-ramp lithofacies. This is represented by crudely stratified balanid floatstone to rudstone, structureless pebbly packstone, and skeletal packstone to floatstone. The latter is composed of well-rounded and usually micritized mollusc, barnacle and echinoid fragments, bryozoans, and red algal fragments. The middle-ramp facies consists of medium to coarse-grained packstone with red algae, molluscs, echinoid fragments and larger foraminifera and rhodolithic floatstone to rudstone. These rocks are crudely stratified with subhorizontal bedding planes and alternate with tabular, massive oyster floatstone beds (0.4 m to 2 m thick). The floatstone matrix is represented by a coralline algae packstone similar to that of the red algae packstone facies. The rhodolith interval studied in detail starts at 54 m from the base of the section (Fig. 3).

The rhodolith interval

The rhodolith interval is 7 m thick and it is organized in tabular, massive beds (0.4 m to 1 m thick). Four facies can be distinguished based on component distribution and textural characteristics (Fig. 3). In all facies, compaction features are very common and breakage of skeletal components is evident; pressure solution is widespread, with abundant sutured grain contacts and grain-interpenetration features. The four facies are described below:

Coralline branch rudstone (samples LR4, LR10)

This is composed of unfragmented coralline algal branches, rhodoliths and their detritus (Figs. 4 and 5). The branches have a maximum size of 2 cm. The rhodoliths are ellipsoidal to subspherical in shape with longer axes ranging from 3–6 cm. Their nuclei consist of medium-grained skeletal grains and their structure is branching with a branching density ascribed to group III, *sensu* Bosence (1983a). The other sedimentary components are represented by foraminifera, molluscs and echinoid remains, as well as bryozoan colonies.

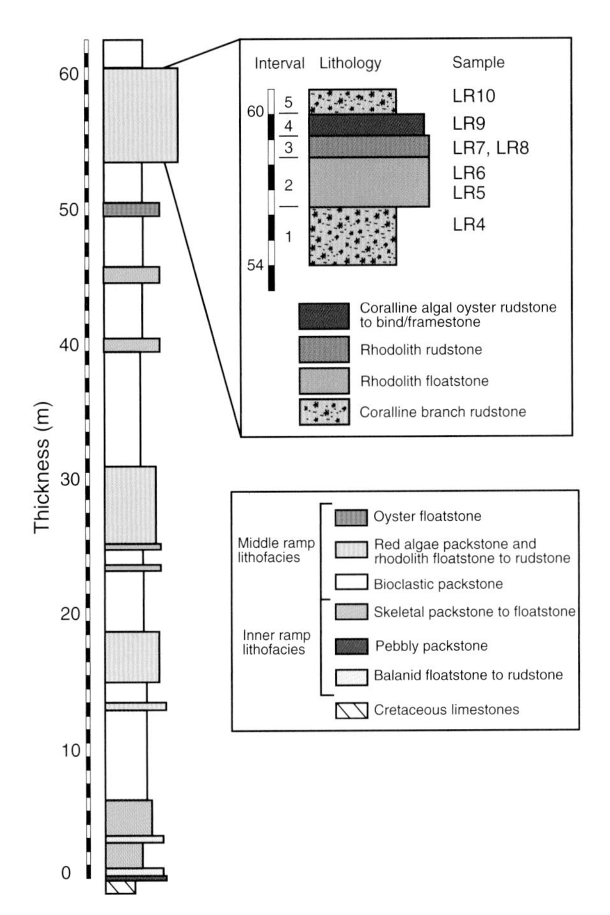

Fig. 3. The Monte Lungo Mandrella section showing lithofacies composition and the position of the analyzed rhodolith interval (expanded). Sample levels (LR) are indicated.

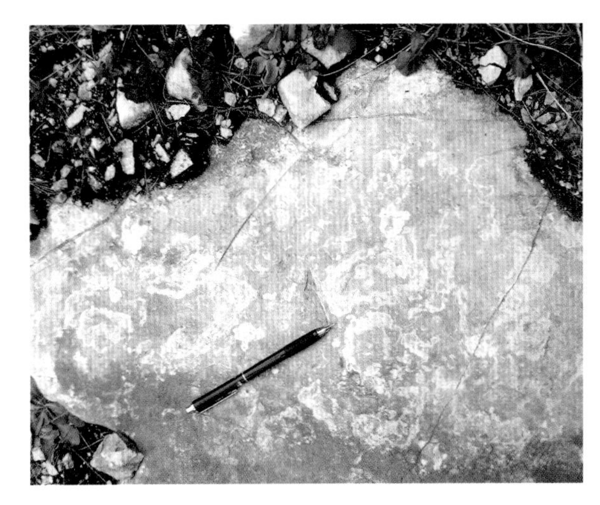

Fig. 5. Coralline branch rudstone. The rhodoliths are ellipsoidal to subspherical in shape ranging from 3–6 cm across (Fig. 3, level 1; sample LR4). Pen is 14 cm long.

Rhodolith floatstone (samples LR5, LR6)

Most rhodoliths are 4–15 cm in diameter and represented by different shapes (Fig. 6). Irregular and regular ellipsoidal shapes make up 25% each, subspherical shapes dominate with 50%. Most rhodoliths start with a laminar-columnar structure grading outward into laminar growth. Branching rhodoliths are also present. Nuclei frequently consist of oysters (neopycnodonts) or their fragments (up to 7 cm in size), but in some cases the space of the original nucleus is filled by bioclastic sediment. The subspherical rhodoliths have bryozoan colonies (up to 5 cm in size) as nuclei. Some of them seem to be non-nucleated. Bryozoan colonies

Fig. 4. Coralline branch rudstone. The sediment is composed of unfragmented coralline algal branches, rhodoliths and their detritus (Fig. 3, level 5; sample LR10). Coin is 2.4 cm diameter.

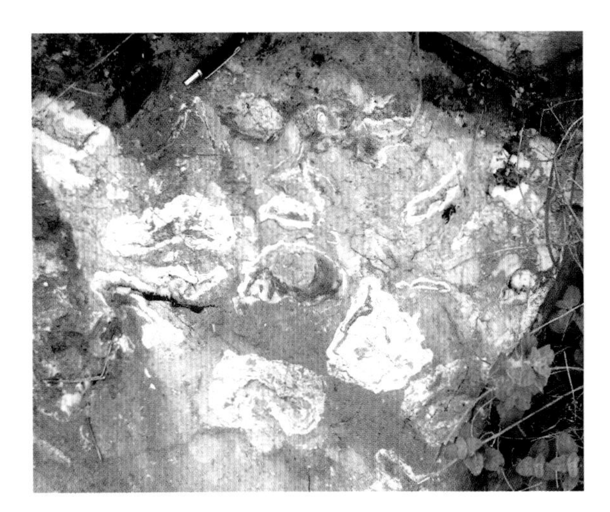

Fig. 6. Rhodolith floatstone. Rhodoliths are 4–15 cm in diameter, with irregular and regular ellipsoidal shapes. Nuclei frequently consist of oysters (neopycnodonts) or their fragments (Fig. 3, level 2; sample LR6).

and acervulinid foraminifera represent an important percentage of the total biogenic carbonate composing the rhodoliths.

The matrix is formed by unsorted bioclastic sediment. The skeletal components are represented mainly by echinoid spines, neopycnodont remains, bryozoans, larger foraminifera (*Amphistegina*, *Heterostegina* and rare miogypsinids) and red algae fragments; balanid fragments and serpulids are subordinate. Bioerosion by clionid sponges and bivalves (ichnogenera *Entobia*, *Gastrochaenolites*) is very abundant on the red algae thalli and on the oysters of the nuclei. The bioerosion cavities are frequently filled by silty sediment.

Rhodolith rudstone (samples LR7, LR8)

The rhodoliths in this layer are generally subspherical in shape with diameters ranging between 6 and 15 cm, although the larger sizes are dominant (Fig. 7). Most rhodoliths show a laminar structure, while branched forms are more rare. Most rhodoliths also seem to be non-nucleated, but some rhodoliths show bryozoan colonies as nuclei. The matrix is represented by a poorly sorted packstone with echinoid spines, neopycnodont remains, bryozoans, larger foraminifera and red algae fragments.

Coralline algal–oyster rudstone to bind/framestone (sample LR9)

Thick laminar layers (up to 5 cm) of coralline algae encrust mostly double-valved neopycnodont oyster shells (up to 8 cm; Fig. 8). The sediment in between is a poorly sorted bioclastic grainstone rich in bryozoans and echinoid remains. In addition, benthic and encrusting foraminifera, serpulids and algal detritus are present. Oysters may be encrusted only on one side or may form nuclei of mostly irregular rhodoliths (Fig. 9). The oysters and coralline algal crusts are heavily bioeroded. Borings are filled by bioclastic sediment. The texture is overprinted by pressure-solution features.

The rhodolith interval shows a clear vertical facies sequence (Fig. 3). The base is characterized by 2.4 m thick coralline branch rudstone, followed by 2 m of rhodolith floatstone. Above, 0.8 m is composed of rhodolith rudstone and another 0.8 m of coralline algal–oyster rudstone to bind/framestone. This sequence is overlain by a further level of coralline branch rudstone (1 m).

Fig. 8. Coralline algal–oyster rudstone to bind/framestone. (Fig. 3, level 4; sample LR9). Pen is 14 cm long.

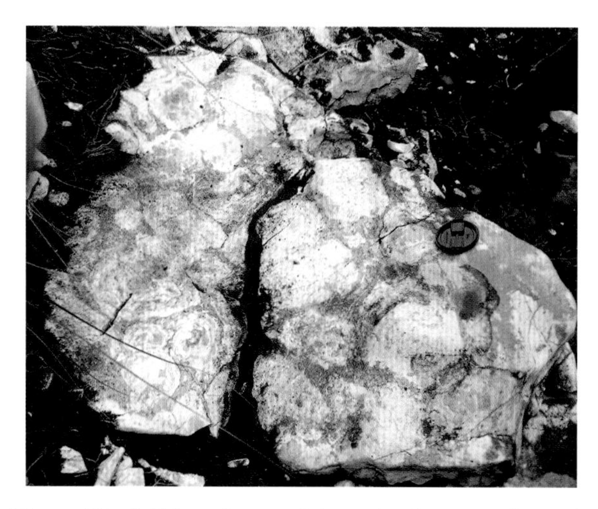

Fig. 7. Rhodolith rudstone. (Fig. 3, level 3; sample LR8). Lens cap is 5 cm diameter.

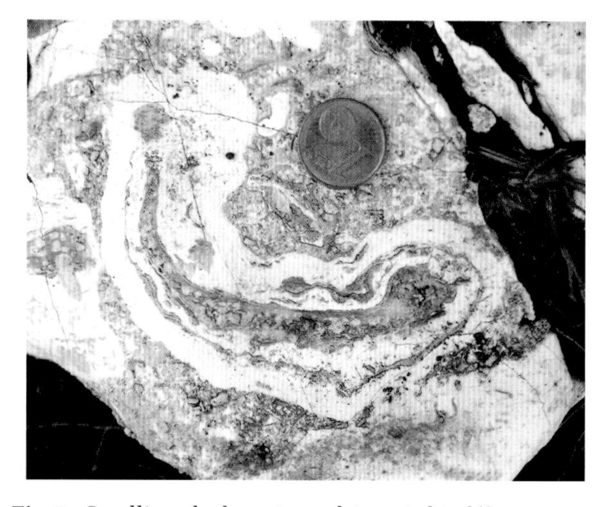

Fig. 9. Coralline algal–oyster rudstone to bind/framestone. (Fig. 3, level 4; sample LR9). Coin is 1.6 cm diameter.

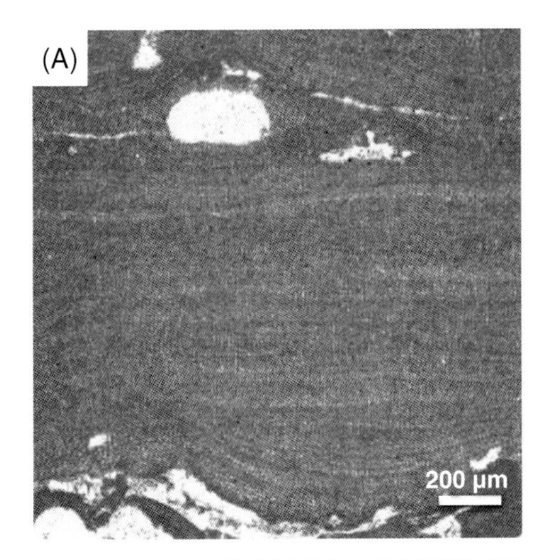

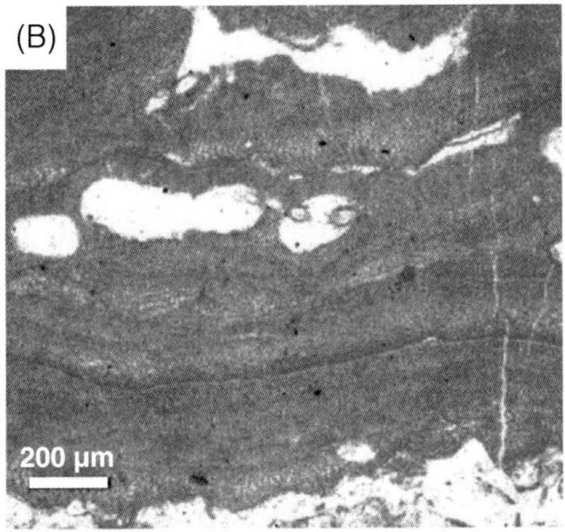

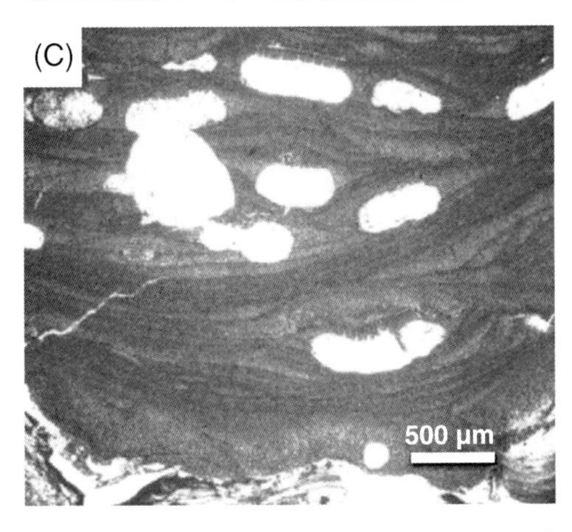

Fig. 10. Coralline algal internal structure I. (A) The rare species *Phymatolithon* sp. 1, with a thin basal core and a thick crustose peripheral portion with a multiporate conceptacle, thin-section LR5-2a. (B) A sequence of repeated thin-crustose thalli of *Lithothamnion* sp. 1 with conceptacles raised above the thallus surface, thin-section LR5-2a. (C) *Lithothamnion* cf. *macrosporangicum* (Mastrorilli, 1950), the most abundant coralline algal species, showing its characteristic large conceptacles, thin-section LR5-1a. All images plain polarized light.

Systematics of coralline algae

Division **Rhodophyta** (Wettstein, 1901)
Class **Rhodophyceae** (1863)
Order **Corallinales** (Silva & Johansen, 1986)
Family **Corallinaceae** (Lamouroux, 1812)
Subfamily **Melobesioideae** (1885)

Phymatolithon sp. 1 (Fig. 10A)

Growth form encrusting to warty. Sometimes thalli form only thin crusts (150–350 μm), they can, however, reach up 2 mm in thickness.

Thallus monomerous, core filaments are non-coaxial, predominantly curving towards the dorsal thallus surface. Core portion 60–120 μm thick, with cell length from 8–26 μm and diameter 7–10 μm.

The peripheral portion is in the thin crusts between 70 and 170 μm thick. Growth bands are indistinctly present. Cell length 9–16 μm, diameter 8–12 μm. The epithallium is one cell-layer thick, its cells are rarely well preserved.

Tetra/bisporangial conceptacles are multiporate, with diameters from 244–346 μm and heights from 116–171 μm.

The thalli are frequently intergrown with bryozoan and other coralline algae.

Lithothamnion sp. 1 (Fig. 10B)

Growth form encrusting to layered. Thickness of thalli up to 500 μm; sometimes superimposed.

Thallus monomerous; core filaments are non-coaxial, cell fusions are present. The core portion is usually 80–100 μm thick, in places up to 140 μm. Filaments curve predominantly toward the dorsal, sometimes towards the ventral thallus surface. Cell length 10–28 μm, diameter 7–13 μm.

Peripheral portion is 200–300 μm thick and no growth bands are developed. Cell length is 10–15 μm, diameter 8–12 μm. Subepithallial initials were not observed and epithallial cells are 6–10 μm in length and 6–14 μm in diameter.

Tetra/bisporangial conceptacles are multiporate and raised above thallus surface; their diameter ranges from 182–485 μm and height from 130–190 μm.

Lithothamnion cf. macrosporangicum (Mastrorilli, 1950; Fig. 10C)

Growth form encrusting to fruticose. Several plants superimposed reaching a thickness of several centimetres.

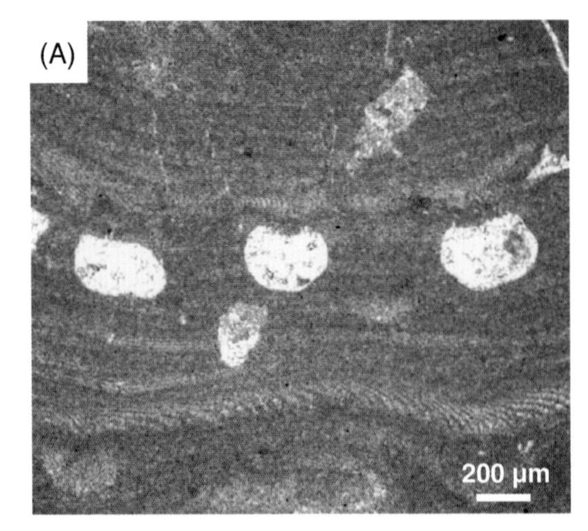

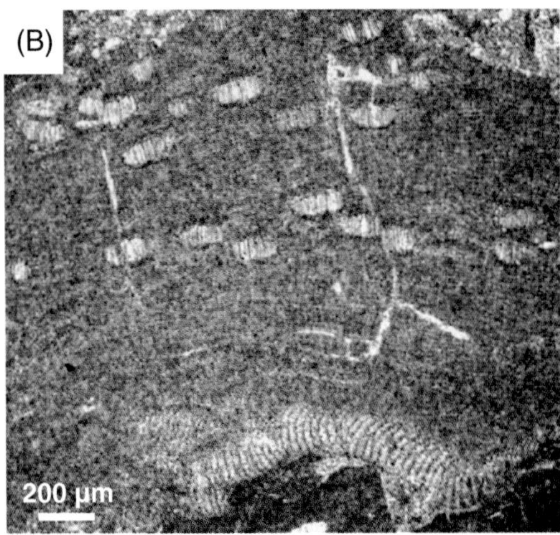

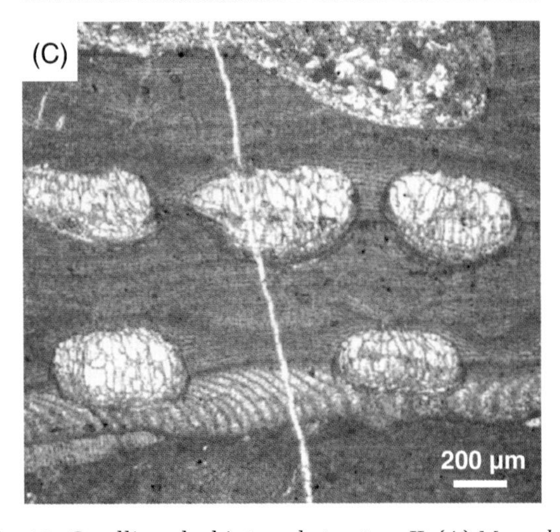

Fig. 11. Coralline algal internal structure II. (A) *Mesophyllum gignouxi* (Lemoine, 1939); thin-section LR7-3d. (B) *Mesophyllum* cf. *inaspectum* (Airoldi, 1932; Ogniben, 1958); thin-section LR7-3e. (C) *Mesophyllum roveretoi* (Conti, 1943); thin-section LR8-1. All images plain polarized light.

Thallus monomerous, core filaments non-coaxial and predominantly curve outward (dorsal); cell fusions are present. The core portion is 105–223 μm thick, cell length is 11–32 μm, diameter 4–14 μm.

Peripheral filaments show distinct growth bands of 50–105 μm in thickness; cell fusions are present, cell length is 5–14 μm, the diameter 6–11 μm. Subepithallial initials and ephitallial cells were not observed.

The tetra/bisporangial conceptacles are multiporate; their diameter is 403–1176 μm, their height 124–259 μm. Sexual conceptacles were also observed, are monoporate and reach 618–725 μm in diameter and 211–239 μm in height.

This is the most abundant coralline algal species in the studied material and also the primary rhodolith builder.

Mesophyllum gignouxi (Lemoine, 1939; Fig. 11A)

Growth form encrusting and crusts reach a total thickness up to 1.2 mm.

Thallus monomerous, core filaments are coaxial and curve to both sides. Cell fusions occur. The core portion is 60–275 μm thick and cores occur repeatedly in thick crusts; repeated cores are thinner as those at the very base of thallus. Cell length is 11–30 μm, diameter 5–15 μm.

Peripheral portion is highly varible in thickness, reaching from 190 μm up to 800 μm; cells range from 6–18 μm in length and 5–11 μm in diameter. Epithallial cells have not been observed.

Tetra/bisporangial conceptacles are multiporate. Conceptacles are 229–608 μm in diameter and 132–236 μm high.

Mesophyllum cf. inaspectum (Airoldi, 1932; Ogniben, 1958; Fig. 11B)

Growth form encrusting to warty with a total thickness up to 6 mm.

Thallus monomerous, core filaments coaxial, curving to both sides. Thickness of core ranges from 100–175 μm, cell fusions occur. Cell length is 14–34 μm, the diameter 5–10 μm.

The region is several mm thick, with cells 6–15 μm long and 6–10 μm in diameter. Growth rhythms are slightly evident. Ephitallium and vegetative initials were not observed.

Tetra/bisporangial conceptacles are multiporate and 153–277 μm in diameter and 74–163 μm high.

Mesophyllum cf. roveretoi (Conti, 1943; Fig. 11C)

Growth form encrusting with a total thickness of crusts up to 2 mm.

Thallus monomerous, core filaments coaxial, curving to both sides. Cell fusions occur. The core reaches 195–435 μm in thickness; cell length is 17–38 μm, their diameter 5–12 μm.

The peripheral portion is 490–1230 μm thick and growth bands are present (intervals from 80–115 μm). Cell length is 6–17 μm, their diameter 5–9 μm. Subepithallial initials and epithallial cells were not observed.

Tetrasporangial conceptacles are multiporate. The old conceptacles are positioned close to or directly on the core portion, in some cases even sunken. Conceptacle diameter is 398–605 μm, their height 195–327 μm.

Subfamily Mastophoroideae (Setchell, 1943)

Neogonioltihon sp. 1 (Fig. 12A)

Growth form encrusting with a total crust thickness of 2.5 mm. Crusts are built by superimposed monomerous thalli between 250 and 900 μm thickness each.

Core 80–120 μm thick and filaments in coaxial arrangement. Cell length 15–22 μm, diameter 5–8 μm. Cell fusions are present.

Postigenous filaments are irregular and show cells of 5–14 μm length and 4–8 μm diameter.

Tetra/bisporangial conceptacles triangular in shape, uniporate with a conical pore channel. One conceptacle in a vertical/central section: height 167 μm, diameter 363 μm.

Spongites albanensis (Lemoine, 1924; Braga *et al.*, 1993, Fig. 12C)

Growth form encrusting with a crust thickness up to 4 mm; thallus monomerous.

Core filaments non-coaxial, curving to both sides. Cell length 16–25 μm (mean: 19.8 μm), diameter 9–13 μm (mean: 11.2 μm).

The peripheral portion is 100–135 μm thick.

Postigenous filaments are well developed forming a zoned grid. Cell length 10–18 μm (mean: 14.3 μm), diameter 8–16 μm (mean: 10.3 μm). Cell fusions abundant.

Tetra/bisporangial conceptacles are uniporate, bean-shaped in vertical section with a conical pore. Conceptacle diameter 336–440 μm, height 140–185 μm.

Spongites sp. 1 (Fig. 12D)

Growth form encrusting with total crust thickness of up to 2 mm; crusts are built by several superimposed monomerous thalli (thickness 250–400 μm). Core filaments non-coaxial, core portion between 150–200 μm. Filaments curved to both sides. Cell fusions are present; cell length 9–16 μm, diameter 6–8 μm.

The peripheral portion is 150–200 μm thick, indistinctly zoned and filaments are rather well developed. Cell length 6–9 μm, diameter 5–9 μm; irregularly shaped cells present. Subepithallial initials and epithallial cells were not observed.

Tetra/bisporangial conceptacles are uniporate with a cylindrical pore. The few observed conceptacles are raised above respective thallus surface. Conceptacle diameter 282–460 μm, height 109–121 μm.

Subfamily Lithophylloideae (Setchell, 1943)

Lithophyllum nitorum (Adey & Adey, 1973; Fig. 12E)

Growth form encrusting with maximum observed thallus thickness of 2 mm.

Thallus dimerous. Primigenous filaments of non-palisade, squarish cells of 12.3–16.5 μm in height and 10.5–14.7 μm in length.

Postigenous filaments are well developed with good lateral cell alignment. Cells are subrectangular in shape and measure 6–20 μm in length and 5–12 μm in diameter. Epithallial cells have not been observed.

Conceptacles are uniporate with conical to cylindrical pore canal. Some have a raised floor. Longer cells in conceptacle roofs are not unambiguously identified. Tetra/bisporangial conceptacles measure 126–161 μm in diameter and 53–96 μm in height.

Lithophyllum racemus (Lamarck, 1836; Foslie, 1901, Fig. 12F)

Growth form encrusting with thallus thickness up to 1 mm.

Thallus dimerous or monomerous with a coaxial arrangement. Cells of primigenous filaments not observed.

Postigenous filaments are well developed with good lateral cell alignment. Cells are 9–20 μm long and 9–13 μm in diameter. No epithallial cells have been observed in the studied samples.

Conceptacles are uniporate with a cylindrical to conical pore canal. Tetra/bisporangial conceptacles are bean-shaped with fairly flat bottom.

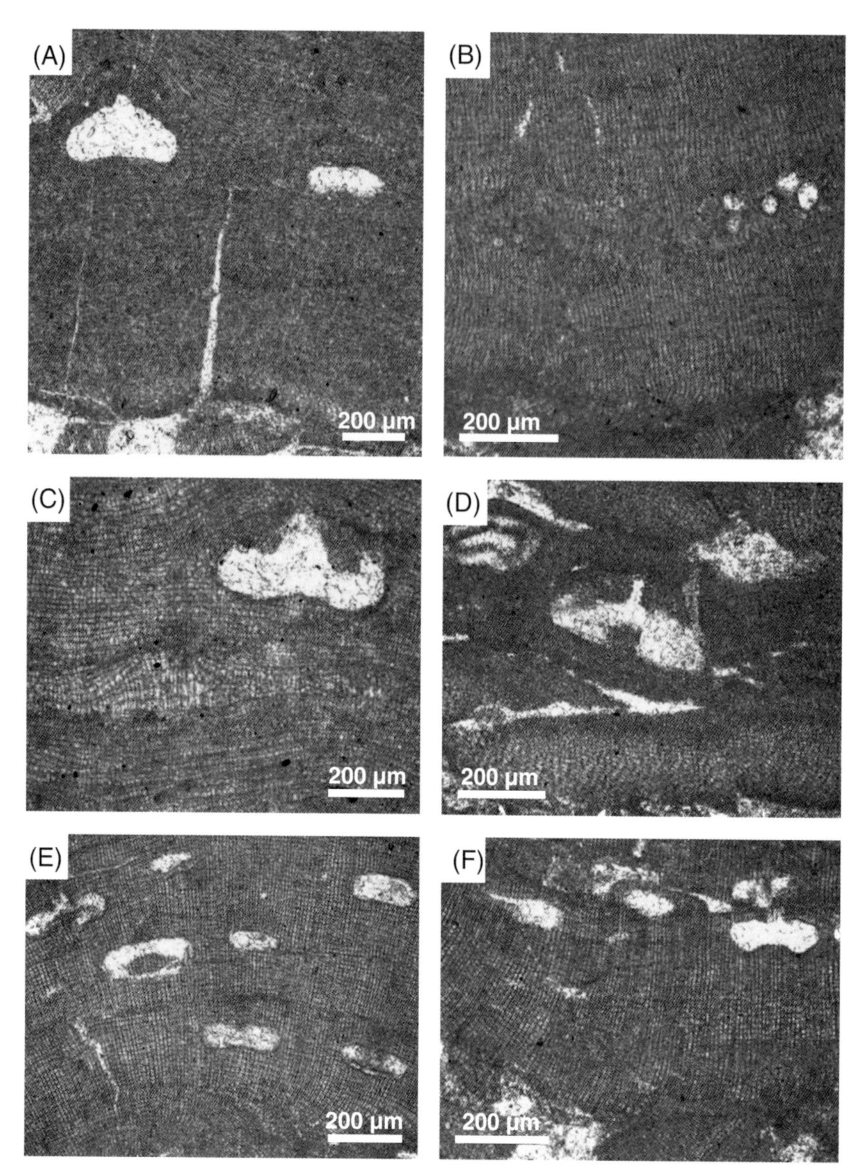

Fig. 12. Coralline algal internal structure III. (A) An encrusting thallus of *Neogoniolithon* sp. 1 with a uniporate conceptacle, thin-section LR8a. (B) An extremely rare example of *Sporolithon* with a cluster of conceptacles, thin-section LR6c. (C) *Spongites albanensis* sp. 1 (Lemoine, 1924; Braga *et al.*, 1993), thin-section LR4-1. (D) *Spongites* sp. 1 (thin-section LR5-2a). Both *Spongites* images (C and D) show the characteristic conceptacles. (E) *Lithophyllum nitorum* (Adey & Adey, 1973), thin-section LR7-5e. (F) *Lithophyllum racemus* (Lamarck, 1836; Foslie, 1901), thin-section LR4-0. All images plain polarized light.

Chambers are 214–252 µm in diameter and 67–87 µm in height. Male (spermatangial) conceptacles are triangular in section and small (72–124 µm in diameter, 19–24 µm in height).

Family sporolithaceae (Verheij, 1993)

Sporolithon sp. 1 (Fig. 12B)

Only two small encrusting fragments classified with this genus have been observed, one 1.7 mm thick, the other 260 µm.

Thallus monomerous. Core filaments curve outwards (thickness and cell dimensions are not measurable).

Peripheral filaments are well developed, rows are less obvious. Cell length range from 9.5 to 16 µm, diameters from 8 to 11 µm. Some rows show a greater cell length. Epithallial cells not observed.

In one of the specimens three groups of conceptacles (sori) are preserved, which measure up to 70 µm in height and 50 µm in diameter (orientation of sections is, however, not fully vertical).

Distribution of coralline algae

A total of 12 non-geniculate corallinacean taxa have been identified, out of which, two taxa (*Phymatolithon* sp. 1, *Neogoniolithon* sp. 1) occur only rarely in one single interval. Generally, the algal flora is dominated by *Lithothamnion* cf. *macrosporangicum*, which is abundant in all samples. This species is highly variable in its growth form ranging from encrusting to lumpy and was also able to grow on loose sediment (Fig. 13). *Mesophyllum* cf. *inaspectum* ranks second and *Mesophyllum gignouxi* third in terms of abundance. *Lithothamnion* sp. 1 and *Spongites albanensis* are common in one sample each. The relative abundance of coralline algal taxa in the investigated intervals is given in Table 1.

The coralline branch rudstone (samples LR4 and LR10) is dominated by *Lithothamnion* cf. *macrosporangicum* in both samples. In LR4, *Mesophyllum gignouxi* and *Spongites albanensis* are also common. In addition, in sample LR10,

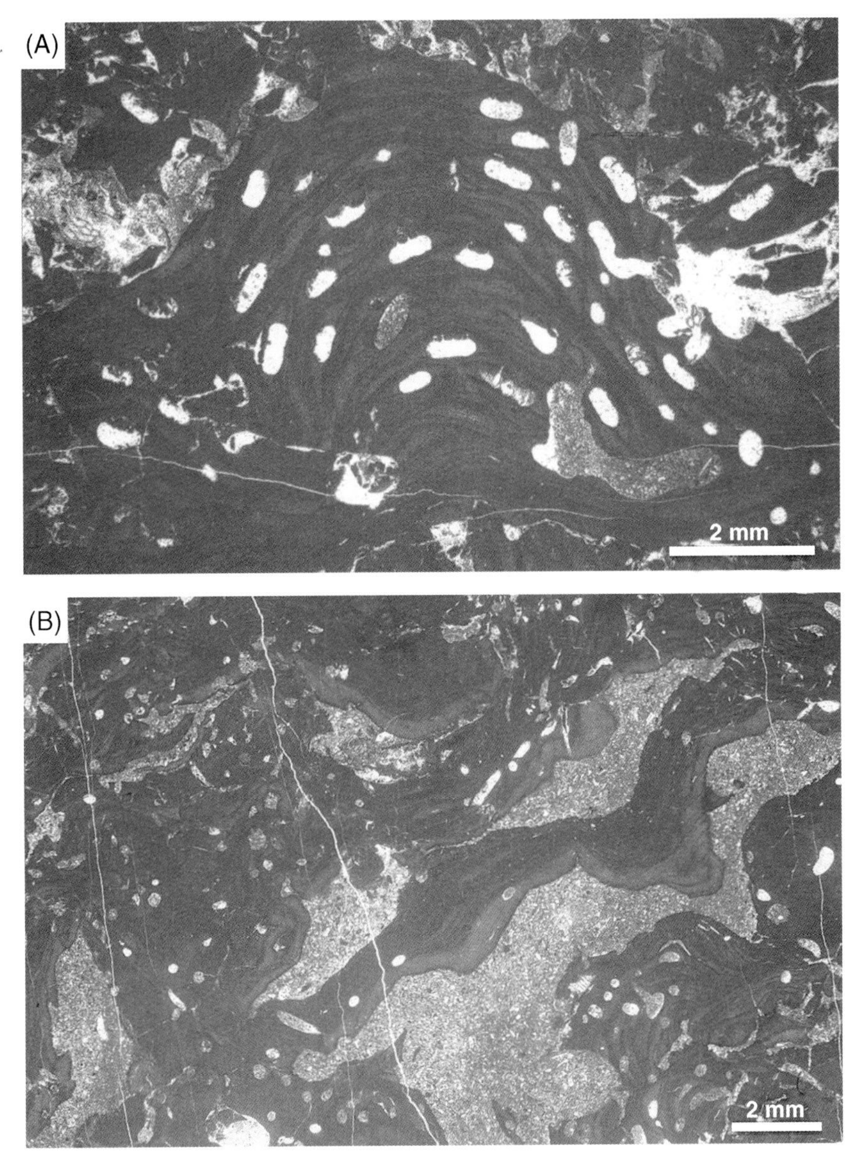

Fig. 13. The two different growth forms of *Lithothamnion* cf. *macrosporangicum* (Mastrorilli, 1950). (A) A protuberance of the fruticose growth form predominating in rhodoliths; thin-section LR5-2b. (B) Bridging growth form typical for bindstone/framestone construction, thin-section LR9-2f. Both images plain polarized light.

Table 1. Relative abundance of coralline algal taxa in the investigated intervals. *L.* = *Lithothamnion*, *P.* = *Phymatolithon*, *M.* = *Mesophyllum*, *Lph.* = *Lithophyllum*, *N.* = *Neogoniolithon*, *S.* = *Spongites*, *Spr.* = *Sporolithon*. Samples LR4-10 (Fig. 3)

	Algal taxa											
Facies	*L.* sp.1	*L. macrospor.* cf.	*P.* sp.1	*M. gignouxi*	*M. inaspectum* cf.	*M. roveretoi* cf.	*Lph. nitorum*	*Lph. racemus*	*N.* sp.1	*S.* sp.1	*S. albanensis*	*Spr.* sp.1
LR 10		A		VR	A		R	R		R	VR	
LR 9	C	A										
LR 8		A		A	C	R	VR	R	R			
LR 7		A		A	A		R	VR				
LR 6		A			C			VR				VR
LR 5	R	A	R	R				VR		VR		VR
LR 4		A		C	R		R	R		VR	C	

VR very rare R rare C common A abundant

Mesophyllum cf. *inaspectum* is very abundant in some thin-sections. Seven taxa are recorded in both levels. The rhodoliths are made up mainly by *Lithothamnion* cf. *macrosporangicum*, alternating with *Lithothamnion* sp. 1, *Mesophyllum gignouxi*, *M.* cf. *inaspectum* and *Spongites albanensis*. Branches and nodules are formed principally by *Lithothamnion* cf. *macrosporangicum*. The other species occur as thin thalli in the rhodoliths and fragmented thalli in the sediment.

The rhodolith floatstone reveals a conspicuous dominance of *Lithothamnion* cf. *macrosporangicum* in sample LR5. All other taxa are rare or very rare. In sample LR6, *Mesophyllum* cf. *inaspectum* is also common besides *Lithothamnion* cf. *macrosporangicum*. In both samples together, 8 taxa have been recorded. *Lithothamnion* cf. *macrosporangicum* may form monospecific rhodoliths. In the multispecific rhodoliths *Lithothamnion* cf. *macrosporangicum* is abundant and associated mainly with *Mesophyllum* cf. *inaspectum* and subordinately with *Lithothamnion* sp. 1, *Phymatolithon* sp. 1 and *Mesophyllum gignouxi*. The other species form debris, small thalli in the rhodoliths or thin crusts on oyster shells.

The rhodolith rudstone (samples LR7 and LR8) is characterized by an almost equal abundance of *Lithothamnion* cf. *macrosporangicum*, *Mesophyllum gignouxi* and *M.* cf. *inaspectum*. LR8 is the only interval in which *Mesophyllum* cf. *roveretoi* and *Neogoniolithon* sp. 1 occur. These species together with *Lithophyllum nitorum* are rare, while *Lithophyllum racemus* is very rare. The total number of taxa is 7. The three dominant species form multispecific rhodoliths, with a complex coating sequence. The other species are accessory and often form the first crust on the oyster shells.

In the coralline algal-oyster rudstone to bind/framestone (sample LR9) only the two species of *Lithothamnion* are present, out of which *Lithothamnion* cf. *macrosporangicum* is abundant and *Lithothamnion* sp. 1 is common. The two species form crusts on oysters and bryozoans and they may form irregular rhodoliths (boxwork). Due to its plasticity, *Lithothamnion* cf. *macrosporangicum* is also able to grow on loose sediment

DISCUSSION

The most obvious and important rock constituents in the studied interval (Fig. 3) are coralline algae, particularly in the form of rhodoliths. In the lowermost and uppermost beds (intervals 1 and 5, samples LR4 and LR10), coralline algae occur both as branches or their fragments, and as crustose or branched rhodoliths of small size (coralline branch rudstone; Figs. 3 and 4). This is in contrast to intervals 2 (rhodolith floatstone, samples LR5 and LR6) and 3 (rhodolith rudstone, samples LR7 and LR8), in which large-sized, predominantly crustose rhodoliths occur in association with neopycnodont oysters that serve as a substrate for the corallines (Figs. 5–7).

In interval 4 (sample LR9), besides large rhodoliths, a bindstone to framestone is developed involving neopycnodont oysters (Figs. 8 and 9) and *Lithothamnion* cf. *macrosporangicum* as well as subordinately, *Lithothamnion* sp. 1. *Lithothamnion* cf. *macrosporangicum* encrusts hard substrate (and therefore is the rhodolith builder in this interval) but is also able to bridge and build a bind/framestone. This species covers a wide morphospace, showing a thin basal core when encrusting as well as a thick one (more than twice the thickness) when bridging (Fig. 13). The rhodoliths, as well as the framework, have a boxwork texture.

The growth form of coralline algae is generally interpreted to be controlled by ecological conditions (e.g. hydrodynamic energy, substrate, bioturbation). However, the growth forms of coralline algal species are influenced also by the general ability of taxa to form certain growth features (Rasser & Piller, 2004). In the studied succession, *Lithothamnion* cf. *macrosporangicum* is abundant in all intervals and acts either as the main rhodolith former by encrustation of various nuclei, or as a settler on loose substrates, where it forms complex crustal frameworks (boxwork texture). Consequently, the high abundance and success of *Lithothamnion* cf. *macrosporangicum* seems to be due to its independence of a specific substrate type, showing a great ability to change its growth form from encrusting to lumpy.

From a general taxonomic view, the rhodoliths are mainly composed by melobesioids (*Lithothamnion*, *Mesophyllum*). Lithophylloids (*Lithophyllum*) and mastophoroids (*Spongites*) occur only subordinately. This assemblage, and in particular the dominance of the genera *Lithothamnion* and *Mesophyllum*, indicates a deeper water setting as is also reported from other Mediterranean Miocene examples (see review in Braga *et al.*, this volume, pp. 167–184). In the studied outcrop it also coincides with the general reconstruction of the platform architecture and location of the coralline locality on the middle ramp (Brandano, 2001; Brandano & Corda, 2002).

The diversity of coralline algae in the five intervals shows a similar species number (7–8) for all intervals, except for interval 4 (LR9), where only two species are recorded. This reflects a drop in species richness and can be related to the distinct change in substrate availability. Although oysters form the dominant substrates in intervals 2 and 3 (LR5–LR8), their size and density is much higher in interval 4 (LR9). In most cases the sediment is therefore not made up predominantly by rhodoliths, but represents a bind/framestone-type. Even the rhodoliths there show a different texture, of boxwork-type.

Initiation of most rhodoliths studied is by encrustation of a nucleus. Nuclei are predominantly by neopycnodont oysters, in some intervals bryozoans are also important. Due to their relatively small sizes, nodular bryozoan nuclei play no essential role in the final shape of rhodoliths. However, the very large size of neopycnodont oysters, controls the shapes of the final nodules (Fig. 9).

All-side growth of rhodoliths requires a periodic movement of the initially encrusted nuclei (incipient rhodoliths). The main agent for this movement is considered to be wave or current energy. However, it is known that benthic organisms (nekto-benthic, epifaunal and mesobenthic to semi-infaunal) may also have a large effect on rhodolith movement (Steneck, 1986; Marrack, 1999). Such organisms may be ground-stirring fish, crustaceans, or echinoderms (echinoids, asterozoans, holothurians). Rhodolith formation due to overturning by high water energy can be excluded in the current example, since apical abrasion of columns and branches does not occur. The high amount and all-around occurrence of echinoid remains, particularly coronal plates and spines of regular echinoids, indicates that the rhodolith fields were densely settled by these echinoderms, and that they also may have caused or essentially supported the formation of rhodoliths by overturning.

The very intense bioerosion of all rhodoliths and oyster shells clearly indicates a long exposure time for these components and, consequently, a low sedimentation rate. Major bioeroders are clionid sponges, molluscs, echinoids and a variety of microborers. For the production of fine sediment, this bioerosion is a very important process. The primary sediment particles <2 mm are dominated by bryozoans, foraminifera and echinoderms (echinoids, ophiuroids). The bioerosion of oyster shells and coralline algae produces angular bioclasts and fine sand to silt that is also composed of angular grains. These angular components indicate a lack of currents and the absence of substantial sediment transport.

The component composition reflects a decrease of coralline algal fragments and larger foraminifera upsection, and an increase in bryozoan colonies, particularly in level 4 (LR9). The degree of bioerosion mostly remains constant for all intervals, but the amount of fragmentation decreases upwards. This change in biotic composition may either reflect a deepening upward trend, or an increase in turbidity up-section. This turbidity increase could be attributed to an increasing nutrient input.

CONCLUSION

The Monte Lungo Mandrella section of the Early Miocene middle-ramp deposits of the Latium-Abruzzi carbonate platform includes a rhodolith interval. This interval is 7 m thick and is composed of 5 levels reflecting a distinct vertical sequence, including from bottom to top: (1) coralline branch rudstone; (2) rhodolith floatstone; (3) rhodolith rudstone; (4) coralline algal–oyster rudstone to bind/framestone; (5) coralline branch rudstone.

Rhodoliths predominantly show nuclei made by neopycnodont oysters and bryozoan colonies. The non-geniculate coralline algal flora is comprised of 12 taxa, out of which *Lithothamnion* cf. *macrosporangicum* is most abundant in all levels. Besides this species, *Mesophyllum gignouxi* and M. cf. *inaspectum* frequently occur. In addition, *Lithothamnion* sp. 1 is common in level 4 and *Spongites albanensis* is abundant in level 1.

Coralline algal diversity is similar in all levels with the exception of level 4, where only two species constitute the flora. The major reason for this reduction in species richness is considered to be specific substrate availability, namely a dense coverage by large neopycnodont oysters. Since *Lithothamnion* cf. *macrosporangicum* is able to grow not only on hard but also on soft substrates, and even bridge between larger components, it acts as a rhodolith builder as well as a framework constructor. The result is the coralline algal–oyster rudstone to bind/framestone of level 4.

Intensive bioerosion of rhodoliths suggests low sedimentation rates, but there is no indication that a high current regime was present. This excludes movement and formation of rhodoliths by raised water energy, but points to a mainly biogenically induced periodic turning of the rhodoliths. Regular echinoid bioclasts are a major sediment constituent in the rhodolith interval and these may represent the most important candidates for rhodolith movement, supporting their all-sided growth.

ACKNOWLEDGMENTS

We are grateful to Laura Corda and Giacomo Civitelli for helpful discussions. The reviews by André Freiwald, Lucia Simone and Maria Mutti helped to improve the manuscript and are gratefully acknowledged. This work was supported by MIUR (progetti della Facoltà di Scienze M.F.N., Università di Roma "La Sapienza").

REFERENCES

Accordi, B., Devoto, G., La Monica, G.B., Praturlon, A., Sirna, G. and **Zalaffi, M.** (1967) Il Neogene nell'Appennino laziale-abruzzese. Commitee Mediterranean Neogene Stratigraphy, (1969), Proc. IV Session, Bologna, Giorn. Geol., **35**, 235–268.

Adey, W.H. and **Adey, P.J.** (1973) Studies on the biosystematics and ecology of the epilithic crustose corallinaceae of the British Isles. *Brit. Phycol. J.*, **8**, 343–407.

Airoldi, M. (1932) Contributo allo studio delle Corallinacee del terziario italiano. I Le Corallinacee dell'Oligocene ligure-piemontese. *Paleont. Ital. Mem. Paleont.*, **33**, 55–83.

Barbera, C., Simone, L. and **Carannante, G.** (1978) Depositi circalittorali di piattaforma aperta nel Miocene Campano, analisi sedimentologica e paleoecologica. *Boll. Soc. Geol. Ital.*, **97**, 821–834.

Basso, D. (1998) Deep rhodolith distribution in the Pontian Islands, Italy: a model for the paleoecology of a temperate sea. *Palaeoeogr. Palaeoclimatol. Palaeoecol.*, **137**, 173–187.

Bizzozero, G. (1885) *Flora Veneta Crittogramica, Parte II. Seminario*, Padova, 255 pp.

Bosellini, A. and **Ginsburg, R.N.** (1971) Form and internal structure of recent algal nodules (rhodolites) from Bermuda. *J. Geol.*, **79**, 669–682.

Bosence, D.W.J. (1983a) Description and classification of rhodoliths (rhodoids, rhodolites). In: *Coated Grains* (Ed. T.M. Peryt), pp. 217–224. *Springer-Verlag*, Berlin.

Bosence, D.W.J. (1983b) The occurrence and ecology of recent rhodoliths. In: *Coated Grains* (Ed. T.M. Peryt), pp. 225–242. *Springer-Verlag*, Berlin.

Bosence, D.W.J. (1991) Coralline algae: mineralization, taxonomy, and palaeoecology. In: *Calcareous algae and stromatolites* (Ed. R. Riding). *Springer-Verlag*, Berlin, pp. 98–113.

Bosence, D.W.J. and **Pedley, H.M.** (1982) Sedimentology and palaeoecology of a Miocene coralline algal biostrome from the Maltese islands. *Palaeoeogr. Palaeoclimatol. Palaeoecol.*, **38**, 9–43.

Braga, J.C. and **Aguirre, J.** (2001) Coralline algal assemblages in Upper Neogene reef and temperate carbonates in Southern Spain. *Palaeoeogr. Palaeoclimatol. Palaeoecol.*, **175**, 27–41.

Braga, J.C. and **Martín, J.M.** (1988) Neogene coralline-algal growth-form and their palaeoenvironments in Almanzora River Valley (Almeria, S. E. Spain). *Palaeoeogr. Palaeoclimatol. Palaeoecol.*, **67**, 285–303.

Braga, J.C., Bosence, D.W.J. and **Steneck, R.S.** (1993) New anatomical characters in fossil coralline algae and their taxonomic implications. *Palaeontology*, **36**, 535–547.

Brandano, M. (2001) *Risposta fisica delle aree di piattaforma carbonatica agli eventi piu significativi del Miocene nell'Appennino centrale.* Unpubl. PhD thesis, Univ. of Rome "La Sapienza". pp. 180.

Brandano, M. and **Corda, L.** (2002) Nutrients, sea level and tectonics: constrains for the facies architecture of a Miocene carbonate ramp in central Italy. *Terra Nova*, **14**, 257–262.

Carannante, G. and **Simone, L.** (1996) Rhodolith facies in the central-southern Apennines mountains, Italy. In:

Models for Carbonate Stratigraphy from Miocene Reef Complexes of Mediterranean Regions (Eds E. K. Franseen, M. Esteban, W. Ward and J. Rouchy). *SEPM Concepts Sedimentol. Paleontol.*, **5**, 261–275.

Catenacci, V., **Matteucci, R.** and **Schiavinotto, F.** (1982) La superficie di trasgressione alla base dei "Calcari a Briozoi e Litotamni" nelle Maiella Meridionale. *Geol. Romana*, **21**, 559–575.

Civitelli, G. and **Brandano, M.** (2005) Atlante delle litofacies e modello deposizionale dei Calcari a Briozoi e Litotamni nella Piattaforma carbonatica laziale-abruzzese. *Boll. Soc. Geol. Ital.*, **124**, 611–643.

Conti, S. (1943) Contributo allo studio delle Corallinacee del Terziario italiano. II Corallinacee del Miocene ligure-piemontese. *Paleont. Ital. Mem. Paleont.*, **41**, 37–61.

Damiani, A.V. (1990) Studi sulla Piattaforma laziale-abruzzese. Nota II. Contributo alla interpretazione della evoluzione tettonico sedimentaria dei Monti Affilani e "pre-Ernici" e cenni sui rapporti con le adiacenti aree appenniniche. *Mem. Descr. Carta Geol. Ital.*, **38**, 177–206.

Damiani, A.V., **Chiocchini, M.**, **Colacicchi, R.**, **Mariotti, G.**, **Parotto, M.**, **Passeri, L.** and **Praturlon, A.** (1991) Elementi litostratigrafici per una sintesi delle facies carbonatiche meso-cenozoiche dell'Appennino centrale. In: *Studi Preliminari all'Acquisizione Dati del Profilo **CROP 11** Civitavecchia-Vasto* (Eds M. Tozzi, G.P. Cavinato and M. Parotto). *Stud. Geol. Camerti, Vol. Spec.*, **1991/2**, 187–213.

Devoto, G. (1967) Note geologiche sul settore centrale dei Monti Simbruini ed Ernici (Lazio nord-orientale). *Boll. Soc. Nat. Napoli*, **76**, 112 pp.

Foslie, M.H. (1901) New Forms of Lithothamnia. K. Norske Videnskab. Selskabskr. 1901.

Franseen, E.K., **Esteban, M.**, **Ward, W.** and **Rouchy, J.** (1996) Models for Carbonate Stratigraphy from Miocene Reef Complexes of Mediterranean Regions. *SEPM Concepts Sedimentol. Paleontol.*, **5**, 391 pp.

Freiwald, A. (1994) Sedimentological and biological aspects in the formation of branched rhodoliths in northern Norway. *Beitr. Pälaont. Osterr.*, **20**, 7–19.

Gischler, E. and **Pisera, A.** (1999) Shallow water rhodoliths from Belize reefs. *N. Jb. Geol. Paläont. Abh.*, **214**, 71–93.

Lamouroux, J.V.F. (1812) Classification des Ploipyiers coralligènes. *Bull. Phil.*, **3**, 181–188.

Lemoine, M. (1939) Les algues calcaires fossiles de l'Algérie. *Matér. Carte Géol. l Algérie Ser.1 Paléontol.*, **9**, 1–128.

Lemoine, P. (1924) Contribution a l'étude des Corallinacées fossiles, VII, Mélobésiées miocènes recueilles par M. Bougart en Albanie. *CR Somm. Bull. Soc. Géol. Fr.*, **4**, 275–283.

Marrack, E.C. (1999) The relationship between water motion and living rhodolith beds in the southwestern Gulf of California, Mexico. *Palaios*, **14**, 159–171.

Mastrorilli, V.I. (1950) Corallinacee fossili del Calabriano di Miradolo. *Atti. Ist. Geol. Univ. Pavia*, **4**, 57–67.

Mastrorilli, V.I. (1974) Affioramenti neogenici a Corallinacee in Italia. *Bur. Rech. Géol. Min. Mém.*, **78**, 525–532.

Ogniben, L. (1958) Melobesie basso-elveziane di Caiazzo (Caserta). *Palaeontogr. Ital.*, **53**, 49–71.

Parotto, M. and **Praturlon, A.** (1975) Geological summary of the Central Appennines. In: *Structural Model of Italy. CNR Quad. Ric. Sci.*, **90**, 256 pp.

Patacca, E., **Scandone, P.**, **Bellatalla, M.**, **Perilli, N.** and **Santini, U.** (1991) La zona di giunzione tra l'arco appenninico settentrionale e l'arco appenninico meridionale nell'Abruzzo e nel Molise. In: *Studi Preliminari all'Acquisizione Dati del Profilo **CROP 11** Civitavecchia-Vasto* (Eds M. Tozzi, G.P. Cavinato and M. Parotto). *Stud. Geol. Camerti, Vol. Spec.*, **1991/2**, 417–441.

Pomar, L., **Brandano, M.** and **Westphal, H.** (2004) Environmental factors influencing skeletal-grain sediment associations: A critical review of Miocene examples from the Western Mediterranean. *Sedimentology*, **51**, 627–651.

Prager, E.J. and **Ginsburg, R.N.** (1989) Carbonate nodule growth on Florida's outer shelf and its implications for fossil interpretations. *Palaios*, **4**, 310–317.

Rabenhorst, I. (1863) *Kryptogamen-Flora von Sachsen, der Ober-Lausitz, Thuringen und Nordbőhmen.* E. Krummer, Leipzig, 653 pp.

Rasser, M.W. and **Piller, W.E.** (1999) Application of neontological taxonomic concepts to Late Eocene coralline algae (Rhodophyta) of the Austrian Molasse Zone. *J. Micropalaeontol.*, **18**, 67–80.

Rasser, M.W. and **Piller, W.E.** (2004) Crustose algal frameworks from the Eocene Alpine Foreland (Austria). *Palaeogeogr. Palaeoclimatol. Palaeoecol.*, **206**, 21–39.

Riegl, B. and **Piller, W.E.** (1999) Framework revisited: Reefs and coral carpets of the northern Red Sea. *Coral Reefs*, **18**, 305–316.

Setchell, W.A. (1943) Mastophora and the Mastophoreae: genus and subfamily of Corallinaceae. *Proc. Natl Acad. Sci. USA*, **29**, 127–135.

Silva, P.C. and **Johansen, H.W.** (1986) A reappraisal of the order Corallinales (Rhodophyceae). *Brit Phycol. J.*, **21**, 245–254.

Steneck, R.S. (1986) The ecology of coralline algal crusts: convergent patterns and adaptive strategies. *Ann. Rev. Ecol. Syst.*, **17**, 273–303.

Vannucci, G., **Piazza, M.**, **Fravega, P.** and **Arnera, V.** (1993) Le rodoliti del Miocene inferiore del settore SW del Bacino Terziario del Piemonte (Spigno Monteferrato – Alessandria). *Atti Soc. Tosc. Sc. Nat. Mem.*, **100**, 93–117.

Vannucci, G., **Piazza, M.**, **Fravega, P.** and **Abate, C.** (1996) Litostratigrafia e paleoecologia di successioni a rodoliti della «Pietra da Cantoni» (Monferrato orientale, Italia Nord-occidentale). *Atti Soc. Tosc. Sci. Nat., Mem., Ser. A.* **103**, 69–86.

Verheij, E. (1993) The genus Sporolithon (Sporolithaceae fam. nov., Corallines, Rhodophyta) from the Spermonde Archipelago, Indonesia. *Phycologia*, **32**, 184–196.

Wettstein, R.R. (1901) *Handbuch der systematischen Botanik*, **1**. Deuticke, Leipzig, 201 pp.

Woelkerling, W.J. (1988) *The Coralline Red Algae: an Analysis of the Genera and Subfamilles of Nongeniculate Corallinaceae.* British Museum (Natural History), London and *Oxford University Press*, Oxford, 268 pp.

Woelkerling, W.M.J., **Irvine, L.M.** and **Harvey, A.S.** (1993) Growth forms in non-geniculate coralline red algae (Corallinales, Rhodophyta). *Aust. Syst. Bot.*, **6**, 277–293.

Int. Assoc. Sedimentol. Spec. Publ. (2010) **42**, 165–182

Palaeoenvironmental significance of Oligocene–Miocene coralline red algae – a review

JUAN C. BRAGA*, DAVIDE BASSI† and WERNER E. PILLER‡

**Departamento de Estratigrafía y Paleontología, Universidad de Granada, Campus de Fuentenueva, E-18002 Granada, Spain (E-mail: jbraga@ugr.es)*
†Dipartimento di Scienze della Terra Università degli Studi di Ferrara, Via Saragat 1, I-44122 Ferrara, Italy
‡Institut für Erdwissenschaften – Geologie und Paläontologie, Karl-Franzens-Universität Graz, Heinrichstrasse 26, A-8010 Graz, Austria

ABSTRACT

Coralline red algae are common in Oligocene and Miocene marine shallow-water carbonate and siliciclastic rocks, as well as deep-water re-deposited sediments containing particles removed from platforms. Corallines are mostly reported in reef-related carbonates but are also the main components in shallow-water heterozoan carbonates from temperate regions. The known distribution of Oligocene coralline assemblages does not suggest any palaeobiogeographical differentiation. In contrast, for the Miocene, the occurrence of taxa still living today with restricted geographic distribution supports an actualistic approach and the rough differentiation of palaeobiogeographic regions as follows: (a) a tropical region (characterised by thick *Hydrolithon* plants and *Aethesolithon*); (b) a subtropical Mediterranean (with common *Spongites* and *Neogoniolithon* species); and (c) a temperate region with shallow-water assemblages dominated by *Lithophyllum*.

In a few examples from the northern margin of the western Tethys, the correlation of Oligocene carbonate-facies and algal assemblages indicates a dominance of *Lithothamnion* species in the shallower environments, while *Mesophyllum* is most abundant in deeper platform settings. The taxonomic composition of Miocene coralline assemblages and growth forms changes with depth, having patterns similar to those in the algal associations in present-day marine platforms. Mastophoroids and lithophylloids (*Aethesolithon, Hydrolithon, Neogoniolithon, Spongites,* and *Lithophyllum* species) characterize the shallower assemblages, whereas the melobesioids *Lithothamnion* and *Mesophyllum* and the sporolithacean *Sporolithon* are dominant in deeper-water settings. The common algal nodules (rhodoliths) comprise thick plants of few species in the shallowest palaeoenvironments, while in deeper platform areas they are composed of more diverse algal assemblages, with thin encrusting and protuberant-branching growth forms.

Keywords Coralline red algae, Oligocene, Miocene, palaeobiogeography, palaeoenvironment.

INTRODUCTION

Coralline red algae are common components in Oligocene and Miocene platform carbonates (Halfar & Mutti, 2005). They are also frequent in shallow-water marine siliciclastic rocks, and in deep sediment gravity deposits as reworked particles from shallow-water environments. Present-day coralline algae occur from the equator to subpolar areas and extend from intertidal environments to as deep as 270 m on marine bottoms (Adey & Macintyre, 1973; Littler *et al.*, 1985). They thrive in a wide range of trophic conditions, from oligotrophic reef environments, such as reef crests in which they can be the major builders (Bosence, 1984; Borowitzka & Larkum, 1986), to mesotrophic waters on marine platforms in diverse latitudes (Adey & Macintyre, 1973). Coralline algae can encrust soft plants (algae, seagrass) or hard substrates (rocks, coral skeletons, shells) throughout their depth range. They can also grow unattached on the sea floor as loose individual plants or forming nodules (rhodoliths). In present-day seas, rhodoliths with living algal covers are found in intertidal and shallow subtidal environments (Bosellini & Ginsburg, 1971; Bosence, 1976; Scoffin *et al.*, 1985; Marrack, 1999; Steller *et al.*, 2003) to depths of more than 100 metres (Bosence, 1983b; Matsuda & Tomiyama, 1988; Prager & Ginsburg, 1989; Marshall *et al.*, 1998; Lund *et al.*, 2000; Steller *et al.*, 2003).

The ubiquitous and cosmopolitan character of coralline algae is reflected in their widespread occurrence throughout a large array of facies and biogeographical settings in the fossil record, since the appearance of the group in the Early Cretaceous (Arias *et al.*, 1995). Their skeletons, composed of high-Mg calcite that impregnates the cell walls, confer on these algae a high preservation potential (Bosence, 1991). Oligocene–Miocene coralline red algae are important components in photozoan and heterozoan associations from tropical to cool-water carbonates and have been reported from the Pacific and Southern Oceans to the Atlantic Ocean over a large interval of latitude (Tables 1 and 2, Figs 1 and 2). However, the taxonomic components of living coralline algal assemblages are distinct in different geographical regions and they and their growth forms also vary along palaeoenvironmental gradients within a given region. The genera and species of coralline algae change from warm to cool waters as well as with depth and other environmental gradients. These variations in assemblage composition and plant morphology can be used to interpret certain palaeoecological parameters of the settings in which fossil coralline algae grew.

This potential use in palaeoenvironmental reconstruction is nonetheless seriously limited by the long-lasting disparity between the botanical and palaeontological criteria for identification of genera and species of coralline algae. The diagnostic characters traditionally used to establish many fossil taxa are ambiguous or meaningless according to the modern criteria used for delimiting living taxa from the species to family level (Bosence, 1983b; Braga *et al.*, 1993; Braga & Aguirre, 1995; Aguirre *et al.*, 1996). In addition, the validity of hundreds of fossil species names is questionable since they have been based only on the cell sizes of thallus fragments or other features of uncertain taxonomic significance (Bosence, 1983b; Braga & Aguirre, 1995). The illustrations and original diagnoses of many fossil genera and species are inadequate from a modern perspective and the interpretations of particular taxa by later authors do not always coincide. Moreover, many palaeoalgologists working on coralline algae have

Table 1. Country/region and age of reported Oligocene coralline algal assemblages

Country/Region	Age	Reference
N Spain (1)	Oligocene	Miranda, 1935
SW France (Aquitaine) (2)	Oligocene	Lemoine, 1938 Poignant, 1972
Algeria (3)	Oligocene	Lemoine, 1939
NW Italy/ Ligurian-Piedmont Basin (4)	Oligocene	Airoldi, 1930 Airoldi, 1932 Conti, 1950 Mastorilli, 1968 Giammarino *et al.*, 1970 Fravega *et al.*, 1988 Fravega *et al.*, 1993a, b Fravega *et al.*, 1994 Vannucci *et al., 1997*
	E Oligocene	Fravega Vannucci, 1980a
	Rupelian	Fravega & Vannucci, 1980b
	M-L Oligocene	Fravega *et al.*, 1987
	L Oligocene	Vannucci *et al.*, 1993
NE Italy/ Venetian area (5)	Oligocene	Mastrorilli, 1973
	Rupelian	Francavilla *et al.*, 1970
Northern Calcareous Alps (6)	Oligocene	Rasser and Nebelsick, 2003
N Slovenia (7)	Rupelian	Bassi & Nebelsick, 2000; Nebelsick *et al.*, 2000
S Italy (Salento) (8)	Oligocene	Bosellini and Russo, 1992
Maltese Islands (9)	Oligocene	Budai & Wilson, 1980
Macedonia (10)	Oligocene	Lemoine, 1977
Greece (11)	Oligocene	Johnson, 1965
Romania/NW Transylvania (12)	E Oligocene	Bucur *et al.*, 1989
N Iraq (13)	Oligocene	Elliott, 1960
Oman (14)	Oligocene	Elliott, 1960
W India (15)	Oligocene	Tandon *et al.*, 1977 Misra *et al.*, 2001
Andaman Archipelago (16)	Oligocene	Badve and Kundal, 1989
Borneo (17)	Oligocene	Johnson, 1966
Sulu Seas, Philippines (18)	Oligocene	Wiedicke, 1987
Indonesia (18)	Oligocene	Saller & Vijaya, 2000
Japan (19)	Oligocene	Ishijima, 1954, 1956, 1979 Hiroshi, 1998
Guam (20)	Oligocene	Johnson, 1964b
New Zealand (21)	Oligocene	MacGregor, 1983 Burgess and Anderson, 1983 Hood *et al.*, 2003
S Georgia/N Florida (22)	E Oligocene	Manker & Carter, 1987
Puerto Rico (23)	Oligocene	Frost *et al.*, 1983
Trinidad (24)	Oligocene	Johnson, 1955

Numbers in parentheses indicate geographic locations in Fig. 1. E, Early; M, Mid-; L, Late.

Table 2. Country/region and age of reported Miocene coralline algal assemblages

Country/Region	Age	Reference
SW Spain (1)	L Miocene	Civis *et al.*, 1994
SE Spain (2)	M-L Miocene	Segonzac, 1990
	Langhian	Braga *et al.*, 1996
	Tortonian	Braga & Martín, 1988
	L Miocene	Segonzac, 1972
		Mankiewicz, 1996
		Braga & Aguirre, 2001
S Aquitaine (3)	Serravallian	Buge *et al.*, 1973
		Boulanger and Poignant, 1975
		Orszag-Sperber *et al.*, 1977
Algeria (4)	Burdigalian–Tortonian	Lemoine, 1939
Mallorca (5)	L Miocene	Pomar, 1991
		Perrin *et al.*, 1995
Menorca (5)	Tortonian	Brandano *et al.*, 2005
N Corsica (6)	late Burdigalian–Tortonian	Orszag-Sperber and Poignant, 1972
		Orszag-Sperber *et al.*, 1977
S Corsica (6)	Miocene	Bellini and Mastrorilli, 1975
W Sardinia (7)	Messinian	Fravega *et al.*, 1988
NW Italy/Ligurian-Piedmont Basin (8)	Miocene	Conti, 1942, 1943, 1946a–c
	E Miocene	Scuttenhelm, 1976
		Vannucci *et al.*, 1994
	Aquitanian–Serravallian	Vannucci *et al.*, 1996
	Burdigalian	Fravega *et al.*, 1988
		Fravega *et al.*, 1993b
	Burdigalian–Tortonian	Fravega *et al.*, 1993a
	Serravallian–Tortonian	Fravega and Vannucci, 1987
	Serravallian	Vannucci, 1980
		Fravega & Vannucci, 1982
		Fravega *et al.*, 1984a
NE Italy (8)	Aquitanian	Scudeller Bacceller *et al.*, 1988
Tuscany (9)	E Messinian	Fravega *et al.*, 1994
Central Italy (10)	E Miocene	Simone & Carannante, 1985
		Carannante & Simone, 1996
S Italy (10)	Messinian	Bosellini *et al.*, 2001
Malta (11)	Tortonian	Bosence and Pedley, 1982
		Bosence, 1983a
Island of Lampedusa (11)	L Miocene	Fravega *et al.*, 1998
Albania (12)	E Miocene	Lemoine, 1924
Austria (13)	Miocene	Piller, 1993
	Badenian	Conti, 1946a, 1946b
	Sarmatian	Kamptner, 1942
		Piller & Harzhauser, 2005
Slovakia (13)	Miocene	Schaleková, 1969
S Poland (14)	Badenian	Pisera, 1985
	M Miocene	Studencki, 1988; Pisera & Studencki, 1989
W Ukraine (14)	Badenian	Maslov, 1956, 1962
Rumania (15)	Badenian	Bucur and Filipescu, 1994
Crimea (16)	L Miocene	Belokrys, 1981
S Turkey (17)	Aquitanian	Poisson & Poignant, 1974
		Orszag-Sperber *et al.*,1977
	E-M Miocene	Hayward *et al.*,1996
Libya (Cyrenaica) (18)	Miocene	Raineri, 1924
	M Miocene	Fravega *et al.*, 1984b
NE Egypt-Gulf of Suez (19)	Miocene	Souaya, 1963a, 1963b
	E-M Miocene	Burchette, 1988
		James *et al.*, 1988
NW Saudi Arabia (20)	M Miocene	Dullo *et al.*,1983
Lebanon (21)	Miocene	Edgell & Basson, 1975

(*Continued*)

Table 2. (*Continued*)

Country/Region	Age	Reference
Israel (21)	Langhian	Buchbinder, 1977, 1996
N Iraq (22)	Burdigalian–M Miocene	Johnson, 1964a
	E Miocene	Elliott, 1960
SW Iran (23)	M Miocene	Elliott, 1970
Somalia (24)	Miocene	Airoldi, 1937
		Bosellini *et al.*,1987
W India-Pakistan (25)	Miocene	Pal & Gosh, 1974
		Ishijima, 1975
		Kundal & Humane, 2002
		Kundal & Mude-Shyam, 2004
Andaman Islands (26)	Miocene	Chatterji & Gururaja, 1972
Indonesia (27)	Miocene	Johnson & Ferris, 1949
Philippines (27)	E Miocene	Wiedicke, 1987
Borneo (27)	Miocene	Johnson, 1966
Malaysia (27)	E Miocene	Ishijima, 1978
Philippines (27)	E Miocene	Ishijima, 1943, 1970
Ryukyu Islands	Miocene	Ishijima, 1941, 1954, 1965a
Taiwan (28)		Johnson, 1961
Japan (29)	Miocene	Ishijima, 1933, 1942, 1944, 1954, 1960, 1968, 1969
Saipan (30)	Miocene	Johnson, 1957
Guam (30)	E-L Miocene	Johnson, 1964b
NE Australia (31)	E-M Miocene	Martín & Braga, 1993
New Caledonia (32)	Miocene	Poignant & Bourrouilh-le-Jan, 1986
Vanuatu (33)	Miocene	Lemoine *et al.*, 1981
Bikini (34)	Miocene	Johnson, 1954
Eniwetok (34)	Miocene	Johnson, 1961
Keeling Islands (35)	E Miocene	Ishijima, 1965b
Guatemala (36)	Miocene	Johnson & Kaska, 1965
Cuba (37)	E-M Miocene	Beckmann & Beckmann, 1966
Martinique (38)	Miocene	Lemoine, 1918
Trinidad (39)	E Miocene	Howe, 1919

Numbers in parentheses indicate geographic locations in Fig. 2. E, Early; M, Mid-; L, Late.

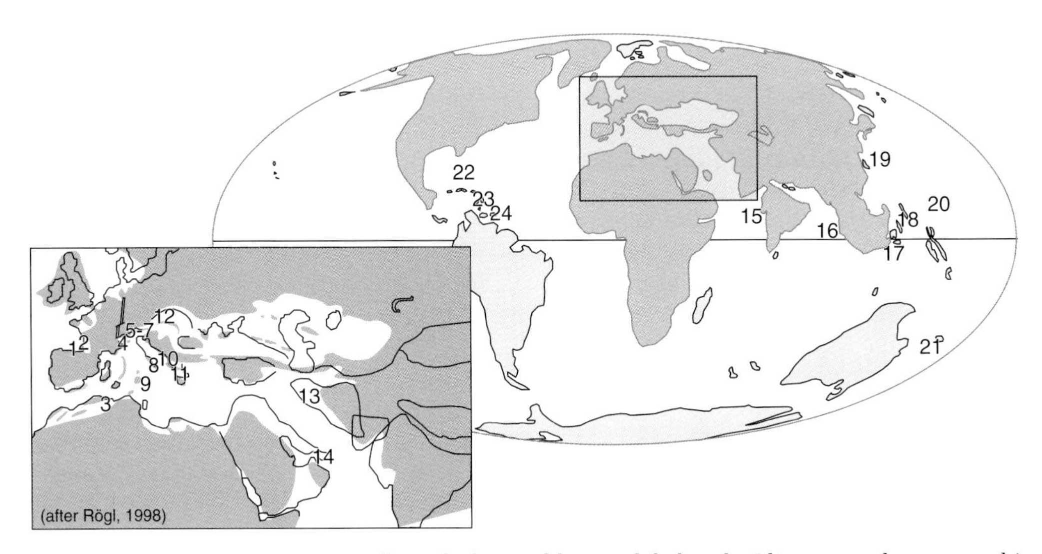

Fig. 1. Distribution of reported Oligocene coralline algal assemblages. Global Early-Oligocene palaeogeographic map after Perrin (2002); inset of the western Tethys after Rögl (1998). Numbers refer to those of Table 1.

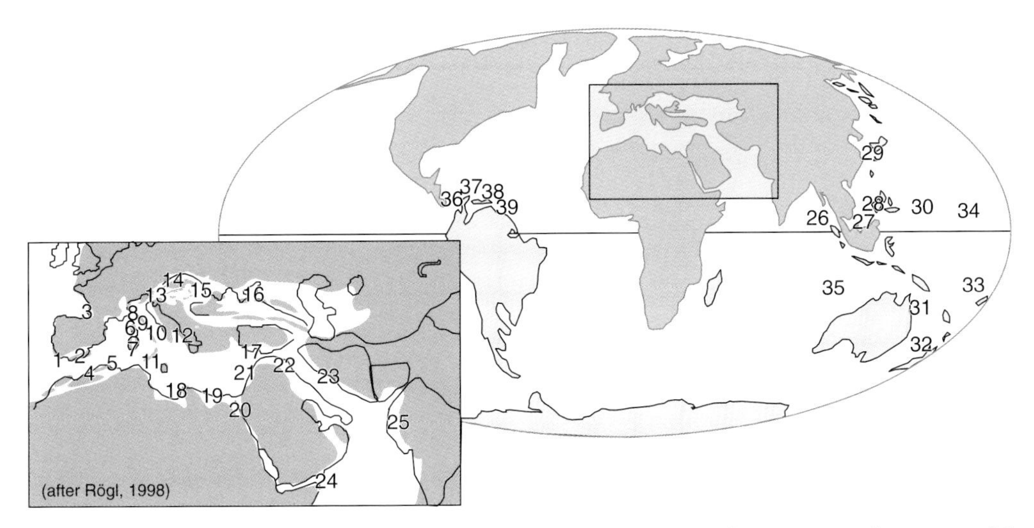

Fig. 2. Distribution of reported Miocene coralline algal assemblages. Global Early-Miocene palaeogeographic map after Perrin (2002); inset of the western Tethys after Rögl (1998). Numbers refer to those of Table 2.

preferred to establish new species based on features of questionable taxonomic value instead of attempting to attribute specimens to previously described species (Aguirre & Braga, 2005). This lack of a consistent and reliable taxonomy of fossil corallines at the species level is a major obstacle for the study of this group and its application to interpreting ancient deposits. In addition, it has also prevented attempts to use coralline algae taxa as biostratigraphic markers (Mastorilli, 1968; Moussavian, 1984).

Data from literature surveys suggest that the Oligocene–Miocene interval was the period of maximum richness in coralline algal species since the Early Cretaceous. Coralline algal diversity peaked in the Early Miocene, with 245 species reported in the Aquitanian (Aguirre *et al.*, 2000). However, it is very difficult to assess the taxonomic meaning of most of these species names, and to evaluate the accuracy of their use by later authors after being established. As a result of these taxonomic uncertainties at the species level, palaeobiogeographic and palaeoenvironmental studies of Oligocene–Miocene coralline algae must mainly be based upon the distribution of genera and higher-level taxa. Table 3 presents an identification key to Oligocene and Miocene genera following botanical taxonomy and using characters preservable in fossil plants.

OLIGOCENE CORALLINE ALGAE

Oligocene coralline algae have primarily been reported from central and southern Europe, palaeogeographically corresponding to the northern margin of the western Tethys (Table 1, Fig. 1). Most studies were carried out in the Piedmont and Ligurian basins and in the Venetian area by authors from the University of Genoa (Airoldi, 1930; Conti, 1950; Mastorilli, 1968; Fravega *et al.*, 1987; further references in Table 1). Very large numbers of species of Oligocene coralline algae have been described in these regions; for example, 180 species have been recorded in the coralline assemblages in the Piedmont Basin since they were first studied in the 1930s. Several other studies report Oligocene species (Johnson, 1965; Lemoine, 1977; Bucur *et al.*, 1989) or refer to the occurrence of coralline algae (Geel, 2000; Stoklosa & Simo, 2000; Kaiser *et al.*, 2001) in various localities in southern Europe. In recent years, a few papers have applied new taxonomical concepts to document Oligocene algal assemblages (Bassi & Nebelsick, 2000; Rasser & Nebelsick, 2003). The occurrence of coralline algae is also recorded at the southern margin of the western Tethys, from Oman to Algeria (Lemoine, 1939; Elliott, 1960; Budai & Wilson, 1980; Bosellini & Russo, 1992).

Many species of non-geniculate and geniculate corallines have been recognized in the Oligocene rocks of Western India (Ishijima, 1954; Tandon *et al.*, 1977; Misra *et al.*, 2001; Kundal & Humane, 2002). However, taxonomic descriptions and reports on the occurrence of coralline algae are relatively scarce in the eastern Tethys, in the large area corresponding to SE Asia and the Pacific Ocean (Ishijima, 1956, 1979; Johnson, 1964b,

Table 3. Key to Oligocene and Miocene coralline algal families, subfamilies and genera

1. Calcified sporangial compartments, sori	**Sporolithaceae** *Sporolithon*
1. Sporangial conceptacles	2
2. Sporangial conceptacles uniporate	Corallinaceae
2. Sporangial conceptacles multiporate	Hapalidiaceae
Corallinaceaea	
3. Secondary pits as predominant interfilamental cell connections	Lithophylloideae
3. Cell fusions as predominant interfilamental cell connections	4
4. Genicula present	Corallinoideae
4. Genicula absent	Mastophoroideae
Lithophylloideae	
5. Genicula present	*Amphiroa*
5. Genicula absent	*Lithophyllum*
Mastophoroidea	
6. Thallus with a conspicuous ventral layer of palisade-cell filaments	7
6. Thallus lacking a conspicuous ventral layer of palisade-cell filaments	8
7. Sporangial conceptacles with a central columella, roofs formed only by filaments peripheral to the sporangial initials	(*Mastophora*)
7. Sporangial conceptacles lacking a central columella, roofs formed by filaments peripheral and interspersed amongst the sporangial initials	*Lithoporella*
8. Sporangial conceptacle pore canals lined by cells orientated more or less perpendicularly to the roof surface.	*Hydrolithon*
8. Sporangial conceptacle pore canals lined by cells orientated more or less parallel to the roof surface.	9
8. Ventral layer of irregularly shaped, large cells. Cells in peripheral filaments strongly fused.	*Karpathia*
8. Coaxial core? Cells in peripheral filaments strongly fused (large polygonal cells)	*Aethesolithon*
* 9. Coaxial core (spermatangia formed both on the floors and roofs of male conceptacle chambers)	*Neogoniolithon*
* 9. Non-coaxial core (spermatangia only formed on the floors of male conceptacle chambers)	*Spongites*
Hapalidiaceae	
Melobesioideae	
10. Thallus dimerous	*Melobesia*
10. Thallus monomerous	11
11. Distal walls of terminal epithallial cells flattened and flared	*Lithothamnion*
11. Distal walls of terminal epithallial cells rounded or flattened but not flared	12
12. Subepithallial initials as short as or shorter than the cells immediately subtending them	*Phymatolithon*
12. Subepithallial initials as long as or longer than the cells immediately subtending them	13
* 13. Coaxial core (male conceptacles with only unbranched spermatangial filaments)	*Mesophyllum*
* 13 Non coaxial core	(*Synarthrophyton* *Clathromorphum*)

Key follows the supra-generic taxonomy proposed by Harvey *et al.* (2003). In parentheses: extant genera not recorded in Oligocene and Miocene deposits but possessing thallus structures that can be preserved as fossils. Other genera not recorded (small and weakly calcified) have been excluded. Asterisks indicate ancillary characteristics used in fossil corallines (Braga *et al.*,1993; Braga, 2003), as the actual diagnostic characteristics (in parentheses) have very low preservation potential. *Subterraniphyllum* Elliott has not been included, as its reproductive structures are unknown. See Braga *et al.* (1993) and Braga (2003) for further information on terms of coralline anatomy.

1966; Wiedicke, 1987; Badve & Kundal, 1989; Hiroshi, 1998; Saller & Vijaya, 2000). Several coralline genera have been reported in the Oligocene algal limestones from New Zealand (Burgess & Anderson, 1983; Mac Gregor, 1983; Hood *et al.*, 2003). Miranda (1935) and Lemoine (1938) described coralline species from the Oligocene deposits in the Atlantic-linked western Pyrenean Basin. The occurrence of Oligocene corallines in the western central Atlantic has also been documented (Johnson, 1955; Frost *et al.*, 1983; Manker & Carter, 1987) (Table 1, Fig. 1).

In most of the reported localities, Oligocene coralline algae occur in coral reef deposits or laterally associated sediments, or in basins within the Oligocene reef belt (see Perrin, 2002). In contrast, the coralline algae in the Oligocene of New Zealand are major components of cool-water carbonates

(MacGregor, 1983; Hood *et al.*, 2003). The geographical distribution of sampling of Oligocene corallines is uneven, with a high concentration of study sites and publications in the Mediterranean region. However, current knowledge of these corallines prevents an assessment as to whether this sampling bias is the only reason for the large number of species reported in the western Tethys in relation to other palaeogeographical domains, or whether there was in fact a true diversity hotspot (even if exaggerated by taxonomic practice) at the western end of the Tethys. Similarly, despite the widespread geographic distribution of coralline algae during the Oligocene and their appearance in both tropical and temperate seas, current knowledge of algal assemblages does not allow the differentiation of palaeobiogeographical regions. The uneven taxonomic approaches to the study of algal assemblages from different basins, and uncertainties concerning the meaning of the species and genus names referred to in the accounts, prevent comparisons of coralline floras from different areas in order to test similarities or divergences in their components.

Oligocene coralline algae appear both in carbonate and mixed siliciclastic-carbonate deposits. However, only papers from the last few decades have closely considered the sedimentology and palaeoenvironmental interpretation of the rocks containing the coralline algae (Table 4). Older work mainly focused on the taxonomic description of the algal taxa. Approaches to palaeoenvironmental interpretation from the algal assemblages are mainly based on comparisons with the taxonomic composition and growth forms of modern algal assemblages. Distribution models of assemblage components and growth morphologies along palaeoenvironmental gradients in deposits have only been developed in the cases of the Oligocene carbonates of the Gornji Grad Beds (North Slovenia; Nebelsick & Bassi, 2000) and the Lower Inn Valley in the Northern Calcareous Alps (Austria; Rasser & Nebelsick, 2003).

The occurrence of genera such as *Sporolithon* (as *Archaeolithothamnium*) and *Lithoporella* has been used to infer warm-water conditions by analogy with the modern distribution of these genera (e.g. Manker & Carter, 1987). In addition, the shape and size of rhodoliths have been applied to interpret water depth and turbulence (MacGregor, 1983; Kaiser *et al.*, 2001) in accordance with present-day models (Adey & Macintyre, 1973; Bosence, 1983a, 1983b; Scoffin, 1988). The correlation of carbonate-facies types and defined algal assemblages has been pointed out in a few cases (MacGregor, 1983; Bassi & Nebelsick, 2000; Nebelsick & Bassi, 2000; Nebelsick *et al.*, 2000). Cluster analysis of the relative abundance of coralline species shows that shallower facies with coarser grain-size are dominated by *Lithothamnion* species, whereas *Mesophyllum* is most abundant in deeper and finer-grained substrates, as in the case of the Lower Inn Valley. This distribution pattern has also been used as an indicator of the provenance of re-deposited carbonates (Rasser & Nebelsick, 2003).

MIOCENE CORALLINE ALGAE

Most palaeontological studies on Miocene coralline algae have concentrated on the Mediterranean region, particularly southern European localities (Table 2, Fig. 2). However, there is a large number of monographs dealing with corallines from islands in the Pacific Ocean and reports of coralline algae are distributed globally (Halfar & Mutti, 2005), in localities from the Indo-Pacific islands to southern Asia, Africa, central America, and the Caribbean islands (Table 2).

A high number of species of nongeniculate corallines have been described in Mediterranean Lower Miocene deposits, where reported Middle Miocene algae are less numerous. Long lists of species were identified in the Piedmont Basin in northern Italy (see Table 2). Recorded species are usually fewer in Lower and Middle Miocene rocks in other localities around the Mediterranean Sea (Table 2). However, Bellini & Mastrorilli (1975) reported up to forty-nine species in the Lower Miocene of southern Corsica. Long lists of species have also been documented in the Middle Miocene (Badenian) deposits of the Paratethys from Austria to Ukraine (Table 2). The number of described species decreases substantially in Sarmatian rocks (late Mid-Miocene), with two species noted in the Central Paratethys (Kamptner, 1942; Piller & Harzhauser, 2005) and one on the eastern margin of this epeiric sea (Maslov, 1956). The number of species listed from Upper Miocene deposits in Mediterranean basins is comparatively low with respect to older Miocene rocks. However, up to twenty-six species have been reported in a single basin (Tuscany; Fravega *et al.*, 1994). Detailed species inventories and descriptions have been published from western Mediterranean areas (Lemoine, 1939; Bosence, 1983a; Braga &

Table 4. Selected cases of palaeoenvironmental interpretation of coralline algal occurrences in Oligocene and Miocene deposits

Palaeoenvironment	Coralline occurrence type	Region, age	Reference
Reef-rimmed shelf			
Back reef, lagoon	Rhodoliths in channels, rhodoliths	Mallorca, L Miocene	Pomar, 1991; Pomar *et al.*, 1996;
		SE Spain, Messinian	Braga & Aguirre, 2001
Reef core	Crusts and small branching frameworks on corals, rhodoliths in pockets and channels	SE Italy, Chattian	Bosellini & Russo, 1992
		Gulf of Suez, Burdigalian–Langhian	Burchette, 1988; James *et al.*, 1988
		Mallorca, L Miocene	Pomar, 1991; Perrin *et al.*, 1995; Pomar *et al.*, 1996
		SE Spain, Messinian	Riding *et al.*, 1991; Mankiewicz, 1996; Braga & Aguirre, 2001
		SE Italy, Messinian	Bosellini *et al.*, 2001
Fore-reef slope	Crusts on bioclasts, rhodoliths	Mallorca, L Miocene	Pomar, 1991; Pomar *et al.*,1996
		SE Spain, Messinian	Braga & Aguirre, 2001
Ramp			
Beach, cliff	Rhodoliths, reworked rhodoliths and fragments	SE Spain, L Tortonian	Braga & Martín, 1988 Betzler *et al.*, 2000
Shoals	Reworked rhodoliths and fragments	N Slovenia, Rupelian	Nebelsick *et al.*, 2000
		SE Spain, L Tortonian	Braga & Martín, 1988
Patch reefs	Crusts and small branching frameworks on corals, rhodoliths	NE Italy, Oligocene	Fravega *et al.*, 1993a, 1994
		SE Spain, Langhian	Braga *et al.*, 1996
		S Turkey, Langhian–Tortonian	Hayward *et al.*, 1996
		S Poland, Badenian	Pisera, 1985
		SE Spain, L Tortonian	Braga & Martín, 1988; Braga and Aguirre, 2001
Outer ramp	Rhodoliths, encrusting coralline frameworks, loose branching plants	S Georgia/N Florida, Oligocene	Manker & Carter, 1987
		NE Italy, Rupelian	Francavilla *et al.*, 1970
		Central and southern Italy, E-M Miocene	Simone & Carannante, 1985; Carannante & Simone, 1996
		S Sardinia, Aquitanian–Burdigalian	Bassi *et al.*, 2006
		S Poland, Badenian	Studencki, 1988; Pisera & Studencki, 1989
		Maltese Islands, Tortonian	Bosence & Pedley, 1982; Bosence, 1983a;
		SE Spain, L Tortonian	Braga & Martín, 1988; Braga & Aguirre, 2001
		Menorca, Tortonian	Brandano *et al.*,2005
Slope and basin	Re-deposited rhodoliths and fragments	Austria	Nebelsick *et al.*, 2001; Rasser & Nebelsick, 2003
		S Sardinia, Aquitanian–Burdigalian	Bassi *et al.*, 2006
		SE Spain, L Tortonian	Braga *et al.*, 2001

Martín, 1988; Fravega *et al.*, 1988, 1994; Segonzac, 1990; Braga & Aguirre, 2001), and the occurrence of coralline algal genera has been noted for several localities (Perrin *et al.*, 1995; Bosellini *et al.*, 2001; Brandano *et al.*, 2005).

A lack of temporal resolution prevents the assignment of coralline species reported in several peri-Mediterranean basins to chronostratigraphic units within the Miocene, such as in the case of the corallines in the Miocene of Egypt

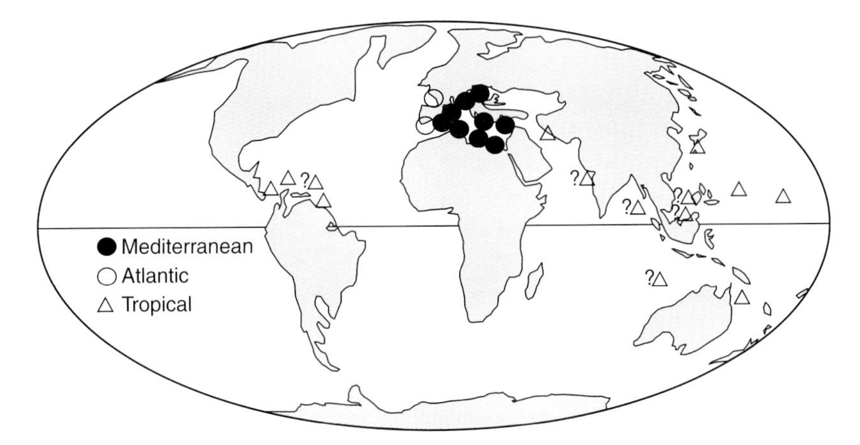

Fig. 3. Palaeogeographic distribution of tropical, Mediterranean and Atlantic Miocene coralline algal assemblages. Base map modified from Perrin (2002). Question marks refer to the uncertain biogeographical character of the coralline algal assemblages recorded.

(Souaya, 1963a, b). This lack of chronostratigraphic resolution also affects most of the species inventories from Indo-Pacific Islands (Johnson & Ferris, 1949; Johnson, 1954, 1957, 1961, 1966; Chatterji & Gururaja, 1972; Ishijima, 1975 and several other publications; Lemoine *et al.*, 1981; Poignant & Bourrouilh-le-Jan, 1986), continental India (Pal & Gosh, 1974), and Somalia (Airoldi, 1937). Only in Guam and Saurashtra has the chronostratigraphic resolution allowed the long list of recorded algae to be assigned to the Early Miocene (Johnson, 1964b; Kundal & Mude-Shyam, 2004). Similarly, the algal species documented from the Caribbean area can only be attributed to the Miocene, with no further stratigraphic precision (Lemoine 1918; Howe, 1919; Johnson & Kaska, 1965; Beckmann & Beckmann, 1966). Miocene coralline species have been reported from very few localities in North Atlantic basins: only the Serravallian from the Aquitanian Basin in France (Poignant, 1977) and the Tortonian from the Guadalquivir Basin in southern Spain (Civis *et al.*, 1994). In addition, non-exhaustive accounts of coralline genera and the occurrence of coralline algal limestones have been documented from globally widespread localities (Halfar & Mutti, 2005).

It is difficult to assess the taxonomic meaning of the majority of the species names reported in the literature on coralline algae. As is the case with the Oligocene algae, this prevents the confident assessment of patterns of biodiversity distribution and hampers comparisons of coralline floras from different localities by simple similarity tests of species inventories. However, the occurrence/absence of a few taxa points to the existence of a roughly defined phytogeographical differentiation of coralline algal floras in the Miocene (Fig. 3). In particular, plants attributed in the past to *Paraporolithon* and *Goniolithon* (Johnson, 1957, 1961, 1964b), with horizontal rows of trichocytes, seem to be restricted to low-latitude localities. The rows of trichocytes are typical of plants (included in *Hydrolithon* in modern botanical taxonomy) that are characteristic of present-day tropical waters. In a few localities these plants occur together with the genus *Aethesolithon* Johnson (Guam, Johnson, 1964b; Guatemala, Johnson & Kaska, 1965). Thick *Hydrolithon* plants also occur in the Miocene of the Marion Plateau (Martín *et al.*, 1993), and together with *Aethesolithon* in the Middle Miocene of the Queensland Plateau off NE Australia (Martín & Braga, 1993). The latter genus continued to inhabit the Great Barrier Reef region at least until the Holocene (Braga & Aguirre, 2004).

The above kinds of corallines do not occur in the shallow-water Miocene Mediterranean algal assemblages that are instead characterized by the common occurrence of other mastophoroids such as *Spongites* and *Neogoniolithon* (reported as *Lithophyllum* species in most papers; Braga *et al.*, 1993). Separation of a subtropical Mediterranean flora (Fig. 3) probably occurred from the Early Miocene onwards and was accentuated by the isolation of the Mediterranean from the Indian Ocean in the early Middle–Miocene (Harzhauser & Piller, 2007). The number of species of the tropical genus *Sporolithon* in the Mediterranean decreased to two before the Messinian Salinity Crisis, after a maximum of species richness in the Langhian, coincident with the Miocene climatic optimum. This decline follows the global cooling that began at around 14 Ma, but probably also reflects the isolation of the Mediterranean region from the tropical oceans since the Langhian (Braga & Bassi, 2007).

Paratethyan assemblages seem to be similar to Mediterranean ones until the late Middle Miocene. In contrast, Sarmatian corallines in the Paratethys are represented by very few species of thin plants. Thin corallines probably continued to live in the Paratethys when it became an inland sea, only sporadically connected to the Mediterranean Sea during the Late Miocene (Clauzon *et al.*, 2005), and they still occur in the Caspian Sea (*Melobesia caspica* Foslie, considered a younger heterotypic synonym of *Lithophyllum pustulatum* (Lamouroux) Foslie by Woelkerling & Campbell, 1992).

At least in the Late Miocene, the few coralline assemblages documented in North Atlantic basins include only scarce mastophoroid species, with lithophylloids (*Lithophyllum*) as the major components in shallow-water deposits (Civis *et al.*, 1994). These *Lithophyllum*-dominated assemblages are also characteristic of the Upper Miocene shallow-water temperate carbonates in the westernmost Mediterranean basins (Braga & Aguirre, 2001).

Most coralline algal genera identified in Miocene rocks are still living and have a relatively well-known geographical and ecological distribution. Even certain species recognized in the Miocene coralline assemblages are still extant (Braga & Aguirre, 1995). An actualistic approach to interpret the palaeonvironments in that the Miocene algae grew in, at least at the generic level, therefore seems appropriate. The occurrence of several genera such as *Lithoporella* and *Sporolithon*, mainly restricted to tropical waters in present-day oceans, can be used to infer warm-water conditions in past environments (Fravega *et al.*, 1989; Brandano *et al.*, 2005).

Coralline algae have been reported in a wide array of depositional settings and both the composition of algal assemblages and the occurrence type have been used to infer palaeoenvironmental parameters (Table 4). Studies on coralline algal distribution along palaeoenvironmental gradients in several Mediterranean localities suggest that Miocene coralline algal assemblages can be used as palaeodepth proxies. The taxonomic composition of assemblages and growth forms probably changed with variations in light and turbulence, usually correlated with changes in depth in marine platforms. In the analyzed cases, shallow-water assemblages are dominated by representatives of the subfamilies Mastophoroideae and Lithophylloideae, namely *Spongites* and *Neogoniolithon* (both commonly identified as *Lithophyllum*, Braga *et al.*, 1993) and *Lithophyllum* species. With increasing palaeodepth, the melobesioids *Lithothamnion* (including in most cases *Phymatolithon)* and *Mesophyllum* became the main components locally together with the sporolithacean *Sporolithon* (= *Archaeolithothamnium*) (Braga & Martín, 1988; Perrin *et al.*, 1995; Braga & Aguirre, 2001; Brandano *et al.*, 2005; Bassi *et al.*, 2006). These changes in composition of Miocene algal assemblages mimic the ones reported along depth gradients in present-day environments (Adey & Macintyre, 1973; Adey, 1979, 1986; Adey *et al.*, 1982; Bosence, 1983b; Rasser & Piller, 1997; Lund *et al.*, 2000). Applying the known depth distribution of extant genera, rough estimates can be attempted of the depositional palaeodepth in which the coralline assemblages grew (Martín *et al.*, 1993; Betzler *et al.*, 2000; Brandano *et al.*, 2005).

Downslope changes in algal growth forms and rhodolith internal structure have been documented in a Tortonian mixed carbonate-siliciclastic ramp in the Almanzora Corridor in SE Spain. Rhodoliths in the shallowest settings are small and made up of thick encrusting plants with low protuberances belonging to a few taxa (mainly *Lithophyllum*). The usually detrital nucleus is relatively large in relation to the algal cover and it strongly determines the general rhodolith shape. In deeper settings, rhodoliths are ellipsoidal to spheroidal and the algal covers are thicker relative to the nuclei (Fig. 4). Several coralline species occur as thin encrusting and protuberant, branching plants intergrown with epizoans such as foraminifera and bryozoans (Braga & Martín, 1988). However, in other localities large rhodoliths occur in very shallow environments, such as the reef crest of the Upper Miocene reefs in Mallorca (Pomar *et al.*, 1996).

Miocene coralline algal crustose frameworks (Studencki, 1979; Bosence & Pedley, 1982; Bosence, 1983a) can also be interpreted in the light of Holocene analogues that have been reported in the Mediterranean from 20 to 160 m (Pérès, 1967; Bosence, 1985) and in the Pacific Ocean (Davies *et al.*, 2004).

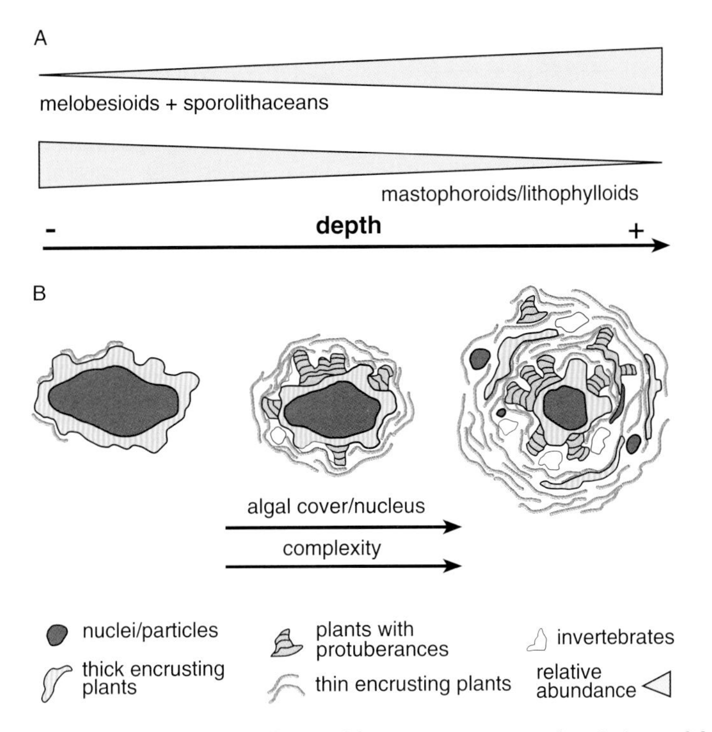

Fig. 4. Changes in relative abundance of coralline algal assemblage components and variations of rhodolith internal structure and composition with depth (after Braga & Martín, 1988).

CONCLUSIONS

Although most studies have focused on the Mediterranean region, Oligocene coralline algae have been reported from the western Pacific to the Caribbean. They mainly occur in coral reef deposits and related carbonates and mixed siliciclastic-carbonate rocks. They also appear in temperate settings in the heterozoan carbonates from New Zealand. However, scanty knowledge of the coralline algae from different regions and, particularly, uneven taxonomic approaches, have hampered any delineation of palaeogeographical domains in coralline algal distribution.

The few attempts to interpret palaeoenvironmental conditions from the Oligocene algal assemblages use the occurrence of genera restricted to tropical latitudes in modern oceans, such as *Sporolithon* (as *Archaeolithothamnium*) and *Lithoporella*, to infer warm-water conditions. The correlation of carbonate-facies and algal assemblages in a few examples from the northern margin of the western Tethys shows that the shallower environments are dominated by *Lithothamnion* species, whereas *Mesophyllum* is most abundant in deeper areas.

Miocene coralline algae have been reported from New Zealand, the western Pacific, southern and central Asia, peri-Mediterranean regions, the eastern North Atlantic and the Caribbean. Most of the records correspond to platform chlorozoan carbonates or deep-water sediments with reworked shallow-water carbonate particles. Coralline algae in carbonates from temperate areas occur in New Zealand and southern Spain. Comparison of the coralline floras from different localities is difficult due to taxonomic problems, but the occurrence of a few taxa suggests the existence of roughly defined phytogeographical regions.

The genus *Aethesolithon* and thick plants with horizontal rows of trichocytes attributable to *Hydrolithon* seem to be restricted to low-latitude localities defining a tropical phytogeographical region. These kinds of plants occur in present-day and Holocene tropical waters. In contrast, shallow-water Miocene Mediterranean algal assemblages include other mastophoroids such as *Spongites* and *Neogoniolithon*, delimiting a subtropical Mediterranean region that was probably fully established by the isolation of the Mediterranean from the Indian Ocean in the early Middle Miocene. Sarmatian corallines in the Paratethys, represented by a few species of thin plants, continued to inhabit the Paratethyan inland sea and one species can still be found in the Caspian Sea. Late Miocene coralline assemblages in North

Atlantic basins and shallow-water heterozoan carbonates in the western Mediterranean are dominated by lithophylloids (*Lithophyllum*), suggesting the existence of a temperate phytogeographic region.

Most coralline algal genera identified in Miocene rocks are still living and they can be used to interpret palaeoenvironments with an actualistic approach. The taxonomic composition of assemblages and growth forms changed with depth in marine platforms. Shallow-water assemblages are dominated by Mastophoroideae and Lithophylloideae (*Hydrolithon*, *Spongites*, *Neogoniolithon* and *Lithophyllum* species), whereas the melobesioids *Lithothamnion* and *Mesophyllum* and the sporolithacean *Sporolithon* (= *Archaeolithothamnium*) are the most common genera in deeper-water settings.

Regarding growth forms, rhodoliths in the shallowest settings are made up of thick encrusting plants belonging to a few taxa; rhodoliths in deeper settings, on the other hand, contain several coralline species that occur as thin encrusting and branching plants.

ACKNOWLEDGMENTS

This work was funded by "Ministerio de Educación y Ciencia (Spain)" Project CGL2004-04342/BTE, by a local research fund from the University of Ferrara, and by the International Inter-university Collaboration fund at the University of Ferrara (Italy). We are very grateful to J. Halfar and A. Henrich for their helpful comments. Christine Laurin is thanked for correcting the English text.

REFERENCES

Adey, W.H. (1979) Crustose coralline algae as microenvironmental indicators in the Tertiary. In: *Historical Biogeography, Plate Tectonics and the Changing Environment* (Eds J. Gray and A.J. Boucot), pp. 459–464. *Oregon State Univ. Press*, Corvallis.

Adey, W.H. (1986) Coralline algae as indicators of sea-level. In: *Sea-level Research: a Manual for the Collection and Evaluation of Data* (Ed. O. van de Plassche), pp. 229–279. Free University Amsterdam, Amsterdam.

Adey, W.H. and **Macintyre, I.G.** (1973) Crustose coralline algae: a re-evaluation in the geological sciences. *Geol. Soc. Am. Bull.*, **84**, 883–904.

Adey, W.H., Townsend, R.A. and **Boykins, W.T.** (1982) The crustose coralline algae (Rhodophyta: Corallinaceae) of the Hawaiian Islands. *Smithson. Contrib. Mar. Sci.*, **15**, 1–74.

Aguirre, J. and **Braga, J.C.** (2005) The citation of nongeniculate fossil coralline red algal species in the twentieth century literature: an analysis with implications. *Rev. Esp. Micropaleontol.*, **37**, 57–62.

Aguirre, J., Braga, J.C. and **Piller, W.E.** (1996) Reassessment of *Palaeothamnium* Conti 1946 (Corallinales, Rhodophyta). *Rev. Paleobot. Palynol.*, **94**, 1–9.

Aguirre, J., Riding, R. and **Braga, J.C.** (2000) Diversity of coralline red algae: origination and extinction patterns from the Early Cretaceous to the Pleistocene. *Paleobiology*, **26**, 651–667.

Airoldi, M. (1930) Su di un nuovo genere di Corallinacea fossile dell'Oligocene Ligure. *Atti. Real. Accad. Naz. Lincei Ser. 6*, **12**, 681–684.

Airoldi, M. (1932) Contributo allo studio delle corallinacee del terziario italiano. 1. Le Corallinacee dell'Oligocene Ligure-Piemontese. *Paleontogr. Ital., Mem. Paleont.* **33**, 55–83.

Airoldi, M. (1937) Le Corallinacee del Miocene della Somalia Italiana. *Paleontogr. Ital.*, **32** (Suppl. 2), 25–43.

Arias, C., Masse, J.P. and **Vilas, L.** (1995) Hauterivian shallow marine calcareous biogenic mounds: SE Spain. *Palaeogeogr. Palaeoclimatol. Palaeoecol.*, **119**, 3–17.

Badve, R.M. and **Kundal, P.** (1989) Studies on crustose coralline algae from Palaeocene to Oligocene rocks of Baratang Island, Andaman, India. In: *Proceedings of the 12th Indian Colloquium on Micropaleontology and Stratigraphy of the Shelf Sequences of India* (Ed. P. Kalia), pp. 253–265. *Papyrus Publ. House*, New Delhi.

Bassi, D. and **Nebelsick, J.H.** (2000) Calcareous algae from the Lower Oligocene Gornji Grad Limestones of Northern Slovenia. *Riv. Ital. Paleontol. Stratigr.*, **106**, 99–122.

Bassi, D., Carannante, G., Murru, M., Simone, L. and **Toscano, F.** (2006) Rhodalgal/bryomol assemblages in temperate type carbonate, channelised depositional systems: the Early Miocene of the Sarcidano area (Sardinia, Italy). In: *Cool-water Carbonates: Depositional Systems and Palaeoenvironmental Control* (Eds H.M. Pedley and G. Carannante), *Geol. Soc. London Spec. Publ.*, **255**, 35–52.

Beckmann, J.P. and **Beckmann, R.** (1966) Calcareous algae from the Cretaceous and Tertiary of Cuba. *Schweiz. Paläontol. Abh.*, **85**, 1–45.

Bellini, A. and **Mastrorilli, V.I.** (1975) Les corallinacées des coupes basales du Miocène de Bonifacio. *Extr. Bull. Soc. Sci. Hist. Nat. Corse*, **615/616**, 33–59.

Belokrys L.S. (1981) Meotian red algae of the Crimen. *Paleontol. J.*, **2**, 117–125.

Betzler, C., Martín, J.M. and **Braga, J.C.** (2000) Non-tropical carbonates related to rocky submarine cliffs (Miocene, Almería, Southern Spain). *Sed. Geol.*, **131**, 51–65.

Borowitzka, M.A. and **Larkum, A.W.D.** (1986) Reef Algae. *Oceanus*, **29**, 49–54.

Bosellini, A. and **Ginsburg, R.N.** (1971) Form and internal structure of Recent algal nodules (rhodolites) from Bermuda. *J. Geol.*, **79**, 669–682.

Bosellini, A., Russo, A., Arush, M.A. and **Cadbulqadir, M.M.** (1987) The Oligo-Miocene of Eil (N.E. Somalia): a prograding coral-*Lepidocyclina* system. *J. African Earth Sci.*, **6**, 583–593.

Bosellini, F.R. and **Russo, A.** (1992) Stratigraphy and facies of an Oligocene fringing reef (Castro Limestone, Salento Peninsula, southern Italy). *Facies*, **26**, 146–166.

Bosellini, F.R., **Russo, A.** and **Vescogni, A.** (2001) Messinian reef-building assemblages of the Salento Peninsula (southern Italy): palaeobathymetric and palaeoclimatic significance. *Palaeogeogr. Palaeoecol. Palaeoclimatol.*, **175**, 7–26.

Bosence, D.W.J. (1976) Ecological studies on two unattached coralline algae from western Ireland. *Palaeontology*, **19**, 365–395.

Bosence, D.W.J. (1983a) Coralline algae from the Miocene of Malta. *Palaeontology*, **26**, 147–173.

Bosence, D.W.J. (1983b) The occurrence and ecology of recent rhodoliths. In: *Coated Grains* (Ed. T.M. Peryt), pp. 225–242. Springer-Verlag, Berlin.

Bosence, D.W.J. (1984) Construction and preservation of two recent algal reefs, St. Croix, Caribbean. *Palaeontology*, **27**, 549–574.

Bosence, D.W.J. (1985) The "Coralligène" of the Mediterranean – a Recent analog for Tertiary coralline algal limestones. In: *Paleoalgology: Contemporary Research and Applications* (Eds D.F. Toomey and M.H. Nitecki), pp. 217–225. Springer–Verlag, New York.

Bosence, D.W.J. (1991) Coralline Algae: mineralization, taxonomy, and palaeoecology. In: *Calcareous Algae and Stromatolites* (Ed. R. Riding), pp. 98–113. Springer-Verlag, Berlin.

Bosence, D.W.J. and **Pedley, H.M.** (1982) Sedimentology and palaeoecology of a Miocene coralline algal biostrome from the Maltese Islands. *Palaeogeogr. Palaeoclimatol. Palaeoecol.*, **38**, 9–43.

Boulanger, D. and **Poignant, A.F.** (1975) Les nodules algaires du Miocene d'Aquitaine Meridionale. *Bull. Centre Etud. Rech. Sci. Biarritz*, **10**, 685–691.

Braga, J.C. (2003) Application of botanical taxonomy to fossil coralline algae (Corallinales, Rhodophyta). *Acta Micropaleontol. Sinica*, **20**, 47–56.

Braga, J.C. and **Aguirre, J.** (1995) Taxonomy of fossil coralline algal species: Neogene Lithophylloideae (Rhodophyta, Corallinaceae) from Southern Spain. *Rev. Paleobot. Palynol.*, **86**, 265–285.

Braga, J.C. and **Aguirre, J.** (2001) Coralline algal assemblages in upper Neogene reef and temperate carbonates in Southern Spain. *Palaeogeogr. Palaeoclimatol. Palaeoecol.*, **175**, 27–41.

Braga, J.C. and **Aguirre, J.** (2004) Coralline algae indicate Pleistocene evolution from deep open platform to outer barrier reef environments in the northern Great Barrier Reef margins. *Coral Reefs*, **23**, 547–558.

Braga, J.C. and **Bassi, D.** (2007) Neogene history of *Sporolithon* Heydrich (Corallinales, Rhodophyta) in the Mediterranean region. *Palaeogeogr. Palaeoclimatol. Palaeoecol.*, **243**, 189–203.

Braga, J.C. and **Martín, J.M.** (1988) Neogene coralline-algal growth-forms and their palaeoenvironments in the Almanzora River Valley (Almeria, S.E. Spain). *Palaeogeogr. Palaeoclimatol. Palaeoecol.*, **67**, 285–303.

Braga, J.C., **Bosence, D.W.** and **Steneck, R.S.** (1993) New anatomical characters in fossil coralline algae and their taxonomic implications. *Palaeontology*, **36**, 535–547.

Braga, J.C., **Jiménez, A.P.**, **Martín, J.M.** and **Rivas, P.** (1996) Middle Miocene coral-oyster reefs, Murchas, Granada, southern Spain. In: *Models for Carbonate Stratigraphy from Miocene Reef Complexes of Mediterranean Regions* (Eds E.K. Franseen, M. Esteban, W.C. Ward and J.M. Rouchy). *SEPM Concepts Sedimentol. Paleontol.*, **5**, 131–139.

Braga, J.C., **Martín, J.M.** and **Wood, J.L.** (2001) Submarine lobes and feeder cannels of redeposited, temperate carbonate and mixed siliciclastic-carbonate platform deposits (Vera Basin, Almeria, southern Spain). *Sedimentology*, **48**, 99–116.

Brandano, M., **Vannucci, G.**, **Pomar, L.** and **Obrador, A.** (2005) Rhodolith assemblages from the lower Tortonian carbonate ramp of Menorca (Spain): environmental and paleoclimatic implications. *Palaeogeogr. Palaeoclimatol. Palaeoecol.*, **226**, 307–323.

Buchbinder, B. (1977) Systematics and palaeoenvironments of the calcareous algae from the Miocene (Tortonian) Tziqlag Formation, Israel. *Micropaleontology*, **23**, 415–435.

Buchbinder, B. (1996) Miocene carbonates of the eastern Mediterranean, the Red Sea and the Mesopotamian Basin: geodynamic and eustatic controls. In: *Models for Carbonate Stratigraphy from Miocene Reef Complexes of Mediterranean Regions* (Eds E.K. Franseen, M. Esteban, W.C. Ward and J.M. Rouchy). *SEPM Concepts Sedimentol. Paleontol.*, **5**, 89–96.

Bucur, I.I. and **Filipescu, S.** (1994) Middle Miocene red algae from the Transylvanian Basin (Romania). *Beitr. Paläontol.*, **19**, 39–47.

Bucur, I.I., **Onac, B.P.** and **Todoran, V.** (1989) Algues calcaires dans les dépots Oligocènes inférieurs de la région Purcăret-Mesteacân-Valea Chioarului (NW du Bassin de Transylvanie). In: *The Oligocene from the Transylvanian Basin* (Petrescu-Iustinian Editions), pp. 141–148. *Univ. Cluj-Napoca*, Cluj-Napoca.

Budai, J.M. and **Wilson, J.L.** (1980) Diagenesis of lower Coralline Limestone (Chattian), Maltese Islands. *AAPG Bull.*, **64**, 682.

Buge, E., **Debourle, A.** and **Deloffre, R.** (1973) Gisement Miocene a nodules algaires (Rhodolithes) a l'ouest de Salies-de-Bearn (Aquitaine sud-ouest). *Bull. Centre Rech. Pau-SNPA*, **7**, 1–51.

Burchette, T.P. (1988) Tectonic control on carbonate platform facies distribution and sequence development: Miocene, Gulf of Suez. *Sed. Geol.*, **59**, 179–204.

Burgess, C.J. and **Anderson, J.M.** (1983) Rhodoids in temperate carbonates from the Cenozoic of New Zealand. In: *Coated Grains* (Ed. T.M. Peryt), pp. 243–258, Springer-Verlag, Berlin.

Carannante, G. and **Simone, L.** (1996) Rhodolith facies in the central-southern Apennines, Italy. In: *Models for Carbonate Stratigraphy from Miocene Reef Complexes of Mediterranean Regions* (Eds E.K. Franseen, M. Esteban, W.C. Ward and J.M. Rouchy). *SEPM Concepts Sedimentol. Paleontol.*, **5**, 89–96.

Chatterji, A.K. and **Gururaja, M.N.** (1972) Coralline algae from Andaman Islands, India. *Rec. Geol. Soc. India*, **99**, 133–144.

Civis, J., **Alonso-Gavilán, G.**, **González Delgado, J.A.** and **Braga, J.C.** (1994) Sédimentation carbonatée transgressive sur la bordure occidentale du couloir nord-bétique pendant le Tortonien supérieur (Fm Calcarenita de Niebla, SW de l'Espagne). *Géol. Méditerr.*, **21**, 9–18.

Clauzon, G., **Suc, J.-P.**, **Popescu, S.-M.**, **Marunteanu, M.**, **Rubino, J.-L.**, **Marinescu, F.** and **Melinte, M.C.** (2005) Influence of Mediterranean sea-level changes on the Dacic Basin (Eastern Paratethys) during the late Neogene: the Mediterranean Lago Mare facies deciphered. *Basin Res.*, **17**, 437–462.

Conti, S. (1942-43) Contributo allo studio delle Corallinacee del terziario italiano. II: Le Corallinacee del Miocene del Bacino Ligure-Piemontese. *Paleontogr. Ital.*, **41**, 37–61.

Conti, S. (1946a) Revisione critica di *Lithothamnium ramosissimum* Reuss. *Publ. Ist. Geol. Univ. Genova*, **Quad. 1 - 2, ser. A**, 3–29.

Conti, S. (1946b) Le Corallinacee del calcare miocenico (Leithakalk) del bacino di Vienna. *Publ. Ist. Geol. Univ. Genova*, **Quad. 1 - 2, ser. A**, 31–68.

Conti, S. (1946c) Su alcune specie di Melobesie di Ponzone (Aqui). *Atti Ist. Univ. Pavia*, **3**, 1–14.

Conti, S. (1950) Alghe Corallinacee fossili. *Publ. Ist. Geol. Univ. Genova*, **Quad. 4 ser. A**, 1–156.

Davies, P.J., **Braga, J.C.**, **Lund M.** and **Webster, J.** (2004) Holocene deep water algal buildups on eastern Australian shelf. *Palaios*, **19**, 598–609.

Dullo, W.C., **Hoetzl, H.** and **Jado, A.R.** (1983) New stratigraphical results from the Tertiary sequence of the Midyan area, NW Saudi Arabia. *Newsl. Stratigr.*, **12**, 75–83.

Edgell, H.S. and **Basson, P.W.** (1975) Calcareous algae from the Miocene of Lebanon. *Micropaleontology*, **21**, 165–184.

Elliott, G.F. (1960) Fossil calcareous algal floras of the Middle East with a note on a Cretaceous problematicum, *Hensonella cylindrica* gen. et sp. nov. *Q.J. Geol. Soc. London*, **3**, 217–232.

Elliott, G.F. (1970) *Pseudoaethesolithon*, a calcareous alga from the Fars (Persian Miocene). *Geol. Rom.*, **9**, 31–46.

Francavilla, F., **Frascari Ritondale Spano, F.** and **Zecchi, R.** (1970) Alghe e macroforaminiferi al limite Eocene-Oligocene presso Barbarano (Vicenza). *Giorn. Geol.*, **36**, 653–686.

Fravega, P. and **Vannucci, G.** (1980a) Segnalazione di una nuova specie di *Lithophyllum*: *Lp. sassellense* n. sp., nel "Rupeliano" superiore di Sassello (Bacino Ligure-Piemontese). *Quad. Ist. Geol. Univ. Genova*, **3**, 31–37.

Fravega, P. and **Vannucci, G.** (1980b) Associazione a corallinacee nella serie di Costa Merlassino (Alessandria) e suo significato ambientale. In: Atti del 1° convegno di ecologia e paleoecologia delle comunità bentoniche. Ferrara, 8-12 Ottobre 1979. *Ann. Univ. Ferrara, N.S.*, sez. 9, **6** suppl., 93–117.

Fravega, P. and **Vannucci, G.** (1982) Significato e caratteristiche degli episodi a rhodoliti al "top" del Serravalliano tipo. *Geol. Romana*, **21**, 705–715.

Fravega, P. and **Vannucci, G.** (1987) Significato delle facies algali delle sequenze tardo serravalliane-tortoniane ad Ovest di Gavi (Bacino Terziario del Piemonte). *Acc. Gioenia Sci. Nat. Catania*, **20**, 317–334.

Fravega, P., **Giammarino, S.** and **Vannucci, G.** (1984a) Episodi ad "algal balls" e loro significato al passaggio Arenarie di Serravalle-Marne di S. Agata fossili a Nord di Gavi (Bacino Terziario del Piemonte). *Atti Soc. Tosc. Sci. Nat. Mem., Ser. A*, **91**, 1–20.

Fravega, P., **Giammarino, S.** and **Vannucci, G.** (1984b) Algal remains and their correlation with microfaunas in the Benghazi limestone (Middle Miocene of Western Cyrenaica). *Boll. Soc. Geol. Ital.*, **103**, 503–514.

Fravega, P., **Giammarino, S.**, **Piazza, M.**, **Russo, A.** and **Vannucci, G.** (1987) Significato paleoecologico degli episodi coralgali a Nord di Sassello. Nuovi dati per una ricostruzione paleogeografica-evolutiva del margine meridionale del Bacino Terziario del Piemonte. *Atti Soc. Tosc. Sci. Nat. Mem. Ser. A*, **94**, 19–76.

Fravega, P., **Piazza, M.** and **Vannucci, G.** (1988) *Leptolithophyllum elongatum* n. sp. del Burdigagliano del Bacino Terziario del Piemonte. In: *Atti 5° Simp. Ecologia e Paleoecologia delle Comunità Bentoniche, Sorrento 1988* (Ed. E. Robba), pp. 265–274. Mus. Reg. Sci. Nat. Torino.

Fravega, P., **Piazza, M.** and **Vannucci, G.** (1989) *Archaeolithothamnium* Rothpletz indicatore ecologico-stratigrafico. In: *Atti del 3° Simposio di Ecologia e Paleoecologia delle Comunità Bentoniche, Catania 1985* (Ed. I. Di Geronimo), pp. 729–743. Univ. Catania.

Fravega, P., **Piazza, M.** and **Vannucci, G.** (1993a) Importance and significance of the rhodoliths bodies in the miocenic sequences of Tertiary Piedmont Basin. In: *Studies on Fossil Benthic Algae* (Eds P. de Castro, F. Barattolo and M. Parente). *Boll. Soc. Paleontol. Ital. Spec. Vol.*, **1**, 197–210.

Fravega, P., **Piazza, M.** and **Vannucci, G.** (1993b) Three new species of coralline algae (genera *Lithothamnion* and *Lithophyllum*) from the Tertiary Piedmont Basin. *Riv. Ital. Paleontol. Stratigr.*, **98**, 453–466.

Fravega, P., **Piazza, M.** and **Vannucci, G.** (1994) Nongeniculate coralline algae associations from the Calcare di Rosignano Formation, lower Messinian, Tuscany (Italy). In: *Studies on ecology and palaeoecology of benthic communities* (Eds R. Matteucci, M.G. Carboni and J.S. Pignatti). *Boll. Soc. Paleontol. Ital. Spec. Vol.* **2**, 127–140.

Fravega, P., **Giammarino, S.**, **Piazza, M.** and **Vannucci, G.** (1998) The rhodolith pavement of the Punta Guitja section, Lampedusa Formation (Upper Miocene, Island of Lampedusa, Pelagian Block). Preliminary note. *Boll. Soc. Paleontol. Ital.*, **36**, 413–416.

Frost, S.H., **Harbour, J.L.**, **Beach, D.K.**, **Realini, M.J.** and **Harris, P.M.** (1983) Oligocene reef tract development. *Sedimenta*, **9**, 1–144.

Geel, T. (2000) Recognition of stratigraphic sequences in carbonate platform and slope deposits: empirical model based on microfacies analysis of Paleogene deposits in southeastern Spain. *Palaeogeogr. Palaeoclimatol. Palaeoecol.*, **155**, 211–238.

Giammarino, S., **Nosengo, S.** and **Vannucci, G.** (1970) Risultanze geologico-paleontologiche sul conglomerato di Portofino (Liguria orientale). *Atti Ist. Geol. Univ. Genova*, **7**, 305–363.

Halfar, J. and **Mutti, M.** (2005) Global dominance of coralline red algal facies: a response to Miocene oceanographic events. *Geology*, **33**, 481–484.

Harvey, A.S., **Broadwater, S.T.**, **Woelkerling, W.J.** and **Mitrovski, P.J.** (2003) *Choreonema* (Corallinales, Rhodophyta): 18S rDNA phylogeny and resurrection of the Hapalidiaceae for the subfamilies Choreonematoideae, Austrolithoideae, and Melobesioideae. *J. Phycol.*, **39**, 988–998.

Harzhauser, M. and **Piller, W.E.** (2007) Benchmark data of a changing sea – Palaeogeography, Palaeobiogeography and Events in the Central Paratethys during the Miocene. *Palaeogeogr. Palaeoclimatol. Palaeoecol.*, **253**, 8–31.

Hayward, A.B., **Robertson, A.H.F.** and **Scoffin, T.P.** (1996) Miocene patch reefs from a Mediterranean marginal terrigenous setting in southwest Turkey. In: *Models for Carbonate Stratigraphy from Miocene Reef Complexes of Mediterranean Regions* (Eds E.K. Franseen, M. Esteban, W.C. Ward and J.M. Rouchy). *SEPM Concepts Sedimentol. Paleontol.*, **5**, 317–332.

Hiroshi, O. (1998) Sedimentary environment and periodicity of coralline algal biostromes in the Nanatsugama Sandstone, Nishisonogi Peninsula, Nagasaki Prefecture, Japan. *Bull. Geol. Surv. Japan*, **49**, 379–394.

Hood, S.D., **Nelson, C.S.** and **Kamp, P.J.J.** (2003) Petrogenesis of diachronous mixed siliciclastic-carbonate megafacies in the cool-water Oligocene Tikorangi Formation, Taranaki Basin, New Zealand. *NZ J. Geol. Geophys.*, **46**, 387–405.

Howe, M.A. (1919) Tertiary calcareous algae from the islands of St. Bartholomew, Antigua, and Anguilla. In: *Contributions to the Geology and Paleontology of the West Indies* (Ed. T.W. Vaughan), pp. 9–19. Carnegie Institution of Washington, Washington, DC.

Ishijima, W. (1933) On three species of Corallinaceae lately obtained from the Megamiyama limestone, Sagara district, Prov. Totomi. *Jpn. J. Geol. Geogr.*, **11**, 27–30.

Ishijima, W. (1941) A new species of *Archaeolithothamnium* from Kotosyo. *Taiwan Chigaku Kizi*, **12**, 1–5.

Ishijima, W. (1942) On the coralline algae from the Ryukyu limestome of Kotosyo (Botel Tobago Island). *Taiwan Chigaku Kizi*, **13**, 78–84.

Ishijima, W. (1943) On several kinds of calcareous algae in the Binangonan limestone. *Trans. Nat. Hist. Soc. Taiwan*, **33**, 643–652.

Ishijima, W. (1944) On some fossil coralline algae from the Ryukyu limestone of the Ryukyu Islands and Formosa (Taiwan). *Mem. Fac. Sci. Taihoku Imp. Univ.*, Ser. 3, **1**, 49–76.

Ishijima, W. (1954) *Cenozoic Coralline Algae from the Western Pacific*. Yuhodo, Tokyo, 87 pp.

Ishijima, W. (1956) On some fossil corraline algae from the Tertiary of Japan. *St. Paul's Rev. Arts Sci. (Nat. Sci.)*, **1**, 59–67.

Ishijima, W. (1960) Notes on some Miocene algae from central Japan. *St. Paul's Rev. Arts Sci. (Nat. Sci.)*, **7**, 1–9.

Ishijima, W. (1965a) Calcareous algae in the Chihko-shan Limestone from Sinchu District, Taiwan. *St. Paul's Rev. Sci.*, **2**, 179–190.

Ishijima, W. (1965b) On some coralline algae from a guyot in the Cocos-Keeling basin, Eastern Indian Ocean. *St. Paul's Rev. Sci.*, **2**, 79–88.

Ishijima, W. (1968) Calcareous algae from Makinogo near Shuzenji, Izu Peninsula, Japan. *St. Paul's Rev. Sci.*, **2**, 245–254.

Ishijima, W. (1969) Calcareous algae from the Miocene Mizunami group, Central Japan, a preliminary note. *St. Paul's Rev. Sci.*, **8**, 255–261.

Ishijima, W. (1970) Tertiary calcareous algae from Marinduque Islands, the Philippines. *Geol. Palaeontol. Southeast Asia*, **2**, 151–164.

Ishijima, W. (1975) On some coralline algae from the Miocene Gaj Formation of West Pakistan. *St. Paul's Rev. Sci.*, **3**, 109–118.

Ishijima, W. (1978) Calcareous algae from Philippines, Malaysia and Indonesia. *Geol. Palaeontol. Southeast Asia*, **19**, 167–190.

Ishijima, W. (1979) On the calcareous algae from Nanatsugama Sandstone of the Nishisonogi Peninsula, Nagasaki Prefecture. *Bull. Nat. Sci. Mus. Ser. C Geol. Paleontol.*, **4**, 131–137.

James, N.P., **Coniglio, M.**, **Aissaoui, D.M.** and **Purser, B.H.** (1988) Facies and geologic history of an exposed Miocene rift-margin carbonate platform: Gulf of Suez, Egypt. *AAPG Bull.*, **72**, 555–572.

Johnson, J.H. (1954) Fossil calcareous algae from Bikini atoll. *US Geol. Surv. Prof. Pap.*, **260(M)**, 536–545.

Johnson, J.H. (1955) Early Tertiary coralline algae from Trinidad, British West Indies. *Eclogae Geol. Helv.*, **48**, 69–78.

Johnson, J.H. (1957) Calcareous algae. Geology of Saipan, Mariana Islands. *US Geol. Surv. Prof. Pap.*, **280(E)**, 209–246.

Johnson, J.H. (1961) Fossil algae from Eniwetok, Funafuti, and Kita-Daito-jima. *US Geol. Surv. Prof. Pap.*, **260-Z**, 907–950.

Johnson, J.H. (1964a) Miocene coralline algae from northern Iraq. *Micropaleontology*, **10**, 477–485.

Johnson, J.H. (1964b) Fossil and Recent calcareous algae from Guam. *US Geol. Surv. Prof. Pap.*, **403-G**, G1–40.

Johnson, J.H. (1965) Coralline algae from the Cretaceous and early Tertiary of Greece. *J. Paleont.*, **39**, 802–814.

Johnson, J.H. (1966) Tertiary red algae from Borneo. *Bull. Br. Mus. (Nat. Hist.) Geol. Ser.*, **11**, 255–280.

Johnson, J.H. and **Ferris, B.J.** (1949) Tertiary coralline algae from the Dutch East Indies. *J. Paleont.*, **23**, 193–198.

Johnson, J.H. and **Kaska, H.V.** (1965) Fossil algae from Guatemala. *Prof. Contrib. Colorado Sch. Min.*, **1**, 1–152.

Kaiser, D., **Rasser, M.W.**, **Nebelsick, J.H.** and **Piller, W.E.** (2001) Late Oligocene algal limestones on a mixed carbonate-siliciclastic ramp at the southern margin of the Bohemian Massif (Upper Austria). In: *Paleogene of the Eastern Alps* (Eds W.E. Piller and M.W. Rasser), pp. 197–223. Akad. Wiss. Schriftenr. Erdwiss. Komm. 14, Wien.

Kamptner, E. (1942) Zwei Corallinaceen aus dem Sarmat des Alpen-Ostrandes und der Hainburger Berge. *Ann. Naturhist. Mus. Wien*, **52**, 5–19.

Kundal, P. and **Humane, S.K.** (2002) Geniculate coralline algae from Middle Eocene to Lower Miocene of Kachchh, Gujarat, India. *Gond. Geol. Mag.*, **17**, 89–101.

Kundal, P. and **Mude-Shyam, N.** (2004) Record of coralline algae from lower Miocene Gaj Formation, Porbander, southwest coast of Saurachtra: paleoenvironment and paleobathymetric implications. *Gond. Geol. Mag.*, **19**, 159–164.

Lemoine, M. (1918) Contribution à l'étude des Corallinacées fossiles. III. Corallinacéees fossiles de la Martinique. *CR Somm. Soc. Géol. Fr., Ser. 4*, **17**, 233–279.

Lemoine, M. (1924) Contribution à l'étude des Corallinacées fossiles. VII. Mélobésiées miocènes recueilles par M. Bourcart en Albanie. *Bull. Soc. Géol. Fr.*, **23**, 275–283.

Lemoine, M. (1938) Les Corallinacées du sondage des Abatilles, prés Arcachon. *CR Somm. Soc. Géol. Fr.*, **137**, 67–72.

Lemoine, M. (1939) Les algues calcaires fossiles de l'Algérie. *Mater. Carte Géol. Algérie Sér. 1 Paléontol.* **9**, 1–131.

Lemoine, M. (1977) Etude d'une collection d'algues corallinacées de la région de Skopje (Yougoslavie). *Rev. Micropaléontol.*, **20**, 10–42.

Lemoine, M., Ricard, M. and **Dugas, F.** (1981) Les algues calcaires fossiles de biomicrites draguées sur l'Arc du Vanuatu (Nouvelles-Hebrides). *CR Hebdom. Séan. Acad. Sci. Ser. 3*, **292**, 669–672.

Littler, M.M., Littler, D.S., Blair, S.M. and **Norris, J.N.** (1985) Deepest known plant life discovered on an uncharted seamount. *Science*, **227**, 57–59.

Lund, M., Davies, P.J. and **Braga, J.C.** (2000) Coralline algal nodules off Fraser Island, Eastern Australia. *Facies*, **42**, 25–34.

MacGregor, A.R. (1983) The Waitakere Limestone, a temperate algal carbonate in the lower Tertiary of New Zealand. *J. Geol. Soc. London*, **140**, 387–399.

Manker, J.P. and **Carter, B.D.** (1987) Paleoecology and paleogeography of an extensive rhodolith facies from the Lower Oligocene of South Georgia and North Florida. *Palaios*, **2**, 181–188.

Mankiewicz, C. (1996) The Middle to Upper Miocene carbonate complex of Níjar, Almería province, South-eastern Spain. In: *Models for Carbonate Stratigraphy from Miocene Reef Complexes of Mediterranean Regions* (Eds E.K. Franseen, M. Esteban, W.C. Ward and J.M. Rouchy). *SEPM Concepts Sedimentol. Paleontol.*, **5**, 141–157.

Marrack, E.C. (1999) The relationship between water motion and living rhodolith beds in the southwestern Gulf of California, Mexico. *Palaios*, **14**, 159–171.

Marshall, J.F., Tsuji, Y., Matsuda, H., Davies, P.J., Iryu, Y., Honada, N. and **Satoh, Y.** (1998) Quaternary and Tertiary sub-tropical carbonate platform development on the continental margin of southern Queensland, Australia. In: *Reefs and Carbonate Platforms in the Pacific and Indian Oceans* (Eds G.F. Camoin and P.J. Davies). *Int. Assoc. Sedimentol. Spec. Publ.*, **25**, 163–195.

Martín, J.M. and **Braga, J.C.** (1993) Eocene to Pliocene coralline algae in the Queensland Plateau (Northestern Australia). *Proc. ODP Sci. Res.*, **133**, 67–74.

Martín, M.M., Braga, J.C., Konishi, K. and **Pigram, C.J.** (1993) A model for the development of rhodoliths on platforms influenced by storms: Middle Miocene carbonates of the Marion Plateau (Northeastern Australia) In: *Proc. ODP Sci. Res.*, **133**, 455–465.

Maslov, V.P. (1956) Fossil calcareous algae of USSR. *Trudy Inst. Geol. Nauk Akad. Nauk SSSR*, **160**, 1–301 (in Russian).

Maslov, V.P. (1962) Fossil red algae of USSR and their connections with facies. *Trudy Inst. Geol. Akad. Nauk SSSR*, **53**, 1–222 (in Russian).

Mastorilli, V.I. (1968) Nuovo contributo allo studio delle Corallinacee dell'Oligocene Ligure-Piemontese: i reperti della tavoletta Ponzone. *Atti Ist. Geol. Univ. Genova*, **5**, 153–406.

Mastrorilli, V.I. (1973) Flore fossili a Corallinacee di alcune località venete tra i Berici e l'Altopiano di Asiago. *Atti Soc. Ital. Sci. Nat., Mus. Civ. St. Nat. Milano*, **114**, 209–292.

Matsuda, S. and **Tomiyama, T.** (1988) An investigation of Recent deepwater rhodoliths from the Ryukyu Islands. *Bull. Coll. Educ. Univ. Ryukyu*, **33**, 343–354 (in Japanese with English fig. captions and abstract).

Miranda, F. (1935) Algas coralináceas fósiles del Terciario de San Vicente de la Barquera (Santander). *Boll. Soc. Esp. Hist. Nat.*, **35**, 279–287.

Misra, P.K., Jauhri, A.K., Singh, S.K., Kishore, S. and **Chowdhury, A.** (2001) Coralline algae from the Oligocene of Kachchh, Gujarat, India. *J. Palaeontol. Soc. India*, **46**, 59–76.

Moussavian, E. (1984) Die Gosau- und Alttertiaer-Geroelle der Angerberg-Schichten (Hoeheres Oligozaen, Unterinntal, Noerdliche Kalkalpen). *Facies*, **10**, 1–86.

Nebelsick, J.H. and **Bassi, D.** (2000) Diversity, growth forms and taphonomy: key factors controlling the fabric of coralline algae dominated shelf carbonates. In: *Carbonate platform systems: components and interactions* (Eds E. Insalaco, P.W. Skelton and T.J. Palmer). *Geol. Soc. London Spec. Publ.*, **178**, 89–107.

Nebelsick, J.H., Bassi, D. and **Drobne, K.** (2000) Microfacies analysis and palaeoenvironmental interpretation of Lower Oligocene, shallow-water carbonates (Gornji Grad Beds, Slovenia). *Facies*, **43**, 157–176.

Nebelsick, J.H., Stingl, V. and **Rasser, M.** (2001) Autochthonous facies and allochthonous debris flows compared: Early Oligocene carbonate facies patterns of the Lower Inn Valley (Tyrol, Austria). *Facies*, **44**, 31–46.

Orszag-Sperber, F. and **Poignant, A.F.** (1972) Corallinacées du Miocène de la plaine orientale corse. *Rev. Micropaléontol.*, **15**, 115–124.

Orszag-Sperber, F., Poignant, A.F. and **Poisson, A.** (1977) Paleogeographic significance of rhodolites: some examples from the Miocene of France and Turkey. In: *Fossil Algae. Recent Results and Developments* (Ed. E. Flügel), pp. 286–294. Springer-Verlag, Berlin.

Pal, A.K. and **Gosh, R.N.** (1974) Fossil algae from the Miocene of Kutch, India. *Palaeobotany*, **21**, 189–192.

Pérès, J.M. (1967) The Mediterranean benthos. *Oceanogr. Mar. Biol. Ann. Rev.*, **5**, 449–533.

Perrin, C. (2002) Tertiary: the emergence of modern reef ecosystems. In: *Phanerozoic Reef Patterns* (Eds W. Kiessling, E. Flügel and J. Golonka). *SEPM Spec. Publ.*, **72**, 587–621.

Perrin, C., Bosence D.W. and **Rosen, B.** (1995) Quantitative approaches to palaeozonation and palaeobathymetry of corals and coralline algae in Cenozoic reefs. In: *Marine Palaeoenvironmental Analysis from Fossils* (Eds D.W.J. Bosence and P.A. Allison). *Geol. Soc. London Spec. Publ.*, **83**, 181–229.

Piller, W.E. (1993) Facies development and coralline algae in the Vienna and Eisenstadt Basins (Miocene). In: *Facial Development of Algae-bearing Carbonate Sequences in the Eastern Alps* (Eds R. Höfling, E. Moussavian and W.E. Piller), pp. 1–24. Field Trip Guidebook, Alpine Algae '93, Munich-Vienna.

Piller, W.E. and **Harzhauser, M.** (2005) The myth of the brackish Sarmatian Sea. *Terra Nova*, **17**, 450–455.

Pisera, A. (1985) Palaeoecology and lithogenesis of the Middle Miocene (Badenian) algal-vermetid reefs from the Roztocze Hills, south-eastern Poland. *Acta Geol. Pol.*, **35**, 89–155.

Pisera, A. and **Studencki, W.** (1989) Middle Miocene rhodoliths from the Korytnica Basin (Southern Poland):

environmental significance and paleontology. *Acta Palaeontol. Pol.*, **34**, 179–209.

Poignant, A. (1972) Microfaciès et microfaunes du Priabonien de l'Oligocène et du Miocène d'Aquitaine méridionale. *Trav. Lab. Micropaléontol.*, **1**, 11 pp.

Poignant, A. (1977) Les algues fossiles, point de vue du géologue. *Bull. Soc. Phycol. Fr.*, **22**, 87–98.

Poignant, A. and **Bourrouilh-le-Jan, F.G.** (1986) Les Corallinacées miocènes de l'Atoll soulevé de Mare (Archipel des Loyaute). Territoire de la Nouvelle-Caledonie. In: *Actes du 111 Congres National des Societes Savantes. CR Congr. Nat. Soc. Sav. Sec. Sci.*, **111**, 119–133.

Poisson, A.A. and **Poignant, A.F.** (1974) La formation de Karabayir, base de la transgression miocène dans la region de Korkuteli (departement d'Antalya-Turquie); *Lithothamnium pseudoramossissimum*, nouvelle espèce d'algue rouge de la formation de Karabayir. *Bull. Min. Res. Expl. Inst. Turkey*, **82**, 66–70.

Pomar, L. (1991) Reef geometries, erosion surfaces and high frequency sea-level changes, Upper Miocene Reef Complex, Mallorca, Spain. *Sedimentology*, **38**, 243–269.

Pomar, L., **Ward, W.C.** and **Green, D.G.** (1996) Upper Miocene reef complex of the Llucmajor area, Mallorca, Spain. In: *Models for Carbonate Stratigraphy from Miocene Reef Complexes of Mediterranean Regions* (Eds E.K. Franseen, M. Esteban, W.C. Ward and J.M. Rouchy). *SEPM Concepts Sedimentol. Paleontol.*, **5**, 191–225.

Prager, E.J. and **Ginsburg, R.N.** (1989) Carbonate nodule growth on Florida's outer shelf and its implications for fossil interpretations. *Palaios*, **4**, 310–317.

Raineri, R. (1924) Alghe fossili mioceniche della Cirenaica raccolte dall'Ing. Crema. *Nuova Notar.*, **25**, 28–46.

Rasser, M.W. and **Nebelsick, J.H.** (2003) Provenance analysis of Oligocene autochthonous and allochthonous coralline algae: a quantitative approach towards reconstructing transported assemblages. *Palaeogeogr. Paleoclimatol. Palaeoecol.*, **201**, 89–111.

Rasser, M. and **Piller, W.E.** (1997) Depth distribution of calcareous encrusting associations in the Northern Red Sea (Safaga, Egypt) and their geological implications. *Proc. 8th Int. Coral Reef Symp.*, **1**, 743–748.

Riding, R., **Martín, J.M.**, and **Braga, J.C.** (1991) Coral stromatolite reef framework, Upper Miocene, Almeria, Spain. *Sedimentology*, **38**, 799–818.

Rögl, F. (1998) Palaeogeographic considerations for Mediterranean and Paratethys seaways (Oligocene to Miocene). *Ann. Naturhist. Mus. Wien*, **99A**, 279–310.

Saller, A.H. and **Vijaya, S.** (2000) Kerendan: an isolated Oligocene carbonate platform in the western Kutai Basin, central Kalimantan, Indonesia. *AAPG Bull.*, **84**, 1484.

Schaleková, A. (1969) Zur näheren Kenntnis der Corallinaceen im Leithakalk des Sandberges bei Devínska Nová Ves (Theben-Neudorf) in der Südwestslowakei. *Acta Geol. Geogr. Univ. Comenianae*, **18**, 93–102.

Scoffin, T.P. (1988) The environments of production and deposition of calcareous sediments on the shelf west of Scotland. *Sed. Geol.*, **60**, 107–134.

Scoffin, T.P., **Stoddart, D.R.**, **Tudhope, A.W.** and **Woodroffe, C.** (1985) Rhodoliths and coralliths of Muri Lagoon, Rarotonga, Cook Islands. *Coral Reefs*, **4**, 71–80.

Scudeller Bacceller, L. and **Reato, S.** (1988) Cenozoic algal biostromes in the eastern Veneto (northern Italy): a possible example of non-tropical carbonate sedimentation. *Sed. Geol.*, **60**, 197–206.

Scuttenhelm, R.T.E. (1976) History and modes of Miocene carbonate deposition in the interior of the Piedmont Basin, N.W. Italy. *Utrecht Micropaleontol. Bull.*, **14**, 1–208.

Segonzac, G. (1972) Nouvelles espèces de Corallinacées (algues calcaires) du Néogène d'Espagne. *Bull. Soc. Nat. Toulouse*, **108**, 280–286.

Segonzac, G. (1990) Algues néogènes betiques. *Doc. Trav. IGAL*, **12–13**, 35–41.

Simone, L. and **Carannante, G.** (1985) Evolution of a Miocene carbonate open shelf from inception to drowning: The case of the southern Apennines. *Rend. Accad. Sci. Fis. Mat. Napoli, IV*, **53**, 1–43.

Souaya, F.J. (1963a) On the calcareous algae (Melobesioideae) of Gebel Gharra (Cairo-Suez road) with a local zonation and some possible correlations. *J. Paleontol.*, **37**, 1204–1216.

Souaya, F.J. (1963b) Micropaleontology of four sections south of Qoseir, Egypt. *Micropaleontology*, **9**, 233–266.

Steller, D.L., **Riosmena-Rodriguez, R.**, **Foster M.S.** and **Roberts, C.** (2003) Rhodolith bed diversity in the Gulf of California: the importance of rhodolith structure and consequences of anthropogenic disturbances. *Aquat. Conserv. Mar. Freshwat. Ecosyst.*, **13**, S5–S20.

Stoklosa, M.L. and **Simo, J.A.T.** (2000) Paleoclimatic and eustatic controls on Oligocene shallow-water environments, SE Spain. *AAPG 2000 Annual Meeting*, p. 143.

Studencki, W. (1979) Sedimentation of algal limestones from Busko-Spa environs (Middle Miocene, Central Poland). *Palaeogeogr. Palaeoclimatol. Palaeoecol.*, **27**, 155–165.

Studencki, W. (1988) Red algae from the Pinczów Limestone (Middle Miocene; Swietokrzyskie Mts., Central Poland). *Acta Palaeontol. Pol.*, **33**, 4–57.

Tandon, K.K., **Gupta, S.K.** and **Saxena, R.K.** (1977) A new species of *Lithophyllum* from Oligocene of southwestern Kutch. *J. Palaeontol. Soc. India*, **21–22**, 74–77.

Vannucci, G. (1980) Prime indagini sulle rodoliti del "Serravalliano" della Vale Scrivia. *Quad. Ist. Geol. Univ. Genova*, **5**, 59–64.

Vannucci, G., **Stockar, R.**, **Piazza, M.** and **Fravega, P.** (1993) La trasgressione tardo-oligocenica nella zona di Milesimo (Savona): caratteristiche e significato delle associazioni faunistiche ed algali. *Boll. Mus. Reg. Sc. Nat. Torino*, **11**, 239–266.

Vannucci, G., **Piazza, M.**, **Fravega, P.** and **Arnera, V.** (1994) Le rodoliti del Miocene inferiore del settore SW del Bacino Terziario del Piemonte (Spigno Monferrato - Alessandria). *Atti Soc. Tosc. Sci. Nat. Mem. ser. A*, **100** (1993), 93–117.

Vannucci, G., **Piazza, M.**, **Fravega, P.** and **Abate, C.** (1996) Litostratigrafia e paleoecologia di successioni a rodoliti della "Pietra da Cantoni" (Monferrato orientale, Italia nord-occidentale). *Atti Soc. Tosc. Sci. Nat. Mem. ser. A*, **103**, 69–86.

Vannucci, G., **Piazza, M.**, **Pastorino, P.** and **Fravega, P.** (1997) Le facies a coralli coloniali e rodoficee calcaree di alcune sezioni basali della Formazione di Molare (Oligocene del Bacino Terziario del Piemonte, Italia Nord-Occidentale). *Atti Soc. Tosc. Sci. Nat. Mem. ser. A*, **104**, 13–39.

Wiedicke, M. (1987) Biostratigraphie, Mikrofazies und Diagenese tertiärer Karbonate aus dem Südchinesischen Meer (Dangerous Grounds-Palawan, Philippinen). *Facies*, **16**, 195–302.

Woelkerling, W.J. and **Campbell, S.T.** (1992) An account of southern Australian species of *Lithophyllum* (Corallinaceae, Rhodophyta). *Bull. Br. Mus. Nat. Hist. (Bot.)*, **22**, 1–107.

Int. Assoc. Sedimentol. Spec. Publ. (2010) **42**, 183–200

Molluscs as a major part of subtropical shallow-water carbonate production – an example from a Middle Miocene oolite shoal (Upper Serravallian, Austria)

MATHIAS HARZHAUSER* and WERNER E. PILLER†

**Natural History Museum Vienna, Burgring 7, A-1014 Vienna, Austria (E-mail: mathias.harzhauser@nhm-wien.ac.at)*
†Institut für Erdwissenschaften, Bereich Geologie und Paläontologie, Universität Graz, Heinrichstrasse 26, A-8010 Graz, Austria

ABSTRACT

Molluscs are usually subordinate contributors to Cenozoic subtropical carbonate factories. A spectacular exception is represented by the shell-carbonate deposits from the Middle Miocene of the Vienna Basin (Austria). These strata consist of up to 81% shells and shell-hash of marine bivalves and gastropods. These locally widespread deposits fill the inlet of an Upper Serravallian (= Sarmatian regional stage) oolite shoal, forming foresets of 10–13 m height and slope angles of 20°.

Medium- to small-sized shell dunes of up to 280 cm height and shell ripples of 10 cm height and up to 190 cm length can be distinguished within the foresets. Due to amalgamation and mechanical nesting of shells, the ripples grew into the direction of the flow and were run over by subsequent ripples. The piling of shells causes stoss sides with high preservation potential within the ripples. The shells and shell-debris involved in the dune formation are interpreted to be derived from the surrounding shoal. The geometry of the foresets, dunes and ripples documents a dominant current entering a shallow lagoon framed by oolite shoals via an inlet. Based on the palaeogeographic position, these bedforms are interpreted to indicate the presence of a subaqueous flood-tidal delta marking the entrance into a shallow lagoon.

The absence of corals and corallinacean algae and a relatively reduced biotic inventory following a major extinction of marine biota in the enclosed Sarmatian Sea, allowed a few pioneer mollusc species to settle at the coasts in considerable numbers. Due to the absence of the classical constituents of a shallow-water subtropical carbonate factory (i.e. a photozoan association), molluscs came to dominate carbonate production.

Keywords Oolite shoal, molluscs, flood-tidal delta, shell dunes, Miocene, Vienna Basin.

INTRODUCTION

The Middle Miocene Sarmatian stage, a regional equivalent of the Upper Serravallian (Fig. 1), coincides with the last marine phase of the European Central Paratethys Sea. Due to the sea-level low coinciding with the glacio-eustatic isotope event MSi-3 at 12.7 Ma (Abreu & Haddad, 1998), strong restrictions of the open ocean connections of the Central Paratethys occurred (Harzhauser & Piller, 2004a). This induced the development of a highly endemic marine fauna that lacks any stenohaline organisms, pointing to shifts of the water chemistry (Piller & Harzhauser, 2005). Simultaneously, changes in ecosystem complexity and food-webs occurred. Many predators preying on molluscs, such as crustaceans, carnivorous gastropods and durophagous fishes disappeared. The fully endemic development is reflected in a regional eco/biostratigraphic zonation based on molluscs and benthic foraminifera (Fig. 1).

The Lower Sarmatian in the western part of the Vienna Basin is generally dominated by fine-grained siliciclastics. Carbonates are rare, except for small bryozoans-serpulid-algal-microbial bioconstructions. During the *Ervilia* Zone, sedimentation switched from a siliciclastic dominated system to a carbonate depositional environment, characterized by more than 20 m thick Upper Sarmatian carbonate platforms with oolites and foraminiferal (nubeculariid)

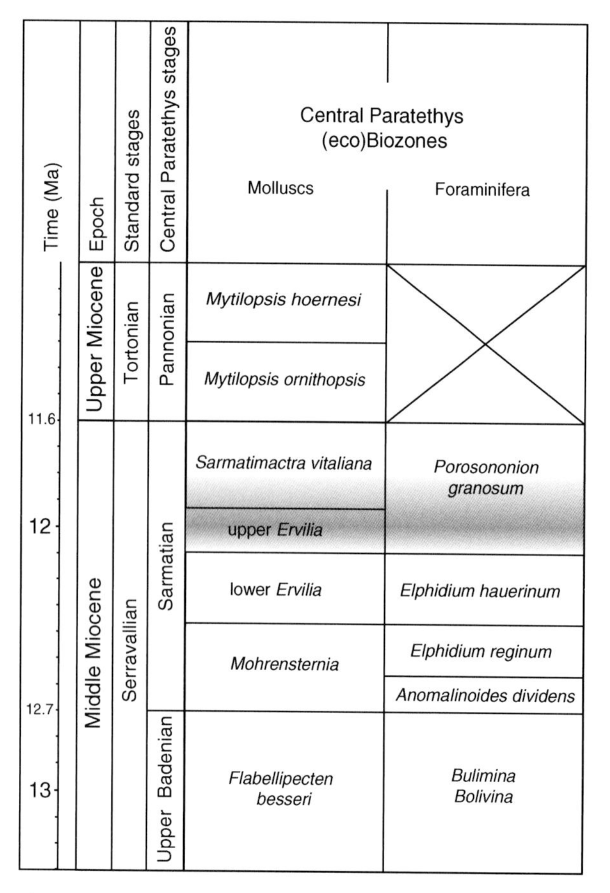

Fig. 1. Miocene chronostratigraphy of Europe. Modified after Rögl (1998) and Harzhauser and Piller (2004a), including the mollusc-based ecostratigraphic zonation of the Sarmatian. The grey-shaded area indicates the stratigraphic position of the investigated deposits.

bioconstructions. Piller & Harzhauser (2005) observed a contemporaneous increase in shell thickness of bivalves. A unique feature related with the oolite shoals are the rock-forming mollusc-shell-hash deposits at Nexing in the Vienna Basin (Lower Austria). This locality, designated as the holostratotype of the Sarmatian Stage (Papp & Steininger, 1974), is characterized by largely biogenic sediments. The sedimentological setting of the section was interpreted by Papp & Steininger (1974) as a strongly agitated shallow sublittoral zone close to the mainland. In the present study these strata are interpreted as subaqueous tidal deposits that formed along the rim of an oolite shoal. The specific environmental framework allows answers to be given concerning the important question why molluscs, characteristic of heterozoan carbonates *sensu* James (1997), can substitute the classical coralgal carbonate factory (heterozoan carbonates) in shallow-water subtropical settings.

GEOLOGICAL SETTING

The study area is situated in the northern Vienna Basin, a pull-apart basin about 200 km long surrounded by the Eastern Alps, the West Carpathians and the western part of the Pannonian Basin (Wessely, 1988; Fig. 2). The Sarmatian portion of the 7000 km thick Neogene basin fill attains more than 1000 m in the central Vienna Basin (Harzhauser & Piller, 2004a, b).

The deposits at the Nexing section (Fig. 2B) are part of the Miocene sediment-cover of the Mistelbach tectonic block. This tectonic unit represents a marginal block of about 60 km length and 18 km width that is separated from the deeper Vienna Basin by the Steinberg fault zone (Kröll & Wessely, 1993). The Steinberg elevation and the bordering fault zone have been the target of many geological studies during the pioneer-phase of Austrian petroleum exploration (e.g. Friedl, 1936). Generally, most of the Sarmatian succession is covered by Upper Miocene (Pannonian) fluvial and limnic deposits or by Pleistocene loess, obscuring the facies relationships of the Sarmatian. Nevertheless, age-equivalent deposits are exposed in several outcrops that have been studied during the current project (Fig. 2).

According to the regional eco/biostratigraphic zonation, the Nexing section encompasses the Upper *Ervilia* Zone and lowermost *Sarmatimactra* Zone in terms of mollusc zonation (Fig. 1), and the Lower *Porosononion granosum* Zone of the benthic foraminiferal zonation. The boundary between the Upper *Ervilia* Zone and the *Sarmatimactra vitaliana* Zone is situated at the base of Unit 3 (Fig. 3). The dating is based on the occurrence of the cardiid *Plicatiforma latisulca* (Münster) and the evolutionary levels of the gastropod *Duplicata duplicata* (Sowerby) and of the bivalves *Ervilia dissita podolica* (Eichwald), *Venerupis gregarius* (Partsch) and *Sarmatimactra eichwaldi* Laskarev (see Papp, 1956, for details).

Material and methods

The outcrop covers an area of about 0.2 km^2 (Fig. 2C). It is currently divided into an active pit in the east and an abandoned quarry in the west. The mined bioclastic sediments have been used as bird food for a long time. The quarry walls are

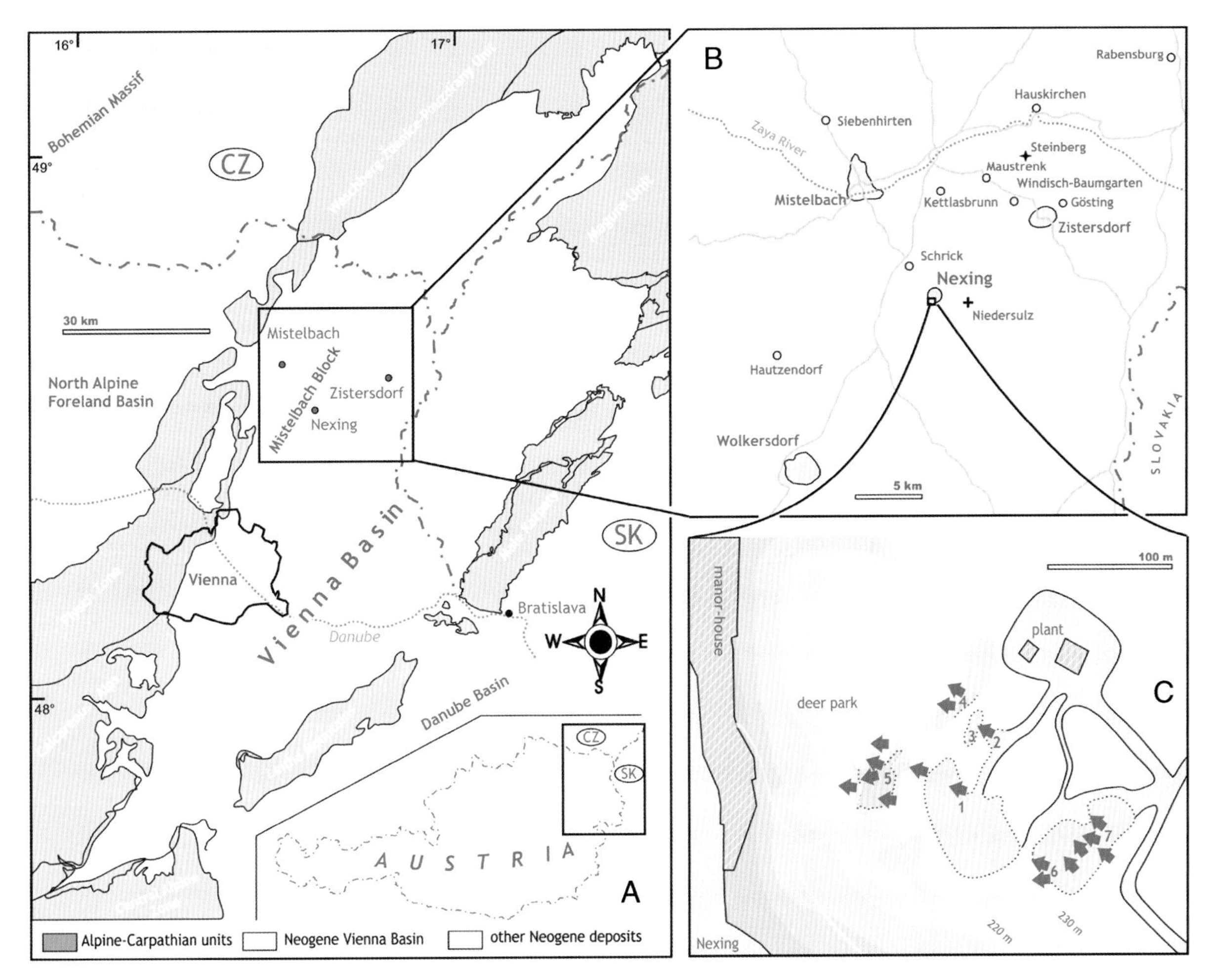

Fig. 2. Location of the study sections. (A) Regional location maps of the investigated sections on the Mistelbach block along the margin of the Vienna Basin. (B) Geographic position of the outcrop at Nexing (Austria). (C) Overview of the outcrop with the position of the investigated stratigraphic sections (Nexing 1–7). Arrows represent dip directions of the foresets. CZ = Czech Republic; SK = Slovakia.

roughly oriented WNW–ESE and SW–NE. Seven sections have been logged, providing an excellent insight into the facies architecture and the texture of the deposits. Six of these profiles are illustrated in Fig. 3. The total thickness attains about 19 m, representing a small part of the Upper Sarmatian Skalica Formation (Harzhauser & Piller, 2004a). The stratigraphic succession has been divided into four main units on the basis of lithology, sedimentary structure and biotic content (Fig. 3). Logs can be correlated laterally by a marker interval (Unit 3 in Fig. 3). A granulometric analysis has been performed for eight samples. To achieve a reliable ranking of quantitatively important mollusc species, which are the main constituents of the foresets and shell dunes (see below), five bulk samples (same numbers as granulometric samples in Fig. 3) were taken at the section covering all important lithological units. The samples were sieved through a 1 mm screen, divided into four splits, and all taxa of each split were counted. Additional data about the lateral variation and character of the Sarmatian carbonates are provided by nine well logs drilled by the OMV-AG oil company (see Harzhauser & Piller, 2004a).

RESULTS

Lithofacies units

Unit 1

The lower part of the deposits only crops out at section Nexing 1. Green silt with scattered plant debris (1.5 m thick) forms the lowermost unit. The silt bed dips 15°–16° toward the NW. Any

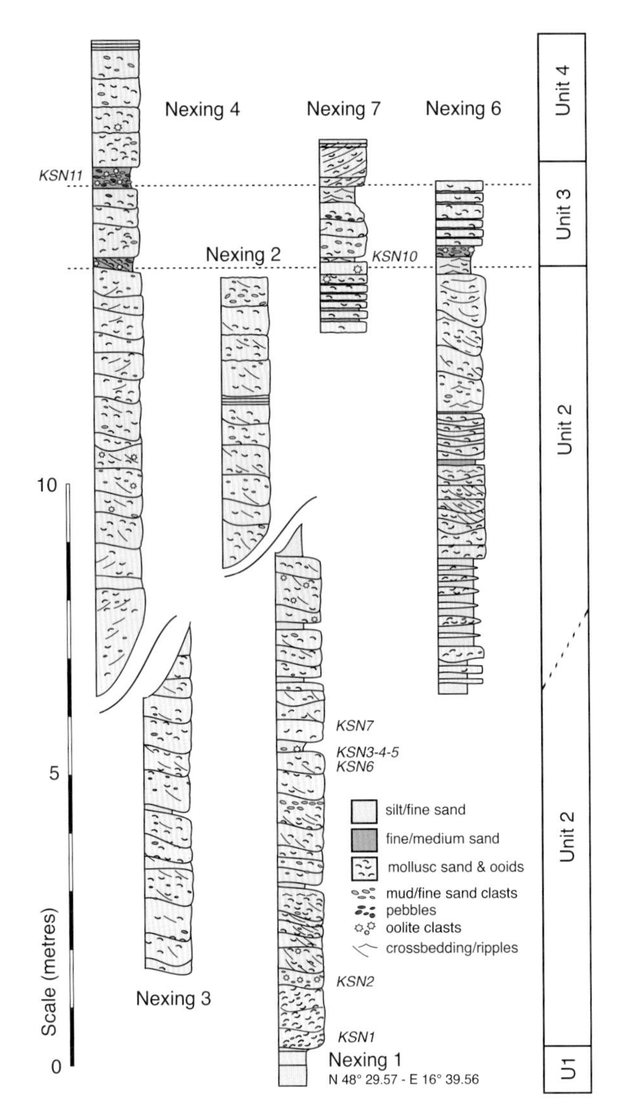

Fig. 3. Logs of the sections Nexing 1–4 and 6–7. The sandy intercalations at the base and the top of Unit 3 are used as reference levels to correlate the sections (dashed lines). KSN refers to samples mentioned in the text.

Fig. 4. Steeply inclined planar-bedded foresets of coarse mollusc shell-hash (Unit 2). (A) Lower part of the succession, representing a lateral equivalent of Nexing 1 (total height of outcrop approximately 15 m). Up to 80% of the foresets are made up of mollusc shells. The dip of the foresets is roughly toward the WNW. (B) Lower bounding of a shell ripple; viewer looks from the lee-side. Most of the shells are strongly fragmented and abraded to various degrees. The large shells are *Venerupis gregarius*. Scale bar units: 1 cm.

macrofauna is missing. The scarce microfauna consist of foraminifera, such as *Porosononion granosum* and various elphidiids.

Unit 2

The silty sediment of Unit 1 is overlain by about 14 m of steeply inclined planar-bedded foresets of coarse mollusc shell-hash (Unit 2; Fig. 4A). The sediment is a bioclast-supported, polytaxic skeletal concentration (Figs 4A–B and 5A–F). The carbonate content of the sediment, such as that illustrated in Fig. 5A–D, ranges from 78–81%. This content may decrease to 60 % in poorly sorted layers with higher amounts of siliciclastics (e.g. Fig. 5E–F). Aside from the predominance of biogenic components, the poorly sorted sediment consists of medium to coarse quartz sand, associated with ooids, scattered pebbles of Cretaceous sandstone and rare reworked oolite clasts. These coquinas are very poorly sorted because the bioclasts (~2–30 mm) are generally larger than the bulk of the siliciclastic components, whilst the pebbles and oolite clasts surpass the bioclasts in size (up to 80 mm).

The thickness of the foresets ranges from 80 to 280 cm, being separated by fine to medium sand intercalations of 1–30 cm thickness. Occasionally, this siliciclastic intercalation is missing, and

Fig. 5. Outcrop photographs of various shell ripples and shell dunes showing the eye-catching imbrication of shells. (A) Detail of two shell ripples separated by a layer of better sorted and planar-bedded shell-hash, indicated by arrow (flow from right; length of picture = 3.2 cm). (B) Distal part of a shell dune, separated from underlying and overlying dunes by poorly developed shell drapes indicated by arrow (flow from right; length of picture = 8.7 cm). (C) Shell ripple in the centre of the picture forming the crest of a shell dune. A drape of sand is well developed (flow from left). Scale bar units: 1 cm. (D) Similar situation as (C); shell ripple in the centre of the picture with drape separating two dunes. Note the pebble on the right which follows the imbrication of the shell along the stoss (flow from left). Scale bar units: 1 cm. (E) Proximal part of shell ripple showing the increasing imbrication (flow from left). Scale bar units: 1 cm. (F) Centre of an isolated shell ripple that developed within two sand drapes. Note the steep-angled imbrication and the tendency of large valves to lie with the convex side in the lee direction (flow from left; length of picture = 13 cm).

reworked lithoclasts occur in the shell bed. Lower parts of that unit, as exposed in the Nexing 1 and 3 sections, display planar bedding of more or less parallel foresets. Towards the top, exposed in sections Nexing 2, 4 and 6, the sets are divided by shallow and broad channel-like structures with thick silt to fine sand drapes of up to 40 cm thickness. The foresets dip at angles ranging from 21° to 36° in a west to NW direction. A general steepening of the foresets from base to top occurs, with angles from 21° to 28° in the lower part of the unit, and 30° to 36° in the upper portion. However, the dip-angles distinctly decrease in the uppermost parts of the shell-hash foresets, where values of 15–25° are measured.

Unit 3

This unit represents a marker interval allowing a lateral correlation across the outcrop area due to the marked lithological change. At section Nexing 4, it consists of a 20–180 cm thick layer of silty fine sand bearing well-rounded pebbles of Cretaceous sandstone, oolite pebbles (Fig. 6A–C) and mud to fine sand clasts of up to 5 cm diameter. Large-sized molluscs of up to 6 cm diameter (e.g. *Sarmatimactra*) are embedded within these deposits. Laterally, sections Nexing 6 and 7 show a decrease in the amount of pebbles and 10–30 cm of silt and fine sand with well-developed small-scale wave-ripples of 6–14 cm length (Fig. 6D). These ripples are

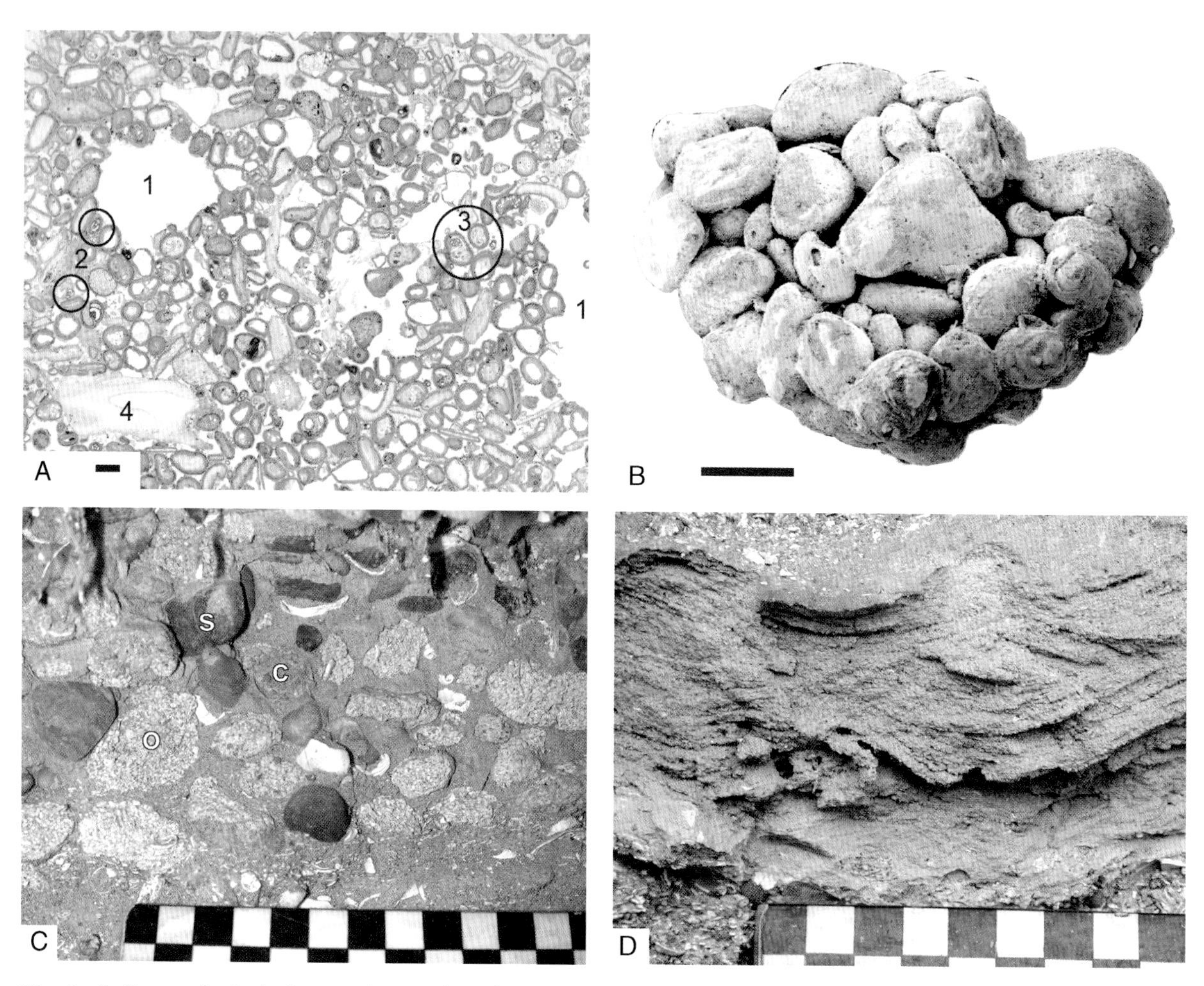

Fig. 6. Sedimentological characteristics of marker Unit 3. (A) Thin-section of an oolite pebble in Unit 3. (1) Large dissolution vugs are frequent. In addition to quartz grains and shell-hash, (2) miliolid foraminifera and (3) hydrobiid gastropods are the most abundant nuclei in the ooids. Larger bioclasts such as (4) cardiid shells lack the coatings (scale bar: 1 mm). (B) SEM photograph of an oolite clast from Unit 2; an ooid in the front bears a small-sized gastropod as a nucleus (scale bar: 1 mm). (C) Sand-supported conglomerate in the upper part of Unit 3 with well-rounded sandstone pebbles (s), carbonate concretions (c) and reworked oolites (o). Scale bar units: 1 cm. (D) Wave ripples made up of fine to medium sand in the lower part of Unit 3, close to sample KSN10. Shell-hash stringers are frequent. Scale bar units: 1 cm.

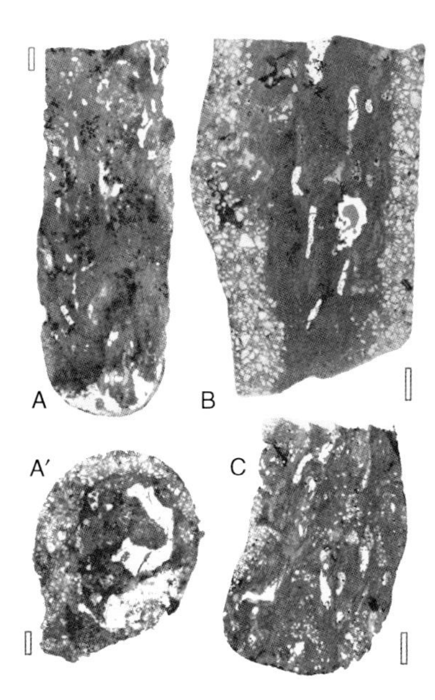

Fig. 7. Rhizoliths from Unit 3. (A–C) Tangential sections. (A′) Cross-section. Note the characteristic halo of quartz sand. This decalcified zone is interpreted to result from dissolution by the acidic microhabitat surrounding the roots (scale bars: 1 mm).

overlain by 100–150 cm inclined planar-bedded mollusc shell sand with sand intercalations and frequent pebbles and reworked rhizoliths. The latter are irregular, tube-like, glossy brown calcite structures representing reworked root-horizons (Fig. 7). The dip-angle of this bed ranges from 15° to 4°. At Nexing 7, another layer of 30–50 cm of silt and fine sand with small-scale ripples follows; the dip-angle is around 4°. Again, this layer of well-sorted sediment changes within 150 m towards the NW and is replaced by fine sand with well-rounded oolite pebbles and Cretaceous sandstone pebbles at section Nexing 4 (Fig. 6C).

Unit 4

Unit 4 has a sharp lower boundary, which is overlain by a 2–3 m thick, steeply inclined, planar-bedded shell-hash. In terms of composition and sedimentary structures, deposits are similar to Unit 2, but dip-angles are lower, with values between 12° and 25°. This Miocene unit grades into various silty, sandy and gravelly layers with shell-hash, which are interpreted as reworked deposits of Pleistocene age, as supported by the relation of the deposits to the adjoining Pleistocene loess.

Sedimentary structures

Dune and related terms such as sandwave and ripple are frequently used in the literature with a broad spectrum of meanings (Allen, 1980; Boersma & Terwindt, 1981; Dalrymple, 1984; Galloway & Hobday, 1996; Leclair, 2002). *Subaqueous dune* is used herein according to Ashley (1990), who pointed out the preference of that term over terms such as megaripple or sandwave. Many authors, such as Allen (1980), Galloway & Hobday (1996) interpreted dunes/sandwaves as several metres long, flow-transverse bar macroforms triggered by currents of tidal origin. As the structures described herein are characterized by a composition dominated by mollusc shells, the bedform is regarded as being a shell dune. Consequently, the low-order bedforms superimposed on or constituting the structures are termed shell ripples. A closer examination of the planar-bedded foresets reveals a complex internal geometry. Three different bedform types of different hierarchical order can be defined on the basis of size and texture:

Bedform type 1

The Bedform Type 1 is a composite bedform. It comprises foresets (Fig. 4A) that range in thickness from 80 to 280 cm. When optimally preserved, single sets are separated by silt/sand-drapes of 1–30 cm thickness. These drapes became frequently eroded during the deposition of the subsequent set, but they are usually traceable as clasts in the basal part of the overlying set. The drapes consist mainly of silty fine sand with planar bedding and rare cm-scale cross bedding at the top, whilst pure clay and silt layers are very rare. In some cases, they are replaced by well-sorted, fine shell-hash.

Bedform type 2

Bedform Type 2 consists of shelly dunes, which are bundled into Bedform Type 1 (Fig. 8A). Individual shell dunes may reach a length of several metres, and more than 1 m in height, They are separated by thin drapes of fine to medium sand (sometimes replaced by shell-hash). Figure 8 illustrates a part of such a structure being composed of ripples, which represent the third bedform type. The separating layers of shell-hash and sand are well developed in Fig. 5B and D.

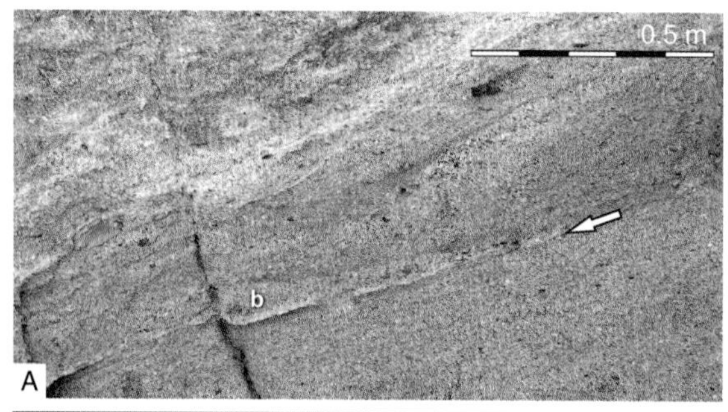

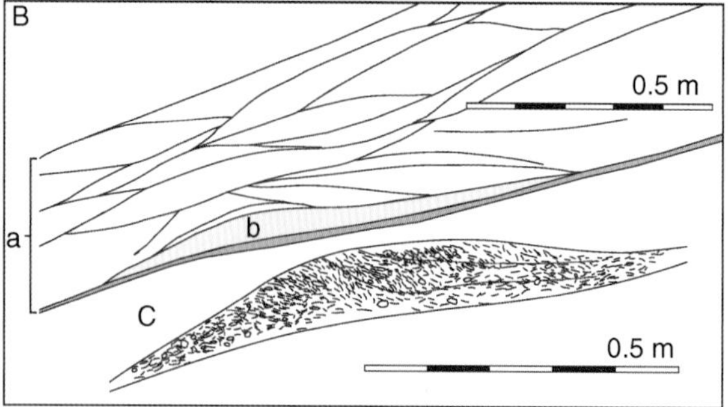

Fig. 8. Bedform Type 2 shelly dunes. (A) Outcrop photograph showing Bedform Type 2, a shell dune with mud-drape at the base (arrow). (B) Sketch illustrating the internal structure of the dune (a), formed by stacked shell ripples (b; Bedform Type 3). (C) The orientation of shells and fragments in one ripple. Steepest angles and maximum imbrication is achieved in the middle part of the ripple, overlying a basal layer of more or less planar-bedded shell hash. A further, faint, subdivision of the ripple is indicated by the dashed lines.

Bedform type 3

The shell dunes are subdivided into a smaller-scaled bedform type, termed herein *shell ripple* (Figs 5F and 11.8B). Individual ripples are up to 190 cm long and 10 cm high. Hence, the shell ripples attain heights ranging from 3.5 to 10% of the total height of the foresets. Internally, the ripples are composed of shell-hash and mollusc shells. Fossils are predominantly abraded fragments. Well-preserved valves are only a subordinate component. Shells and shell debris are associated with up to 8 cm large pebbles of well-rounded Cretaceous sandstone and oolites. Fine to coarse sand and reworked single ooids are common.

Due to cementation, only few ripples are appropriate for granulometric analysis (Fig. 9). The grain-size distributions of the measured samples display two patterns. Samples KSN3, 4, 5 and 6 show a distinct bimodal distribution with peaks at 0–1Φ and at 3–4Φ. In contrast, samples KSN1, 2 and 7 have a unimodal distribution with a peak around 0–1Φ. The sorting (method of Folk & Ward, 1957) ranges between 1.82 and 2.0 for bimodal samples and from 1.1 to 1.36 for unimodal samples, thus in both cases representing poorly sorted sediments. The shell ripples are usually separated by densely packed, roughly planar-bedded shell-hash, but lack silt – fine sand drapes (Fig. 5A).

The elongated shell ripples display a complex internal texture: the basal layer of a shell ripple is formed by planar-bedded shell-hash and shells with admixed sand, ooids and scattered pebbles. At the leeward side of the ripple, the shells and fragments are arranged in a rather chaotic pattern but soon start to become imbricated on the stoss-side (Fig. 5F). This imbrication starts with a low-angle amalgamation of shells with a tendency for orientation of the convex sides towards the lee-direction. This "convex-side down-forward" pattern is most obvious on the bedding planes of the ripples (Fig. 4B).

The accretion of shells culminates in imbrications with steep dip-angles of up to 50–70°. This texture is best developed in the middle part of the ripple and starts to degrade towards the stoss-side due to gradual overtilting (Fig. 5E). Thus, shells and fragments tend to be arranged rather randomly or more or less parallel to the angle of the stoss.

Biota

Up to 81% of the deposits consist of skeletal grains derived almost exclusively from molluscs. Bivalves are only found with disarticulated valves. Abraded shells are typical for both bivalves and gastropods. Based on the collections of the Natural History Museum Vienna, 21 gastropod species and 11 bivalve species are recorded from the current outcrop (Table 1). In contrast to the 32 species known from museum collections, only 17 species were identified in the bulk samples (Fig. 10). A total of 1837 mollusc specimens could be identified in these five bulk samples (Fig. 10). The quantitatively most important species are the same in all standardized samples from five different shell beds. In all samples, a very small number of species contribute up to 92–98% of the biogenes. These are *Granulolabium bicinctum* (batillariid gastropod), *Venerupis gregarius* (venerid bivalve), *Obsoletiforma obsoleta vindobonensis* (cardiid bivalve), *Hydrobia frauenfeldi* (rissoid gastropod), *Sarmatimactra vitaliana* (mactrid bivalve), *Ervilia dissita podolica* (mesodesmatid bivalve), and *Cerithium rubiginosum* (cerithiid gastropod). Species-

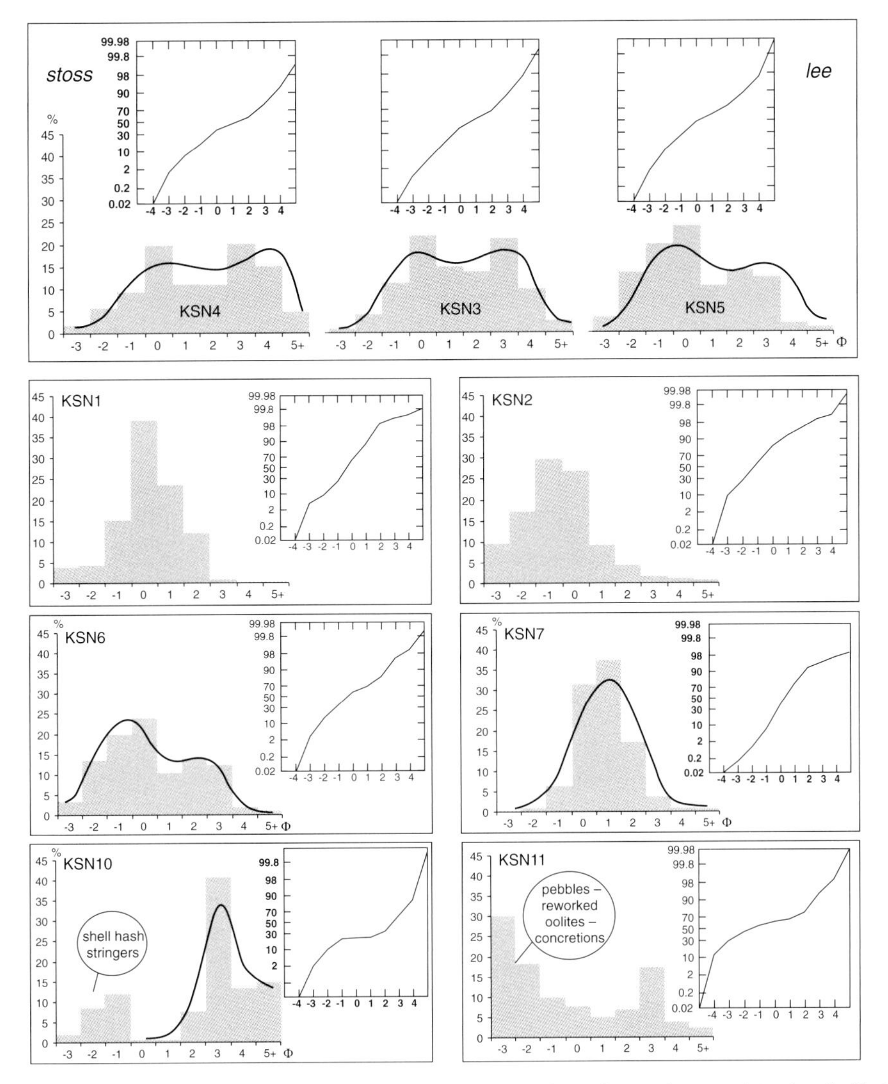

Fig. 9. Grain size distributions and cumulative probability curves of selected samples. Within the shell ripples, a bimodal grain-size distribution of samples KSN3–6 contrasts with a unimodal distribution of samples KSN1, 2 and 7. KSN10 derives from the sandy wave ripples illustrated in Fig. 6D. The coarse peak results from admixed shells. The extremely poorly sorted sample KSN11 derives from Unit 3 and yields a mixture of reworked shell dunes, oolites, fluvial pebbles and sand that formed during the renewed transgression in the latest Sarmatian (cf. Fig. 13E).

sampling curves for the splits level off nearly immediately. Thus, the sampling intensity is adequate for the documentation of all important species; even most of the rare taxa are registered.

INTERPRETATION

Lithofacies

The high angles of the slipfaces of the foresets in Unit 2 seem to be tectonically accentuated. In fact, a back-tilting of about 15° is required if one assumes that the thin-bedded, plant-debris bearing silt of Unit 1 was originally horizontally bedded (Fig. 11). Consequently, the slope angles are around 6–15° in the lower part of Unit 2, increase to around 20° in the middle part, and decrease to angles between 10 and 20° towards the top of Unit 2. This pattern results from the gradual progradation in western and north-western directions. However, this back-tilting would cause an impossible "updipping" if applied to the wave ripple-bearing layers in Unit 3. This points to a tectonic phase prior to the formation of Unit 3, causing a 15° tilt of Units 1 and 2.

This interpretation is strongly supported by the deposition of gravel and reworked oolites at the base of Unit 3. The oolites were weakly lithified

Table 1. Total mollusc fauna recorded from the shell dunes, based on museum collections

Bivalvia
Modiolus subincrassatus (d'Orbigny)
Obsoletiforma obsoleta vindobonensis (Laskarev)
Obsoletiforma ghergutai (Jekelius)
Inaequicostata politioanei (Jekelius)
Plicatiforma latisulca (Münster)
Sarmatimactra eichwaldi (Laskarev)
Solen subfragilis (Eichwald)
Donax dentiger (Eichwald)
Ervilia dissita podolica (Eichwald)
Mytilopsis sp.
Venerupis gregarius (Partsch in Goldfuss)
Gastropoda
Acmaea soceni (Jekelius)
Gibbula podolicum (Dubois)
Gibbula poppelacki (Hörnes)
Gibbula picta (Eichwald)
Theodoxus crenulatus (Klein)
Hydrobia frauenfeldi (Hörnes)
Hydrobia sp.
Granulolabium bicinctum (Brocchi)
Potamides nodosoplicatum (Hörnes)
Potamides disjunctus (Sowerby)
Potamides hartbergensis (Hilber)
Cerithium rubiginosum (Eichwald)
Melanopsis impressa (Krauss)
Euspira helicina sarmatica (Papp)
Ocenebra striata (Eichwald)
Mitrella bittneri (Hörnes & Auinger)
Duplicata duplicata (Sowerby)
Acteocina lajonkaireana (Basterot)
Gyraulus vermicularis (Stoliczka)
Tropidomphalus gigas (Papp)
Cepaea gottschicki (Wenz)

by submarine isopachous fibrous carbonate cement prior to erosion and transportation. Moreover, the frequent rhizoliths (Fig. 7), document an episode of emergence and vegetation of the area. After that phase, the last Sarmatian flooding during the *Sarmatimactra* Zone (Harzhauser & Piller 2004a) initiated the development of a second but less prominent unit of foresets. Therefore, the more or less undisturbed top of the section (Unit 4) was deposited with slope angles fully corresponding to the (back-tilted) foresets of Unit 2.

The current outcrop situation allows estimates of the length and width of the sedimentary bodies to be made. The foresets in the lower part of Unit 2 display more or less straight "crests" and planar slopes. The length of the slopes, based on outcrop observations, can be estimated to attain at least 30–40 m. Taking the calculated dip-angle of 20° into consideration, a height of about 10–13 m can be predicted for that structure. This rough estimation is corroborated by the outcrop at section Nexing 1m where the base the foresets is exposed. Consequently, the lowest dip-angles, observed at the base of the succession, correspond to very distal parts of the slipfaces, indicating early progradation.

The sedimentological interpretation of the deposits strongly relies on the general palaeogeographic framework. As shown in Fig. 12, the locality of Nexing is situated in a seaway or channel framed by ooid shoals which connected the northern Vienna Basin with the lagoon of the Mistelbach Basin. As the stratification indicates a west-directed progradation of sediment bodies, a current entering the lagoon was the main controlling factor for bedform formation. The huge foresets are interpreted, therefore, as flood tidal delta foresets. During the westward migration of the delta, the steep foresets gradually buried the preceding ones. At the time of flood tidal delta growth, the relative sea-level in the Vienna Basin was quite stable (Papp, 1956), and a coinciding loss of accommodation space can therefore be predicted. This assumption is supported by the upsection changing geometry of the foresets.

Towards the top of Unit 2, the regular foreset pattern is replaced by a more wavy and partly channel-like bedding. These structures might either represent large lobate foresets or longitudinal tidal bars, as described by Lesueur *et al.* (1990) from the Miocene of the Rhone Basin. However, their internal structure is identical to that of the planar-bedded foresets. This morphological shift is interpreted to be caused by a shallowing-upward trend. Consequently, current velocity probably increased, being reflected in a succession from lower-speed bedforms towards higher-speed bedforms (Rubin & McCullogh, 1980; Costello & Southard, 1981; Dalrymple, 1984; Terwindt & Brouwer, 1986). Despite the virtual dominance of gravel-sized shells and pebbles, the deposits are mainly composed of the sand grain-size class. Even the higher amount of gravel occasionally observed in the outcrop does not contradict an interpretation as a dune. As demonstrated by Carling (1996), the predominant sediment grain size is not a major control for dune formation in coarse sediment; 2-D and 3-D dunes may even form from coarse gravel.

At first glance the overall geometry is therefore strongly reminiscent of 2-D dunes (*sensu* Ashley, 1990) and resembles the *Class IIIA* category of tidal dunes of Allen (1980). Steep foresets with downslope angles of 20° develop beneath a large-scale separated flow. During phases of

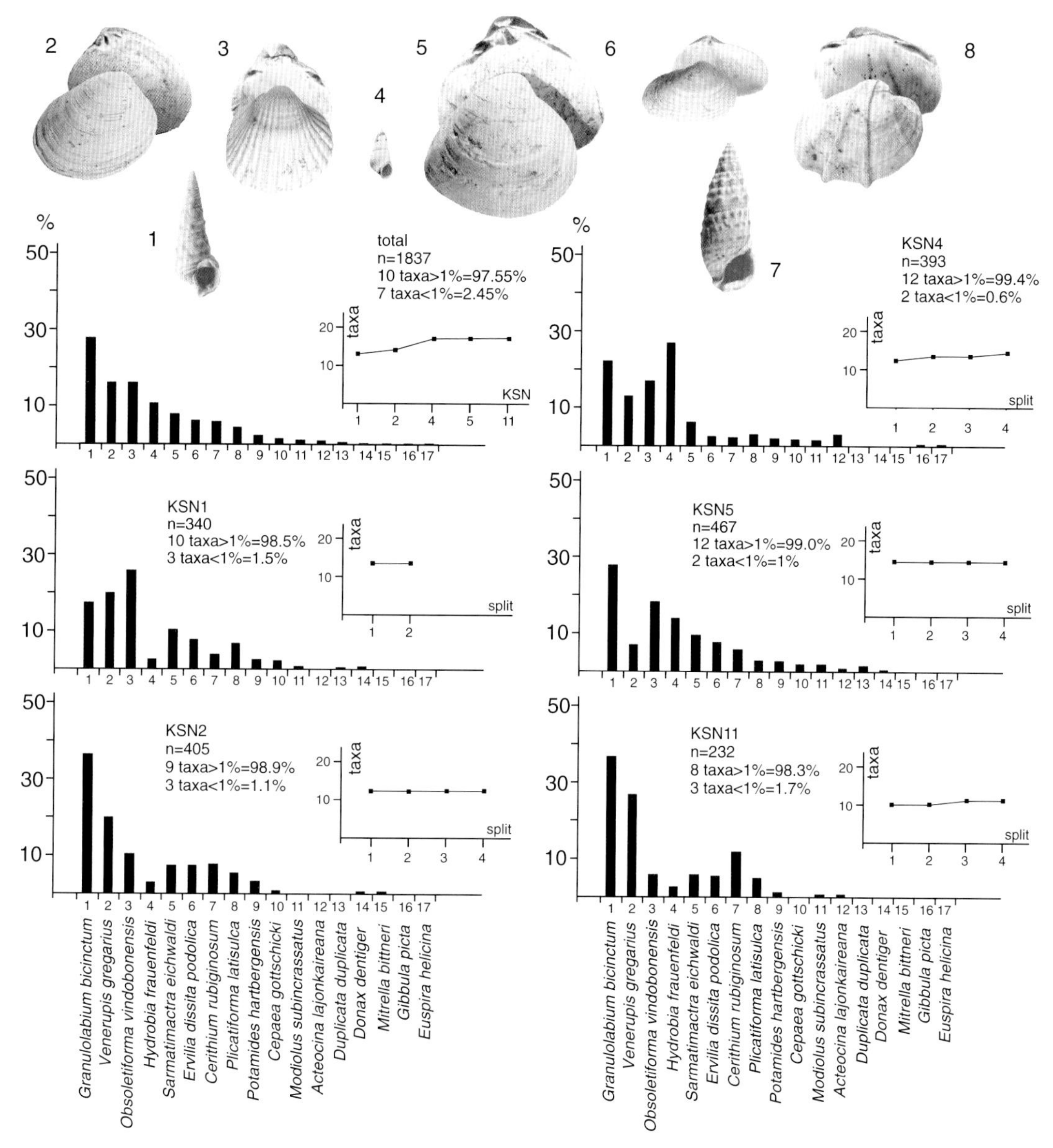

Fig. 10. Taxonomic composition and percentage abundance of mollusc species in five standardized bulk samples and for all samples (total). The eight quantitatively most important taxa are illustrated. Determination of bivalves is based on hinges and umbos; fragments are not included. Species-sampling curves for each sample (KSN1, 2, 4, 5, 11) and for all samples (total) are drawn, documenting a well-balanced sampling quality and quantity.

substantial slackening of currents the foresets became separated by thick mud drapes. However, interpreting the foresets as slipfaces of giant dunes would result in problematic inferences with palaeogeography, namely with the depositional depth.

Although there is considerable doubt about a straightforward correlation between dune height and flow depth (e.g. Stride, 1970; Terwindt & Brouwer, 1986; Allen & Homewood, 1984; Flemming, 2000), several authors including Allen (1980), Yalin (1972), Allen *et al.* (1985) and Mosher & Thomson (2000) have discussed a vague correlation between dune height and total water depth. Following the various rules of thumb presented by the mentioned authors, a total water depth ranging from 40 to 90 m would have to be calculated for the 10–13 m high structures of Nexing. Based on the topographic altitude of the correlative littoral deposits of the shallower oolite shoal, this depth estimation turns out to be much too deep. Especially the marker horizon in Unit 3, suggesting a

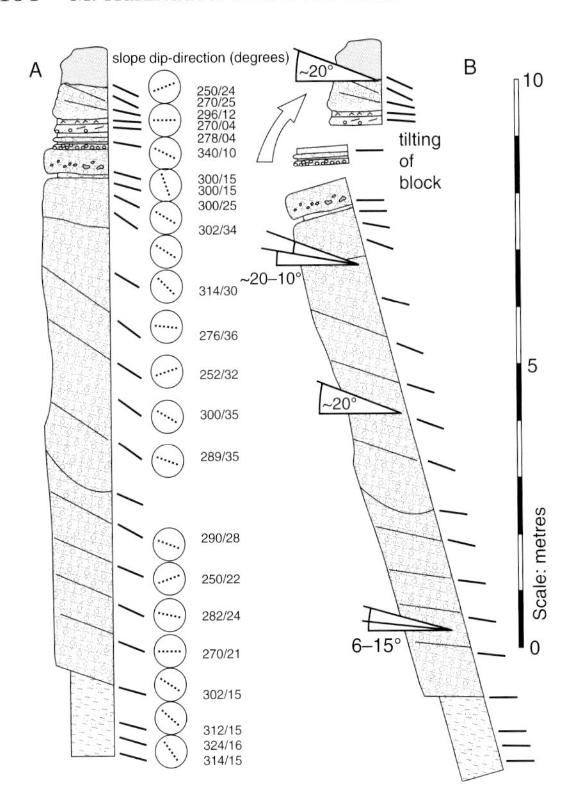

Fig. 11. Composite section for the shoal succession. (A) Idealized composite section. The dip of the foresets is consistently towards the W and NW. (B) Back-tilted log (−15°) and tectonic phase separating Unit 2 from Units 3 and 4 (see text for discussion).

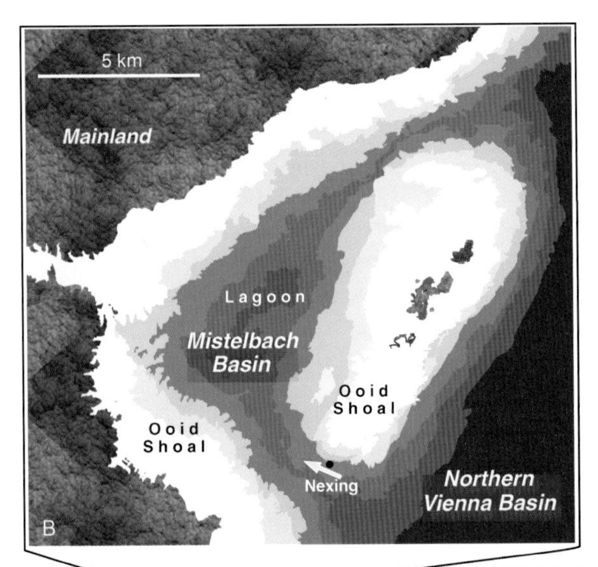

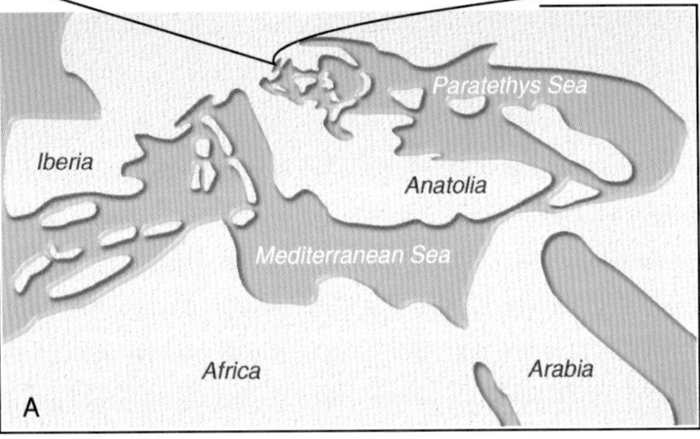

Fig. 12. Palaeogeographic setting. (A) Palaeogeography of Eurasia during the Serravallian, modified from Rögl (1998). The Paratethys Sea is nearly completely disconnected from the Mediterranean Sea. (B) The Mistelbach block during the Late Sarmatian (*Ervilia* Zone) based on sedimentological, palaeontological and tectonic information in Friedl (1936), Grill (1988), Kröll & Wessely (1993), and Harzhauser & Piller (2004a, b). The Steinberg elevation acted as an island with attached shoal about 15 km off the coast. The arrow illustrates the dominant flow direction interpreted from the foresets and subaqueous dunes at Nexing.

phase of emersion of the entire shoal, allows a good correlation of the deposits throughout the Mistelbach block. Hence, a maximum water depth of 10–20 m is most plausible and an interpretation as a dune field is rejected. Nevertheless, bioclastic sand dunes from Ackers Shoal (Torres Strait, NE Australia) reach a height between 3–8 m in a moderate water depth of around 20 m (Keene & Harris, 1995), and therefore an interpretation as "mega"dunes cannot be ruled out completely.

A second possibility is to discuss these structures as washover deposits that formed along the seaward fringe of the shoal. However, this contrasts with the internal architecture of the shell dunes and ripples, which points to regular short-term high-energy conditions rather than to random events. Furthermore, the steep-angled foresets differ distinctly from the subhorizontal to low-angle planar stratification as described by Schwartz (1982) from washover fan deltas.

Biofacies

The statistically important molluscan species are the same in all samples, but the predominance of single taxa varies (Fig. 10). Due to the poor sorting of the shell fragments, differences in the composition of the five samples cannot be explained solely by transport. Thus, the composition might rather reflect differences in the community structures of the source areas. The shell ripple from which sample KSN4 derives is characterized by a conspicuous predominance of hydrobiid and batillariid gastropods. Therefore, the shell concentration seems to have been fed by an intertidal environment where these taxa flourished (Latal *et al.*, 2004). In contrast, samples KSN1 and KSN2, which derive from the base of the succession, contain fewer hydrobiids but yield more foreshore and shoreface taxa such as venerid, mesodesmatid, donacid, and cardiid bivalves.

Small-sized gastropods such as the hydrobiids and *Acteocina lajonkaireana* frequently have an ooid-like coating. This points to a time-averaged accumulation of dead shells, which partly spent a long time in a wave-swept environment prior to their incorporation in the shell ripples. Most of the comparatively rare vertebrate remains also display traces of abrasion.

The supposedly high-energy environment of the shell deposits and the lack of bioturbation within the ripples as well as within the mud/sand drapes assigns these settings as an extremely unlikely natural habitat of the molluscs. Outcrops to the west and south of the Steinberg elevation document an extended area of very shallow-marine settings characterized by oolite shoals and sand. At these localities, many of the bivalves have articulated shells, pointing to an autochthonous occurrence.

Thus, it is suggested that the molluscs are derived from these shallow-marine intertidal to shallow-subtidal environments. The process behind how the shells became transported and finally incorporated into the shell deposits remains unknown. The position of the shell beds of Nexing is, however, immediately south of the topographic high of the Steinberg. This high zone may therefore have acted (temporarily) as island or shoal along which currents transported shells and sediment southwards. At the southern end of this topographic high, a curved island spit probably developed or a tidal inlet connected with a flood-tidal delta. Functionally this topographic high acted like a barrier island (e.g. Oertel, 1985) surrounded by oolite shoals and sand flats. The mass of shells may originate from intertidal flats or shallow-subtidal areas as well as from shore erosion along and around this topographic high (island or/and shoal).

Comparable depositional systems

No modern complex shell accumulation in planar foresets, comparable to the Sarmatian example, has yet been described in detail. However, similar general structures of bioclastic sand dunes with up to 95% biogenic carbonate were reported from the Torres Strait by Keene & Harris, (1995). The cores taken in those dunes, as illustrated by Keene & Harris (1995), seem to lack internal shell ripples. The texture of the shells and fragments in the cores reveals a subhorizontal pattern. Nesting and piling are virtually absent, pointing to a quite different depositional environment. Modern dunes of the German Bight described by Lüders (1929) display some similarities in composition. These dunes are similar in their dominance of cardiid, venerid and mytilid bivalves, but differ considerably in their internal organization. The shell accumulation there is caused by winnowing in the crest area, whilst growth results from avalanching along the lee side.

A slightly reminiscent fossil example was described by Portell *et al.* (1995) from the Pleistocene Leisey shell pits in Florida. There, the so-called upper shell bed (Fort Thompson Formation) is characterized by a high percentage of mollusc shells along with fine-grained quartz sand comparable to the deposits in Austria. Abraded shells and fragments are common and Portell *et al.* (1995) interpreted the depositional environment among other variants as a tidal channel or inlet. Details concerning the geometry and internal architecture of the sedimentary body are missing.

A lack of comparative information makes it difficult to interpret the complex internal composition of the shell dunes and shell ripples in "standard" terms of hydrodynamics. The unimodal and bimodal grain-size distributions detected in the shell ripple samples (Fig. 9) are reminiscent of grain-size distributions recorded by Allen & Homewood (1984) from a Lower Miocene tidal dune in the Swiss Molasse. In that key study, bimodal distributions occurred during the neap tides and at the transition towards the spring tides, whilst unimodal distributions developed during the spring tides. A corresponding trigger might have caused the grain-size distributions of the shell ripples. Indeed, the bimodal patterns of samples KSN3–5 were detected in a ripple that developed within a thick fine sand drape that points to low current energy.

In the following, a model will be presented for the internal structure of the flood-tidal delta foresets, the superimposed shell dunes, and the ripples, but this has to remain hypothetical because of the lack of a modern analogue for these deposits. It is proposed that the shell ripple bedform (Type 3) might be equivalent to a sand dune foreset of Allen & Homewood (1984). This interpretation is supported by the aforementioned grain-size distributions. The intervening and sometimes eroded shell-hash layer separating the shell ripples would thus have formed during the weaker ebb flow. Consequently, the sum of ripples constituting a shell dune represents tidal bundles that formed during one major cycle. This cycle might well correspond to a spring-neap cycle. The separating fine sand drapes developing during phases of low current energy would thus coincide with neap-tide deposition. The formation of the thick sand drapes separating the superior foresets of the flood-tidal delta is thus triggered by a higher-order cyclicity which cannot be predicted within this scheme. The slacking might be seasonally induced, e.g. by storm seasons.

Palaeogeographic setting of the mollusc-carbonate factory

The palaeogeography illustrated in Fig. 12 shows that the deposits formed at the seaward margin of an oolite shoal. In the northeast of the Nexing area, the Steinberg elevation was an island whereas in the west, there was the mainland that formed the coast. The island situation is reflected in the rocky littoral deposits, such as the "Riesenkonglomerat" (giant conglomerate) along the basinward side of the Steinberg elevation (Grill, 1988). This topographic accentuation is also expressed by strongly different sedimentation rates at the distinct sides of the island. About 300 m of Sarmatian deposits on the elevated block (well Mistelbach 1) contrast with a total thickness of up to 1100 m of Sarmatian sediments in close-by basinal settings (Grill, 1988). This, and the different sediment types (marls and sandy intercalations in the basin versus mixed siliciclastic-carbonate systems on the block) document that the Mistelbach block acted as a submarine shoal during the Sarmatian, about 12 Myr ago.

Due to the current pattern, inferred from the dip-directions of the foresets, most of the terrigenous material involved in the shell dune formation could not have been derived from the mainland at the time of deposition. Its origin is rather related to reactivation and reworking of deposits that accumulated along the Steinberg elevation during the earliest Sarmatian. The source for those deposits was the Molasse Basin and the Waschberg Zone. After the retreat of the Badenian Sea from the Molasse Basin, incised valleys developed in the subsequent third order lowstand systems tract encompassing the Badenian/Sarmatian boundary (Harzhauser & Piller, 2004a, b). Drainage from the Molasse Basin via the modern Zaya Valley developed and transported fluvial gravel and various siliciclastics onto the Mistelbach block. During the following Early Sarmatian transgressive systems tract, the sea entered the valley and replaced the fluvial system (Papp, 1950) and the fluvial input ceased.

During the transgression, a lagoon filled the shallow Mistelbach Basin and silts were deposited in the gateway towards the deeper Vienna Basin (Fig. 13A). The shoals surrounding the Steinberg elevation, settled by enormous masses of shallow-marine molluscs, are thought to have supplied the foresets with the striking amount of shells and shell-hash. These shell masses, as well as the reworked oolites, may have been transported by SSW-directed shoal-parallel currents to the inlet where they were deposited in the prograding foresets (Fig. 13B). Thus, the WNW orientation of the downslopes of the foresets and shell dunes is interpreted to result from a dominant flood current entering the shoal via the broad inlet in the area of Nexing. In that inlet, the foresets migrated within a large channel. Similar channels with flow-transverse bedforms occur on modern oolite sand shoals of the Bahamas (Hine, 1977) and the Persian Gulf (Evans *et al.*, 1973). That channel might have been initiated during the lowstand in the earliest Sarmatian by the above-mentioned fluvial system that structured the Mistelbach block prior to the Sarmatian flooding, a mechanism frequently reported (Davis & FitzGerald, 2004).

A gradual decrease in accommodation space coinciding with the westward migration of the delta front is reflected in a shift from planar-bedded foresets towards either lobate foresets or tidal ridges (Fig. 13B and C). Finally, the shoal on which the flood-tidal delta developed became exposed during the Late Sarmatian at the boundary between the *Ervilia* Zone and the *Sarmatimactra* Zone (Fig. 13D). Fluvial gravel was deposited and vegetation developed. The subsequent flooding marking the base of the *Sarmatimactra* Zone reworked parts of the shell dunes, but also involved the fluvial gravel and oolite from the surrounding shoal (Fig. 13E). Roots that penetrated the exposed oolite caused characteristic concretions which also became reworked. Terrestrial vertebrates are most abundant in Units 3 and 4, probably as a result of lag formation and subsequent reworking. In the course of that renewed, Late Sarmatian transgression, a second flood-tidal delta developed (Fig. 13F). It, however, is strongly truncated by post-Sarmatian erosion.

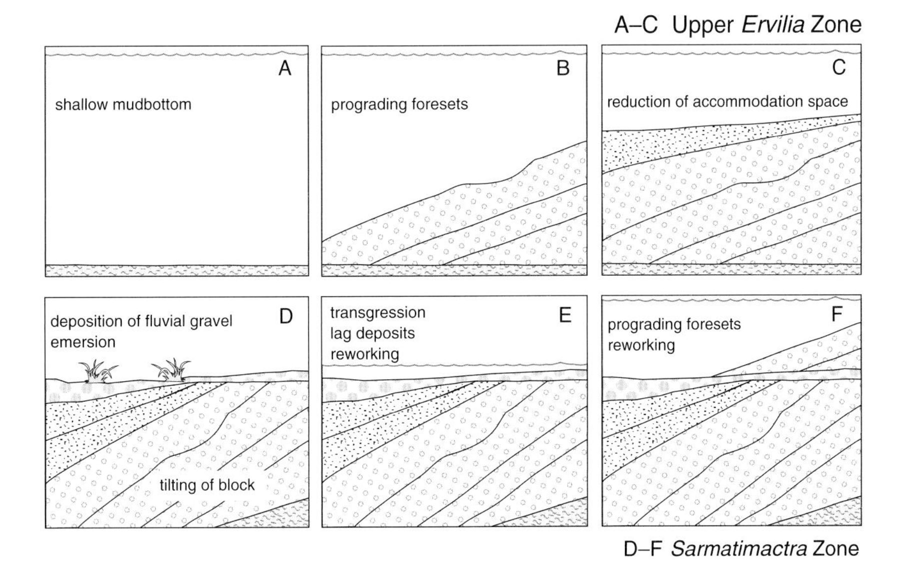

Fig. 13. Simplified evolution of the shoal on the marginal block of the Vienna Basin during the Late Sarmatian. (A–C) The growth of the flood-tidal delta started during the Late *Ervilia* Zone, with a progradation toward the west across shallow-marine clays. With successive infill of accommodation space under stable sea-level conditions, the foreset dip-angle gradually decreased. (D–F) An emergent phase of the delta, also evident in other parts of the shoal, enabled the development of vegetation. A fluvial system, transporting pebbles, spread over the dry shoal. The renewed transgression during the *Sarmatimactra* Zone led to reworking of these deposits and also allowed the re-establishment of a marine environment.

CONCLUSIONS

Mollusc shells are rarely rock-forming constituents of Miocene subtropical shallow-water carbonate factories. In the example described herein, shells of bivalves and gastropods produce large foresets of up to 13 m height with complex internal structures. The consistent dip direction to the W and WNW towards the palaeoshore suggests the flood current as the dominant flow that produced the large-scale foresets and dunes. The weaker counter current cannot be definitively detected in the sedimentary record. The planar-bedded shell-hash separating single shell ripples could be related with that weaker counter flow.

Each large-scale foreset is internally structured by two smaller bedform types. The smallest one is referred to as a *shell ripple,* attaining about 10 cm in height and up to 190 cm in length. A bundle of such ripples constitutes *shell dunes* of up to 1 m height and several metres length. If these structures escaped erosion, a thin layer of sand or shell-hash separates the dunes, being equivalents of the mud drapes of classical sand dunes. The internal texture of single shell ripples indicates that growth was a function of piling and amalgamation of shells into the direction of the dominant current. Thus, the stoss within the ripple gradually steepened. This scenario points to a prevalence of stoss-oriented growth instead of avalanching along the slipface.

The deposits described are an example of the importance of molluscs as part of Miocene subtropical shallow-water carbonate factories. However, the dominance of molluscs evolved within a unique system in which corals and coralline algae were absent due to palaeo(bio)geographic and adverse environmental factors. Instead, molluscs and ooids provided the carbonate, contributing up to 81 % of the dune sediment. Only a very small number of 5–7 mollusc species predominate in the spectrum, representing up to 98 % of the shells.

The unusual accumulation of shelly material with a relatively depleted fauna is most probably a result of the Sarmatian ecosystem that allowed only a small number of specialists to flourish. Hence, high productivity coupled with low diversity seems to support the formation of such shelly deposits. Consequently, mollusc-based carbonate factories and noteworthy shell-accumulations seem to be unusual in normal marine settings with diverse faunas. Aside from that biotic prerequisite, such shell-accumulations are linked to hydrodynamically favourable positions such as tidal inlets and channels providing suitable flow conditions for the transport of the shells.

ACKNOWLEDGEMENTS

We are greatly indebted to Bernhard Krainer and Hanns-Peter Schmid (OMV) for providing access to well log data of the Vienna Basin. Many thanks to Gerhard Niedermayr (NHMW) for the chemical analyses of the carbonate concretions and to Gudrun Höck (NHMW) and Ortwin Schultz (NHMW) for help with the vertebrates. Gerhard Penz (Vienna) kindly provided information on the mollusc fauna. The paper profited greatly by careful reviews of Giovanna Della Porta (Cardiff University), Christian Betzler (University Hamburg) and anonymous reviewers. This work was supported by the Fonds zur Förderung der wissenschaftlichen Forschung in Österreich (FWF, grants P-13745-Bio and P-14366-Bio).

REFERENCES

Abreu, V.S. and **Haddad, G.A.** (1998) Glacioeustatic Fluctuations: The Mechanism linking stable isotope events and sequence stratigraphy from the Early Oligocene to Middle Miocene. In: *Mesozoic and Cenozoic Sequence Stratigraphy of European Basins* (Eds C.-P. Graciansky, J. Hardenbol, T. Jacquin and P.R. Vail). *SEPM Spec. Publ.*, **60**, 245–259.

Allen, J.R.L. (1980) Sand waves: a model of origin and internal structure. *Sed. Geol.*, **26**, 281–328.

Allen, P.A. and **Homewood, P.** (1984) Evolution and mechanics of a Miocene tidal sandwave. *Sedimentology*, **31**, 63–81.

Allen, P.A., **Mange-Rajetzky, M.**, **Matter, A.** and **Homewood, P.** (1985) Dynamic palaeogeography of the open Burdigalian seaway. Swiss Molasse basin. *Eclogae Geol. Helv.*, **78**, 351–381.

Ashley, G.M. (1990) Classification of large-scale subaqueous bedforms: a new look at an old problem. *J. Sed. Petrol.*, **60**, 160–172.

Boersma, J.R. and **Terwindt, J.H.J.** (1981) Neap-spring tide sequences of intertidal shoal deposits in a mesotidal estuary. *Sedimentology*, **28**, 151–170.

Carling, P.A. (1996) Morphology, sedimentology and paleohydraulic significance of large gravel dunes, Altai Mountains. Siberia. *Sedimentology*, **43**, 647–664.

Costello, W.R. and **Southard, J.B.** (1981) Flume experiments on lower flow regime bed forms in coarse sand. *J. Sed. Petrol.*, **51**, 849–864.

Dalrymple, R.W. (1984) Morphology and internal structure of sandwaves in the Bay of Fundy. *Sedimentology*, **31**, 365–382.

Davis, R.A. Jr. and **FitzGerald, D.M.** (2004) Beaches and Coasts. Blackwell Science, Oxford, 419 pp.

Evans, G., **Murray, J.W.**, **Biggs, H.E.J.**, **Bate, R.** and **Bush, P. R.** (1973) The oceanography, ecology, sedimentology and geomorphology of parts of the Trucial Coast barrier island complex, Persian Gulf. In: *The Persian Gulf: Holocene Carbonate Sedimentation and Diagenesis in a Shallow Epicontinental Sea* (Ed. B.H. Purser), pp. 233–277. *Springer*, Berlin.

Flemming, B.W. (2000) The role of grain size, water depth and flow velocity as scaling factors controlling the size of subaqueous dunes. In: *Marine Sandwave Dynamics, International Workshop, March 23-24, 2000* (Eds A. Trentesaux and T. Garlan), pp. 131–142. University of Lille 1, France.

Folk, R.L. and **Ward, W.** (1957) Brazos River bar: a study in the significance of grain size parameters. *J. Sed. Petrol.*, **27**, 3–26.

Friedl, K. (1936) Der Steinberg-Dom bei Zistersdorf und sein Ölfeld. *Mitt. Geol. Ges. Wien*, **29**, 21–290.

Galloway, W.E. and **Hobday, D.K.** (1996) *Terrigeneous Clastic Depositional Systems*. 2nd Edn. Springer, Berlin, 489 pp.

Grill, R.(1968) *Erläuterungen zur Geologischen Karte des nordöstlichen Weinviertels und zu Blatt Gänserndorf. Flyschausläufer, Waschbergzone mit angrenzenden Teilen der flachlagernden Molasse, Korneuburger Becken, Inneralpines Wiener Becken nördlich der Donau.* Geologische Bundesanstalt, Wien, 155 pp.

Harzhauser, M. and **Piller, W.E.** (2004a) Integrated Stratigraphy of the Sarmatian (Upper Middle Miocene) in the western Central Paratethys. *Stratigraphy*, **1**, 65–86.

Harzhauser, M. and **Piller, W.E.** (2004b) The Early Sarmatian – hidden seesaw changes. *Cour. Forsch.-Inst. Senckenb*, **246**, 89–112.

Hine, A.C. (1977) Lily Bank. *Bahamas: History of an active oolite sand shoal. J. Sed. Petrol.*, **47**, 1554–1581.

James, N.P. (1997) The cool-water carbonate depositional realm. In: *Cool-water Carbonates* (Eds N.P. James and J.A.D. Clarke). *SEPM Spec. Publ.*, **56**, 1–20.

Keene, J.B. and **Harris, P.T.** (1995) Submarine cementation in tide-generated bioclastic sand dunes: epicontinental seaway, Torres Strait, north-east Australia. In: *Tidal Signatures in Modern and Ancient Sediments* (Eds B. W. Flemming and A. Bartholomä). *Int Assoc. Sedimentol. Spec. Publ.*, **24**, 225–236.

Kröll, A. and **Wessely, G.** (1993) Wiener Becken und angrenzende Gebiete. Strukturkarte-Basis der tertiären Beckenfüllung. *Geologische Themenkarten der Republik Österreich 1:200.000.* Geologische Bundesanstalt Wien.

Latal, C., **Piller, W.E.** and **Harzhauser, M.** (2004) Paleoenvironmental Reconstruction by stable isotopes of Middle Miocene Gastropods of the Central Paratethys. *Palaeogeogr, Palaeoclimatol. Palaeoecol.*, **211**, 157–169.

Leclair, S.F. (2002) Preservation of cross-strata due to the migration of subaqueous dunes: an experimental investigation. *Sedimentology*, **49**, 1157–1180.

Lesueur, J.-P., **Rubino, J.-L.** and **Giraudmaillet, M.** (1990) Organisation et structures internes des dépots tidaux du Miocène rhodanien. *Bull. Soc. Géol. Fr.*, **6**, **1**, 49–65.

Lüders, K. (1929) Entstehung und Aufbau von Großrücken mit Schillbedeckung in Flut, bzw. *Ebbetrichtern der Außenjade. Senckenbergiana*, **11**, 123–142.

Mosher, D.C. and **Thomson, R.E.** (2000) Massive submarine sand dunes in the eastern Juan de Fuca Strait, British Columbia. In: *Marine Sandwave Dynamics, International Workshop, March 23–24, 2000* (Eds A. Trentesaux and T. Garlan), pp. 131–142. University of Lille 1, France.

Oertel, G.F. (1985) The barrier island system. *Mar. Geol.*, **63**, 1–18.

Papp, A. (1950) Das Sarmat von Hollabrunn. *Verh. Geol. Bundesanst.*, **1948**, 110–112.

Papp, A. (1956) Fazies und Gliederung des Sarmats im Wiener Becken. *Mitt. Geol. Ges. Wien*, **47**, 1–97.

Papp, A. and **Steininger, F.** (1974) Holostratotypus Nexing N.Ö. In: *M5. Sarmatien* (Eds A. Papp, F. Marinescu and J. Senes), *Chronostratigr. Neostratotyp.*, **4**, 162–166.

Piller, W.E. and **Harzhauser, M.** (2005) The Myth of the Brackish Sarmatian Sea. *Terra Nova*, **17**, 450–455.

Portell, R.W., **Schindler, K.S.** and **Nicol, D.** (1995) Biostratigraphy and Paleoecology of the Pleistocene invertebrates from the Leisey shell pits, Hillsborough County. *Florida. Bull. Florida Mus. Nat. Hist.*, **37**, 127–164.

Rögl, F. (1998) Palaeogeographic Considerations for Mediterranean and Paratethys Seaways (Oligocene to Miocene). *Ann. Naturhist. Mus. Wien*, **99A**, 279–310.

Rubin, D.M. and **McCullogh, D.S.** (1980) Single and superimposed bedforms of San Francisco Bay and flume observations. *Sed. Geol.*, **26**, 207–231.

Schwartz, R.K. (1982) Bedform and stratification characteristics of some modern small-scale washover sand bodies. *Sedimentology*, **29**, 835–849.

Stride, A.H. (1970) Shape and size trends for sandwaves in a depositional zone of the North Sea. *Geol. Mag.*, **107**, 469–477.

Terwindt, J.H.J. and **Brouwer, M.J.N.** (1986) The behaviour of intertidal sandwaves during neap-spring tide cycles and the relevance for palaeoflow reconstructions. *Sedimentology*, **33**, 1–31.

Yalin, M.S. (1972) *Mechanics of Sediment Transport*. Pergamon Press Oxford 290 pp..

Wessely, G. (1988) Structure and Development of the Vienna BasIin in Austria. In: *The Pannonian System: A study in Basin Evolution* (Eds L.H. Royden and F. Horvath). *AAPG Mem.*, **45**, 333–346.

Int. Assoc. Sedimentol. Spec. Publ. (2010) **42**, 201–228

Echinoderms and Oligo-Miocene carbonate systems: potential applications in sedimentology and environmental reconstruction

ANDREAS KROH* and JAMES H. NEBELSICK†

**Naturhistorisches Museum Wien, Geologisch-Paläontologische Abteilung, Burgring 7, A-1010 Wien, Austria (E-mail: andreas.kroh@nhm-wien.ac.at)*

†Institut für Geowissenschaften, Universität Tübingen, Sigwartstrasse 10, D-72076 Tübingen, Germany

ABSTRACT

Echinoderms represent a major ecological component and contribute considerably to Oligocene–Miocene carbonate sediments, both as macrofossils and as skeletal grains. The skeletal morphology of all five extant echinoderm classes (echinoids, asteroids, ophiuroids, crinoids, holothuroids) is reviewed. Disarticulated skeletal elements are much more common in sediments than articulated specimens for all echinoderm classes except for echinoids; studies relying on complete specimens alone may be severely biased. The reproduction and growth of echinoderms, the composition of the skeleton, and the crystallography and diagenesis of echinoderm ossicles are reviewed. The echinoderm skeleton consists of high-Mg calcite with 3–18.5 wt% Mg. The skeleton exhibits strong interlacing of microcrystalline calcite with organic material and non-random orientation of crystals, achieving considerable hardness and durability. Echinoderm biostratinomy and the identification of disarticulated material are considered. The echinoderm origin of sediment particles can usually be recognised by their characteristic microstructure. Due to the high degree of specialisation, disarticulated remains can often be identified to family or genus level, leading to a more accurate picture of spatial and temporal echinoderm distributions.

Echinoderm geochemistry is reviewed with respect to the Mg-content of the skeleton as a palaeotemperature proxy, and the Mg/Ca ratio as a monitor of ancient seawater composition; Sr/Ca ratios and carbon and oxygen stable-isotopes are considered. The echinoderm skeleton is altered during diagenesis and is transformed to low-Mg calcite. The microstructure of the skeleton is largely unaffected by this process, but changes in the isotopic signature and minor/trace-element contents may occur. These factors, together with physiological effects of isotope intake, hamper geochemical applications. However, echinoderms have been used successfully in studies of Phanerozoic seawater chemistry: the Mg and Sr contents of echinoderm skeletons apparently strongly correlate with temperature. Asteroids and ophiuroids are probably best suited for palaeotemperature reconstructions because of the lack of known fractionation within the skeleton and because genetic effects are less pronounced than in echinoids. Controlled laboratory experiments are needed to establish calibrations.

Echinoderm remains may account for 5–30 % of the particles within specific Oligocene and Miocene carbonate facies. They seem to be more abundant in temperate shelf carbonates than in tropical settings. Diagenetic changes associated with echinoderm ossicles strongly affect the embedding sediment and promote lithification. Bioerosion by grazing echinoids is important for carbonate budgets in coral reefs and influences the modal size-distribution of sediments by the production of carbonate mud. Burrowing echinoderms may cause intensive bioturbation and reworking of sediments. Echinoderms provide valuable evidence for palaeoenvironmental reconstructions. Ecological information can both be gained by actualistic comparisons with modern echinoderms and by a functional morphological approach, allowing the detailed assessment of general life habits, substrate conditions, nutrient availability and hydrodynamic regimes.

Keywords Echinodermata, taphonomy, palaeoecology, skeletal chemistry, Mg/Ca ratio, Sr/Ca ratio, stable isotopes.

INTRODUCTION

Echinoderms are a group of skeleton-bearing organisms that radiated in the early Palaeozoic, where they showed considerable diversity and morphological plasticity. The Permo-Triassic extinction event strongly affected echinoderms leaving only five groups, all of which are still extant today. These groups are the echinoids (sea urchins, sand dollars, heart urchins), asteroids (sea stars), ophiuroids (brittle stars and basket stars), crinoids (sea lilies and feather stars), and holothuroids (sea cucumbers). In each group, a variety of different life styles and feeding habits are realized, allowing their survival in a broad range of marine settings. With few exceptions among ophiuroids (Talbot & Lawrence, 2002) echinoderms are strictly stenohaline organisms.

Due to their peculiar high-Mg calcite skeleton, echinoderms exhibit a high fossilization potential, and are among the most common macrofossils found in Oligocene and Miocene carbonates. This paper presents a review of the current knowledge on the echinoderm skeleton and taphonomy, as well as their potential use in Oligo–Miocene carbonate sedimentology. Possible applications include actualistic and palaeoecological approaches to palaeoenvironmental reconstruction, taphonomic studies for facies analysis, and investigation of their geochemical signature as palaeoclimate proxies and/or ancient seawater archives.

ECHINODERM SKELETON

Morphology

The echinoderm skeleton is a highly specialized, three dimensional meshwork of high-Mg calcite, called stereom. Its high porosity with an average of 50–60 % volume (the full range is c. 10–70 %: Weber, 1969a; Weber *et al.*, 1969) makes it a lightweight structure, conserving energy resources and facilitating rapid growth and regeneration. The morphology of the stereom network, including the thickness of individual rods and pore size, varies strongly according to position and function of the ossicles (Smith, 1980; Fig. 1). Although only covered by a thin layer of epithelium, the echinoderm skeleton is an endoskeleton of mesodermal origin. During life, the pores within the stereom are filled with tissue, which is termed stroma. This design accounts for high structural integrity and regeneration potential.

Echinoderms possess a multi-element skeleton. The skeleton of *Paracentrotus lividus*, the common stone urchin of the Mediterranean and East Atlantic coasts, for example, encompasses more than 110,000 ossicles (Fig. 2). Most crinoids possess

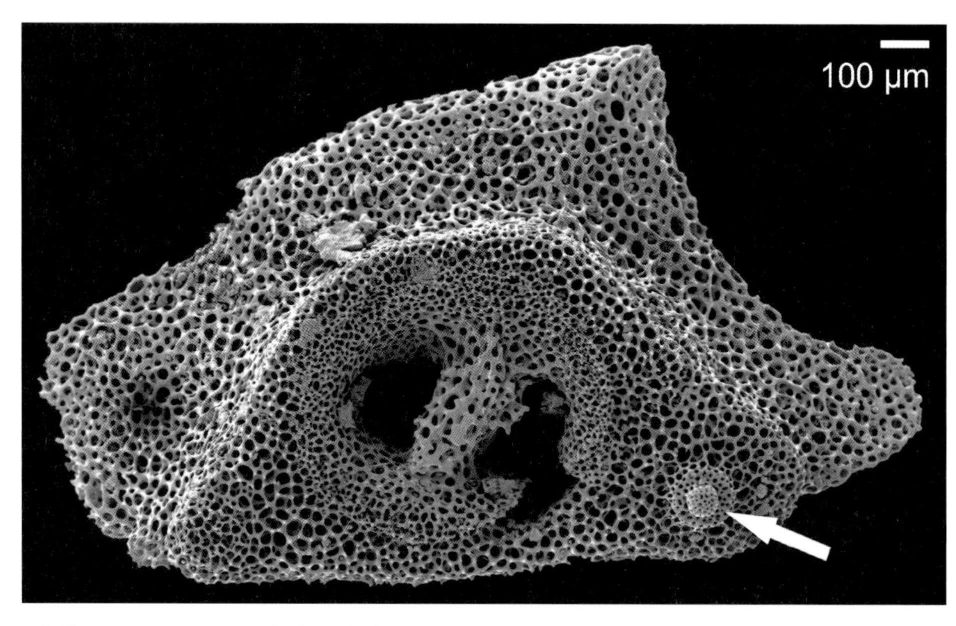

Fig. 1. Stereom differentiation on an ambulacral plate of an echinoid (*Asthenosoma ijimai*). Pore and trabecle size changes according to the attached soft tissue and its function: e.g. very small pores in the stereom forming the small tubercle on the lower right-hand side, marked by the arrow; rather coarse meshwork on the flanks.

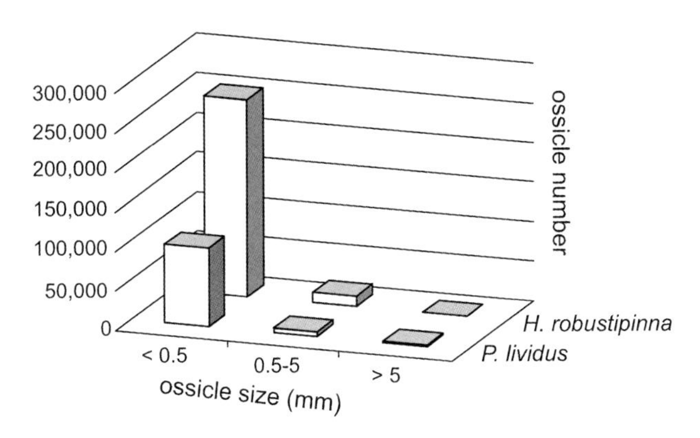

Fig. 2. Number of ossicles and their size in the skeleton in an individual regular echinoid (*Paracentrotus lividus*) and a comatulid crinoid (*Himerometra robustipinna*; data from Meyer & Meyer, 1986).

more than twice as many elements (e.g. Fig. 2: *Himerometra robustipinna*, Meyer & Meyer, 1986). Ossicle shape is in most cases strongly functionally controlled. Due to the fact that the echinoderm skeleton is highly repetitive (with a pentaradiate symmetry in most groups), the inventory of different ossicle types is limited (see Figs. 3–5 for characteristic ossicles of Cenozoic echinoderms). Despite this high degree of functional control, the shapes of individual groups are usually well differentiated and allow a rather straightforward identification (e.g. Fig. 6, showing spine cross-sections). The degree of taxonomic resolution that may be reached when working with echinoderm remains depends on ossicle type and echinoderm group involved, ranging from phylum level at the worst, to species level in the best case. Oligocene, Miocene and Pliocene echinoderm faunas show a modern composition, many genera ranging back in time to the Late Eocene. This facilitates recognition of individual skeletal ossicles, particularly when extant material is used for comparative purposes. This method has been successfully employed in a number of studies dealing with Neogene echinoderms (e.g. Gordon & Donovan, 1993; Donovan *et al.*, 1993; Donovan, 2001; Kroh, 2003a, b, 2004; Kroh & Nebelsick, 2003).

Echinoid skeleton

Echinoids are characterized by a corona made up of 20 columns of perforate (ambulacral) and imperforate (interambulacral) plates. Attached to this corona are spines of variable shape, length and function, as well as tube feet (for locomotion and feeding) and pedicellariae (tiny claw-like defence and sanitary organs). While these appendages are generally lost before fossilization, the coronae often remain intact and are commonly found in Cenozoic sediments. This is due to the presence of organic and skeletal structures tightly interlocking the plates (except in cidaroids, diadematoids and echinothurioids, which are thus usually found as isolated plates or spines only).

Fig. 3. Examples of Neogene echinoids. (a) Spine of the cidaroid *Stylocidaris? polyacantha* (Reuss, 1860); (b) Test fragment of the cidaroid *Stylocidaris? schwabenaui* (Laube, 1869). (c–e) Lantern elements of the "regular" echinoid *Schizechinus hungaricus* (Laube, 1869). (f) and (g) *Parmulechinus hoebarthi* (Kühn, 1936), a sand dollar. (h) and (i) Corona of the "regular" echinoid *Echinometra mathaei* (de Blainville, 1825). (j) and (k) *Schizaster eurynotus* (Sismonda, 1841), a heart urchin. Scale bars equal 10 mm (upper left scale bar valid for a and b, lower left bar for c–e, upper right bar for f–k). Modified from Kroh (2005).

Fig. 4. Preservation of fossil Neogene asteroids. (a–e) Disarticulated ossicles. (f) A whole specimen. Arrows indicate possible positions of the disarticulated material in the body of an asteroid (specimens shown here are not conspecific). (a) and (b) marginal ossicles of a goniasterid. (c) Astropectinids ambulacral ossicle. (d) Paxilla (abactinal ossicle) of *Luidia* sp. (e) Abactinal ossicle of an asteroid. (f) "*Goniaster*" *muelleri* (Heller, 1858). The scale bar equals 1 mm (for a–e), and 1 cm (f). Modified from Kroh & Harzhauser (1999).

Based on the gross morphology, three different types of echinoids can be identified in Oligo-Miocene deposits: (1) "regular" echinoids, characterized by their radial symmetry (Cidaroida, Diadematoida, Echinacea; Fig. 3a–i); (2) sand dollars and sea biscuits (Clypeasteroida) which are often flat and disc-like with a strong corona (Fig. 3f and g); and (3) heart urchins (Spatangoida), usually with a thin-walled corona and sunken ambulacra (Fig. 3j and k). Each of these groups is adapted to a different environment and can be used in facies analysis and palaeoenvironmental reconstruction (see below).

The identification of (fossil) echinoids is based mainly on corona morphology. Although a number of additional features are used in extant forms, identification using literature on recent genera is rather straightforward, at least in Neogene strata. A monograph of utmost importance in this respect is Mortensen's (1928, 1935, 1940, 1943a, b, 1948a, b, 1950, 1951) "*Monograph of the Echinoidea*". In most cases, identification of fragments and isolated spines is possible by comparison with extant material. For further information on echinoids the reader is referred to Smith (1984), a highly useful textbook on echinoid palaeobiology.

Asteroid skeleton

The asteroid skeleton is composed of large, often block-shaped marginal ossicles (Fig. 4a and b) forming a pentagonal or pentastellate frame. The dorsal (aboral) side is covered by smaller polygonal plates (abactinal plates, Fig. 4d and e) set in a flexible membrane, but may also include large ossicles in some forms. The ambulacral grooves are located on the ventral (oral) side, formed by ambulacral (Fig. 4c) and adambulacral ossicles. The region of the mouth is framed by another set of specialized ossicles. After death, the asteroid skeleton disarticulates rapidly resulting in the common occurrence of isolated ossicles in the fossil record. Despite this fact, research has mainly focused on more or less complete asteroids (with the exceptions of Müller, 1953; Hess, 1955; Blake,

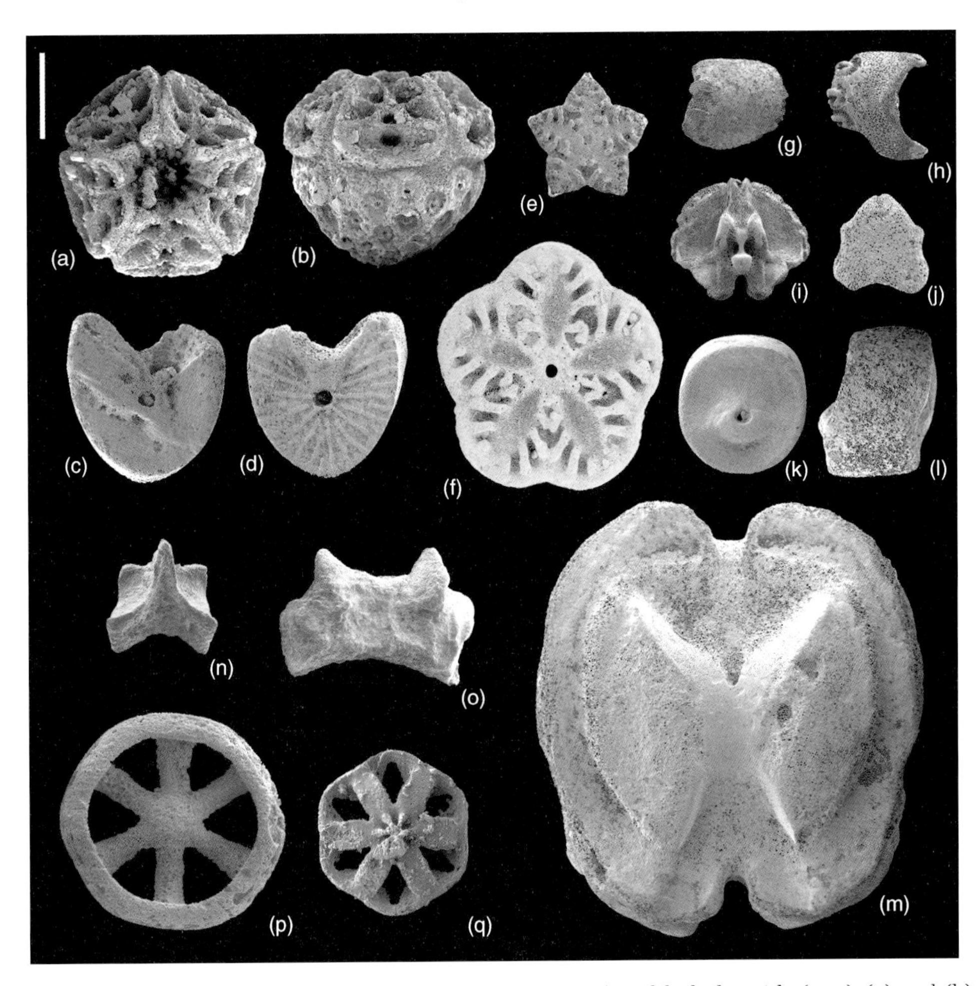

Fig. 5. Characteristic ossicles of crinoids (a–f, k–l), ophiuroids (g–j, m) and holothuroids (n–q); (a) and (b) Calyx of the comatulid crinoid *Conometra hungarica* (Vadász, 1915). (c) and (d) Crinoid brachials. (e) and (f) Isocrinoid columnals. (g) and (h) Ophiuroid lateral shields. (i) Ophiuroid vertebra. (j) Ophiuroid ventral shield. (k) and (l) Crinoid cirrals. (m) Vertebra of a gorgonocephalid ophiuroid. (n) and (o) Elements of the calcareous ring of an apodid holothuroid. (p) and (q) Wheel-shaped holothuroid sclerites: (p) *Theelia kutscheri* (Reich, 2003); (q) *Palaeotrochodota jagti* Reich, 2003. Scale bar equals: 1 mm (a–m); 0.5 mm (n–o); and 0.1 mm (p–q). Modified from Kroh (2003a), Kroh & Jagt (2006), Reich (2001, 2003) and Reich and Frenzel (2002).

1973; Breton, 1992; Villier, 1999; Jagt, 2000b). Identification of isolated ossicles is difficult, mainly due to the lack of comparative literature. Detailed descriptions of the ossicle morphology of some, more common forms can be found in Blake (1973).

Asteroids are very common in Oligocene and Miocene marine sediments, occurring mainly in form of isolated ossicles (e.g. Kaczmarska, 1987; Jagt, 1991; Kroh & Harzhauser, 1999) and, more rarely, as whole skeletons (Nosowska, 1997). In cases these ossicles can constitute a large part of the fossil biomass recovered from sediment bulk samples (e.g. phosphate beds in the Lower-Middle Miocene *Globigerina* Limestone of the Maltese Islands).

Ophiuroid skeleton

Ophiuroids are characterized by their body shape with a usually well delimited disc and five arms. In some forms, the so-called basket stars, the arms may be branched, forming a catch apparatus for filter feeding. The disc is composed of a stout mouth frame, jaw apparatus, bursal ossicles, and a high number of polygonal platelets set in a flexible dorsal membrane. The arms are built of a central structure of longitudinally aligned vertebrae and protective platelets (dorsal, ventral and lateral shields). The structure of the ophiuroid arms is thus highly repetitive, with proximal elements very similar in shape to distal ones, only differing in their size and elongation.

Fig. 6. Cross-sections of fossil and recent echinoid spines illustrating the potential use of spine microstructure for classification in thin section. (a) Cidaroid *Plegiocidaris*? *peroni*. (b) Cidaroid *Eucidaris zeamays*. (c) Diadematoid. (d) Echinacean *Paracentrotus lividus*. (e) Echinacean *Tripneustes ventricosus*. (f) Cassiduloid *Echinolampas crassa*. (g) Echinoneoid *Echinoneus cyclostomus*. (h) Spatangoid *Brissopsis ottnangensis* (i) Spatangoid.

As in asteroids, palaeontological research has mainly focused on complete specimens, with the exception of some workers (Hess, 1962a, b, 1963, 1965, 1966, 1975a, b; Jagt, 2000a; Kutscher & Jagt, 2000; Kroh, 2003b, 2004; Kroh & Jagt, 2006) who studied disarticulated ophiuroid material, in particular lateral shields and vertebrae. Identification of these isolated remains is usually difficult, necessitating numerous specimens for evaluation of morphological variation and extant material for comparison. Despite these difficulties, identification to genus level is often possible.

In comparison with asteroid ossicles, ophiuroid remains tend to be less common in Oligocene and Miocene sediments. Locally, however, they can be quite abundant and characteristic (e.g. basket star vertebrae in the Middle Miocene Hartl Formation of Austria; Kroh, 2003a; Kroh *et al.*, 2003).

Crinoid skeleton

The crinoid skeleton often consists of a very large number of ossicles (see Fig. 2), in spite of being composed of a small number of ossicle types. The majority of the ossicles are found in the arms, which basically consist of just two ossicle types, the larger brachials and the smaller pinnulae (Fig. 5c and d). The calyx in contrast, is made up of a small number of ossicles, which are often tightly fused (5 radials, 5 basals, 1 centrodorsale). Attached to the lower side of the calyx there may be a number of cirrae, being composed of numerous individual cirrals (Fig. 5k and l). In stalked forms, a stem consisting of numerous columnals is also present. Considerable information on fossil crinoids can be found in Hess *et al.* (1999), although primarily dealing with Palaeozoic and

Mesozoic forms. A review on comatulids (stalkless crinoids) has been published by Messing (1997).

In Oligo-Miocene sediments, crinoids play only a subsidiary role. Stalked crinoids (isocrinids and bourgueticrinids) are rare, except in deep-water settings (Oji, 1990; Eagle, 1993; Donovan, 1995; Donovan & Veltkamp, 2001; Kroh, 2003a). Stalkless forms, the comatulids, are more common, especially in reefal environments (Meyer & Macurda, 1997). Identification of these forms, however, is severely limited by the lack of adequate illustrations of the individual ossicles of most extant forms. Isolated brachial and cirral ossicles, as well as pinnulae, in most cases cannot be related to a specific taxon. Useful compendia on Cenozoic crinoids have been published by Gislén (1924), Biese & Sieverts-Doreck (1939), and Rasmussen (1972).

Holothuroid skeleton

In contrast to all other echinoderm groups discussed above, holothuroids have a highly reduced skeleton composed from numerous, minute sclerites (<1 mm) embedded in the body wall, tube feet and tentacles (Fig. 5p and q). Apart from these tiny ossicles, the only hard parts present are those of the calcareous ring, a structure usually composed of 10 elements (Fig. 5n and o). While widespread and common in extant settings, fossil holothuroids are much less abundant, particularly in the Cenozoic. Holothuroid body fossils are especially rare. Usually only isolated sclerites are found in micropalaeontological samples. In most cases, an artificial classification (parataxonomy) is used for the identification of fossil sclerites (Gilliland, 1993). A wide range of sclerites may occur in a single animal and individual sclerite types may occur in more than one species. A review of holothuroid skeletal morphology, systematics and evolutionary history was provided by Gilliland (1993) and the reader is referred to that work.

Reproduction and growth

Echinoderms, like many other invertebrates, usually reproduce by external fertilization, i.e. by the release of vast numbers of sperm and eggs into the water column. Fertilization is followed by a planktonic larval stage (except in direct developing forms) and finally metamorphosis and settlement. After a short lag-phase in early juvenile growth, a period of rapid growth followed by a long interval of declining growth can be observed in most echinoderms. Growth in echinoderms, unlike most vertebrates, does not seem to stop at maturity although rather strongly decrease in rate (Lawrence, 1987; Pearse & Pearse, 1975; Russell & Meredith, 2000).

Growth in echinoderms is realized not only by enlargement of individual skeletal elements, but also by the continuous addition of new ossicles (Lawrence, 1987; Smith, 1984). Due to this, the position of an individual element may change considerably during growth (e.g. an interambulacral plate of an echinoid may migrate from a position near the apex to the underside of the animal). As a consequence, ossicle shape also changes during growth, both through lateral deposition of new skeletal material and by resorption. This change of shape is well illustrated by the growth lines present in many echinoderm ossicles (Fig. 7). These growth lines form by episodic growth breaks, injury and regeneration, as well as by other, unexplained causes. An interpretation of these growth lines as annual increments might seem tempting, but recent studies, employing tagging experiments, showed that this is not necessarily true (Russell & Meredith, 2000; Ebert, 2001, 2007). Slow growth rates in older individuals are especially problematic in this respect, and it now appears that the age of echinoids and, possibly most other echinoderms may have been seriously underestimated (Ebert, 1998, 2001; Russell *et al.*, 1998; Russell & Meredith, 2000).

For more information on echinoderm growth and mathematical models, in particular concerning echinoids, the reader is referred to Ebert (2001) and references therein. Further details on echinoderm biomineralization can be found in the comprehensive review provided by Smith (1990).

Mineralogy of the echinoderm skeleton

The skeleton of all known echinoderms is composed of high-Mg calcite, with $MgCO_3$ contents ranging from ~3 to 18.5 weight % (wt%; Weber, 1969b; Ebert, 2007). In the so-called "stone zone" of echinoid teeth, a primary protodolomite with Mg-values of up to 40 mol% occurs (Schroeder *et al.*, 1968, 1969; Märkel *et al.*, 1971). Transmission electron microscopic (TEM) studies of crinoid stereom revealed that Mg incorporation into the calcite structure of the echinoderm stereom is random and homogeneous to at least the 20 nm level (Blake & Peacor, 1981). Iron and strontium are incorporated as minor elements

Fig. 7. Growth lines in interambulacral plates of an extant *Strongylocentrotus pallidus* as revealed in thin section. Modified from Raup (1966a). Scale bar equals approx. 10 mm (no scale given in original paper).

with contents <1 wt% (e.g. $SrCO_3$ content ranging from 0.296 to 0.354 wt% in the sand dollar *Dendraster excentricus*, Pilkey & Hower, 1960; 1329–1645 $\mu g\,g^{-1}$ (parts per million) Sr, and 32–164 $\mu g\,g^{-1}$ Fe in stalked crinoids, Roux *et al.*, 1995). Manganese, aluminium and silicon are trace elements that occur at concentrations of <100 $\mu g\,g^{-1}$ (e.g. 0–22 $\mu g\,g^{-1}$ Mn in stalked crinoids, Roux *et al.*, 1995).

Crystallography of echinoderm ossicles

Echinoderm ossicles are peculiar among other invertebrates with calcite skeletons as each echinoderm ossicle behaves optically like a single calcite crystal (West, 1937; Raup, 1966a; Donnay & Pawson, 1969; Nissen, 1969). Although parts of the echinoderm skeleton may be polycrystalline and composed of tiny crystallites with preferred orientation (e.g. cortex of spines in cidaroid echinoids, parts of the echinoid teeth, and the calcareous ring of holothurians), the overwhelming part of the echinoderm skeleton is monocrystalline (Märkel *et al.*, 1971) at a micro-optical scale. The polycrystalline parts are usually associated with structures and elements exposed to considerable wear or strain and are the hardest parts of the echinoderm skeleton. This hardness, which lies distinctly above the mineralogical hardness of calcite and protodolomite, is achieved by strong interlacing of the microcrystalline calcite with organic material (Märkel *et al.*, 1971). Individual skeletal plates of crinoids examined with single-crystal X-ray diffraction analysis by Blake & Peacor (1981) yielded diffuse and imperfect X-ray reflections due to a mosaic structure. Close inspection with TEM revealed crystallites within an order of magnitude of about 1 μm in size, but it remains to be shown if this is true for all echinoderms. The observed mosaic structure together with incorporated organic material might be responsible for the lack of cleavage in fracture surfaces of extant echinoderm skeletal material (Blake & Peacor, 1981). In fossil specimens, by contrast, such cleavage is present due to diagenetic alteration resulting in (slight) mineralogical and crystallographical transformation.

Crystallographic orientation is non-random in most ossicles of the echinoderm skeleton, with empirical data showing clear genetic control in the echinoderms studied so far (echinoids: Raup, 1959, 1960, 1962a, 1966a, b; blastoids: Bodenbender, 1996). How exactly this control is achieved by the echinoderms is not completely understood, and several mechanisms have been discussed (see Smith, 1990; Bodenbender, 1996). Crystallographic data may be employed to address palaeobiological questions (e.g. mode of larval development in echinoids, see Raup, 1965; Raup & Swan, 1967; Emlet, 1985, 1989), classification and phylogeny (Raup, 1962a, b; Fisher & Cox, 1988; Bodenbender, 1996), as well as to assess diagenetic alteration.

GEOCHEMISTRY OF THE ECHINODERM SKELETON

As outlined above, echinoderms possess a relatively durable skeleton of high-Mg calcite and are common in many marine deposits. Due to these favourable attributes they represent a potential tool for the investigation of palaeotemperatures, geochemical cycles and ancient seawater composition. Despite their potential usefulness, only a small number of geochemical studies have actually employed echinoderm material (e.g. Bill *et al.*, 1995). This is probably related to the generally incomplete knowledge of the formation of the echinoderm skeleton in relation to seawater chemistry. Although a number of studies on this topic have been published, many of the basic principles are still poorly understood. Furthermore, detailed studies on the vital effects and intra- and interspecific variation are mostly missing (with a few exceptions: e.g. Pilkey & Hower, 1960; Weber & Raup, 1966a, b; Weber, 1968, 1969b, 1973). Nevertheless, echinoderms represent an unexploited source of original data on ancient seawater chemistry and palaeotemperature, as emphasized by Dickson (2004).

Magnesium content as a palaeotemperature proxy

Initial studies on the chemistry of recent echinoderm skeletons (Clarke, 1911; Clarke & Wheeler, 1914, 1915, 1917, 1922) suggested a good correlation between Mg content and water temperature in crinoids, asteroids, echinoids, and ophiuroids (listed according to the quality of the fit). The investigation of fossil echinoderms, in contrast, resulted in highly ambiguous and unexpectedly low values. Nevertheless, the potential of the apparent magnesium-temperature correlation for "studies of climatology" (Clarke & Wheeler, 1917, p. 56) was cautiously highlighted. Chave (1954), in a major paper on the biogeochemistry of a wide range of marine organisms, identified three major factors influencing the Mg content in the skeleton: mineralogy; water temperature; and phylogenetic level. However, the results of these studies (Clarke & Wheeler, 1922; Chave, 1954) were based on a limited dataset, and intraspecific as well as individual variation remain to be investigated. The first systematic investigation in this direction was undertaken by Pilkey & Hower (1960) on a large number of sand dollars (*Dendraster excentricus*) from the US west coast. They found that although a magnesium-temperature correlation existed, Mg content was also influenced by salinity. Furthermore, they showed that the slope of the magnesium-temperature regression line in *Dendraster* differed strongly from that of the entire class and that the different species living at the same temperature showed considerable difference in skeletal $MgCO_3$ concentration (larger differences than the range shown by *Dendraster* over its entire temperature range). It could thus be deduced that a genetic control of Mg-uptake exists, an inference substantiated by subsequent studies (e.g. Raup, 1966b; Weber, 1969b, 1973). New data, albeit preliminary, indicate a negative correlation between growth rate (expressed by the Brody-Bertalanffy growth constant) and Mg content in echinoids (Ebert, 2007).

Weber (1969b) was the first to investigate the variation of $MgCO_3$ content within individual animals on a larger scale. Although it was known from early on that there were differences between spines, coronal plates and lantern elements in echinoids (Clarke & Wheeler, 1915) this had not been investigated in detail. Weber (1969b) showed that variation within individual skeletal elements was low (with the exception of echinoid spines; see below). The existence of a systematic variation of Mg content in the different skeletal elements of echinoids was also confirmed (Fig. 8). These differences were small in some species, but could

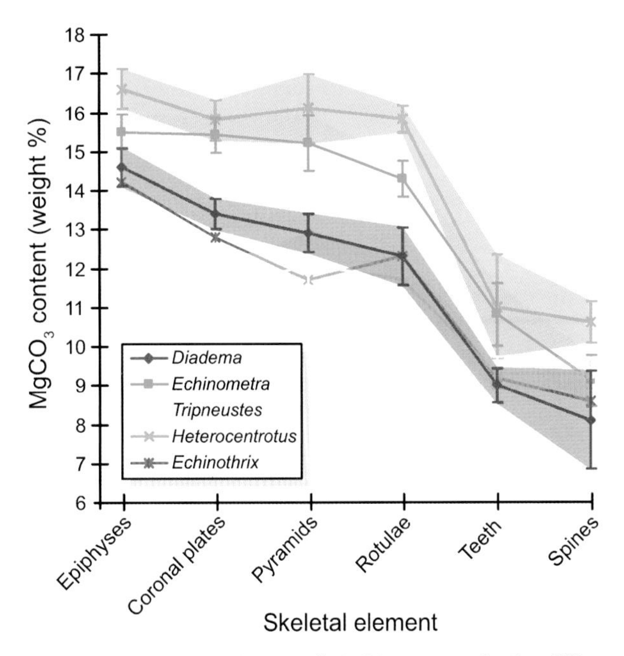

Fig. 8. Systematic variation of $MgCO_3$ content in the different skeletal elements of echinoids (data from Weber, 1973).

amount of up to 8.7 wt% in others (spines vs. coronal plates in *Echinometra mathaei*). In contrast to echinoids, no significant variation was detected in ophiuroids and asteroids. Magnesium content variation within individual populations was found to be rather low (Weber 1969b, Table 7), which confirmed earlier results by Pilkey & Hower (1960, Table 2). Surprisingly, asteroids and ophiuroids, in contrast to echinoids, showed only small differences between different species collected from the same locality (Weber, 1969b, Table 8) and thus genetic control of Mg-uptake seems to be less pronounced in these groups (compare also Weber, 1973, Fig. 4). Additionally, echinoid skeletal elements (even the coronal plates) are generally less rich in Mg than asteroids and ophiuroids from the same sample locations and seem, as a class, to be displaced towards lower Mg values at given temperatures (Weber, 1969b, 1973).

Data on crinoids, and even more so holothurians, are scarce. Roux *et al.* (1995) showed that $MgCO_3$ content in crinoids decreases with depth (Fig. 9) and provided also the first evidence for fractionation of Mg between different parts of the crinoid skeleton. They concluded that the incorporation of Mg in crinoid calcite was not necessarily a direct function of temperature, but is influenced also by metabolic rate and energy allocated to biomineralization in a given adaptive strategy. This is corroborated by newer data presented by Ebert (2001), that show that there is an inverse relationship between growth rate and $MgCO_3$ content. Additionally, regeneration may also have an important impact on Mg-uptake in the echinoderm skeleton.

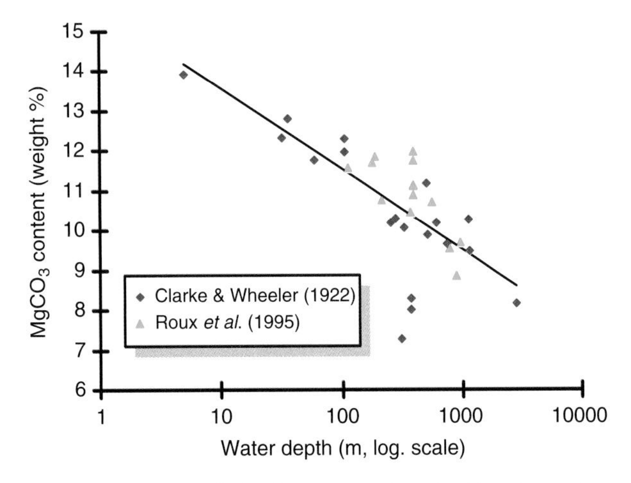

Fig. 9. Decreases of $MgCO_3$ content in crinoids with water depth.

Based on the analysis of regenerated echinoid spines, Davies *et al.* (1972) and Weber (1973) demonstrated that Mg is not uniformly incorporated during regeneration and that the original $MgCO_3$ content is only reached in the final stage of regeneration.

In light of the data available at present, the use of $MgCO_3$ content as a palaeotemperature proxy is complicated by several factors, and the interpretation of its variation is not as simple as once thought. Particular problems are the poorly understood physiological effects causing strong scatter in echinoid data, the influence of factors other than temperature (including food availability, growth rate and regeneration), and difficulties in calibration between measured $MgCO_3$ content and temperature (as growth is temperature dependent and temperature varies seasonally). Slight variations of salinity may also pose a problem. Studying the geochemistry of recent tests of *Echinocyamus* from North Atlantic European localities, Richter & Bruckschen (1998) noted that the Mg content of the skeletons strongly correlated with the average annual temperature but not with salinity. Regarding the different echinoderm groups, asteroids and ophiuroids may be best suited for palaeotemperature reconstructions because fractionation between different skeletal elements is not known and genetic effects are apparently less pronounced than in echinoids (Weber, 1969b, 1973). In any case, controlled laboratory experiments are needed to establish well-founded calibrations for any echinoderm group that is to be used.

There is potentially a wide range of other research issues addressing the use of Mg content in echinoderm skeletons, in addition to palaeoclimatic research. An innovative approach utilizing the Mg content of fossil echinoderm skeletal material, for example, was employed by Richter (1974), who made use of the Mg-loss under subaerial diagenesis for the relative dating of Pleistocene carbonate terraces.

Mg/Ca-ratio as a monitor of ancient seawater composition

Initially, the major ionic composition of seawater was presumed to have been constant throughout the Phanerozoic (Holland, 1978). More recently, however, major changes in ancient seawater chemistry have been discovered by investigation of sediment mineralogy (Sandberg, 1983), geochemical modelling (Wilkinson & Algeo, 1989; Hardie,

1996) and fluid inclusion studies (Lowenstein *et al.*, 2001; Horita *et al.*, 2002).

Recently, Dickson (1995, 2002, 2004) and Ries (2004) employed fossil echinoderm skeletal material as an independent monitor for Mg/Ca oscillation of ancient seawater. Based on 29 specimens, Cambrian to Eocene in age, Dickson (2002, 2004) was able to reconstruct a Phanerozoic Mg/Ca-curve closely resembling first-order Mg/Ca seawater oscillations derived from earlier studies (Wilkinson & Algeo, 1989; Hardie, 1996; Horita *et al.*, 2002). At a finer scale, however, (<100 Myr) considerable differences exist and additional data, as well as improved understanding of Mg partitioning in echinoderms is required. These studies have demonstrated that, despite the uncertainties involved, echinoderms have a high potential as a "seawater archive" and represent an underused resource in this context.

Sr/Ca-ratio

Like magnesium, the strontium/calcium ratio may be used as palaeotemperature proxy (Lea, 2003) and has been successfully utilized in living and fossil corals (e.g. Beck *et al.*, 1992; Linsley *et al.*, 2000). Unlike corals, data on the Sr-uptake and content in echinoderms is largely missing. Pilkey & Hower (1960) found an inverse relationship between Sr and temperature in the sand dollar *Dendraster excentricus* with Sr/Ca ratios decreasing at roughly 0.3×10^{-4} per °C. Contrary to Turekian (1955), they could not detect any consistent influence of salinity.

Vital effects affecting the Sr-uptake in the echinoderm skeleton are, so far, unknown. In corals, at least, growth rate and symbiont activity appear to influence Sr-uptake from seawater (Cohen *et al.*, 2001, 2002). Diagenetic alteration, likewise, is poorly investigated in echinoderms, but it seems likely that, as in corals, subaerial exposure and vadose diagenesis result in Sr-loss from the skeletal calcite (compare McGregor & Gagan, 1994). When employing Sr/Ca-ratio in fossil echinoderms, secular changes in seawater strontium content have to be taken into account. As with Mg, there are data suggesting that the Sr/Ca-ratio changed through time, in particular due to glaciation events (Stoll & Schrag, 1998; Stoll *et al.*, 1999).

Carbon and oxygen isotopes

One of the most powerful methods for reconstructing palaeoenvironmental parameters is the study of stable isotopes of oxygen ($^{18}O/^{16}O$) and carbon ($^{13}C/^{12}C$) in carbonate shell material. Oxygen isotopes are used as temperature and salinities proxies, and carbon isotopes for the reconstruction of water bodies and palaeoproductivity (e.g. Wefer & Berger, 1991; Hoefs, 1997; Brenchley & Harper, 1998).

Despite this, stable-isotope studies on echinoderm skeletons are relatively rare. There is only a small number of research papers dating back to the beginning of biogenic stable isotope research dealing with echinoderm material (e.g. Weber & Raup, 1966a, b, 1968; Weber, 1968). In these initial studies, a wide range of interspecific, intraspecific and even intraindividual variability was detected. In general, echinoderms are enriched in ^{12}C and ^{16}O with respect to inorganically precipitated calcium carbonates under the same environmental conditions. Ophiuroids however, are an exception to this "rule" and show similar isotopic values to calcium carbonates precipitated abiotically at ambient seawater conditions (Weber, 1968).

According to Weber & Raup (1966a) echinoids show considerable fractioning of stable isotopes, especially in the coronal plates. The isotopic values of spines, in contrast, coincide better with data of other marine invertebrates than do other elements of the echinoid skeleton (Weber & Raup, 1966a; Ebert, 2007). During growth, the variation of the isotopic values seems rather small. Similar results were obtained for asteroids, ophiuroids and crinoids (Weber & Raup, 1968; Roux *et al.*, 1995; Baumiller, 2001), which also show large and systematic variation for different skeletal elements. Weber & Raup (1966b, 1968) found both $\delta^{13}C$ and $\delta^{18}O$ values to be largely genetically controlled. Oxygen-isotope values and, to a lesser extent, $\delta^{13}C$ values, are also controlled by temperature (Weber & Raup 1966b; Weber, 1968). If present, the correlation of carbon with temperature is positive, while in oxygen isotopes the correlation is negative. In asteroids, crinoids and echinoids $\delta^{13}C$ is negatively correlated with depth, while a positive correlation has been found for $\delta^{18}O$ (Weber & Raup 1966b; Weber, 1968; Roux *et al.*, 1995; Baumiller, 2001). According to Roux et al. (1995) the $\delta^{13}C$ depth correlation is related to the metabolic rate in crinoids, which in turn is controlled by the inter-related parameters, depth, temperature, food supply, and changing $\delta^{13}C$ levels of dissolved CO_2 at different depths.

Similar to other geochemical parameters (see above), diagenesis represents a potential severe problem in using echinoderm skeletons because the skeletal elements consist of high-Mg calcite, an unstable carbonate mineral under meteoric conditions. Weber & Raup (1968), based on a study on the isotopic composition of fossil echinoids, concluded that distinct patterns may be found in isotopic signals despite diagenesis changing absolute isotopic values, especially when the extent of diagenetic alteration is known.

Despite all the problems mentioned, carbon-isotope data of crinoid stems and echinoid spines have more recently been used successfully to reconstruct environmental parameters in Upper Jurassic sediments from Switzerland (Bill *et al.*, 1995). Apart from isotope stratigraphy, isotope analysis in echinoderms may be applied to a wide range of palaeontological and (palaeo-)biological issues. Oji (1989), for example, used $\delta^{18}O$ values in living crinoids to identify seasonal temperature variation in bottom seawater and annual stem growth rates, while Baumiller (2001) utilized the isotope signature of the crinoid skeleton to recognize regenerated parts and different soft-tissue types.

ECHINODERM TAPHONOMY

Biostratinomy

Echinoderms have been the subject of a number of studies dealing with taphonomic aspects, as reviewed by Lewis (1980), Donovan (1991), Brett et al. (1997), Ausich (2001) and, most recently, Nebelsick (2004). Echinoderm skeletons can rapidly fall apart after death in a matter of days (Kier, 1977; Allison, 1990; Kidwell & Baumiller, 1990; Donovan, 1991; Greenstein, 1992; Nebelsick & Kampfer, 1994). The degree of disarticulation is related to intrinsic (type of connective tissues, structural integrity of the skeleton) as well as extrinsic factors (temperature, water agitation, scavenging). While holothuroids, ophiuroids, crinoids, and asteroids, as a rule, disarticulate into their separate, single ossicles, many echinoids can be preserved as denuded coronas, devoid of spines. This is related to the presence of interlocking skeletal supports and tight suturing of skeletal plates in some echinoid coronas. The skeletons of clypeasteroids are especially durable due to the added presence of internal supports that connect the oral and aboral surfaces (Seilacher, 1979).

Under certain circumstances (e.g. rapid burial, dysoxic conditions) articulated specimens of even the most fragile echinoderms are preserved. Examples of exceptional preservation include holothuroid body fossils (Smith & Gallemí, 1991; Haude, 2002), articulated crinoid communities (numerous examples discussed in Hess *et al.*, 1999), the famous Devonian Hunsrück slates (Lehmann, 1957; Bartels *et al.*, 1966; Hess, 1999a; Glass & Blake, 2004), and the Upper Cretaceous *Uintacrinus* beds of Kansas (Hess, 1999b; Meyer & Milsom, 2001). In some instances, conditions may be exceptionally favourable, leading to the formation of so-called echinoderm-"Lagerstätten" (Rosenkrantz, 1971; Seilacher *et al.*, 1985) including the famous complete crinoids (*Seirocrinus* and *Pentacrinites*) from the Lower Jurassic Posidonian shales of southwestern Germany. Exceptional preservation of Oligo-Miocene echinoderms seems to be limited to concentrated deposits of echinoids, especially that of clypeasteroids (see below). A rare example of Oligocene crinoids preserved as articulated specimens is the famous Oregon sea-lily fauna (Moore & Vokes, 1953; Hess, 1999c).

Due to the properties of the stereo, the echinoderm skeleton is very durable during life, but is easily abraded and transported after death (Chave, 1964). Small ossicles, in particular, are easily lost due to winnowing or dissolution. Transportation however, can often be recognized by ossicle wear and assessing abrasion thus represents a useful tool to separate autochthonous elements from allochthonous ones. Another application, involving assessment of abrasion and transport, was demonstrated by Meyer & Meyer (1986), who showed that comatulid crinoid ossicle abundance and wear might serve as an indicator for reef proximity (Fig. 10).

Identification of disarticulated material

The echinoderm origin of individual skeletal components can be easily recognized by their peculiar skeletal structure (stereom). Echinoderm components in the mud-fraction are less easily identified, as the stereom structure may not be readily recognized in small fragments. If unaltered, skeletal mineralogy (high-Mg calcite) may aid identification. Due to the high degree of specialization of the individual ossicles, even disarticulated remains can often be identified to family or genus level (Nebelsick, 1992b, c; Donovan, 1996). Sometimes identification to species level is possible by comparison with complete fossil and/or extant specimens (e.g. Kroh, 2005). Inclusion of data from

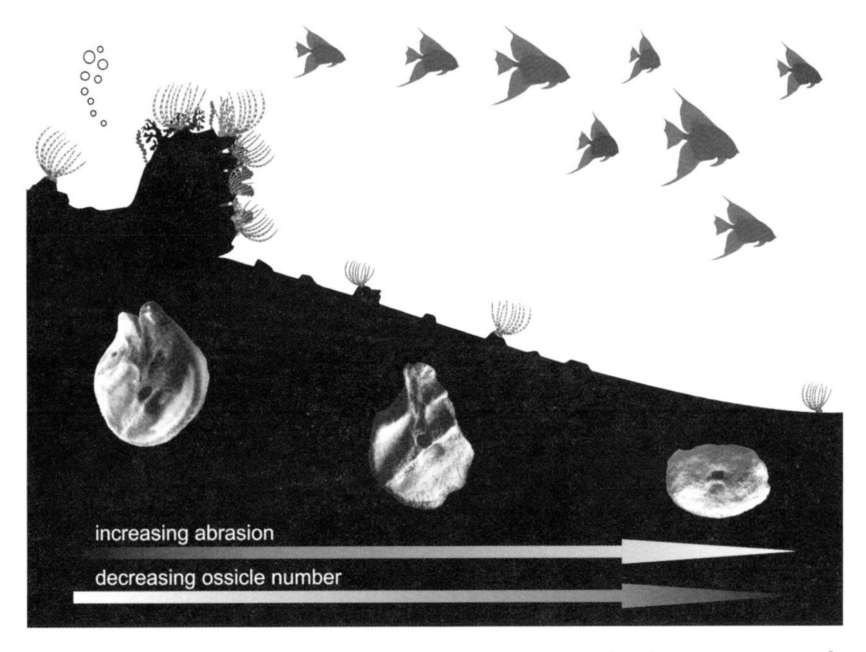

Fig. 10. Comatulid crinoid ossicle abundance and preservation as a potential palaeoenvironmental tool to indicate reef proximity. Modified from Meyer and Meyer (1986).

fragmentary and/or disarticulated specimens leads to a much more accurate picture of spatial and temporal echinoderm distributions (Nebelsick, 1992a, b, 1966; Gordon & Donovan, 1993; Donovan, 2001, 2003; Kroh, 2005). In this respect, the analysis of sediment bulk samples is of special importance. Typical size classes of echinoderm skeletal elements are shown in Fig. 2.

Diagenesis of the echinoderm skeleton

The skeleton of echinoderms consists of high-Mg calcite, which is thermodynamically metastable. Nevertheless, recrystallization of echinoderm calcite during early diagenesis is usually inhibited by organic or inorganic coatings, and stabilizing effects due to the interaction of Mg-ions in the seawater with the skeletal calcite (Berner, 1966, 1996a, b; Bischoff, 1968; summary in Weber, 1969b). The growth of cement crystals, in contrast, usually starts very early, commonly as soon as the soft tissue has decomposed. A peculiarity of calcite cements growing on echinoderm skeletal material is their syntaxial nature (Fig. 11). Depending on Ca-ion availability, pore-water chemistry and available space, individual cement crystals may attain considerable size (Fig. 12A and B), usually completely filling the pores of the stereom. Larger cavities in the echinoderm skeleton may also be filled if they are free from sediment, and eventually so-called "crystal apples" (Donovan & Portell, 2000; Donovan *et al.*, 2005, plate 1, figs. 2–7) may be formed. These are especially common in Palaeozoic forms like *Echinosphaerites*, but also occur in

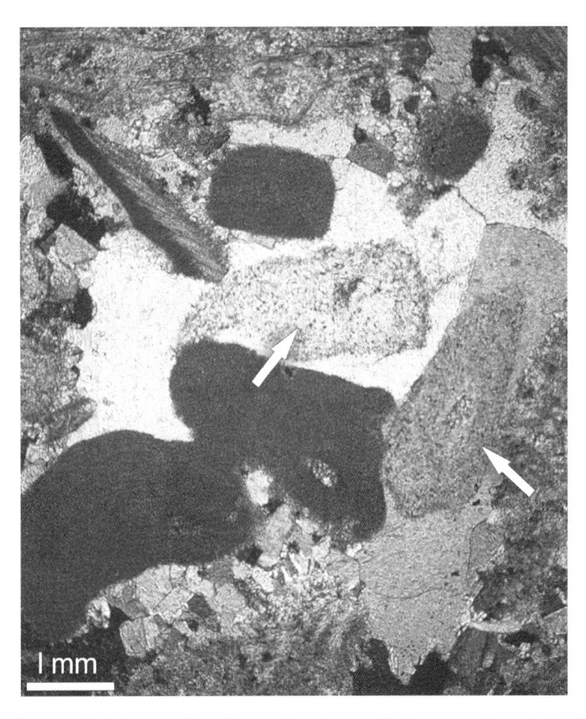

Fig. 11. Thin section from the Lower Miocene Zogelsdorf Formation of Austria showing two crinoid brachials (arrows) with extensive syntaxial cement. Crossed polarized light view (after Nebelsick, 1989).

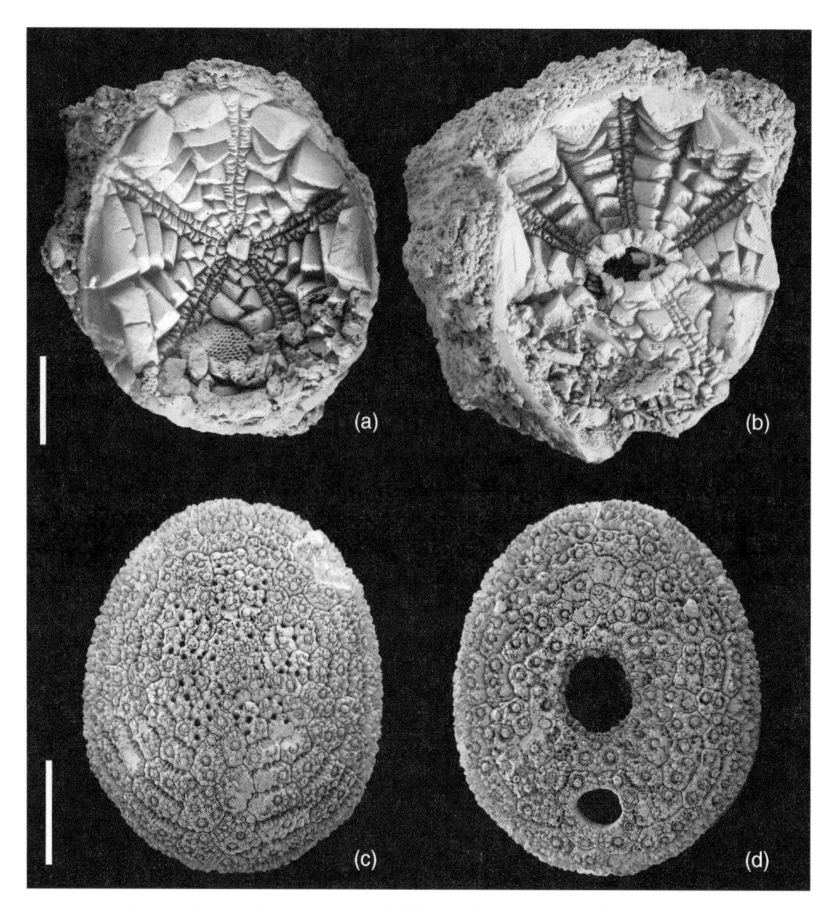

Fig. 12. Syntaxial cement growth on echinoid tests. (a) and (b) On the inside of the corona of *Echinolampas manzoni* (scale bar = 1 cm). (c) and (d) On the outside of *Echinocyamus pseudopusillus* (scale bar = 1 mm).

Cenozoic echinoids. The outlines of these crystals often precisely follow the distribution of skeletal plates (Fig. 12; see also Donovan & Portell, 2000, fig. 1), and may thus sometimes be useful recognizing plating patterns. Neugebauer (1979) recognized two types of cement growth, the dog tooth cement ("Zahnzement" of Neugebauer, 1979a) mentioned above, and rim cement ("Facettenzement" of Neugebauer, 1979a). The development of large cement crystals on the outside of individual ossicles is initiated only after large parts of the pore space within the ossicle are closed.

Apart from cement growth, the mineralogy of the echinoderm skeleton itself changes during diagenesis. In most cases, fossil echinoderms are transformed to low-Mg calcite (Weber, 1969b, 1973; Neugebauer, 1979a, b, c), especially during vadose meteoric diagenesis (Richter, 1974). For example, Weber & Raup (1968) found only one single skeleton consisting of high-Mg calcite (a Pleistocene sand dollar of the genus *Mellita*) in a survey of 83 fossil echinoids from the Carboniferous to Pleistocene. However, echinoderm skeletal material can be altered along different diagenetic pathways (Dickson, 2004). In some cases, high-Mg calcite is preserved (examples dating back to the Silurian are known, e.g. Dickson, 1995, 2004), others are altered to calcite and dolomite, but retain their original bulk composition (Bruckshen *et al.*, 1990; Dickson, 2001b, 2004). Cement coating plays an important role in this respect, as so called "crystal caskets" (Dickson, 2001b) may be formed that prevent ion-exchange with surrounding pore water.

Macro- and microscopic structure of the echinoderm stereom is often preserved, even in mineralogically altered ossicles. Even primary growth banding in individual stereom trabeculae may be preserved (Neugebauer, 1979a; Dickson, 2001b). Transformation however, is usually texturally not perfect, and often a finely mottled or granular texture is observed (contrary to the homogenous texture in unaltered samples). This granular texture resolves into large calcite crystals enclosing many microdolomite blebs (few μm in diameter)

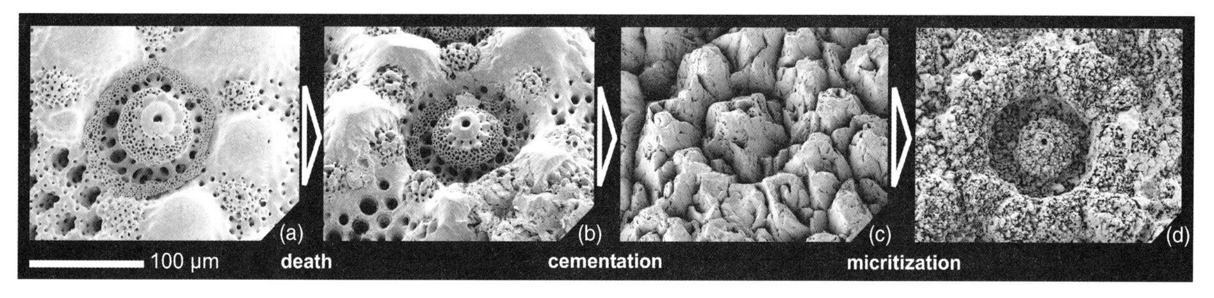

Fig. 13. Diagenetic changes of echinoderm skeletons shown by the example of a tubercle from the dwarf sea urchin *Echinocyamus*. Stages shown: (a) Fresh, unaltered corona; (b) Initial cement growth begins, here in the form of syntaxial rim cement; (c) Strong cementation with dog tooth cement; (d) Marginal micritization.

and micropores (~1 μm in diameter) in the Backscatter Scanning Electron Microscope (BSEM) (compare Dickson, 2001b, 2004). A similar texture can be observed in modern echinoderm ossicles artificially transformed by heating (Gaffey *et al.*, 1991; Dickson, 2001a). More commonly, however, another type of texture consisting of single calcite crystals with homoaxially well-defined microrhombic dolomite crystals occurs in fossil echinoderms (Macqueen & Ghent, 2003; Richter, 1974; Blake *et al.*, 1982). Contrary to the former texture type, which is the result of transformation with little or no involvement from material external to the skeleton, the latter type forms due to open-system transformation.

Similarly to the Mg-loss during vadose diagenesis, changes in the isotopic signature of the echinoderm skeletal remains may occur. Again the microstructure of the skeleton is largely unaffected by the process. Manze & Richter (1979) documented a loss of ^{13}C correlated to Mg-loss under meteoric-vadose conditions. Weber & Raup (1968), by contrast, found no evidence for a correlation of Mg-loss and alteration of the stable-isotope signature in a survey of 191 fossil echinoids spanning an age range from the Devonian to the Pleistocene. The degree of diagenetic alteration of $\delta^{13}C$ and $\delta^{18}O$ values cannot be predicted by Mg content, except in special cases involving conspecific fossil and recent specimens (Weber & Raup, 1968).

Apart from the geochemical alteration discussed above, echinoderm ossicles may also become subject to micritization during diagenesis, be it biogenic (Bathurst, 1990) or abiogenic (Neugebauer, 2001). While complete micritization as described by Neugebauer (2001) is rather rare (at least in the Cenozoic), marginal micritization is commonly observed (Fig. 13). Hand-in-hand with micritization, a considerable loss of volume may occur due to pressure dissolution. This loss may amount to up to 80% of the original skeletal mass in the example of the Cretaceous crinoid *Uintacrinus socialis* investigated by Neugebauer (2001). A similar alteration has been observed in Miocene echinoids (Fig. 14) where the corona is micritized and strongly leached due to pressure solution.

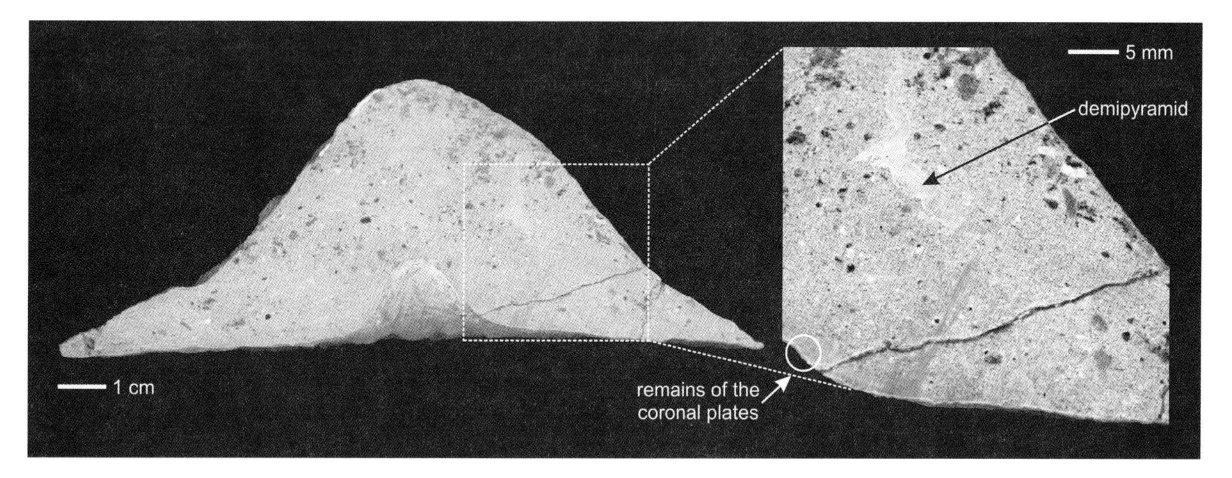

Fig. 14. Vertical cross section through a "Prägesteinkern" of *Clypeaster campanulatus*. The corona is nearly completely micritized and leached due to pressure solution, only traces remain of the original shell. Modified from Kroh (2005).

Diagenetic changes associated with echinoderm ossicles strongly affect the embedding sediment and promote lithification. This especially applies when echinoderm syntaxial cements extend out into the surrounding pore space. Cementation with syntaxial cement is an important factor in sediments where: (1) echinoderms are common; (2) the sediments show high interparticulate porosity (grainstones and rudstones); and (3) there is little or no primary marine cementation. In the Oligo-Miocene, this has been most often described from non-tropical limestones (e.g. Nelson, 1978; James & Bone, 1989; Nebelsick, 1989; Nicolaides, 1995; Nicolaides & Wallace, 1997; Knoerich & Mutti, 2003).

In the Gambier Limestone, for example, syntaxial cements on echinoderm fragments are the only cements of any importance (James & Bone, 1989). In the Oligo-Miocene Clifton Formation of southern Australia, syntaxial calcite overgrowths on echinoderm fragments represent one of three morphological cement types along with scalenohedral and blocky equant calcite spar (Nicolaides & Wallace, 1997). Knoerich and Mutti (2003) noted two phases of syntaxial cements growing on echinoderm fragments as part of a complex diagenetic history in Oligocene heterozoan limestones from Malta: a thin rim (30 μm) of inclusion-rich crystals and a later phase of inclusion-free syntaxial calcite with larger crystals (200 μm). The first phase is attributed to marine diagenesis, the second to possible burial marine cementation. In all cases, high degrees of primary pore space allow syntaxial cement to develop far beyond the primary component, often enclosing surrounding biotic and terrigenous components.

Sediment production, erosion and reworking

Echinoderms play an important role both in the production of carbonate (skeleton, faeces) and the reworking of sediment (bioturbation, erosion). The total production of carbonate by echinoderms and their contribution to the carbonate budget is difficult to ascertain. However, disarticulated remains may account for a significant portion of the sediment particles (up to 45% in some instances) as revealed by modal analysis of components (see below). Carbonate mud from echinoderm faeces contributes to the fine fraction of the sediment.

Echinoids are major bioeroding organisms and are important for carbonate budgets in coral reefs (Otter, 1932, 1937; Russo, 1980; Bak, 1993, 1998; Glynn, 1997; Carreiro-Silva & McClanahan, 2001). The jaw mechanisms of echinoids (Aristotle's lantern) contain hardened teeth which are continuously produced at the base in order to replace the eroded distal tips. Jaws are present in all regular echinoids as well as in many irregulars, including the clypeasteroids. Many regular echinoids are non-specific grazers exploiting plants and animals with both soft and hard skeletal surfaces. The eroded material is then excreted as pellets of fine carbonate mud.

Echinoid bioerosion as a structuring force in the coral reef environments was reported by Mokady *et al.* (1996) from the Red Sea. Typically, the echinoid *Diadema setosum* erodes 310 mg per individual per day, while *Echinometra mathaei* abrades 120 mg per individual per day (see also Muthiga & McClanahan, 2007; McClanahan & Muthiga, 2007). Both echinoids occur in high densities and are thus major converters of carbonate skeletons to carbonate sediments in both reef flats and reef slopes. In a study of echinoid bioerosion and herbivory on Kenyan coral reefs, Carreiro-Silva and McClanahan (2001) showed the importance of regular echinoids (in this case *Echinothrix diadema*, *Diadema setosum*, *Diadema savignyi* and *Echinometra mathaei*) in reef development. Bioerosion was generally found to be higher than herbivory rates. The highest bioerosion rates (1180 ± 230 g $CaCO_3$ $m^{-2}yr^{-1}$) were found in unprotected reefs with highest echinoid densities at 6.2 ± 1.5 individuals m^{-2} due to low fish predation pressures. Intermediate bioerosion rates (711 ± 157 g $CaCO_3$ $m^{-2}yr^{-1}$) were recorded at newly protected reefs with 1.2 ± 0.1 individuals. Lowest bioerosion rates (50.3 ± 25.8 g $CaCO_3$ $m^{-2}yr^{-1}$) were found in protected reefs with 0.06 ± 0.01 individuals m^{-2}.

There has been a dramatic evolution of teeth form and efficiency within echinoids through time (compare Smith, 1984). Steneck (1983) correlated increased bioerosive efficiency of organisms including echinoids during the Mesozoic to adaptive trends in calcareous algal crusts, including the development of sunken conceptacles within the skeletons. Direct evidence of echinoid bioerosion such as five pointed, star-shaped scratch marks (Bromley, 1975) are rare and not conducive to quantification. Although, clypeasteroid teeth do not erode surficial hard substrates as such, they are employed in crushing ingested sand-sized particles (Kampfer & Tertschnig, 1992). This leads to the production of finer sediments, thus influencing the modal size distribution of the sediments.

Echinoids may also produce distinct burrows which can potentially be recognized in the fossil record. The Indopacific echinoid *Echinostrephus molaris* erodes deep burrows in reefal environments (Russo, 1980; Campbell, 1987). Their burrowing activity can also exert an influence on coralline algal growth inducing rhodolith production in the Red Sea (Piller & Rasser, 1996).

Finally, members of the echinoids (especially the irregulars), asteroids, ophiuroids and holothuroids are major infaunal elements and can cause intensive bioturbation of the sediments in calcareous environments (Scheibling, 1982). High maximum individual reworking rates have been recorded for the irregular echinoid *Meoma ventricosa* (8520 $cm^3 day^{-1}$) and the clypeasteroid *Encope michelini* (960 $cm^3 day^{-1}$) from Caribbean lagoons (Thayer, 1983). This bioturbation can be shallow, as for most clypeasteroids, or deep, as for many spatangoids, which can result in characteristic trace fossils (Bromley and Asgaard, 1975; Bertling *et al.*, 2006).

ECHINODERMS AS COMPONENTS IN OLIGO-MIOCENE CARBONATES

Echinoderms are important components in carbonate environments, usually comprising between 5 and 30 % of the sediment particles. They seem to be more abundant in non-tropical than in tropical settings. In temperate shelf carbonates, echinoderms constitute one of the major groups of skeletal fragments together with bryozoans, benthic foraminifera, barnacles, brachiopods, bivalves, and coralline algae (Nelson, 1978). This is not only reflected by quantitative modal analysis of component distributions, but finds its expression also in facies nomenclature as well as in the names of general facies associations (see below). In tropical settings, echinoderms, though common as macrofaunal elements, are rare as components in the sediments. A recent example from the Red Sea shows mean concentrations of <1% and maximum values of <5% within sediment samples (Piller, 1994).

One of the most distinct occurrences of echinoderms in Oligo-Miocene sediments is that of mass occurrences of clypeasteroids in shallow-water environments (Nebelsick & Kroh, 2002; Kroh & Nebelsick, 2003). These mass occurrences are often found within shoreface sequences and can include very large numbers of individuals, leading to sand-dollar beds ranging from a few centimetres to a number of metres in thickness. The origin of these shell beds has been tied to high population densities and concentration processes, including tempestites or winnowing in higher energy environments. Other echinoids can also be included in these special lagerstätten, which can be spectacular in nature and are readily noticed within sedimentary sequences. Encrinites, consisting of mass accumulation of crinoid remains, though common in the Palaeozoic and Mesozoic (Ausich, 1997), are missing in Cenozoic sediments, probably owing to the declining importance of stalked crinoids in shallow-water settings.

Modal distribution of echinoderms

An exact comparison of the modal presence of echinoderms in different Oligo-Miocene carbonates is difficult to accomplish due to the wide variety of qualitative and quantitative assessment methods applied to determine component distribution. Even those studies including multivariate statistical analysis of component distributions differ widely in the exact methods used to accumulate the respective data sets (e.g. by use of semi-quantitative comparative methods or quantitative point-counting methods). However, high numbers of echinoderms are reflected by nomenclature; e.g. "molechfor" facies type of Carannante *et al.* (1988) or "echinofor" facies type of Hayton *et al.* (1995). The ability to distinguish echinoderms in thin section is limited to distinct plates and spines (Fig. 15).

Oligocene carbonates seem to be generally poor in echinoderms, at least in the circum-alpine area. Average concentrations of around 1 % echinoderms are found in the Lower Oligocene Gornji Grad Beds of Slovenia, with maximum values of near 4 % being reached in grainstones of a foraminiferal-coralline algal facies (Nebelsick *et al.*, 2000). Similarly, echinoderm remnants in thin sections from in the Lower Oligocene Werlberg Member of the Paislberg Formation from the Lower Inn Valley show values ranging from 3 to 7 % (Nebelsick *et al.*, 2001). Oligocene carbonates of the Lower Coralline Limestone of Malta consist of packstones to rudstones and are dominated by coralline red algae, bryozoans, echinoids and benthic foraminifera (Knoerich & Mutti, 2003). Modal analysis revealed echinoid dominance between 2 and 29 % within individual samples. These echinoids are especially important for diagenetic pathways including two generations of

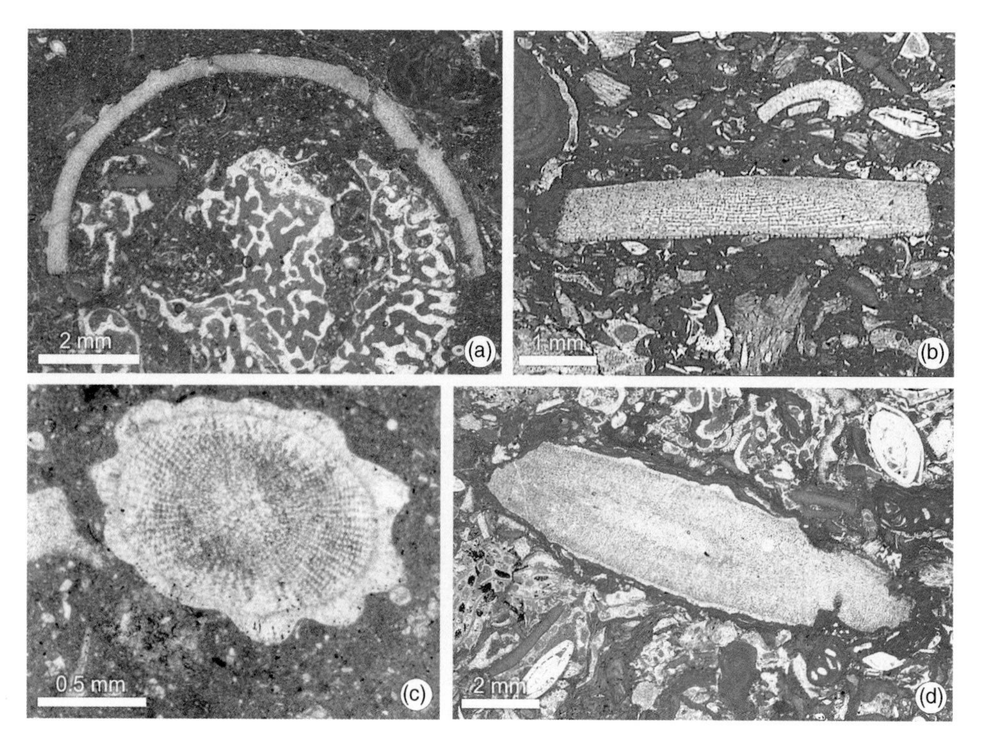

Fig. 15. Echinoderms in thin section. (a) Half of a broken regular echinoid with distinct plates and tubercles cupped around a coral fragment (from Nebelsick *et al.*, 2000). (b) Single regular echinoid plate showing differential stereom construction. (c) Slightly oblique sections through a cidaroid spine (regular echinoid; from Löffler & Nebelsick, 2001). (d) Strongly oblique sections through a cidaroid spine (regular echinoid), note thin crust of coralline algae. (a) Lower Oligocene, Gornji Grad Beds. (b–d) Lower Oligocene, Werlberg Member (Paislberg Formation) from the Lower Inn Valley. All views, plain polarized light.

syntaxial cements. Echinoid content is equally high in samples from Ragusa Island (5–16 %, averaging around 9 %; see Knoerich & Mutti, 2006).

An example from Lower Miocene bryomol carbonates is the "Echinoderm-Foraminiferal Facies" within the Lower Miocene Zogelsdorf Formation (Nebelsick, 1989, 1992a). In this facies, echinoderm remnants comprise more that 30% of the biogenic components. The occurrence of echinoderms is linked to high amounts of bryozoans, which may be due to the fact that dead echinoids represent important sites for incrustation not only for bryozoans, but also for barnacles, serpulids and coralline algae, as they represent relatively large substrates within otherwise highly mobile, particulate skeletal sediment (Nebelsick *et al.*, 1997). Tortonian, nearshore, temperate, carbonate depositional systems from southern Spain contain packstones and rudstones in part rich in echinoderms (e.g. Brachert *et al.*, 1996; Martín *et al.*, 1996). Quantitative data from thin sections show average concentrations of echinoderms near 10 %, with maximum values of 28 % (Betzler *et al.*, 1997); packstones below volcanic palaeocliffs are locally dominated by echinoderm debris, which contributes up to 45 % of the components.

The importance of echinoderms within facies in a non-tropical setting has been reported by Lukasik *et al.* (2000) from the Murray Supergroup in southern Australia. These deposits originated from a shallow, low-energy, mesotrophic, epeiric ramp. Echinoderms are generally common within four facies associations, but especially within the echinoid-bryozoan facies association. This association comprises four echinoderm-dominated subfacies deposited at shallow to moderate depths to the base of the euphotic zone.

Facies associations

Echinoderms are common members of "non-tropical", "warm temperate" (Nelson, 1988) or "heterozoan" communities (James, 1997). They have played an increasing role in facies models used to describe the component associations. These include the "bryomol" (Nelson *et al.*, 1988), the "rhodalgal" (Carannante *et al.*, 1988) and "molechfor" facies types (Carannante *et al.*, 1988).

The latter association is dominated by molluscs, echinoids and (small) benthic foraminifera.

A more detailed classification, based on component dominance, was introduced by Hayton *et al.* (1995) for non-tropical carbonate deposits using data from New Zealand Cenozoic limestones. Seven assemblages were recognized by cluster analysis of abundance data of skeletal components. Of these, two include echinoderms as denominating components: the "echinofor" (echinoderm/benthic foraminiferal) and the "rhodechfor" (calcareous red algal/echinoderm/benthic foraminiferal) assemblages. The echinofor assemblage, which is dominated by echinoid plates and spines with occasional complete sea urchins, is widely distributed in Oligo-Miocene sediments and is interpreted to reflect frequent reworking and high suspension loads that prevented the settlement of bryozoans (Hayton *et al.*, 1995). The rhodechfor assemblage is generally rare in Miocene sediments. Lower light intensity is interpreted to be responsible for an increase in non-algal components compared to a rhodalgal assemblage (Hayton *et al.*, 1995). This terminology has been applied in other studies, for example in Eastern Crete where an echinofor-type lithofacies was recognized in Neogene non-tropical carbonates deposited in a warm-temperate biogeographic province (Pomoni-Papaioannou *et al.*, 2002).

Echinoderms in aphotic sediments

The presence of echinoderms is generally not restricted by depth. Members of all five extant echinoderm classes belong to the most dominant benthic organisms in larger depths below the photic zone where phototrophic (e.g. coralline algae) and symbiont-bearing organisms (e.g. larger foraminifera) cannot exist due to the lack of light. This is demonstrated by Miocene rhodalgal and molechfor sediments from the "Briozoi e Litotamni" Formation from the Latium-Abruzzi Platform of the central Apennines (Brandano & Corda, 2002; Corda & Brandano, 2003). While carbonate production and accumulation in the inner to middle ramp is dominated by coralline algae with rhodoliths, larger benthic foraminifera and molluscs; sediments of the outer ramp below the photic zone are co-dominated by echinoderms (a bryozoan–echinoid unit, a benthic foraminiferal–echinoid unit, and a planktonic foraminifera–echinoid unit). Deep-water, basinal sediments also show a large number of echinoderms in calcarenites, together with siliceous sponge spicules and planktonic foraminifera.

ECHINODERMS AS PALAEOECOLOGICAL TOOLS IN OLIGO-MIOCENE SEDIMENTS

Echinoderms have great potential as palaeoecological tools in Oligo-Miocene carbonate environments. Ecological information can not only be gained by actualistic comparisons with recent echinoderms, but also by using a functional morphological approach (e.g. Kier, 1972; Boggild & Rose, 1984; Kroh & Harzhauser, 1999; Néraudeau *et al.*, 1979c; Kroh & Nebelsick, 2003). Both approaches allow the detailed assessment of general life habits, substrate relationships (grain-size parameters, stability and sorting), nutrient factors (food availability and organic content) and hydrodynamic regime including current velocity and water turbulence (Boggild & Rose, 1984; Kroh & Nebelsick, 2003). Studies of echinoderms in Oligo-Miocene habitats have largely focussed on echinoids (Kier, 1972; Boggild & Rose, 1984; McNamara, 1970; Kroh & Harzhauser, 1999; Carter, 2003; Kroh & Nebelsick, 2003) due to their higher preservation potential, but other echinoderms, especially comatulid crinoids, if present, should also be included.

A combination of morphological description, taxonomic treatment, functional morphological interpretation, and interpretative assignment to habitat occupation, were shown in the classic study of Kier (1972) from the Yorktown Formation of eastern North America (originally held to be Late Miocene in age, now attributed to the Pliocene; Dowsett & Wiggs, 1992). The existence of various habitats was demonstrated, based on the ecological demands of two regular echinoids: *Arbacia imporcera*, *Psammechinus philanthropus* (both occupying an intertidal nearshore habitat); and three irregular echinoids: *Echinocardium orthonotum* (burrowing deeply in a nearshore habitat), *Mellita aclinensis* (burrowing shallowly in a nearshore habitat) and *Spatangus glenni* (burrowing shallowly echinoid in an offshore habitat).

Numerous echinoid biofacies were recognized in Upper Oligocene to Upper Miocene sediments of Malta by Boggild & Rose (1984). These contain a rich echinoid fauna with at least 47 echinoid species. The echinoid biofacies are each characterized by a distinctive echinoid assemblage, associated biota and lithology and were related to the standard facies belts and standard microfacies types as defined by Wilson (1975). The 13 different echinoid biofacies are found in restricted platform lagoons (one echinoid biofacies), open platform lagoons (three biofacies), winnowed edge sands (one biofacies), organic reefs (one biofacies), foreslope (three biofacies) and in the deep shelf margin to open shelf (four biofacies). The Lower Globigerina Limestones, for example, are characterized by a wackestone with a *Schizaster parkinsoni*/*Ditremaster*/*Pecten* assemblage and were ascribed to Standard Facies Belt Zone 3-2 of Wilson (1975) due to its massive, fine-grained, wackestone to mudstone lithology with pelagic microfossils. Both *Schizaster* and *Ditremaster* are irregular echinoids adapted to burrowing into fine, soft substrates as found in the Globigerina Limestone (see also Rose & Watson, 1998). These echinoid biofacies have also been recognized in other Oligo-Miocene successions of the Mediterranean (e.g. Crete, Libya; Rose, 1974; Marcopoulou-Diacantoni, 1979; Boggild & Rose, 1984) and Caribbean region (Boggild & Rose, 1984; Poddubiuk & Rose, 1984).

Late Miocene Messinian irregular echinoids from the Sorbas Basin of southeast Spain were analysed by Néraudeau *et al.* (1979c) with respect to morphological and palaeoecological gradients including detailed biometrical analysis. An ecomorphological gradient from shallow to deeper water was postulated following the succession of: (1) a *Clypeaster altus* association from the upper infralittoral zone; (2) an association from the lower infralittoral zone with *Clypeaster marginatus*, *Schizaster* and *Spatangus*; (3) a *Brissopsis* gr. *lyrifera* association from the upper-circalittoral zone; and (4) the deepest association from the medium ciralittoral to bathyal zone with *Brissopsis* gr. *atlantica*. Furthermore, changes in echinoid diversity are used to record sea-level changes and discuss the existence of marine connections between different marine basins across the Messinian salinity crisis.

Seven different echinoid assemblages were recognized in Lower Miocene siliciclastic and carbonate-dominated sediments from the Gebel Gharra section, Egypt (Kroh & Nebelsick, 2003). These assemblages were based on a wide variety of echinoderms representing distinct ecological habitats. The dominating echinoids include both epibenthic as well as various endobenthic forms from different burrowing depths. Comatulid crinoids and rare marginal ossicles of the sea star *Astropecten* were also present. Changes in echinoderm diversity were related to substrate variations, burrowing depths, taphonomic factors as well as the general deepening and subsequent shallowing of the depositional environment. These findings, based on echinoderms, complemented those based on microfacies analysis and other macrobenthic invertebrate studies from the same section (Abdelghany & Piller, 1999; Mandic & Piller, 2001).

An example of using the complete echinoderm record in palaeoenvironmental reconstruction was shown by Kroh & Harzhauser (1999) from Lower Miocene sediments of Austria. Here, palaeoecological interpretations were based on complete and fragmentary material from echinoids, crinoids, ophiuroids and asteroids. Aut- and synecological analysis of the assemblages allowed the differentiation of a shallow, wave-dominated sand bottom fauna of low diversity, and a rich fauna from more sheltered, diverse habitats.

CONCLUSIONS

The study of echinoderm skeletal remains offers a variety of potentially useful applications for sedimentology and palaeoenvironmental reconstruction. Compared with many other macrofossil groups, echinoderms show many advantages including a durable skeleton, moderate diversity and easy recognition. Once the faunal spectrum present in a given area or time slice has been established and documented by a specialist, usage of disarticulated echinoderm material for sedimentological applications is rather straightforward.

Researchers working on the Oligo-Miocene benefit from the high similarity of fossil echinoderm remains to extant faunas, facilitating element and taxon recognition based on comparison to recent specimens. Many of the potential uses outlined in this paper have not been, or have very rarely been, applied to Oligo-Miocene echinoderms and sedimentary systems thus far. This is, in part, due to the limited awareness of the possibilities and limitations of echinoderms in a sedimentological context and in part due to the lack of communication between echinoderm specialists on the one hand and sedimentologists on the other. The present paper is intended to improve this situation.

ACKNOWLEDGEMENTS

This study was supported by the Austrian Science Fund (FWF) via project no. P-13466-Bio to Werner E. Piller (Univ. Graz). The critical reviews and helpful comments of Andrea Knoerich and Maria Mutti greatly improved the paper. Furthermore, we want to express our sincere thanks to the staff of the Geological Department at the Natural History Museum, Vienna, and the Institute of Geosciences at the University of Tübingen.

REFERENCES

Abdelghany, O. and **Piller, W.E.** (1999) Biostratigraphy of Lower Miocene sections in the Eastern Desert (Cairo - Suez district, Egypt). *Rev. Micropaléontol.*, **18**, 607–617.

Allison, P.A. (1990) Variation in rates and decay and disarticulation of Echinodermata: Implications for the application of actualistic data. *Palaios*, **5**, 432–330.

Ausich, W.I. (1997) Regional encrinites; a vanished lithofacies. In: *Paleontological Events; Stratigraphic, Ecological and Evolutionary Implications* (Eds C.E. Brett and G. C. Baird), pp. 509–519. Columbia University Press, New York.

Ausich, W.I. (2001) Echinoderm taphonomy. In: *Echinoderm Studies*, **6**, (Eds M. Jangoux and J.M. Lawrence), pp. 171–227. Balkema, Lisse.

Bak, R.P.M. (1990) Patterns of echinoid bioerosion in two Pacific coral reef lagoons. *Mar. Ecol. Prog. Ser.*, **66**, 267–272.

Bak, R.P.M. (1993) Sea urchin bioerosion on coral reefs: place in the carbonate budget and relevant variables. *Coral Reefs*, **13**, 99–103.

Bartels, C., **Briggs, D.E.G.** and **Brassel, G.** (1998) The fossils of the Hunsrück Slate – Marine Life in the Devonian. *Cambridge Palaeobiol. Ser.*, **3**, Cambridge University Press, Cambridge, 309 pp.

Bathurst, R.C.G. (1966) Boring algae, micrite envelopes and lithification of molluscan biosparites. *Geol. J.*, **5**, 15–32.

Baumiller, T.K. (2001) Light stable isotope geochemistry of the crinoid skeleton and its use in biology and paleobiology. In: *Echinoderms 2000* (Ed. M.F. Barker), pp. 107–112. Swets & Zeitlinger, Lisse.

Beck, J.W., **Edwards, R.L.**, **Ito, E.**, **Taylor, F.W.**, **Recy, J.**, **Rougerie, F.**, **Joannot, P.** and **Henin, C.** (1992) Sea-surface temperature from coral skeletal strontium/calcium ratios. *Science*, **257**, 644–647.

Berner, R.A. (1966) Chemical diagenesis of some modern carbonate sediments. *Am. J. Sci.*, **264**, 1–36.

Berner, R.A. (1996a) Diagenesis of carbonate sediments: interaction of magnesium in sea water with mineral grains. *Science*, **153**, 188–191.

Berner, R.A. (1996b) Chemical diagenesis of some modern carbonate sediments. *Am. J. Sci.*, **264**, 1–36.

Bertling, M., **Braddy, S.J.**, **Bromley, R.G.**, **Demathieu, G. R.**, **Genise, J.**, **Mikuláš, N.J.K.**, **Nielsen, K.S.S.**, **Rindsberg, A.K.**, **Schlirf, M.** and **Uchman, A.** (2006) Names for trace fossils: a uniformitarian approach. *Lethaia*, **39**, 265–286.

Betzler, C., **Brachert, T.C.** and **Nebelsick, J.H.** (1997) The warm temperate carbonate province: a review of the facies, zonations, and delimitations. *Cour. Forsch.-Inst. Senckenb.*, **201**, 83–99.

Biese, W. and **Sieverts-Doreck, H.** (1939) *Crinoidea caenozoica. - Fossilium Catalogus, I: Animalia.* W. Junk., 's-Gravenhage, 151 pp.

Bill, M., **Baumgartner, P.O.** and **Hunziker, J.C.** (1995) Carbon isotope stratigraphy of the Liesberg Beds Member (Oxfordian, Swiss Jura) using echinoids and crinoids. *Eclogae Geol. Helv.*, **88**, 135–155.

Bischoff, J.L. (1968) Kinetics of calcite nucleation: magnesium ion inhibition and ionic strength catalysis. *J. Geophys. Res.*, **73**, 3315–3322.

Blake, D.B. (1973) Ossicle morphology of some recent asteroids and description of some West American fossil asteroids. *Geol. Sci. Univ. Calif. Publ.*, **104**, 1–59.

Blake, D.F. and **Peacor, D.R.** (1981) Biomineralization on crinoid echinoderms. Characterization of crinoid skeletal elements using TEM and STEM microanalysis. *Scan. Electron Microsc.*, **1981**, 321–328.

Blake, D.F., **Peacor, D.R.** and **Wilkinson, B.H.** (1982) The sequence and mechanism of low-temperature dolomite formation: Calcian dolomites in a Pennsylvanian echinoderm. *J. Sed. Petrol.*, **52**, 59–70.

Bodenbender, B.E. (1996) Patterns of crystallographic axis orientation in blastoid skeletal elements. *J. Paleontol.*, **70**, 466–484.

Boggild, G.R. and **Rose, E.P.F.** (1984) Mid-Tertiary echinoid biofacies as palaeoenvironmental indices. *Ann. Géol. Pays Hellén.*, **22**, 57–67.

Brachert, T.C., **Betzler, C.**, **Braga, J.C.** and **Martín, J.M.** (1996) Record of climatic change in neritic carbonates: turnover in biogenic associations and depositional modes (Late Miocene, Southern Spain). *Geol. Rundsch.*, **85**, 327–337.

Brandano, M. and **Corda, L.** (2002) Nutrient, sea level and tectonics, constraints: for the facies architecture of Miocene carbonate ramp in Central Italy. *Terra Nova*, **14**, 257–262.

Brenchley, P.J. and **Harper, D.A.T.** (1998) *Palaeoecology: Ecosystems, Environments and Evolution.* Chapman & Hall, London, 402 pp.

Breton, G. (1992) Les Goniasteridae (Asteroidea, Echinodermata) jurassiques et crétacés de France: taphonomie, systematique, paléogéographie, évolution. *Bull. Trim. Soc. Géol. Normandie Amis Mus. Havre*, **78**, 1–590.

Brett, C.E., **Moffat, H.A.** and **Taylor, W.L.** (1997) Echinoderm taphonomy, taphofacies, and Lagerstätten. In: *Geobiology of Echinoderms* (Eds J.A. Waters and C.G. Maples). *Paleont. Soc. Pap.*, **3**, 147–190.

Bromley, R.G. (1975) Comparative analysis of fossil and recent echinoid bioerosion. *Palaeontology*, **18**, 725–739.

Bromley, R.G. and **Asgaard, U.** (1975) Sediment structures produced by a spatangoid echinoid: a problem of preservation. *Bull. Geol. Soc. Denmark*, **24**, 261–281.

Bruckshen, P., **Noeth, S.** and **Richter, D.K.** (1990) Tempered microdolomites in crinoids: a new criterion for high-grade diagenesis. *Carbonates Evaporites*, **5**, 197–207.

Campbell, A.C. (1987) Echinoderms of the Red Sea. In: *Red Sea (Key Environments)* (Eds A.J. Edwards and S.M. Head), pp. 215–232. Pergamon Press, Oxford.

Carannante, G., **Esteban, M.**, **Milliman, J.D.** and **Simone, L.** (1988) Carbonate lithofacies as paleolatitude indicators: problems and indicators. *Sed. Geol.*, **60**, 333–346.

Carreiro-Silva, M. and **McClanahan, T.R.** (2001) Echinoid bioerosion and herbivory on Kenyan coral reefs: the role of protection from fishing. *J. Exp. Mar. Biol. Ecol.*, **262**, 133–153.

Carter, B.D. (2003) Diversity Patterns in Eocene to Oligocene Echinoids. In: *From Greenhouse to Icehouse - The Marine Eocene-Oligocene Transition* (Eds D.R. Prothero, L.C. Ivany and E.A. Nesbitt), pp. 354–365. Columbia University Press, New York.

Chave, K.E. (1954) Aspects of the biogeochemistry of magnesium: 1. Calcareous marine organisms. *J. Geol.*, **62**, 266–283.

Chave, K.E. (1964) Skeletal durability and preservation. In: *Approaches to Paleoecology* (Eds J. Imbrie and N. Newell), pp. 377–382. Wiley, New York.

Clarke, A.H. (1911) The inorganic constituents of the skeleton of two recent crinoids. *Proc. US Nat. Mus.*, **39**, 487–488.

Clarke, F.W. and **Wheeler, W.C.** (1914) The composition of crinoid skeletons. *US Geol. Surv. Prof. Pap.*, **90D**, 33–37.

Clarke, F.W. and **Wheeler, W.C.** (1915) The inorganic constituents of echinoderms. *US Geol. Surv. Prof. Pap.*, **90L**, 191–196.

Clarke, F.W. and **Wheeler, W.C.** (1917) The inorganic constituents of marine invertebrates. *US Geol. Surv. Prof. Pap.*, **102**, 1–56.

Clarke, F.W. and **Wheeler, W.C.** (1922) The inorganic constituents of marine invertebrates. *US Geol. Surv. Prof. Pap.*, **124**, 1–62.

Cohen, A., **Layne, G.**, **Hart, S.** and **Lobel, P.** (2001) Kinetic control of skeletal Sr/Ca in a symbiontic coral: implications for the paleotemperature proxy. *Paleoceanography*, **16**, 20–26.

Cohen, A., **Owens, K.E.**, **Layne, G.** and **Shimizu, N.** (2002) The effect of algal symbionts on the accuracy of Sr/Ca paleotemperatures from coral. *Science*, **296**, 331–333.

Corda, L. and **Brandano, M.** (2003) Aphotic zone carbonate production on a Miocene ramp, Central Apennines. Italy. Sed. Geol., **161**, 55–70.

Davies, T.T., **Crenshaw, A.** and **Heatfield, B.M.** (1972) The effect of temperature on the chemistry and structure of echinoid spine regeneration. *J. Paleontol.*, **46**, 874–883.

de Blainville, H.M.D. (1825) Oursins. In: *Dictionnaire des Sciences Naturelles*, **37**, pp. 59–245. F.G. Levrault, Strasbourg, Paris.

Dickson, J.A.D. (1995) Paleozoic Mg calcite preserved: Implications for the Carboniferous ocean. *Geology*, **23**, 535–538.

Dickson, J.A.D. (2001a) Diagenesis and crystal caskets: Echinoderm Mg calcite transformation, Dry Canyon, New Mexico, U.S.A. *J. Sed. Res.*, **71**, 764–777.

Dickson, J.A.D. (2001b) Transformation of echinoid Mg calcite skeletons by heating. *Geochim. Cosmochim. Acta*, **65**, 443–454.

Dickson, J.A.D. (2002) Fossil echinoderms as monitor of the Mg/Ca ratio of Phanerozoic oceans. *Science*, **298**, 1222–1224.

Dickson, J.A.D. (2004) Echinoderm skeletal preservation: Calcite-aragonite seas and the Mg/Ca ratio of Phanerozoic oceans. *J. Sed. Res.*, **74**, 355–365.

Donnay, G. and **Pawson, D.L.** (1969) X-ray diffraction studies of echinoderm plates. *Science*, **166**, 1147–1150.

Donovan, S.K. (1991) The taphonomy of echinoderms: calcareous multi-element skeletons in the marine environment. In: *The Processes of Fossilisation* (Ed. S.K. Donovan), pp. 241–269. Belhaven Press, London.

Donovan, S.K. (1995) Isocrinid crinoids from the late Cenozoic of Jamaica. *Atlantic Geol.*, **30**, 195–203.

Donovan, S.K. (1996) Use of the SEM in interpreting ancient faunas of sea urchins. *Eur. Microsc. Anal.*, **1996**, 29.

Donovan, S.K. (2001) Evolution of Caribbean echinoderms during the Cenozoic: Moving towards a complete picture using all of the fossils. *Palaeogeogr. Palaeoclimatol. Palaeoecol.*, **166**, 177–192.

Donovan, S.K. (2003) Completeness of a fossil record: the Pleistocene echinoids of the Antilles. *Lethaia*, **36**, 1–7.

Donovan, S.K. and **Portell, R.W.** (2000) Incipient "crystal apples" from the Miocene of Jamaica. *Caribb. J. Sci.*, **36**, 168–170.

Donovan, S.K. and **Veltkamp, C.J.** (2001) The Antillean Tertiary Crinoid Fauna. *J. Paleontol.*, **75**, 721–731.

Donovan, S.K., **Gordon, C.M.**, **Veltkamp, C.J.** and **Scott, A.D.** (1993) Crinoids, asteroids, and ophiuroids in the Jamaican fossil record. In: *Biostratigraphy of Jamaica* (Eds R.M. Wright and E. Robinson). *Geol. Soc. Am. Mem.*, **182**, 125–130.

Donovan, S.K., **Portell, R.W.** and **Veltkamp, C.J.** (2005) Lower Miocene echinoderms of Jamaica. West Indies. Scripta Geol., **129**, 91–135.

Dowsett, H.J. and **Wiggs, L.B.** (1992) Planktonic Foraminiferal Assemblage of the Yorktown Formation, Virginia, USA. *Micropaleontology*, **38**, 75–86.

Eagle, M.K. (1993) A new fossil isocrinid crinoid from the Late Oligocene of Waitete Bay, northern Coromandel. Rec. Auckland Inst. Mus., **30**, 1–12.

Ebert, T.A. (1998) An analysis of the importance of Allee effects in management of the red sea urchin *Strongylocentrotus franciscanus*. In: *Echinoderms: San Francisco* (Eds R. Mooi and M. Telford), pp. 619–627. A.A. Balkema, Rotterdam.

Ebert, T.A. (2001) Growth and survival of post-settlement sea urchins. In: *Sea Urchins: Biology and Ecology Edible Developments in Aquaculture and Fisheries Science* (Ed. J.M. Lawrence), pp. 79–102. *Elsevier*, Amsterdam.

Ebert, T.A. (2007) Growth and Survival of Postsettlement Sea Urchins. In: *Edible Sea Urchins: Biology and Ecology. 2nd Edn. Developments in Aquaculture and Fisheries Science* (Ed. J.M. Lawrence), pp. 95–134. Elsevier, Amsterdam.

Emlet, R.B. (1985) Crystal axes in recent and fossil adult echinoids indicate trophic mode in larval development. *Science*, **230**, 937–940.

Emlet, R.B. (1989) Apical skeletons of sea urchins (Echinodermata: Echinoidea): two methods for inferring mode of larval develop-ment. *Paleobiology*, **15**, 223–254.

Fisher, D.C. and **Cox, R.S.** (1988) Application of skeletal crystallography to phylogenetic inference in fossil echinoderms. In: *Echinoderm Biology* (Eds R.D. Burke, P.V. Mladenov, P. Lambert and R.L. Parsley), pp. 797. A.A. Balkema, Rotterdam.

Gaffey, S.J., **Kolak, J.J.** and **Bronnimann, C.E.** (1991) Effects of drying, heating, annealing, and roasting on carbonate skeletal material, with geochemical and diagenetical implications. *Geochim. Cosmochim. Acta*, **55**, 1627–1640.

Gilliland, P.M. (1993) The skeletal morphology, systematics and evolutionary history of holothurians. *Spec. Pap. Palaeontol.*, **47**, 1–147.

Gislén, T. (1924) Echinoderm studies. *Zool. Bidr. Uppsala*, **9**, 1–316.

Glass, A. and **Blake, D.B.** (2004) Preservation of tube feet in an ophiuroid (Echinodermata) from the Lower Devonian Hunsrück Slate of Germany and a redescription of *Bundenbachia beneckei* and *Palaeophiomyxa grandis*. *Paläontol. Z.*, **78**, 73–95.

Glynn, P.W. (1997) Bioerosion and coral reef growth: a dynamic balance. In: *Life and Death on Coral Reefs* (Ed. C. Birkeland), pp. 68–95. Chapman & Hall, New York.

Gordon, C.M. and **Donovan, S.K.** (1992) Disarticulated echinoid ossicles in paleoecology and taphonomy: The last interglacial Falmouth Formation of Jamaica. *Palaios*, **7**, 157–166.

Greenstein, B.J. (1993) Is the fossil record of regular echinoids really so poor? A comparison of living and subfossil assemblages. *Palaios*, **8**, 597–601.

Hardie, L.A. (1996) Secular variation in seawater chemistry: An explanation for the coupled secular variation in the mineralogies of marine limestones and potash evaporates over the past 600 my. *Geology*, **24**, 279–283.

Haude, R. (2002) Origin of holothurians (Echinodermata) derived by constructional morphology. *Mitt. Mus. Naturk. Berlin Geowiss. Reihe*, **5**, 141–153.

Hayton, S., **Nelson, C.S.** and **Hood, S.D.** (1995) A skeletal assemblage classification system for non-tropical carbonate deposits based on New Zealand Cenozoic limestones. *Sed. Geol.*, **100**, 123–141.

Heller, C. (1858) Über neue fossile Stelleriden. *Sitzber. Kaiserl. Akad. Wiss. Math.-naturwiss. Cl. Abt. 1.* **28**, 155–170.

Hess, H. (1955) Die fossilen Astropectiniden (Asteroidea). Neue Beobachtungen und Übersicht über die bekannten Arten. *Mém. Soc. Paléontol. Suisse*, **71**, 1–113.

Hess, H. (1962a) Mikropalaeontologische Untersuchungen an Ophiuren – I. Einleitung. *Eclogae Geol Helv.*, **55**, 595–608.

Hess, H. (1962b) Mikropalaeontologische Untersuchungen an Ophiuren. II. Die Ophiuren aus dem Lias (Pliensbachien – Toarcien) von Seewen (Kt. Solothurn). *Eclogae Geol. Helv.*, **55**, 609–656.

Hess, H. (1963) Mikropalaeontologische Untersuchungen an Ophiuren. III. Die Ophiuren aus dem Callovien-Ton von Liesberg (Berner Jura). *Eclogae Geol. Helv.*, **56**, 1141–1164.

Hess, H. (1965) Mikropalaeontologische Untersuchungen an Ophiuren. IV. Die Ophiuren aus dem Renggeri-Ton (Unter-Oxford) von Chapois (Jura) und Longecombe (Ain). *Eclogae Geol. Helv.*, **58**, 1059–1082.

Hess, H. (1966) Mikropalaeontologische Untersuchungen an Ophiuren. V. Die Ophiuren aus dem Argovien (unteres Ober-Oxford) vom Guldenthal (Kt. Solothurn) und von Savigna (Dépt. Jura). *Eclogae Geol. Helv.*, **59**, 1025–1063.

Hess, H. (1975a) Mikropaläontologische Untersuchungen an Ophiuren. VI. Die Ophiuren aus den Günsberg-Schichten (oberes Oxford) von Guldenthal (Kt. Solothurn). *Eclogae Geol. Helv.*, **68**, 591–601.

Hess, H. (1975b) Mikropaläontologische Untersuchungen an Ophiuren. VII. Die Ophiuren aus den Humeralis-Schichten (Ober-Oxford) von Raedersdorf (Ht-Rhin). *Eclogae Geol. Helv.*, **68**, 603–612.

Hess, H. (1999a) Lower Devonian Hunsrück Schales of Germany. In: *Fossil Crinoids* (Eds H. Hess, W.I. Ausich, C.E. Brett and M.J. Simms), pp. 111–121. Cambridge University Press, Cambridge.

Hess, H. (1999b) *Unitacrinus* beds of the Upper Cretaceous Niobrara Formation, Kansas, USA. In: *Fossil Crinoids* (Eds H. Hess, W.I. Ausich, C.E. Brett and M.J. Simms), pp. 225–232. Cambridge University Press, Cambridge.

Hess, H. (1999c) Tertiary. In: *Fossil Crinoids* (Eds H. Hess, W.I. Ausich, C.E. Brett and M.J. Simms), pp. 233–236. Cambridge University Press, Cambridge.

Hess, H., **Ausich, W.I.**, **Brett, C.E.** and **Simms, M.J.** (Eds) (1999) *Fossil Crinoids*. Cambridge University Press, Cambridge, 275 pp.

Hoefs, J. (1997) *Stable Isotope Geochemistry. 4th Edn.* Springer-Verlag, Berlin-Heidelberg, 201 pp.

Holland, H.D. (1978) *The Chemistry of the Atmosphere and Oceans*. John Wiley, New York, 351 pp.

Horita, J., **Zimmermann, H.** and **Holland, H.D.** (2002) Chemical evolution of seawater during the Phanerozoic: Implications from the record of marine evaporates. *Geochim. Cosmochim. Acta*, **66**, 3733–3756.

Jagt, J.W.M. (1991) Early Miocene luidiid asteroids (Echinodermata, Asteroidea) from Winterswijk-Miste (The Netherlands). *Contrib. Tert. Quatern. Geol.*, **28**, 35–43.

Jagt, J.W.M. (2000a) Late Cretaceous-Early Palaeogene echinoderms and the K/T boundary in the southeast Netherlands and northeast Belgium – Part 3: Ophiuroids. With a chapter on: Early Maastrichtian ophiuroids from Rügen (northeast Germany) and Møn (Denmark) by M. Kutscher and J.W.M. Jagt. *Scripta Geol.*, **121**, 1–179.

Jagt, J.W.M. (2000b) Late Cretaceous-Early Palaeogene echinoderms and the K/T boundary in the southeast Netherlands and northeast Belgium – Part 5: Asteroids. *Scripta Geol.*, **121**, 377–503.

James, N.P. (1997) The cool-water carbonate depositional realm. In: *Cool-water Carbonates* (Eds N.P. James and J. A.D. Clarke). *SEPM Spec. Publ.*, **56**, 1–22.

James, N.P. and **Bone, Y.** (1989) Petrogenesis of Cenozoic, temperate water calcarenites, South Australia: a model for meteoric/shallow burial diagenesis of shallow water calcite sediments. *J. Sed. Petrol.*, **59**, 191–203.

Kaczmarska, G. (1987) Asteroids from the Korytnica Basin (Middle Miocene; Holy Cross Mountains, Central Poland). *Acta Geol. Polonica*, **37**, 131–144.

Kampfer, S. and **Tertschnig, W.** (1992) Feeding biology of *Clypeaster rosaceus* (Echinoidea, Clypeasteroidea) and its impact on shallow logoon sediments. *Echinoderm Research 1991 – Proc. 3rd Eur. Echino. Conf., Lecce, Italy, 1991* (Eds L. Scalera-Liaci and C. Canicatti), pp. 197–200. A.A. Balkema, Rotterdam.

Kidwell, S.M. and **Baumiller, T.** (1990) Experimental disintegration of regular echinoids: roles of temperature, oxygen and decay thresholds. *Paleobiol.*, **16**, 247–271.

Kier, P.M. (1972) Upper Miocene echinoids from the Yorktown Formation of Virginia and their environmental significance. *Smithson. Contrib. Paleobiology*, **13**, 1–41.

Kier, P.M. (1977) The poor fossil record of the regular echinoids. *Paleobiology*, **3**, 168–174.

Knoerich, A.C. and **Mutti, M.** (2003) Controls of facies and sediment composition on the diagenetic pathway of shallow-water Heterozoan carbonates: the Oligocene of the Maltese Islands. *Int. J. Earth Sci.*, **92**, 494–510.

Knoerich, A.C. and **Mutti, M.** (2006) Missing aragonitic biota and the diagenetic evolution of heterozoan carbonates: a case study from the Oligo-Miocene of the Central Mediterranean. *J. Sed. Res.*, **76**, 871–888.

Kroh, A. (2003a) First record of gorgonocephalid ophiuroids (Echinodermata) from the Middle Miocene of the Central Paratethys. *Cainoz. Res.*, **2**, 143–155.

Kroh, A. (2003b) The Echinodermata of the Langhian (Lower Badenian) of the Molasse Zone and the northern Vienna Basin (Austria). *Ann. Naturhist. Mus. Wien*, **104A**, 155–183.

Kroh, A. (2004) First fossil record of the family Euryalidae (Echinodermata: Ophiuroidea) from the Middle Miocene of the Central Mediterranean. In: *Echinoderms: München* (Eds T. Heinzeller and J.H. Nebelsick), pp. 447–452. Taylor & Francis, London.

Kroh, A. (2005) *Catalogus Fossilium Austriae,* **2**. *Echinoidea neogenica*. Österreichische Akademie der Wissenschaften, Wien, 210 pp.

Kroh, A. and **Harzhauser, M.** (1999) An echinoderm fauna from the Lower Miocene of Austria: Paleoecology and implications for Central Paratethys paleobiogeography. *Ann. Naturhist. Mus. Wien*, **101A**, 145–191.

Kroh, A. and **Jagt, J.W.M.** (2006) Notes on North Sea Basin Cainozoic echinoderms, Part 3. Pliocene gorgonocephalid ophiuroids from borehole IJsselmuiden-1 (Overijssel, the Netherlands). *Cainoz. Res.*, **4**, 67–70.

Kroh, A. and **Nebelsick, J.H.** (2003) Echinoid assemblages as a tool for palaeoenvironmental reconstruction – an example from the Early Miocene of Egypt. *Palaeogeogr. Palaeoclimat. Palaeoecol.*, **201**, 157–177.

Kroh, A., **Harzhauser, M.**, **Piller, W.E.** and **Rögl, F.** (2003) The Lower Badenian (Middle Miocene) Hartl Formation (Eisenstadt - Sopron Basin, Austria). In: *Stratigraphia Austriaca* (Ed. W.E. Piller). *Schriftenr. Erdwiss. Komm.*, **16**, 87–109.

Kühn, O. (1936) Eine neue Burdigalausbildung bei Horn. *Sitzber. Akad. Wiss. Math.-naturwiss. Kl. Abt. 1*, **145**, 35–45.

Kutscher, M. and **Jagt, J.W.M.** (2000) Early Maastrichtian ophiuroids from Rügen (northeast Germany) and Møn (Denmark). In: *Late Cretaceous-Early Palaeogene Echinoderms and the K/T Boundary in the Southeast Netherlands and Northeast Belgium – Part 3: Ophiuroids* (Ed. J.W.M. Jagt). *Scripta Geol.*, **121**, 45–107.

Laube, G.C. (1869) Die Echinoiden der österreichisch-ungarischen oberen Tertiärablagerungen. *Verh. k.-k. Geol. R.-A.*, **3**, 182–184.

Lawrence, J.M. (1987) *A Functional Biology of Echinoderms. Croom Helm*, London & Sydney, 340 pp.

Lea, D.W. (2003) Elemental and isotopic proxies of marine temperatures. In: *The Oceans and Marine Geochemistry* (Ed. H. Elderfield), *Treat. Geochem.*, **6**, 365–390.

Lehmann, W.M. (1957) Die Asterozoen in den Dachschiefern des rheinischen Unterdevons. *Abh. Hessisch. Landes. Bodenforsch.*, **21**, 1–160.

Lewis, R. (1980) Taphonomy. In: *Echinoderms, Notes for a Short Course* (Eds T.W. Broadhead and J.A. Waters). *Univ. Tenn. Stud. Geol.*, **3**, 27–39.

Linsley, B.K., **Wellington, G.M.** and **Schrag, D.P.** (2000) Decadal sea surface temperature variability in the subtropical South Pacific from 1726 to 1997 A.D. *Science*, **290**, 1145–1148.

Löffler, S.-B. and **Nebelsick, J.H.** (2001) Palaeoecological Aspects of the Lower Oligocene "Zementmergel" Formation based on Molluscs and Carbonates. In: *The Paleogene of the Eastern Alps* (Eds W.E. Piller and M.W. Rasser), pp. 641–670. Verlag der Österreichischen Akademie der Wissenschaften, Vienna.

Lowenstein, T.K., **Timofeeff, M.N.**, **Brennan, S.T.**, **Hardie, L.A.** and **Demicco, R.V.** (2001) Oscillations in Phanerozoic seawater chemistry: evidence from fluid inclusions. *Science*, **294**, 1086–1088.

Lukasik, J.L., **James, N.P.**, **Mcgowran, B.** and **Bone, Y.** (2000) An epeiric ramp: low-energy, cool-water carbonate facies in a Tertiary inland sea, Murray Basin, South Australia. *Sedimentology*, **47**, 851–881.

McClanahan, T.R. and **Muthiga, N.A.** (2007) Ecology of *Echinometra*. In: *Edible Sea Urchins: Biology and Ecology. 2nd Edn. Developments in Aquaculture and Fisheries Science* (Ed. J.M. Lawrence), pp. 297–317. Elsevier, Amsterdam.

McGregor, H.V. and **Gagan, M.K.** (2003) Diagenesis and geochemistry of *Porites* corals from Papua New Guinea: implications for palaeoclimate reconstruction. *Geochim. Cosmochim. Acta*, **67**, 2147–2156.

McNamara, K.J. (1994) Diversity of Cenozoic marsupiate echinoids as an environmental indicator. *Lethaia*, **27**, 257–268.

Macqueen, R.W. and **Ghent, E.D.** (1970) Electron microprobe study of magnesium distribution in some Mississippian echinoderm limestones from Western Canada. *Can. J. Earth Sci.*, **7**, 1308–1316.

Mandic, O. and **Piller, W.E.** (2001) Pectinid coquinas and their palaeoenvironmental implications – examples from the early Miocene of northeastern Egypt. *Palaeogeogr. Palaeoclimatol. Palaeoecol.*, **172**, 171–191.

Manze, U. and **Richter, D.K.** (1979) Die Veränderung des C^{13}/C^{12} - Verhältnisses in Seeigelcoronen bei der Umwandlung von Mg-Calcit in Calcit unter meteorisch-vadosen Bedingungen. *Neues Jb. Geol. Paläontol. Abh.*, **158**, 334–345.

Marcopoulou-Diacantoni, A. (1979) Biofacies au moyen des echinides du Miocene superieur dans l'Ile de Crete (Grece) (Recherche biostratigraphique et paleontologique). *Ann. Géol. Pays Hellén, Tome hors Sér.*, **2**, 745–854.

Märkel, K., **Kubanek, F.** and **Willgallis, A.** (1971) Polykristalliner Calcit bei Seeigeln (Echinodermata, Echinoidea). *Z. Zellforsch. Mikroskop. Anat.*, **119**, 355–377.

Martín, J.M., **Braga, J.C.**, **Betzler, C.** and **Brachert, T.C.** (1996) Sedimentary model and high-frequency cyclicity in a Mediterranean, shallow-shelf, temperate-carbonate environment (uppermost Miocene, Agua Amarga Basin, Southern Spain). *Sedimentology*, **43**, 263–277.

Messing, C.G. (1997) Living Comatulids. In: *Geobiology of Echinoderms* (Eds J.A. Waters and C.G. Maples). *Paleont. Soc. Pap.*, **3**, 3–30.

Meyer, D.L. and **Macurda, D.B. Jr.** (1997) Adaptive radiation of the comatulid crinoids. *Paleobiology*, **3**, 74–82.

Meyer, D.L. and **Meyer, K.B.** (1986) Biostratinomy of recent crinoids (Echinodermata) at Lizard Island, Great Barrier Reef, Australia. *Palaios*, **1**, 294–302.

Meyer, D.L. and **Milsom, C.V.** (2001) Microbial sealing in the biostratinomy of Uintacrinus Lagerstätten in the upper Cretaceous of Kansas and Colorado, USA. *Palaios*, **16**, 535–546.

Mokady, O., **Lazar, O.** and **Loya, Y.** (1996) Echinoid Bioerosion as a Major Structuring Force of Red Sea Coral Reefs. *Biol. Bull.*, **190**, 367–372.

Moore, R.C. and **Vokes, H.E.** (1953) Lower Tertiary crinoids from northwestern Oregon. *US Geol. Surv. Prof. Pap.*, **233-E**, 113–148.

Mortensen, T. (1928) *A Monograph of the Echinoidea. I. Cidaroidea*. C.A. Reitzel & Oxford Univ. Press, Copenhagen & London, 551 pp.

Mortensen, T. (1935) *A Monograph of the Echinoidea. II. Bothriocidaroida, Melonechinoida, Lepidocentroida, and Stirodonta*. C.A. Reitzel & Oxford Univ. Press, Copenhagen & London, 647 pp.

Mortensen, T. (1940) *A Monograph of the Echinoidea. III, 1. Aulodonta, with Additions to Vol. II (Lepidocentroida and Stirodonta)*. C.A. Reitzel, Copenhagen, 370 pp.

Mortensen, T. (1943a) *A Monograph of the Echinoidea. III, 2. Camarodonta. I. Orthopsidæ, Glyphocyphidæ, Temnopleuridæ and Toxopneustidæ*. C.A. Reitzel, Copenhagen, 553 pp.

Mortensen, T. (1943b) *A Monograph of the Echinoidea. III, 3. Camarodonta. II. Echinidæ, Strongylocentrotidæ, Parasaleniidæ, Echinometridæ*. C.A. Reitzel, Copenhagen, 446 pp.

Mortensen, T. (1948a) *A Monograph of the Echinoidea. IV, **1**, Holectypoida, Cassiduloida*. C.A. Reitzel, Copenhagen, 371 pp.

Mortensen, T. (1948b) *A Monograph of the Echinoidea. IV, **2**. Clypeasteroida. Clypeasteridæ, Arachnoidæ, Fibulariidæ, Laganidæ and Scutellidæ*. C.A. Reitzel, Copenhagen, 471 pp.

Mortensen, T. (1950) *A Monograph of the Echinoidea, **1**. Spatangoida I. Protosternata, Meridosternata, Amphisternata I. Palæopneustidæ, Palæostomatidæ, Aëropsidæ, Toxasteridæ, Micrasteridæ, Hemiasteridæ*. C.A. Reitzel, Copenhagen, 432 pp.

Mortensen, T. (1951) *A Monograph of the Echinoidea, **2**. Spatangoida II. Amphisternata II. Spatangidæ, Loveniidæ, Pericosmidæ, Schizasteridæ, Brissidæ*. C.A. Reitzel, Copenhagen, 593 pp.

Müller, A.H. (1953) Die isolierten Skelettelemente der Asteroidea (Asterozoa) aus der obersenonen Schreibkreide von Rügen. *Beih. Z. Geol.*, **8**, 1–66.

Muthiga, N.A. and **McClanahan, T.R.** (2007) Ecology of *Diadema*. In: *Edible Sea Urchins: Biology and Ecology. 2nd Edn. Developments in Aquaculture and Fisheries Science* (Ed. J.M. Lawrence), pp. 205–225. Elsevier, Amsterdam.

Nebelsick, J.H. (1989) Temperate water carbonate facies of the Early Miocene Paratethys (Zogelsdorf Formation, Lower Austria). *Facies*, **21**, 11–40.

Nebelsick, J.H. (1992a) Components analysis of component distribution in Lower Miocene temperate carbonates of the Zogelsdorf Formation, Lower Austria. *Palaeogeogr. Palaeoclimatol. Palaeoecol.*, **91**, 59–69.

Nebelsick, J.H. (1992b) Echinoid distribution by fragment identification in the Northern Bay of Safaga, Red Sea, Egypt. *Palaios*, **7**, 316–328.

Nebelsick, J.H. (1992c) The northern Bay of Safaga (Red Sea, Egypt): an actuopalaeontological approach III. Distribution of echinoids. *Beitr. Paläontol. Österr.*, **17**, 5–79.

Nebelsick, J.H. (1996) Biodiversity of shallow-water Red Sea Echinoids: implications for the fossil record. *J. Mar. Biol. Assoc. UK*, **76**, 185–194.

Nebelsick, J.H. (2004) Taphonomy of Echinoderms: introduction and outlook. In: *Echinoderms: München* (Eds T. Heinzeller and J.H. Nebelsick), pp. 471–477. Taylor & Francis, London.

Nebelsick, J.H. and **Kampfer, S.**(1994) Taphonomy of *Clypeaster humilis* and *Echinodiscus auritus* from the Red Sea. In: *Echinoderms Through Time* (Eds B.A. David, J.-P. Guille and M. Roux), pp. 803–808. Rotterdam, Balkema.

Nebelsick, J.H. and **Kroh, A.** (2002) The stormy path from life to death assemblages: The formation and preservation of mass accumulations of fossil sand dollars. *Palaios*, **17**, 378–394.

Nebelsick, J.H., **Schmid, B.** and **Stachowitsch, M.** (1997) The encrustation of fossil and recent sea-urchin tests: ecological and taphonomic significance. *Lethaia*, **30**, 271–284.

Nebelsick, J.H., **Drobne, K.** and **Bassi, D.** (2000) Microfacies analysis and Palaeoenvironmental Interpretation of Lower Oligocene, shallow water carbonates (Gornji Grad Beds, Slovenia). *Facies*, **43**, 157–176.

Nebelsick, J.H., **Stingl, V.** and **Rasser, M.** (2001) Facies and component distribution in and autochthonous and allochthonous Lower Oligocene Carbonates. *Facies*, **44**, 31–46.

Nelson, C.S. (1978) Temperate shelf carbonates in the Cenozoic of New Zealand. *Sedimentology*, **29**, 737–772.

Nelson, C.S. (1988) An introductory perspective on non-tropical shelf carbonates. *Sed. Geol.*, **60**, 3–14.

Nelson, C.S., **Keane, S.L.** and **Head, P.S.** (1988) Non-tropical deposits on the New Zealand Shelf. *Sed. Geol.*, **60**, 71–94.

Néraudeau, D., **Goubert, E.**, **Lacour, D.** and **Rouchy, J.M.** (2001) Changing biodiversity of Mediterranean irregular echinoids from the Messinian to Present-Day. *Palaeogeogr. Palaeoclimatol. Palaeoecol.*, **175**, 43–60.

Neugebauer, J. (1978) Micritization of crinoids by diagenetic dissolution. *Sedimentology*, **25**, 267–283.

Neugebauer, J. (1979b) Drei Probleme der Echinodermendiagenese: Innere Zementation, Mikroporenbildung und der Übergang von Magnesiumcalcit zu Calcit. *Geol. Rundsch.*, **68**, 856–875.

Neugebauer, J. (1979b) Echinodermen-Diagenese. *Neues Jb. Geol. Paläontol. Abh.*, **157**, 193–195.

Neugebauer, J. (1979c) Fossil-Diagenese in der Schreibkreide: Echinodermen. *Clausthaler Geol. Abh.*, **30**, 198–229.
Nicolaides, S. (1995) Cementation in Oligo-Miocene non-tropical shelf limestones. Otway Basin, Australia. *Sed. Geol.*, **95**, 97–121.
Nicolaides, S. and **Wallace, M.W.** (1997) Pressure-dissolution and cementation in an Oligo-Miocene non-tropical limestone (Clifton Formation) Otway Basin, Australia. In: *Cool-water Carbonates* (Eds N.P. James and J.A.D. Clarke). *SEPM Spec. Publ.*, **56**, 249–261.
Nissen, H.-U. (1969) Crystal orientation and plate structure in echinoid skeletal units. *Science*, **166**, 1150–1152.
Nosowska, E. (1997) Asteroids from the Nawodzice Sands (Middle Miocene); Holy Cross Mountains, Central Poland). *Acta Geol. Pol.*, **47**, 225–241.
Oji, T. (1989) Growth rate of stalk of *Metacrinus rotundus* (Echinodermata: Crinoidea) and its functional significance. *J. Fac. Sci. Univ. Tokyo, Sect. II*, **22**, 39–51.
Oji, T. (1990) Miocene Isocrinidae (stalked crinoids) from Japan and their biogeographic implication. *Trans. Proc. Palaeontol. Soc. Japan New Ser.*, **157**, 412–429.
Otter, G.W. (1932) Rock-burrowing echinoids. *Biol. Rev. Cambr. Phil. Soc.*, **7**, 89–107.
Otter, G.W. (1937) Rock-destroying organisms in relationship to coral reefs. *Sci. Rep. Gt. Barrier Reef Exped., 1928–1929; Brit. Mus. Nat. Hist.*, **1**, 232–352.
Pearse, J.S. and **Pearse, V.B.** (1975) Growth zones in the echinoid skeleton. *Am. Zool.*, **15**, 731–753.
Pilkey, O.H. and **Hower, J.** (1960) The effect of environment on the concentration of skeletal Magnesium and Strontium in *Dendraster*. *J. Geol.*, **68**, 203–216.
Piller, W.E. (1994) The Northern Bay of Safaga (Red Sea, Egypt): an actuopalaeontological approach. IV. Thin section analysis. *Beitr. Paläontol.*, **18**, 1–73.
Piller, W.E. and **Rasser, M.** (1996) Rhodolith formation induced by reef erosion in the Red Sea, Egypt. *Coral Reefs*, **15**, 191–198.
Poddubiuk, R.H. and **Rose, E.P.F.** (1984) Relationships between Mid-Tertiary echinoid faunas from the Central Mediterranean and eastern Carribean and their palaeobiogeographic significance. *Ann. Géol. Pays Hellén.*, **31**, 115–127.
Pomoni-Papaioannou, F., **Drinia, H.** and **Dermitzakis, M.D.** (2002) Neogene non-tropical carbonate sedimentation in a warm temperate biogeographic province (Rethymnon Formation, Eastern Crete, Greece). *Sed. Geol.*, **154**, 147–157.
Rasmussen, H.W. (1972) Lower Tertiary Crinoidea, Asteroidea and Ophiuroidea from northern Europe and Greenland. *Biol. Skr. Kong. Dan. Vidensk. Selskab.*, **19**, 1–83.
Raup, D.M. (1959) Crystallography of echinoid calcite. *J. Geol.*, **67**, 661–674.
Raup, D.M. (1960) Ontogenetic variation in the crystallography of echinoid calcite. *J. Paleontol.*, **34**, 1041–1050.
Raup, D.M. (1962a) Crystallographic data in echinoderm classification. *Syst. Zool.*, **11**, 97–108.
Raup, D.M. (1962b) The phylogeny of calcite crystallography in echinoids. *J. Paleontol.*, **36**, 793–810.
Raup, D.M. (1965) Crystal orientations in the echinoid apical disc. *J. Paleontol.*, **39**, 934–951.
Raup, D.M. (1966a) Crystal orientations in the echinoid coronal plates. *J. Paleontol.*, **40**, 555–568.
Raup, D.M. (1966b) The endoskeleton. In: *Physiology of Echinodermata* (Ed. R.A. Boolootian), pp. 379–395. Interscience Publishers, New York.
Raup, D.M. and **Swan, E.F.** (1967) Crystal orientation in the apical plates of aberrant echinoids. *Biol. Bull.*, **133**, 618–629.
Reich, M. (2001) Holothurians from the Late Cretaceous of the Isle of Rügen (Baltic Sea). In: *Echinoderms 2000* (Ed. M. Barker), pp. 89–92. A.A. Balkema Publishers Lisse/Abingdon/Exton (PA)/Tokyo.
Reich, M. (2003) Holothurien (Echinodermata) aus der Oberkreide des Ostseeraumes: Teil 3. Chiridotidae Östergren, 1898. *Neues Jb. Geol. Paläontol., Abh.*, **228**, 363–397.
Reich, M. and **Frenzel, P.** (2002) Die Fauna und Flora der Rügener Schreibkreide (Maastrichtium, Ostsee). *Arch. Geschiebek.*, **3**, 73–284.
Reuss, A.E. (1860) Die marinen Tertiärschichten Böhmens und ihre Verseinerungen. *Sitzber. Kaiserl. Akad. Wiss. Math.-naturwiss. Cl. Abt. 1*, **39**, 207–288.
Richter, D.K. (1974) Zur subaerischen Diagenese von Echinidenskeletten und das relative Alter pleistozäner Karbonatterrassen bei Korinth (Griechenland). *Neues Jb. Geol. Paläontol. Abh.*, **146**, 51–77.
Richter, D.K. and **Bruckschen, P.** (1998) Geochemistry of recent tests of *Echinocyamus pusillus*: Constraints for temperature and salinity. *Carbonates Evaporites*, **13**, 157–167.
Ries, J.B. (2004) Effect of ambient Mg/Ca ratio on Mg fractionation in calcareous marine invertebrates: A record of the oceanic Mg/Ca ratio over the Phanerozoic. *Geology*, **32**, 981–984.
Rose, E.P.F. (1974) Stratigraphical and facies distribution of irregular echinoids in Miocene limestones of Gozo, Malta, and Cyrenaica. Libya. Bur. Rech. Géol. Min. Mém., **78**, 349–355.
Rose, E.P.F. and **Watson, A.C.** (1998) Burrowing adaptions of schizasteroid echinoids from the Globigerina Limestone (Miocene) of Malta and their evolutionary significance. In: *Echinoderms: San Francisco* (Eds R. Mooi and M. Telford), pp. 811–817. A.A. Balkema, Rotterdam.
Rosenkranz, D. (1971) Zur Sedimentation und Ökologie von Echinodermen-Lagerstätten. *Neues Jb. Geol. Paläontol. Abh.*, **138**, 221–258.
Roux, M., **Renard, M.**, **Améziane, N.** and **Emmanuel, L.** (1995) Zoobathymetrie et composition chimique de la calcite des ossicules du pedoncule des crinoides. *CR Acad. Sci. Paris Sér. IIa*, **321**, 675–680.
Russell, M.P. and **Meredith, R.W.** (2000) Natural growth lines in echinoid ossicles are not reliable indicators of age: a test using *Strongylocentrotus droebachiensis*. *Inv. Biol.*, **119**, 410–420.
Russell, M.P., **Ebert, T.A.** and **Petraitis, P.S.** (1998) Field estimates of growth and mortality of the green sea urchin *Strongylocentrotus droebachiensis*. *Ophelia*, **48**, 137–153.
Russo, A.R. (1980) Bioerosion by two rock boring echinoids (*Echinometra mathaei* and *Echinostrephus aciculatus*) en Enewetak Atoll, Marshall Islands. *J. Mar. Res.*, **38**, 99–110.
Sandberg, P.A. (1983) An oscillating trend in Phanerozoic nonskeletal carbonate mineralogy. *Nature*, **305**, 19–22.
Scheibling, R.E. (1982) Habitat utilization and bioturbation by *Oreaster reticulatus* (Asteroidea) and *Meoma*

ventricosa (Echinoidea) in a subtidal sand patch. *Bull. Mar. Sci.*, **32**, 624–629.

Schroeder, J.H., **Papike, J.J.**, **Dwornik, E.J.** and **Lindsay, J.R.** (1968) Compositional and crystallographic aspects of the echinoid skeleton. *Trans. Am. Geophys. Union*, **49**, 219.

Schroeder, J.H., **Dwornik, E.J.** and **Papike, J.J.** (1969) Primary protodolomite in echinoid skeletons. *Geol. Soc. Am. Bull.*, **80**, 1613–1618.

Seilacher, A. (1979) Constructional morphology of sand dollars. *Paleobiology*, **5**, 191–221.

Seilacher, A., **Reif, W.E.** and **Westphal, F.** (1985) Sedimentological, ecological and temporal patterns of fossil Lagerstätten. In: *Extraordinary Fossil Biotas: Their Ecological and Evolutionary Significance* (Eds H.B. Whittington and M.S. Conway Morris). *Phil. Trans. Roy. Soc. London Ser. B: Biol. Sci.*, **311**, 5–24.

Sismonda, E. (1841) Monografia degli Echinidi Fossili del Piemonte. *Mem. R. Accad. Sci. Torino Ser., 2*, **4**, 1–54.

Smith, A.B. (1980) Stereom microstructure of the Echinoid test. *Spec. Pap. Palaeontol.*, **25**, 1–81.

Smith, A.B. (1984) *Echinoid Palaeobiology*. George Allen and Unwin, London, 191 pp.

Smith, A.B. (1990) Biomineralization in echinoderms. In: *Skeletal Biomineralization: Patterns, Processes and Evolutionary Trends, I.* (Ed. J.G. Carter), pp. 170–175. Nostrand Reinhold, New York.

Smith, A.B. and **Gallemí, J.** (1991) Middle Triassic Holothurians from Northern Spain. *Palaeontology*, **34**, 49–76.

Steneck, R.S. (1983) Escalating herbivory and resulting adaptive trends in calcareous algal crusts. *Paleobiology*, **9**, 44–61.

Stoll, H.M. and **Schrag, D.P.** (1998) Effects of quaternary sea level cycles on strontium in seawater. *Geochim. Cosmochim. Acta*, **62**, 1107–1118.

Stoll, H.M., **Schrag, D.P.** and **Clemens, S.C.** (1999) Are seawater Sr/Ca variations preserved in quaternary foraminifera. *Geochim. Cosmochim. Acta*, **63**, 3535–3547.

Talbot, T.D. and **Lawrence, J.M.** (2002) The effect of salinity on respiration, excretion, regeneration and production in *Ophiophragmus filograneus*. *J. Exp. Mar. Biol. Ecol.*, **275**, 1–14.

Thayer, C.W. (1983) Sediment-mediated biological disturbance and the evolution of marine benthos. In: *Biotic Interactions with Recent and Fossil Benthic Communities* (Eds M.J.S. Tevesz and P.L. McCall), pp. 478–625. *Plenum*, New York.

Turekian, K.K. (1955) Paleoecological significance of the strontium-calcium ratio in fossils and sediments. *Geol. Soc. Am. Bull.*, **66**, 155–158.

Vadász, E. (1915) Die mediterranen Echinodermen Ungarns. *Geol. Hung.*, **1**, 79–253.

Villier, L. (1999) Reconstitution du squelette d'astérides fossiles à partir d'ossicules isolés: intérêt taxinomique et phylogénétique. *CR Acad. Sci. Paris*, **328**, 353–358.

Weber, J.N. (1968) Fractionation of stable isotopes of carbon and oxygen in calcareous marine invertebrates – the Asteroidea, Ophiuroidea, and Crinoidea. *Geochim. Cosmochim. Acta*, **32**, 33–70.

Weber, J.N. (1969a) Origin of concentric banding in the spines of the tropical echinoid Heterocentrotus. *Pacific Sci.*, **23**, 452–466.

Weber, J.N. (1969b) The incorporation of magnesium into the skeletal calcite of echinoderms. *Am. J. Sci.*, **267**, 537–566.

Weber, J.N. (1973) Temperature dependence of magnesium in echinoid and asteroid skeletal calcite: a reinterpretation of its significance. *J. Geol.*, **81**, 543–556.

Weber, J.N. and **Raup, D.M.** (1966a) Fractionation of stable isotopes of carbon and oxygen in marine calcareous organisms – the Echinoidea. Part I. Variation of C^{13} and O^{18} content within individuals. *Geochim. Cosmochim. Acta*, **30**, 681–703.

Weber, J.N. and **Raup, D.M.** (1966b) Fractionation of stable isotopes of carbon and oxygen in marine calcareous organisms – the Echinoidea. Part II. Environmental and genetic factors. *Geochim. Cosmochim. Acta*, **30**, 705–736.

Weber, J.N. and **Raup, D.M.** (1968) Comparison of C^{13}/C^{12} and O^{18}/O^{16} in skeletal calcite of recent and fossil echinoids. *J. Paleontol.*, **42**, 37–50.

Weber, J., **Greer, R.**, **Voight, B.**, **White, E.** and **Roy, R.** (1969) Unusual strength properties of echinoderm calcite related to structure. *J. Ultrastruct. Res.*, **26**, 355–366.

Wefer, G. and **Berger, W.H.** (1991) Isotope paleontology: growth and composition of extant calcareous species. *Mar. Geol.*, **100**, 207–248.

West, C.D. (1937) Note on the crystallography of the echinoderm skeleton. *J. Paleont.*, **11**, 458–459.

Wilkinson, P.A. and **Alego, T.J.** (1989) Sedimentary carbonate record of calcium-magnesium cycling. *Am. J. Sci.*, **289**, 1158–1194.

Wilson, J.L. (1975) *Carbonate Facies in Geologic History*. Springer-Verlag, New York. **471** pp.

ADDENDUM

A major focus of research since the submission of this manuscript has been marine acidification and the ability of marine organisms to cope with dramatic changes in ocean chemistry (e.g. Pörtner *et al.*, 2004; Orr *et al.*, 2005; Fabry, 2008; Fabry *et al.*, 2008; Doney *et al.*, 2009). Published studies have dealt primarily with anthropogenically induced increases in CO_2 and their effect on recent organisms, but this work has profound implications for interpreting the fossil record during past phases of long-term and short-term change in ocean chemistry.

Echinoderms have attracted particular attention during discussions of ocean acidification; they may be especially sensitive to changes in ocean chemistry because: (1) their skeletons are constructed of high-Mg calcite; and (2) various echinoderm larval forms posses extremely delicate skeletal support elements, which may be adversely affected by acidification. Echinoderms are also common in various marine environments with differing temperature and ocean chemistry regimes, where acidification may have different consequences. Studies have been conducted on the effects of ocean acidification on echinoderms in general (e.g. Gooding *et al.*, 2009; Dupont *et al.*, 2010), as well as specifically on aspects of their metabolism (Miles *et al.*, 2007), fertilization and development (Dupont *et al.*, 2008; Havenhand *et al.*, 2008; Kurihara, 2008; Wood *et al.*, 2008; Byrne *et al.*, 2009; Clark *et al.*, 2009; O'Donnell *et al.*, 2009, 2010), and sediment/animal relationships (Dashfield *et al.*, 2008).

There has been little corresponding research published on the implications of ocean acidification for the fossil record of echinoderms. New work has emphasized the role of echinoderm skeletons as carbon sinks, since large amounts of echinoderm remains are committed to sediments (Lebrato *et al.*, 2009). The fact that echinoderms play a significant role in the CO_2 budget is amply demonstrated by the rich fossil record of echinoderms, and the significant contribution they make to ancient sediments (see main text). Longer term, secular changes in ocean chemistry occurred through the Phanerozoic, but short term, drastic changes have also been put forward as contributors to mass extinction events (see e.g. Knoll *et al.*, 2007; Veron, 2008). Such events have significantly affected the evolutionary development of echinoderms in the past (e.g. Twitchett, 2007). To what extent ocean acidification has contributed to these mass extinction events and how echinoderms are affected remain to be studied in detail.

Byrne, M., **Ho, M.**, **Selvakumaraswamy, P.**, **Nguyen, H.D.**, **Dworjanyn, S.A.** and **Davis, A.R.** (2009) Temperature, but not pH, compromises sea urchin fertilization and early development under near-future climate change scenarios. *Proc. Roy. Soc.*, **276**, 1883–1888.

Clark, D., **Lamare, M.** and **Barker, M.** (2009) Response of sea urchin larvae (Echinodermata: Echinoidea) to reduced seawater pH: a comparison among a tropical, temperate and a polar species. *Mar. Biol.*, **156**, 1125–1137.

Dashfield, S.L., **Somerfield, P.J.**, **Widdicombe, S.**, **Austen, M.C.** and **Nimmo, M.** (2008) Impacts of ocean acidification and burrowing urchins on within-sediment pH profiles and subtidal nematode communities. *J. Exp. Mar. Biol. Ecol.*, **365**, 46–52.

Doney, S.C., **Fabry, V.J.**, **Feely, R.A.** and **Kleypas, J.A.** (2009) Ocean acidification: the other CO_2 problem. *Ann. Rev. Mar. Sci.*, **1**, 169–192.

Dupont, S., **Havenhand, J.**, **Thorndyke, W.**, **Peck, L.** and **Thorndyke, M.** (2008) Near-future level of CO_2-driven radically affects larval survival and development in the brittlestar *Ophiothrix fragilis*. *Mar. Ecol. Prog. Ser.*, **373**, 285–294.

Dupont, S., **Ortega-Martinez, O.** and **Thorndyke, M.** (2010) Impact of near-future ocean acidification on echinoderms. *Ecotoxycology*, **19**, 449–462.

Fabry, V.J. (2008) Marine calcifiers in a high-CO_2 ocean. *Science*, **320**, 1020–1022.

Fabry, V.J., **Seibel, B.A.**, **Feely, R.A.** and **Orr, J.C.** (2008) Impacts of ocean acidification on marine fauna and ecosystem processes. *ICES J. Mar. Sci.*, **65**, 414–432.

Gooding, R.A., **Harley, C.D.G.** and **Tang, E.** (2009) Elevated water temperature and carbon dioxide concentration increase the growth of a keystone echinoderm. *PNAS*, **106**, 9316–9321.

Havenhand, J.N., **Buttler, F.R.**, **Thorndyke, M.C.** and **Williams J.E.** (2008) Near-future levels of ocean acidification reduce fertilization success in a sea urchin. *Curr. Biol.*, **18**, R651–R652.

Knoll, A.H., **Bambach, R.K.**, **Payne, J.L.**, **Pruss, S.** and **Fischer, W.W.** (2007) Paleophysiology and end-Permian mass extinction. *Earth Plan. Sci. Lett.*, **256**, 295–313.

Kurihara, H. (2008) Effects of CO_2-driven ocean acidification on the early developmental stages of invertebrates. *Mar. Ecol. Prog. Ser.*, **373**, 275–284.

Lebrato, M., **Iglesias-Rodriguez, D.**, **Feely, R.**, **Greeley, D.**, **Jones, D.**, **Suarez-Bosche, N.**, **Lampitt, R.**, **Cartes, J.**, **Green, D.** and **Alker, B.** (2009) Global contribution of echinoderms to the marine carbon cycle a re-assessment of the oceanic $CaCO_3$ budget and the benthic compartments. *Ecological Monographs*. e-View.

Miles, H., **Widdicombe, S.**, **Spicer, J.I.** and **Hall-Spencer, J.** (2007) Effects of anthropogenic seawater acidification on acid–base balance in the sea urchin *Psammechinus miliaris*. *Mar. Poll. Bull.*, **54**, 89–96.

O'Donnell, M.J., **Hammond, L.** and **Hofmann G.E.** (2009) Predicted impact of ocean acidification on a marine invertebrate: elevated CO_2 alters response to thermal stress in sea urchin larvae. *Mar. Biol.*, **156**, 439–446.

O'Donnell, M.J., **Todgham, A.E.**, **Sewell, M.A.**, **Hammond, L.M.**, **Ruggiero, K.**, **Fangue, N.A.**, **Zippay, M.L.** and **Hofmann, G.E.** (2010) Ocean acidification alters skeletogenesis and gene expression in larval sea urchins. *Mar. Ecol. Prog. Ser.*, **398**, 157–171.

Orr, J.C., **Fabry, V.J.**, **Aumont, O.**, **Bopp**, L. and others (2005) Anthropogenic ocean acidification over the twenty-first century and its impact on calcifying organisms. *Nature*, **437**, 681–686.

Pörtner, H.O., **Langenbuch, M.** and **Reipschläger, A.** (2004) Biological impact of elevated ocean CO_2 concentrations: lessons from animal physiology and earth history. *J. Oceanogr.*, **60**, 705–718.

Twitchett, R.J. (2007) The Late Permian mass extinction event and recovery: biological catastrophe in a greenhouse world. In: *Advances in Earth Science - from earthquakes to global warming* (Eds P.R. Sammonds and J.M. T. Thompson). *Roy. Soc. Ser. Adv. Sci.*, 2, 69–90. London: *Imperial College Press*.

Veron, J.E.N. (2008) Mass extinctions and ocean acidification: biological constraints on geological dilemmas. *Coral Reefs*, **27**, 459–472.

Wood, H.L., **Spicer, J.I.** and **Widdicombe, S.** (2008) Ocean acidification may increase calcification rates, but at a cost. *Proc. R. Soc., B* **275**, 1767–1773.

Int. Assoc. Sedimentol. Spec. Publ. (2010) **42**, 229–244

Coral diversity and temperature: a palaeoclimatic perspective for the Oligo-Miocene of the Mediterranean region

FRANCESCA R. BOSELLINI* and CHRISTINE PERRIN†

**Dipartimento di Sciere della Terra, Università di Modena e Reggio Emilia, Largo S. Eufemia 19, 41100 Modena, Italy (E-mail: francesca.bosellini@unimore.it)*

†Laboratoire des Mécanismes et Transferts en Géologie, Université Paul Sabatier, 14 avenue Edouard Belin, 31400 Toulouse & Muséum National d'Histoire Naturelle, Paléobiodiversité et Paléoenvironnements (UMR 5143 - USM 203), Département Histoire de la Terre, 8 rue Buffon, 75005 Paris, France

Present address: UMR7207 du CNRS, Muséum National d'Histoire Naturelle, CP38, 8 rue Buffon 75231 Paris Cedex 05, France (E-mail: cperrin@mnhn.fr)

ABSTRACT

Zooxanthellate-coral (z-coral) generic richness values from 102 Oligocene–Miocene localities from the Mediterranean and adjacent areas are tested in this paper as a proxy for relative palaeotemperatures, essentially using the quantitative relationship between taxonomic richness and prevailing seawater temperature underlined by the so-called "energy hypothesis". Patterns of generic richness and inferred palaeotemperatures are examined for each stage of the Oligocene–Miocene time interval and compared with some global palaeoclimatic curves based on oxygen stable-isotopes. Except for the Mid-Miocene Climatic Optimum, which is not recorded in the generic richness of the Mediterranean z-coral communities, the coral-richness-derived palaeotemperatures correlate well with the general palaeoclimatic trends shown by the isotope curves. Results also show: (1) a gradual increase of temperature from the early to the late Rupelian; and (2) a gradual widening of the temperature range after the Burdigalian. The latter indicates that z-coral communities were able to maintain themselves in the region by progressively adapting to a wider temperature range from the mid-Miocene onwards to the Messinian, as the Mediterranean was migrating northward. Although limitations and biases are underlined and discussed in the paper, the "energy hypothesis" applied to fossil z-coral faunas appears to offer a powerful method when used at global or regional scales to estimate changes in palaeotemperatures. It holds particular promise for time intervals and regions where conventional isotopic data are lacking, such as the Oligocene and Miocene of the Mediterranean.

Keywords Corals, diversity, palaeotemperatures, Oligocene, Miocene, Mediterranean.

INTRODUCTION

Coral reefs and reef corals (or, strictly, zooxanthellate corals or "z-corals") are considered as particularly relevant and suitable for palaeoenvironmental and palaeoclimatic studies because of their well-known ecological and biogeographical sensitivity, and because they are often preserved as almost intact communities. Research in this field is very extensive, and ranges from detailed case studies for particular time slices, to general overviews and models produced at a global scale and based on the analysis and interpretation of reef patterns through time (Kiessling *et al.*, 2002). However, together with global schemes and the universal need of better understanding biotic response to global change in Earth history, attention must also be focused on crucial periods of the geological record. Such studies provide evidence of how reef organisms and carbonate systems reacted to specific environmental changes and help identify times in the past that parallel the present climate-warming trend.

The Oligocene–Miocene time interval corresponds to a period of extensive reef development and is characterized by numerous important changes influencing carbonate-producing biota and the architecture of reefs and carbonate platforms (for a review, see Perrin, 2002). This provides an excellent framework and basis to investigate changes in reef ecosystems and coral assemblages in particular, and to elucidate the complex dynamics of interactions and feedbacks of abiotic and biotic factors in carbonate depositional systems. As a first step within the process to achieve these goals, a detailed database is currently being produced that collates information about spatio-temporal coral distribution from Oligo-Miocene localities of the Mediterranean region. The decision to focus on this region (that includes here the Mediterranean Sea, Paratethys, Middle East and some European Atlantic regions), is based on the following facts:

(1) This time period is characterized by a complex palaeogeographic and climatic history with a stepwise transition towards cooler climates, biotic turnover, and the almost complete disappearance of zooxanthellate corals at the end of the Miocene (Chevalier, 1962);
(2) Well-developed Oligo-Miocene coral reefs crop out in many localities with excellent exposures, and occur in a wide variety of structural and depositional settings; specific references can be found in the compilations of Franseen *et al.* (1996) and Perrin (2002);
(3) Oligo-Miocene coral reef sites have been well explored, providing for a huge body of literature (Esteban, 1996);
(4) The regional scale that is focused on here allows a relatively good control on taxonomy and stratigraphy.

Within this background, the present database is used here to obtain patterns of zooxanthellate coral richness (at the genus level) through time. The scope of this paper is to test these patterns as a proxy for relative palaeotemperatures, essentially using the quantitative relationship between taxonomic richness and prevailing temperature underlined by the so-called "energy hypothesis" (Fraser & Currie, 1996; Rosen, 1999), and also to discuss their palaeobiogeographic significance.

CORAL TAXONOMIC RICHNESS AND TEMPERATURE

Fossil corals and coral reefs have commonly been used as indicators of past tropical warm and shallow-water marine conditions simply by using an uniformitarian approach. Climatic changes, which imply variations in global seawater temperature, have therefore been widely considered to be one of the key factors controlling the development of Phanerozoic reefs, producing changes in size, composition and constructional reef types (Copper, 1994, 2002).

The latitudinal distribution of dominant reef biota (Adams et al., 1990; Kiessling *et al.*, 1999; Kiessling, 2001; Perrin & Kiessling, this volume, pp. 17–34) and patterns of coral taxonomic richness (Veron, 1995; Wilson & Rosen, 1998; Rosen, 1999;Bosellini & Russo, 2000) have been analyzed through database approaches for underlining their link with palaeoclimate. Concerning taxonomic richness, in particular, it is not claimed here that an exhaustive picture of its complex relationship with temperature will be given; rather the focus will be on those aspects that can be readily used in the fossil record to investigate quantitatively the sensitivity of reef corals to climate change.

A relatively good correlation between sea surface temperature (SST) and taxonomic richness has been established for many years (Wells, 1955; Rosen, 1971, 1984; Stehli & Wells, 1971; Veron & Minchin, 1992; Veron, 1995), and clearly shows the general attenuation of taxonomic richness with increasing latitude and decreasing temperature. It is also well known that the global geographical pattern of taxonomic richness of z-corals, and many other warm shallow-water organisms, shows a clear positive correlation with prevailing mean seawater temperatures. This is directly associated with incident solar energy, as shown by the so-called "energy hypothesis" (Fraser & Currie, 1996), which essentially relies on these assumptions and correlations.

This relationship has been used to infer minimum seawater palaeotemperatures from the Miocene coral record of the Mediterranean region (Rosen, 1999) and Palaeogene of Italy (Bosellini & Russo, 2000), thus being an independent approach to the now dominant geochemical technique of stable-isotope palaeothermometry. Adams *et al.* (1990) underlined discrepancies between palaeotemperature values derived from palaeogeographical and palaeoecological patterns and those derived from stable isotopes. However, the study of Rosen (1999) showed that, at the regional level, the only area that closely matches the Oligocene–Miocene cooling trend of relative temperatures obtained from stable isotopes, is the Mediterranean region. This suggests that climate change was a possible first-order control on the observed diversity of patterns (Rosen, 2002).

Before testing the "energy hypothesis" with the Oligo-Miocene reef coral database, it is necessary to emphasize some limitations of this approach. First, as previously mentioned, the response of z-corals to climate changes is complex, with effects occurring at different spatio-temporal scales. This is particularly well shown by the present-day increasing global warming and its effects on the physiology of z-corals (e.g. mortality, coral diseases, bleaching) that may cause local or regional disappearances if they cannot adapt rapidly enough or migrate to colonize possible refugia in deeper waters or higher latitude regions (e.g. Veron, 1995; Hughes *et al.*, 2003; Riegl & Piller, 2003; Hallock, 2005).

Another important aspect is that diversity patterns of zooxanthellate corals are not consistently related to climate because of potential interactions with other environmental factors such as depth and light, turbidity and nutrient supply. In addition, it is noteworthy that global patterns related to coral richness during the Cenozoic are derived from the biogeographical context and history of regional pools of reef corals. For example, Wilson & Rosen (1998) suggested that the low diversity of the Indo-West Pacific Palaeogene corals was mainly due to a strong tectonic control on the distribution of habitats suitable for reef coral fauna. The Western Tethys region was in fact the centre of diversity during the Eocene and Oligocene, with the modern Indo-West Pacific centre emerging only in the Miocene, after the tectonic collision between the cratons of Australia and mainland Asia. Global cooling and biogeographical isolation of the Mediterranean faunas led to their gradual demise during the Miocene (Chevalier, 1962; Rosen, 1999; Perrin, 2002), while the Indo-West Pacific centre became richer during the Neogene, and is the richest at the present time (Veron, 1995; Wilson & Rosen, 1998; Rosen, 2002).

Finally, estimates of biodiversity patterns, mostly reconstructed at a global scale and thus mainly based on literature compilations of coral-bearing deposits, are heterogeneously influenced by uncertainties in taxonomic identifications and in stratigraphic resolution. Therefore it is suggested that there is a strong need for more specific studies focusing on crucial periods of coral history and based on detailed and updated biostratigraphy, systematics and palaeoecology.

TESTING THE "ENERGY HYPOTHESIS"

Methods and dataset

The "energy hypothesis" links first-order global patterns of taxonomic richness of reef corals with prevailing seawater temperatures (Fraser & Currie, 1996; Fig. 1). With this empirical method, it is possible to obtain palaeotemperatures from richness values of a given fauna at a given place, as shown by Rosen (1999), who used richness data from 15 Miocene localities of the Mediterranean.

Richness values from 102 localities ranging in age from the early Rupelian to the early Messinian have been extracted from the database, the geographical distribution of which is considered to be satisfactorily representative of the whole "Mediterranean" region (Table 1). Almost all selected localities refer to relatively recent studies dealing with the systematics and/or palaeoecology of their coral fauna and associated with up-to-date stratigraphy. The classic Miocene localities of the western Mediterranean studied by Chevalier (1962) are also included, for which direct access to the original specimens in the collections deposited in the Muséum National d'Histoire Naturelle (MNHN) Paris was granted.

Data concerning the number of genera already include some systematic revisions that were made directly at the genus level within the database. Richness of z-corals is specifically indicated for each locality. The attribution of fossil genera to the category of zooxanthellate corals (z-genera in Table 1), includes z-corals *sensu* Wilson & Rosen (1998, table 2, p.170: genera that are still living and are known to be zooxanthellate), and those designated as "z-like" (i.e. extinct genera having morphology and skeletal characters typical of living z-corals). Although only z-corals have been used to obtain palaeotemperatures, the total number of genera (which includes non-zooxanthellate corals and also those whose symbiotic status is particularly uncertain) is also given for comparison.

Palaeotemperatures, as indicated in Table 1, have been calculated from the richness values of each locality, using the temperature-richness plot of the "energy hypothesis" (Fig. 1). The method essentially consists of the square root of generic z-coral richness from each locality; a detailed explanation of the method was provided by Rosen (1999). The minimum possible temperature is then read back on the envelope curve of maximum

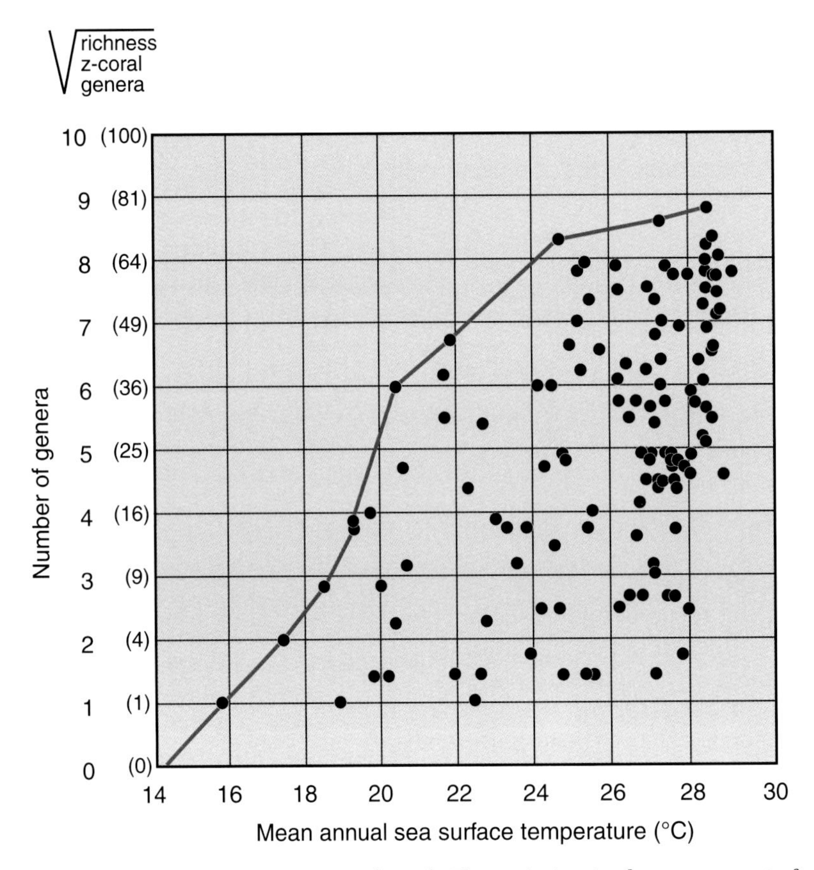

Fig. 1. The temperature-richness plot of Fraser & Currie (1996). The variation in the square root of modern coral generic richness in 130 sites worldwide is expressed as a function of mean annual ocean-surface temperature (°C) at each site. The red curve of maximum richness values has been added, following Rosen (1999), in order to facilitate the application of this plot to the incomplete fossil coral record.

richness values (see Fig. 1) that was added to the original plot in order to facilitate its application to the fossil record. This allows minimizing biases caused by the undoubtly incomplete fossil record that would otherwise lead to values not representative of the original richness.

Results

Diversity patterns

Figure 2A displays generic richness of z-corals for each locality, the stratigraphic range of which is represented by the length of the bar. In order to analyze the general trend, it seems inappropriate to take into account those localities having generic richness values very different from others for the same time interval (localities 23 in the Chattian, 25 and 27 in the Aquitanian, and 57 in the Langhian–Serravallian). The anomalies related to the highest values likely reflect excessive taxonomic subjectivity, whereas those related to lowest values may be due either to sampling and poor specimen preservation or to ecological constraints limiting coral diversity.

The Rupelian shows a gradual increase of z-coral diversity per locality through time. The good stratigraphic resolution of many z-coral sites allows fine-scale variations to be seen (Bosellini & Russo, 2000). The maximum richness values of the Rupelian are broadly maintained during the Chattian. By contrast, the Chattian–Aquitanian boundary is marked by a sharp decrease from about 30 to 15 genera. Generic richness increases again to more than 30 in the Burdigalian. This is followed by two main decreases at the Burdigalian–Langhian and at the Serravallian–Tortonian boundaries. During the early Messinian the diversity of the z-coral assemblages does not reach more than 5 genera.

Patterns of inferred palaeotemperatures

In Fig. 2B, palaeotemperatures inferred from the generic richness of each locality are plotted for

Table 1. Oligocene–Miocene coral localities of the Mediterranean region selected for this study. For each locality, up-dated age, number of z-coral genera (Gen (z)), total number of genera (including non-z corals: Gen (z + nz)), and inferred seawater palaeotemperature (T°C), calculated from the z-coral richness, are indicated. References used for the coral fauna of each locality are listed.

	Locality	Stage	Gen (z)	Gen (z + nz)	T°C	Reference
1	Mallorca, Balearic Islands (Spain)	lower Rupelian	14	15	19.20	Alvarez *et al.*1989; Ramos-Guerrero *et al.*, 1989
2	Marosticano, Vicentin (N Italy)	lower Rupelian	11	11	18.90	Pfister, 1980a
3	Gornji Grad (Slovenia)	lower-middle Rupelian	15	18	19.30	Nebelsick *et al.*, 2000; Bosellini, pers. data
4	Eastern Lessini, Vicentin (N Italy)	middle Rupelian	23	23	19.80	Frost, 1981
5	Marosticano, Vicentin (N Italy)	middle Rupelian	17	18	19.45	Frost, 1981
6	Eastern Lessini, Vicentin (N Italy)	middle Rupelian	16	16	19.40	Bosellini, 1988
7	Ovada, Piedmont (N Italy)	middle Rupelian	13	14	19.15	Fantini Sestini, 1960
8	Marosticano, Vicentin (N Italy)	middle Rupelian	20	20	19.60	Pfister, 1980a
9	Sassello, Liguria (N Italy)	upper Rupelian	18	18	19.50	Fravega *et al.*,1987
10	Aqui, Piedmont (N Italy)	upper Rupelian	12	12	19.00	Pfister, 1980b
11	Cairo Montenotte, Liguria (N Italy)	upper Rupelian	22	22	19.75	Pfister, 1985
12	Mesohellenic Basin (Greece)	upper Rupelian	22	24	19.75	Schuster, 2002c
13	Abadeh (central Iran)	upper Rupelian	31	36	20.20	Schuster, 2002a; Schuster & Wielandt, 1999
14	Gaas, Aquitaine (France)	Rupelian	11	13	18.90	Chevalier, 1956
15	Sirt Basin (Libya)	Rupelian	21	22	19.70	Hladil *et al.*, 1991
16	Salento Peninsula, Apulia (S Italy)	lower Chattian	26	27	19.95	Bosellini & Russo, 1992; Bosellini & Perrin, 1994; Bosellini, pers.data
17	Malta	Chattian	21	21	19.70	Chaix & Saint Martin, 1994
18	Landes, Aquitaine (France)	Chattian	11	17	18.90	Chaix & Cahuzac, 2001
19	Dax, Aquitaine (France)	Chattian	16	21	19.40	Chevalier, 1962
20	Mesohellenic Basin (Greece)	Chattian	15	15	19.30	Schuster, 2002b, 2002c
21	Peyrehorade, Aquitaine (France)	Chattian	24	30	19.90	Cahuzac & Chaix, 1994
22	Sirt Basin (Libya)	Chattian	9	9	18.70	Hladil *et al.*, 1991
23	Aquitaine region (France)	Chattian	45	73	21.70	Cahuzac & Chaix, 1993; Cahuzac & Chaix, 1996;
24	La Nerthe, Provence (France)	upper Chattian	29	32	20.10	Chevalier, 1962

Table 1 (*Continued*)

	Locality	Stage	Gen (z)	Gen (z + nz)	T°C	Reference
25	Aquitaine region (France)	Aquitanian	28	37	20.10	Cahuzac & Chaix, 1993; Cahuzac & Chaix, 1996
26	Gironde, Aquitaine (France)	Aquitanian	15	18	19.30	Chevalier, 1962
27	Qom (central Iran)	Aquitanian	2	2	16.50	Schuster, 2002d; Schuster & Wielandt, 1999
28	Dolianova, Sardinia (Italy)	Aquitanian	6	6	18.00	Governato, 1993; Cherchi *et al.*, 2000
29	Makran Mountains (S Iran)	Aquitanian	8	8	18.55	McCall *et al.*, 1989
30	Sirt Basin (Libya)	Aquitanian–Serravallian	16	17	19.40	Hladil *et al.*, 1991
31	Esfahan (central Iran)	Aquitanian–Burdigalian	8	8	18.55	Schuster, 2002d; Schuster & Wielandt, 1999
32	Paros Island (Greece)	Aquitanian–Burdigalian	13	13	19.15	Collections MNHN
33	Cyprus	Aquitanian–Burdigalian	15	15	19.30	Follows, 1992; Follows et *al.*,1996
34	Aquitaine region (France)	Aquitanian–Burdigalian	16	18	19.40	Oosterban, 1988
35	NW Gulf of Suez (Egypt)	Burdigalian	12	15	19.00	Schuster, 2002b, 2002e
36	Qom (central Iran)	Burdigalian	16	17	19.40	Schuster, 2002d; Schuster & Wielandt, 1999
37	Zagros Mountains (SW Iran)	Burdigalian	7	7	18.30	Schuster, 2002d; Schuster & Wielandt, 1999
38	Makran Mountains (S Iran)	Burdigalian	32	38	20.25	McCall *et al.*, 1989
39	Aquitaine region (France)	Burdigalian	33	45	20.30	Cahuzac & Chaix, 1993; Cahuzac & Chaix, 1996
40	Gironde, Aquitaine (France)	Burdigalian	22	24	19.75	Chevalier, 1962
41	Dax, Aquitaine (France)	Burdigalian	24	27	19.90	Chevalier, 1962
42	Valencia region (Spain)	Langhian	5	5	17.80	Calvet *et al.*, 1994
43	Aquitaine region (France)	Langhian	18	27	19.50	Cahuzac & Chaix, 1993; Cahuzac & Chaix, 1996
44	Manciet, Aquitaine (France)	Langhian	4	6	17.50	Chevalier, 1962
45	Saubrigues, Aquitaine (France)	Langhian	4	5	17.50	Chevalier, 1962
46	Styrian Basin (Austria)	lower Badenian (Langhian)	3	3	17.00	Friebe, 1991
47	Budapest area (Hungary)	lower Badenian (Langhian)	10	10	18.80	Oosterban, 1990
48	Murchas, Granada Basin (S Spain)	Late Langhian	3	3	17.00	Braga *et al.*,1996
49	Duplek (Slovenia)	Badenian (Langhian–L. Serrav.)	4	4	17.50	Baron Szabo, 1997
50	Budapest area (Hungary)	Badenian (Langhian–L. Serrav.)	4	4	17.50	Oosterban, 1990
51	Vienna Basin (Austria)	Badenian (Langhian–L. Serrav.)	11	19	18.90	Piller & Kleemann, 1991; Riegl & Piller, 2000
52	Holy Cross Mountains (Poland)	Badenian (Langhian–L. Serrav.)	7	18	18.30	Roniewicz & Stolarski, 1991
53	North Hurghada (Egypt)	Langhian–L.Serravallian	14	17	19.20	Perrin *et al.*, 1998
54	South Quseir area (Egypt)	Langhian–L.Serravallian	10	10	18.80	Perrin *et al.*,1998

55	Hérault, Languedoc (France)	Langhian–Serravallian	10	12	18.80	Chevalier, 1962
56	Loire Basin (France)	Langhian–Serravallian	10	16	18.80	Chevalier, 1962
57	Torino Hills, Piedmont (N Italy)	Langhian–Serravallian	35	76	20.40	Chevalier, 1962
58	Djebel Chott, Dahara region (Algeria)	Langhian–Serravallian	2	3	16.50	Belkebir *et al.*, 1994
59	Aquitaine region (France)	Serravallian	5	13	17.80	Cahuzac & Chaix, 1993; Cahuzac & Chaix, 1996
60	Midyan region (Saudi Arabia)	Tortonian	3	3	17.00	Dullo *et al.*, 1983
61	Cyprus	Tortonian	1	1	15.90	Follows, 1992; Follows et *al.*,1996
62	Crete (Greece)	Tortonian	5	7	17.80	Buchbinder, 1996
63	Granada Basin (S Spain)	Tortonian	4	4	17.50	Braga *et al.*,1990
64	Almanzora River Valley, Almeria (SE Spain)	Tortonian	1	1	15.90	Martín *et al.*, 1994
65	Kasaba area (Turkey)	Tortonian	1	1	15.90	Hayward, 1982; Hayward *et al.*, 1996
66	NW Crete (Greece)	Tortonian	3	3	17.00	Baron Szabo, 1995
67	Modena area (N Italy)	Tortonian	7	16	18.30	Montanaro, 1929
68	Modena area (N Italy)	Tortonian	6	19	18.00	Chevalier, 1962; Chevalier collection, MNHN
69	Tortona area, Piedmont (N Italy)	Tortonian	3	17	17.00	Chevalier, 1962
70	Sangonera, Murcia (SE Spain)	upper Tortonian	2	2	16.50	Lopez Buendia, 1992
71	Caltanisetta region, Sicily (S Italy)	upper Tortonian	5	5	17.80	Catalano, 1979
72	Isparta region (Turkey)	Tortonian–Messinian	5	5	17.80	Karabiyikoglu *et al.*, 1999
73	Mallorca, Balearic Islands (Spain)	Tortonian–Messinian	3	3	17.00	Pomar, 1991
74	Menorca, Balearic Islands (Spain)	Tortonian–Messinian	2	2	16.50	Obrador et *al.*, 1992
75	Gozo Island	Tortonian–Messinian	1	1	15.90	Pedley, 1979
76	Sirt Basin (Libya)	Tortonian–Messinian	3	3	17.00	Hladil *et al.*, 1991
77	Calabria (S Italy)	Tortonian–Messinian	4	8	17.50	Chevalier, 1962
78	Vigoleno, Piacenza (N Italy)	lower Messinian	2	2	16.60	Barrier *et al.*, 1994
79	Maiella Mountain (Central Italy)	lower Messinian	1	1	15.90	Danese, 1999
80	Cabo de Gata, Almeria (Spain)	lower Messinian	2	2	16.50	Esteban & Giner, 1980
81	Rosignano, Tuscany (Italy)	lower Messinian	5	5	17.80	Bossio et *al.*, 1996; Chevalier, 1962
82	Vibo Valenzia, Calabria (S Italy)	lower Messinian	3	3	17.00	Romano *et al.*, 2005
83	Salento Peninsula, Apulia (S Italy)	lower Messinian	3	3	17.00	Bosellini *et al.*, 2001, 2002

Table 1 (*Continued*)

	Locality	Stage	Gen (z)	Gen (z + nz)	T°C	Reference
84	Caltanisetta region, Sicily (S Italy)	lower Messinian	2	2	16.50	Grasso & Pedley, 1989; Pedley, 1996
85	Melilla (N Morocco)	Messinian	2	2	16.50	Saint-Martin & Cornèe, 1996
86	Sorbas Basin, Almeria (SE Spain)	Messinian	1	1	15.90	Riding *et al.*, 1991
87	San Miguel de Salinas Basin (SE Spain)	Messinian	1	1	15.90	Reinhold, 1995
88	Alicante Basin (SE Spain)	Messinian	1	1	15.90	Calvet et *al.*, 1996
89	Orania (Algeria)	Messinian	2	2	16.50	Saint-Martin, 1996
90	Sicily (Italy)	Messinian	2	2	16.50	Catalano, 1979
91	Malta and Gozo Islands	Messinian	2	2	16.50	Bosence & Pedley, 1982
92	Malta	Messinian	3	3	17.00	Chaix & Saint Martin, 1994
93	Bou Meriem (Morocco)	Messinian	3	3	17.00	Elhamzaoui & Lachkhem, 1994
94	Cabo de Gata, Almeria (SE Spain)	Messinian	1	1	15.90	Esteban & Giner, 1980
95	Fes (Morocco)	Messinian	4	4	17.50	Moissette & Saint Martin, 1995
96	Nijar, Almeria (SE Spain)	Messinian	1	1	15.90	Dabrio et *al.*, 1981
97	Lorca Basin, Murcia (SE Spain)	Messinian	4	4	17.50	Rouchy *et al.*, 1986
98	Orania (N Algeria)	Messinian	1	1	15.90	Rouchy *et al.*, 1986
99	Orania (N Algeria)	Messinian	5	5	17.80	Saint-Martin, 1996
100	Melilla (N Morocco)	Messinian	2	2	16.50	Saint-Martin & Cornèe, 1996
101	Rif (N Morocco)	Messinian	1	1	15.90	Rouchy *et al.*, 1986
102	Santa Pola, Alicante (SE Spain)	Messinian	1	1	15.90	Rouchy *et al.*, 1986

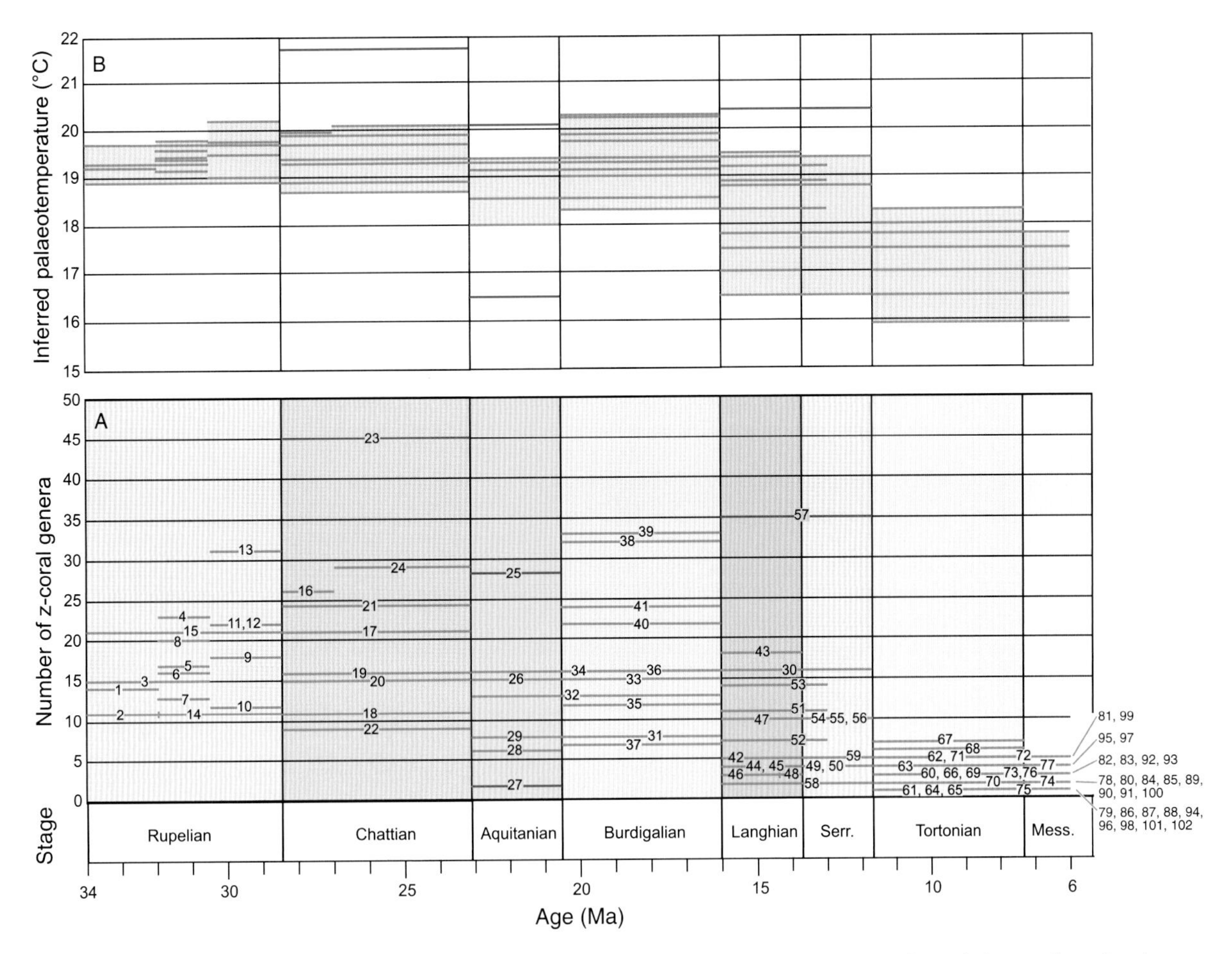

Fig. 2. Stratigraphic ranges and generic richness of Mediterranean Oligo-Miocene z-corals and their inferred palaeotemperatures. (A) Generic richness of z-corals occurring at the different numbered localities listed in Table 1. Bars represent the stratigraphic range of each locality. (B) Palaeotemperatures inferred from generic richness at each locality. Time-scale of Gradstein *et al.* (2004).

each stage, with bars representing the stratigraphic range of z-coral sites. Whereas a trend similar to the one described for generic diversity can be traced also for the palaeotemperatures, this plot shows a gradual widening of the temperature range, particularly evident after the Burdigalian. This means that z-corals were able to adapt to a larger temperature range from the mid-Miocene onwards to the Messinian.

Figure 3 compares the changes of inferred palaeotemperatures and diversity patterns through time with global palaeoclimatic fluctuations derived from oxygen stable-isotopes (Dodd & Stanton, 1990, fig. 3.26; Abreu & Haddad, 1998; Zachos *et al.*, 2001). Diversity patterns are estimated by averaging the generic richness of localities of the same stage (Fig. 3A) and also by considering the richness values of the northernmost localities of each stage (Fig. 3B). When several z-coral sites occur at similar palaeolatitudes for the same stage, the average of their diversity values was taken.

The gradual increase of inferred palaeotemperatures in the Rupelian, not observed in the isotope curves, may be due to the low stratigraphic resolution of the latter. By contrast, data prescribed here are consistent with isotope values showing a cooling from the Chattian–Aquitanian boundary. The diversity patterns record a drastic decline in the average generic richness and a moderate fall at the northernmost localities. A good correlation can also be seen for the warming trend in the Burdigalian, especially with the isotope curves of Dodd & Stanton (1990) and Zachos *et al.* (2001). This warm peak corresponds to the highest values at the northernmost z-coral site and to a relatively high value of average richness.

The general cooling trend, which begins drastically from the Middle Miocene and smooths in the

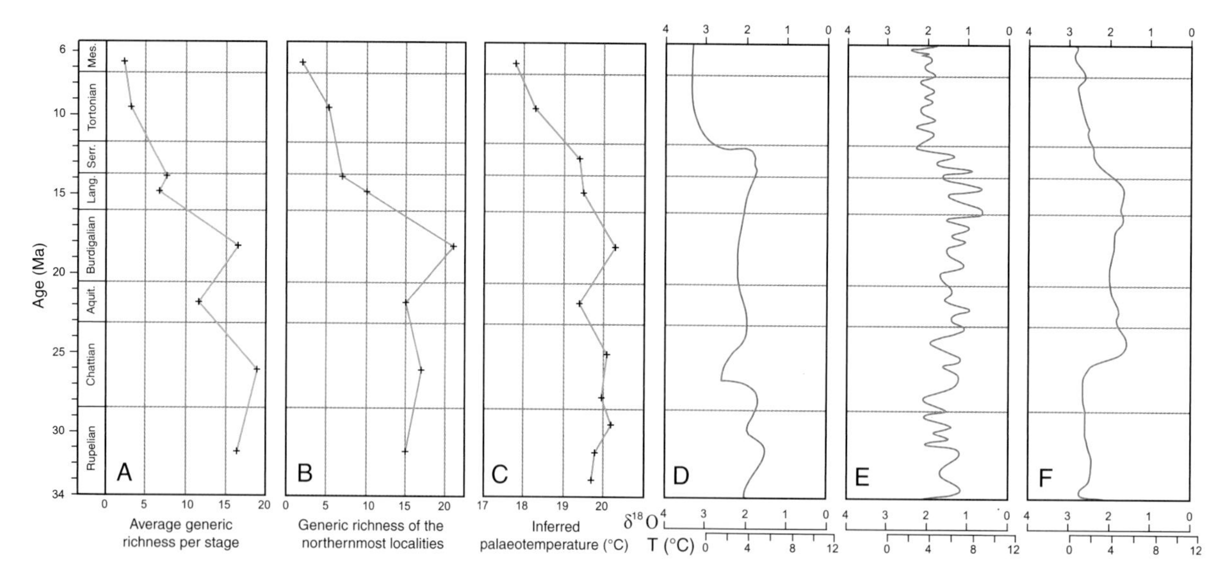

Fig. 3. Trends in z-coral generic richness and estimates of seawater palaeotemperatures through the Oligo-Miocene. (A) Change of diversity pattern through time, estimated by averaging the z-coral generic richness of localities of the same stage. (B) Change of the z-coral generic richness of the northernmost localities through time. (C) Temporal change in palaeotemperature derived from z-coral generic richness. (D) Oxygen-isotope composition of multispecies assemblages of benthic foraminifera and inferred corresponding palaeotemperature (modified from Savin *et al.*, 1975; Dodd & Stanton, 1990). (E) Composite smoothed $\delta^{18}O$ and palaeotemperature curve (modified from Abreu & Haddad, 1998). (F) Global deep-sea oxygen isotope record and palaeotemperature based on data compiled from more than 40 DSDP and ODP sites (modified from Zachos *et al.*, 2001). Time-scale of Gradstein *et al.* (2004).

Upper Miocene from the richness data, differs from the isotopic temperature curves, in particular because the Mid-Miocene Climatic Optimum is not recorded by taxonomic richness of Mediterranean z-corals. However, a good correlation exists after the middle Miocene, as also underlined by the diversity patterns (Fig. 3A and B)

DISCUSSION

The relationship between taxonomic diversity of z-coral communities and abiotic and biotic factors determining the composition and richness of a given community at a given place is highly complex and strongly influenced by many interactions and feedbacks. One of the main explanations of this complexity is that diversity patterns result from both ecological and historical causes, and hence, they cannot be considered solely in terms of stress response (Rosen, 1988; Jokiel & Martinelli, 1992; Perrin *et al.*, 1995). As emphasized by Rosen (1999), the "energy hypothesis" is an empirical approach linking z-coral taxonomic richness and incident solar energy, for which prevailing seawater temperature is considered as a reasonable proxy (Fraser & Currie, 1996). Although Fraser & Currie (1996) clearly stated that seawater temperature is the dominant first-order control of taxonomic richness of z-corals at any given site today, other factors may locally become dominant at particular localities, hence causing anomalies in the inferred temperature plot of the region concerned. In this study, the number of localities examined should minimize, or at least smooth, the anomalies related to local conditions at particular sites.

The "energy hypothesis" approach is also based on the assumption that the relationship between generic richness of z-corals at a given site and seawater temperature has remained the same through time, thus being similar to the "nearest-living-relative" method and to the coexistence approach, both commonly applied by palaeobotanists for quantitative reconstructions of Cenozoic palaeoclimates (Chaloner & Creber, 1990; Mosbrugger & Utescher, 1997; Uhl *et al.*, 2003; Mosbrugger *et al.*, 2005). Although it is clear that for fossil fauna (and flora) uniformitarianism may introduce some biases, these can be minimized using the temperature-generic richness plot (Fig. 1) to discuss relative palaeotemperatures through time, rather than absolute values (see also Rosen, 1999, p. 320).

Similar to other methods (nearest living relative and coexistence approach) that use fossil organisms to obtain quantitative estimates of palaeotemperatures (Mosbrugger *et al.*, 2005), the comparison between the inferred temperature–z-coral richness plot presented here and published isotopic palaeotemperature curves (Fig. 3), demonstrates the consistency and reliability of these results. However, some discrepancies exist for the Rupelian and for the Mid-Miocene. Palaeotemperatures inferred from z-coral richness increase gradually through the Rupelian. This progressive change is not recorded by the isotope data, showing that either it is related to a regional feature and hence does not appear on the global isotope curve, or the stratigraphic resolution, upon which the isotope curves are based, is not precise enough to highlight this trend. Several Rupelian z-coral localities occur in northern Italy (Liguria, Piedmont and the Vicentin Lessini shelf), for which the stratigraphic setting is well constrained (Bosellini & Russo, 2000). The gradually increasing taxonomic richness of coral fauna of the Eastern Lessini shelf throughout the entire Rupelian has been attributed to increasing seawater temperature (Bosellini & Russo, 1988).

No clear indication of the Mid-Miocene Climatic Optimum appears in the palaeotemperatures inferred from the generic richness of Mediterranean z-coral assemblages. While the Mid-Miocene Climatic Optimum is clearly recorded in Mediterranean shallow-water faunas and facies (e.g. Bojar *et al.*, 2004), it seems that the generic diversity of z-coral communities of this age, and consequently its related inferred palaeotemperature, is lower than expected relative to their Burdigalian counterparts.

It is well known that the diversity of the Mediterranean coral fauna declined during the Miocene (Chevalier, 1962). This is explained by the isolation of the Mediterranean basin from the late-Early Miocene to the Middle Miocene when seaway connections with the Indo-Pacific Ocean through the Middle East began to be disrupted by emergent areas and northward rotation of the Arabian Peninsula (Rögl, 1998; Harzhauser et al., 2002). Several reopenings and closures of the Middle East seaway during the Middle Miocene have been proposed by Rögl (1998). However, as far as reef corals are concerned, scleractinian faunas resulting from a mixture of Mediterranean and Indo-Pacific species are not known to occur after the Early Miocene (McCall *et al.*, 1989; Schuster & Wielandt, 1999). Consequently, for these biogeographical reasons, the regional taxonomic pool of z-corals in the Mediterranean began to diminish from the middle towards the Late Miocene, hence limiting species and generic diversity at any localities in the region. The decrease of the regional taxonomic diversity in the Mediterranean may affect the relationship between taxonomic richness and seawater temperature, although the resulting effect may be more or less marked depending on the time considered. The Mid-Miocene Climatic Optimum is not expressed in the z-coral generic richness while, after the Middle Miocene, the diversity-inferred palaeotemperature pattern is well correlated with changes shown by the global isotopic temperature curves.

Several potential biases may alter to some degree the relationship between generic richness of z-coral assemblages and seawater palaeotemperatures, and therefore may produce some anomalies in the general pattern. Although no attempt has yet been made to establish a hierarchy between potential biases, these can be grouped in to five main categories:

(1) Biases related to the heterogeneity of habitats where z-coral communities were able to develop, and to the general heterogeneity of data available, which depends on the level of investigation by geographical area. It is known that the species richness tends to increase with the level of investigation in the area considered (Perrin *et al.*, 1995);
(2) Biases related to data collection and analysis, such as sampling, preservation of coral specimens and taxonomic biases;
(3) Those depending on the precision of stratigraphic resolution, which is highly variable through time and space. From the Mediterranean data, stratigraphic distinction between Langhian and Serravallian, and between Late Tortonian and Messinian, is often problematic;
(4) Those introduced by local environmental factors limiting taxonomic richness: mainly, turbidity affecting light penetration in the water column and local nutrient inputs adjacent to river mouths or produced by land runoff;
(5) Those linked to biogeographical features such as the lack of possible exchange with z-coral faunas of the Indo-Pacific after the late Burdigalian.

Among these potential biases, those having solely a local influence and affecting some localities individually can be minimized by treating a significant number of z-coral sites and by considering the pattern obtained at a regional or a global scale, rather than deducing palaeotemperatures for particular places alone. By exerting a strong control on the regional species pool biogeographical factors influence to some degree the inferred temperature-diversity pattern at a regional scale. However, this effect can be estimated precisely enough by comparing the results obtained with independent data such as global isotope curves of palaeotemperatures.

The "energy hypothesis" applied to fossil z-coral faunas appears to be a powerful and promising method when used at global or regional scales to estimate changes in palaeotemperatures, particularly for time intervals and in regions where conventional isotope data are lacking, such as the Oligocene and Miocene of the Mediterranean.

CONCLUSIONS

The "energy hypothesis" has been applied to Oligocene and Miocene z-coral sites from the Mediterranean and adjacent areas in order to infer minimum seawater palaeotemperatures for this region where isotopic palaeotemperature curves are lacking. The results are in good agreement with known global temperature curves, except for the Mid-Miocene Climatic Optimum which is not recorded in the generic richness of the Mediterranean z-coral communities.

In addition, the results show: (1) a gradual increase in temperature from the Early to the Late Rupelian; and (2) a gradual broadening of the temperature range after the Burdigalian. The latter indicates that z-coral communities were able to maintain themselves in the region by progressively adapting to a wider temperature range from the mid-Miocene onwards to the Messinian, as the Mediterranean was migrating northward.

ACKNOWLEDGMENTS

We would like to sincerely thank Brian Rosen for numerous fruitful discussions on this topic and for his specific advice on this particular project. We also acknowledge Maria Mutti, Werner Piller and Christian Betzler who organized the ESF Exploratory Workshop "Evolution of carbonate Systems during the Oligocene–Miocene climatic Transition", in Potsdam, and edited this volume. We are grateful to Michaela Bernecker and James Nebelsick for their helpful reviews of our manuscript. Studies by Francesca R. Bosellini were funded through grant PRIN 2004.

REFERENCES

Abreu, V.S. and **Haddad, G.A.** (1998) Glacioeustatic fluctuations: the mechanism linking stable isotope events and sequence stratigraphy from the Early Oligocene to Middle Miocene. In: *Mesozoic and Cenozoic Sequence Stratigraphy of European Basins* (Eds P.C. de Graciansky, J. Hardenbol, T. Jacquin and P.R. Vail). *SEPM Spec. Publ.*, **60**, 245–259.

Adams, C.G., Lee, D.E. and **Rosen, B.R.** (1990) Conflicting isotopic and biotic evidence for tropical sea surface temperatures during the Tertiary. *Palaeogeogr. Palaeoclimatol. Palaeoecol.*, **77**, 289–313.

Alvarez, G., Busquets, P., Vilaplana, M. and **Ramos-Guerrero, E.** (1989) Fauna coralina paleogena de las Islas Baleares (Mallorca y Cabrera). *Espana. Batalleria*, **3**, 61–68.

Baron Szabo, R.C. (1995) Taxonomy and palaeoecology of Late Miocene corals of NW Crete (Gramvoússa, Roka and Koukounaras Fms). *Berl. Geowiss. Abh.*, **E 16**, 569–577.

Baron Szabo, R. C. (1997) Miocene (Badenian) corals from Duplek, NE Slovenia. *Razpr. IV, Razr. Sazu*, **38**, 96–115.

Barrier P., Cauquil E., Raffi S., Russo A. and **Tran van Huu M.** (1994) Signification du plus septentrional des récifs messiniens à Algues et *Porites* connus en Méditerranée (Vigoleno, Piacenza, Italie). *Interim Colloqu. RCMNS Marseille (abstract book)* 2–3.

Belkebir L., Mansour B., Bessedik M., Saint-Martin J.P., Belbarbi M. and **Chaix C.** (1994) Présence d'une construction récifale à Djebel Chott (Dahra occidental, Algérie): témoin du maximum transgressif du Miocène moyen en Méditerranée). *Interim Colloqu. RCMNS Marseille (abstract book)* 4.

Bojar, A.-V., Hiden, H., Fenninger, A. and **Neubauer, F.** (2004) Middle Miocene seasonal temperature changes in the Styrian basin, Austria, as recorded by the isotopic composition of pectinid and brachiopod shells. *Palaeogeogr. Palaeoclimatol. Palaeoecol.*, **203**, 95–105.

Bosellini, F.R. (1988) Oligocene corals from Monte Bastia (Vicentin Lessini Mountains, N Italy). *Acc. Naz. Sci. Lett. Arti, Modena*, **VII** (5), 111–157.

Bosellini, F.R. and **Perrin, C.** (1994) The coral fauna of Vitigliano: qualitative and quantitative analysis in a back reef environment (Castro Limestone, Late Oligocene, Salento Peninsula, southern Italy). *Boll. Soc. Paleontol. Ital.*, **33**, 171–181.

Bosellini, F.R. and **Russo, A.** (1988) The Oligocene *Actinacis* coral community of the southern Alps (Italy): temperature vs. terrigenous control. *Proc. 6th Int. Coral Reef Symp. Townsville*, **3**, 385–391.

Bosellini, F.R. and **Russo, A.** (1992) Stratigraphy and facies of an Oligocene fringing reef (Castro Limestone, Salento Peninsula, southern Italy). *Facies*, **26**, 145–166.

Bosellini, F.R. and **Russo, A.** (2000) Biodiversità dei coralli zooxantellati e implicazioni paleoclimatiche. L'esempio del Paleogene Italiano. In: *Crisi Biologiche, Radiazioni Adattative e Dinamica delle Piattaforme Carbonatiche.*

(Eds A. Cherchi and C. Corradini). *Acc. Naz. Sci. Lett. Arti Modena,* **21**, 41–45.

Bosellini, F.R., Russo, A. and **Vescogni, A.** (2001) Messinian reef-building assemblages of the Salento Peninsula (southern Italy): palaeobathymetric and palaeoclimatic significance. *Palaeogeogr. Palaeoclimatol. Palaeoecol.*, **175**, 7–26.

Bosellini, F.R., Russo, A. and **Vescogni, A.** (2002) The Messinian reef complex of the Salento Peninsula (southern Italy): Stratigraphy, facies and palaeoenvironmental interpretation. *Facies*, **47**, 91–112.

Bosence, D.W.J. and **Pedley, H.M.** (1982) Sedimentology and palaeoecology of a Miocene coralline algal biostrome from the Maltese islands. *Palaeogeogr. Palaeoclimatol. Palaeoecol.*, **38**, 9–43.

Bossio, A., Esteban, M., Mazzanti, R., Mazzei, R. and **Salvatorini, G.** (1996) Rosignano reef complex (Messinian), Livornesi Mountains, Tuscany, central Italy. In: *Models for Carbonate Stratigraphy from Miocene Reef Complexes of the Mediterranean Regions* (Eds E.K. Franseen, M. Esteban, W.C. Ward and J.-M. Rouchy). *SEPM Concepts Sedimentol. Paleontol,* **5**, 277–294.

Braga, J.C., Martín, J.M. and **Alcala, B.** (1990) Coral reefs in coarse-terrigenous sedimentary environments (Upper Tortonian, Granada Basin, southern Spain). *Sed. Geol.*, **66**, 135–150.

Braga, J.C., Jimenez, A.P., Martín, J.M. and **Rivas, P.** (1996) Middle Miocene coral-oyster reefs, Murchas, Granada, southern Spain. In: *Models for Carbonate Stratigraphy from Miocene Reef Complexes of the Mediterranean Regions* (Eds E.K. Franseen, M. Esteban, W.C. Ward and J.-M. Rouchy). *SEPM Concepts Sedimentol. Paleontol,* **5**, 131–139.

Buchbinder, B. (1996) Miocene carbonates of the eastern Mediterranean, the Red Sea and the Mesopotamian Basin: geodynamic and eustatic controls. In: *Models for Carbonate Stratigraphy from Miocene Reef Complexes of the Mediterranean Regions* (Eds E.K. Franseen, M. Esteban, W.C. Ward and J.-M. Rouchy). *SEPM Concepts Sedimentol. Paleontol,* **5**, 89–96.

Cahuzac, B. and **Chaix, C.** (1993) Les faunes de coraux (Anthozoaires Scléractiniaires) de la façade atlantique française au Chattien et au Miocène. *Cienc. Terra*, **12**, 57–69.

Cahuzac, B. and **Chaix, C.** (1994) La faune de coraux de l'Oligocène supérieur de la Téoulère (Peyrehorade, Landes). *Bull. Soc. Borda*, 463–483.

Cahuzac, B. and **Chaix, C.** (1996) Structural and faunal evolution of Chattian-Miocene reefs and corals in western France and the northeastern Atlantic Ocean. In: *Models for Carbonate Stratigraphy from Miocene Reef Complexes of the Mediterranean Regions* (Eds E.K. Franseen, M. Esteban, W.C. Ward and J.-M. Rouchy). *SEPM Concepts Sedimentol. Paleontol,* **5**, 105–127.

Calvet F., Esteban M. and **Permanyer A.** (1994) Mid Miocene coral reefs in the Gulf of Valencia, NE Spain. *Interim Colloqu. RCMNS Marseille (abstract book)*, 10.

Calvet, F., Zamarreno, I. and **Vallès, D.** (1996) Late Miocene reefs of the Alicante-Elche Basin, southeast Spain. In: *Models for Carbonate Stratigraphy from Miocene Reef Complexes of the Mediterranean Regions* (Eds E.K. Franseen, M. Esteban, W.C. Ward and J.-M. Rouchy). *SEPM Concepts Sedimentol. Paleontol,* **5**, 177–190.

Catalano, R. (1979) Scogliere ed evaporiti messiniane in Sicilia. Modelli genetici ed implicazioni strutturali. *Lav. Ist. Geol. Univ. Palermo*, **18**, 1–21.

Chaix, C. and **Cahuzac, B.** (2001) Une faune inédite de coraux scléractiniaires dans le gisement chattien d'Escornebéou (Landes, SW France); stratigraphie, systématique et paléoécologie. *Ann. Paléontol.*, **87**, 3–47.

Chaix C. and **Saint Martin J.P.** (1994) Les associations coralliennes oligo-miocènes de Malte, un résumé de l'histoire paléobiogéographique de la Méditerranée. *Interim Colloqu. RCMNS Marseille (abstract book)*, 12.

Chaloner, W.G. and **Creber, G.T.** (1990) Do fossil plants give a climatic signal? *J Geol. Soc. London*, **147**, 343–350.

Cherchi, A., Murru, M. and **Simone, L.** (2000) Miocene carbonate factories in the syn-rift Sardinia Graben Subbasins (Italy). *Facies*, **43**, 223–240.

Chevalier, J.P. (1956) Les polypiers anthozoaires du Stampien de Gaas (Landes). *Bull. Soc. Hist. Nat. Toulouse*, **90**, 375–410.

Chevalier, J.P. (1962) Recherches sur les Madréporaires et les formations récifales miocènes de la Méditerranée occidentale. *Thèse. Mém. Soc. Géol. Fr.*, **93**, 558 pp.

Copper, P. (1994) Reefs under stress: the fossil record. *Cour. Forsch.-Inst. Senckenb.*, **17**, 87–94.

Copper, P. (2002) Silurian and Devonian reefs: 80 million years of global greenhouse between two ice ages. In: *Phanerozoic Reef Patterns* (Eds F W. Kiessling and E. Flügel). *SEPM Spec. Publ*, **72**, 181–238.

Dabrio, C.J., Esteban, M. and **Martín, J.M.** (1981) The coral reef of Nijar, Messinian (uppermost Miocene), Almeria Province, S. E. Spain. *J. Sed. Petrol.*, **51**, 521–539.

Danese, E. (1999) Upper Miocene carbonate ramp deposits from the southernmost part of Maiella Mountain (Abruzzo), central Italy. *Facies*, **41**, 41–54.

Dodd, J.R. and **Stanton, R.J. Jr** (1990) *Paleoecology: Concepts and Applications*. 2nd Edn. *John Wiley & Sons*, New York, 502 pp.

Dullo, W.C., Hötzl, H. and **Jado, A.R.** (1983) New stratigraphical results from the Tertiary sequence of the Midyan area, NW Saudi Arabia; Contributions to the geology of the eastern margin of the Red Sea. *Newsl. Stratigr.*, **12**, 75–83.

Elhamzaoui O. and **Lachkhem H.** (1994) Les affleurements récifaux de Bou Meriem (Région de Tazouta; SE de Fès; Bordure Sud du sillon sud rifain; Maroc). *Interim Colloqu. RCMNS Marseille (abstract book)*, 20.

Esteban, M. (1996) An overview of Miocene reefs from Mediterranean areas: General trends and facies models. In: *Models for Carbonate Stratigraphy from Miocene Reef Complexes of the Mediterranean Regions* (Eds E.K. Franseen, M. Esteban, W.C. Ward and J.-M. Rouchy). *SEPM Concepts Sedimentol. Paleontol,* **5**, 3–54.

Esteban, M. and **Giner, J.** (1980) Messinian coral reefs and erosion surfaces in Cabo de Gata (Almeria, SE Spain). *Acta Geol. Hisp.*, **15**, 97–104.

Fantini Sestini, N. (1960) La fauna oligocenica dei dintorni di Ovada (Alessandria). *Riv. Ital. Paleontol. Stratigr.*, **66**, 403–434.

Follows, E.J. (1992) Patterns of reef sedimentation and diagenesis in the Miocene of Cyprus. *Sed. Geol.*, **79**, 225–253.

Follows, E.J., Robertson, A.H.F. and **Scoffin, T.P.** (1996) Tectonic controls on Miocene reefs and related carbonate facies in Cyprus. In: *Models for Carbonate Stratigraphy from Miocene Reef Complexes of the Mediterranean Regions* (Eds E.K. Franseen, M. Esteban, W.C. Ward and J.-M. Rouchy). *SEPM Concepts Sedimentol. Paleontol,* **5**, 295–315.

Franseen, E.K., **Esteban, M.**, **Ward, W.C.** and **Rouchy, J.-M.** (1996) Models for Carbonate Stratigraphy from Miocene Reef Complexes of the Mediterranean regions. *SEPM Concepts Sedimentol. Paleontol,* **5**, 391 pp.

Fraser, R.H. and **Currie, D.J.** (1996) The species richness-energy hypothesis in a system where historical factors are thought to prevail: coral reefs. *Am. Nat.*, **148**, 138–159.

Fravega, P., **Giammarino, S.**, **Piazza, M.**, **Russo, A.** and **Vannucci, G.** (1987) Significato paleoecologico degli episodi coralgali a nord di Sassello. Nuovi dati per una ricostruzione paleogeografico-evolutiva del margine meridionale del Bacino Terziario del Piemonte. *Atti Soc. Tosc. Sci. Nat.*, **94**, 19–76.

Friebe J.G. (1991) Middle Miocene reefs and related facies in Eastern Austria- II) Styrian Basin. *Excursion Guidebook, 6th Int. Symp. Fossil Cnidaria*, Münster, pp. 29–47.

Frost, S.H. (1981) Oligocene reef coral biofacies of the Vicentin, northeast Italy. In: *European Fossil Reef Models* (Ed. D.F. Toomey). *SEPM Spec. Publ*, **30**, 483–539.

Governato, S. (1993) *La Corallofauna Miocenica della Zona di Dolianova (Cagliari).* Unpubl. Diploma Thesis, University of Modena, 101 pp.

Gradstein, F.M., **Ogg, J.G.**, **Smith, A.G.**, **Agterberg, F.P.**, **Bleeker, W.**, **Cooper, R.A.**, **Davydov, V.**, **Gibbard, P.**, **Hinnov, L.A.**, **House, M.R.**, **Lourens, L.**, **Luterbacher, H. P.**, **McArthur, J.**, **Melchin, M.J.**, **Robb, L.J.**, **Shergold, J.**, **Villeneuve, M.**, **Wardlaw, B.R.**, **Ali, J.**, **Brinkhuis, H.**, **Hilgen, F.J.**, **Hooker, J.**, **Howarth, R.J.**, **Knoll, A.H.**, **Laskar, J.**, **Monechi, S.**, **Plumb, K.A.**, **Powell, J.**, **Raffi, I.**, **Röhl, U.**, **Sadler, P.**, **Sanfilippo, A.**, **Schmitz, B.**, **Shackleton, N.J.**, **Shields, G.A.**, **Strauss, H.**, **van Dam, J.**, **van Kolfschoten, T.**, **Veizer, J.** and **Wilson, D.** (2004) *A Geologic Time Scale 2004.* Cambridge University Press, Cambridge, 589.

Grasso M. and **Pedley H.M.** (1989) Palaeoenvironment of the Upper Miocene coral build-ups along the northern margins of the Caltanisetta Basin (Central Sicily). Atti 3rd Simp. Ecol. Paleoecol. Comunità Bentoniche, Catania, pp. 373–389

Hallock, P. (2005) Global change and modern coral reefs: New opportunities to understand shallow-water carbonate depositional processes. *Sed. Geol.*, **175**, 19–33.

Harzhauser, M., **Piller, W.E.** and **Steininger, F.F.** (2002) Circum-Mediterranean Oligo-Miocene biogeographic evolution: the gastropods' point of view. *Palaeogeogr. Palaeoclimatol. Palaeoecol.*, **183**, 103–133.

Hayward, A.B. (1982) Coral reefs in a clastic sedimentary environment: fossil (Miocene, S.W. Turkey) and modern (Recent, Red Sea) analogues. *Coral Reefs*, **1**, 109–114.

Hayward, A.B., **Robertson, A.H.F.** and **Scoffin, T.P.** (1996) Miocene patch reefs from a Mediterranean marginal terrigenous setting in southwest Turkey. In: *Models for Carbonate Stratigraphy from Miocene Reef Complexes of the Mediterranean Regions* (Eds E.K. Franseen, M. Esteban, W.C. Ward and J.-M. Rouchy). *SEPM Concepts Sedimentol. Paleontol,* **5**, 317–332.

Hladil, J., **Otava, J.** and **Galle, A.** (1991) Oligocene carbonate buildups of the Sirt Basin, Libya. In: *The Geology of Libya* 4 (Eds M.J. Salem, O.S. Hammuda and B.A. Eliagoubi), pp. 1401–1420. Elsevier, Amsterdam.

Hughes, T.P., **Baird, A.H.**, **Bellwood, D.R.**, **Card, M.**, **Connolly, S.R.**, **Folke, C.**, **Grosberg, R.**, **Hoegh-Gouldberg, O.**, **Jackson, J.B.C.**, **Kleypas, J.**, **Lough, J.**, **Marshall, P.**, **Nyström, M.**, **Palumbi, S.R.**, **Pandolfi, J.**, **Rosen, B.R.** and **Roughgarden, J.** (2003) Climate change, human impacts, and the resilience of coral reefs. *Science*, **301**, 929–933.

Jokiel, P. and **Martinelli, F.J.** (1992) The vortex model of coral reef biogeography. *J. Biogeogr.*, **19**, 449–458.

Karabiyikoglu M., **Tuzcu, S.** and **Ciner A.** (1999) Miocene reefs in the Antalya foreland basin, western Taurides, Turkey: Stratigraphy, facies and paleoenvironmental setting. *The 1999 Lyell Meeting, London.*(abstract book).

Kiessling, W. (2001) Paleoclimatic significance of Phanerozoic reefs. *Geology*, **29**, 751–754.

Kiessling, W., **Flügel, E.** and **Golonka, J.** (1999) Paleoreefs Maps: Evaluation of a comprehensive database on Phanerozoic reefs. *AAPG Bull.*, **8**, 1552–1587.

Kiessling W., **Flügel E.** and **Golonka J.** (Eds) (2002). *Phanerozoic Reef Patterns. SEPM Spec. Publ.*, **72**, Tulsa. 275 pp

Lopez Buendia, A.M. (1992) Arrecifes de coral en el Mioceno Superior de la vertiente norte de la Sierra de Carrascoy (Murcia). *Rev. Esp. Paleontol.*, 101–111.

McCall, G.J.H., **Rosen, B.R.** and **Darrell, J.G.** (1994) Carbonate deposition in accretionary prism settings: Early Miocene coral limestones and corals of the Makran mountain range in southern Iran. *Facies*, **31**, 141–178.

Martín, J.M., **Braga, J.C.** and **Rivas, P.** (1989) Coral successions in Upper Tortonian reefs in SE Spain. *Lethaia*, **22**, 271–286.

Moissette, P. and **Saint Martin, J.P.** (1995) Bryozoaires des milieux récifaux miocènes du sillon sud-rifain au Maroc. *Lethaia*, **28**, 271–283.

Montanaro, E. (1929) Coralli Tortoniani di Montegibbio (Modena). *Boll. Soc. Geol. Ital.*, **48**, 107–137.

Mosbrugger, V. and **Utescher, T.** (1997) The coexistence approach – a method for quantitative reconstructions of Tertiary terrestrial palaeoclimate data using plant fossils. *Palaeogeogr. Palaeoclimatol. Palaeoecol.*, **134**, 61–86.

Mosbrugger, V., **Utescher, T.** and **Dilcher, D.L.** (2005) Cenozoic continental climatic evolution of Central Europe. *Proc. Nat. Acad. Sci.*, **102**, 14964–14969.

Nebelsick, J.H., **Bassi, D.** and **Drobne, K.** (2000) Microfacies analysis and palaeoenvironmental interpretation of Lower Oligocene, shallow-water carbonates (Gornji Grad Beds, Slovenia). *Facies*, **43**, 157–176.

Obrador, A., **Pomar, L.** and **Taberner, C.** (1992) Late Miocene breccia of Menorcs (Balearic Islands): a basis for the interpretation of a Neogene ramp deposit. *Sed. Geol.*, **79**, 203–223.

Oosterban, A.F.F. (1988) Early Miocene corals from the Aquitaine basin (SW France). *Meded. Werkgr. Tert. Kwart. Geol. Leiden*, **25**, 247–284.

Oosterban, A.F.F. (1990) Notes on a collection of Badenian (Middle Miocene) corals from Hungary in the National Museum of Natural History at Leiden (The Netherlands). *Contrib. Tert. Quatern. Geol. Leiden*, **27**, 3–15.

Pedley, H.M. (1979) Miocene bioherms and associated structures in the Upper Coralline Limestone of the Maltese Islands: their lithification and palaeoenvironment. *Sedimentology*, **26**, 577–591.

Pedley, H.M. (1996) Miocene reef distributions and their associations in the central Mediterranean region: an overview. In: *Models for Carbonate Stratigraphy from Miocene Reef Complexes of the Mediterranean Regions* (Eds E.K. Franseen, M. Esteban, W.C. Ward and J.-M. Rouchy). *SEPM Concepts Sedimentol. Paleontol*, **5**, 73–87.

Perrin, C. (2002) Tertiary: The emergence of modern reef ecosystems. In: *Phanerozoic Reef Patterns* (Eds W. Kiessling and E. Flügel). *SEPM Spec. Publ.*, **72**, 587–621.

Perrin, C., **Bosence, D.W.J.** and **Rosen, B.R.** (1995) Quantitative approaches to palaeozonation and palaeobathymetry of corals and coralline algae in Cenozoic reefs. In: *Marine Palaeoenvironmental Analysis from Fossils* (Eds D.W.J. Bosence and P.A. Allison). *Geol. Soc. London Spec. Publ*, **83**, 181–229.

Perrin, C., **Plaziat, J.-C.** and **Rosen, B.R.** (1998) The Miocene coral reefs and reef corals of the SW Gulf of Suez and NW Red Sea: distribution, diversity and regional environmental controls. In: *Sedimentation and Tectonics of Rift Basins: Red Sea-Gulf of Aden* (Eds B.H. Purser and D.W.J. Bosence), 296–319. *Chapman & Hall*, London.

Pfister, T. (1980a) *Sistematische und paläokölogische Untersuchungen am Oligozänen Korallen der Umgebung von San Luca (Provinz Vicenza, Norditalien)*. PhD Thesis, Birkhäuser Verlag, Basel, 121 pp.

Pfister, T. (1980b) Paläoökologie des oligozänen Korallenvorkommens von Cascine südlich Acqui (Piemont, Norditalien). *Jb. Naturhist. Mus. Bern*, **7**, 247–262.

Pfister, T. (1985) Coral fauna and facies of the Oligocene fringing reef near Cairo Montenotte (Liguria, Northern Italy). *Facies*, **13**, 175–226.

Piller W.E. and **Kleemann K.** (1991) Middle Miocene reefs and related facies in Eastern Austria I) Vienna Basin. *Excursion Guidebook 6th Int. Symp. Fossil Cnidaria Münster*, 1–28

Pomar, L. (1991) Reef geometries, erosion surfaces and high-frequency sea-level changes, upper Miocene reef complex, Mallorca, Spain. *Sedimentology*, **38**, 243–269.

Ramos-Guerrero, E., **Busquets, P.**, **Alvarez, G.** and **Vilaplana, M.** (1989) Fauna coralina de las plataformas mixtas del Paleogeno de las Baleares. *Boll. Soc. Hist. Nat. Balears*, **33**, 9–24.

Reinhold, C. (1995) Guild structure and aggradation pattern of Messinian *Porites* patch reefs: ecological succession and external environmental control (San Miguel de Salinas, SE Spain). *Sed. Geol.*, **97**, 157–175.

Riding, R., **Martín, J.M.** and **Braga, J.C.** (1991) Coral-stromatolite reef framework, Upper Miocene, Almeria, Spain. *Sedimentology*, **38**, 799–818.

Riegl, B. and **Piller, W.E.** (2000) Biostromal Coral Facies: a Miocene Example from the Leitha Limestone (Austria) and its Actualistic Interpretation. *Palaios*, **15**, 399–413.

Riegl, B. and **Piller, W.E.** (2003) Possible refugia for reefs in times of environmental stress. *Int. J. Earth Sci.*, **92**, 520–531.

Rögl, F. (1998) Palaeogeographic considerations for Mediterranean and Paratethys seaways (Oligocene to Miocene). *Ann. Naturhist. Mus. Wien*, **99**, 279–310.

Romano C., **Neri C.**, **Russo A.** and **Russo F.** (2005) Biofacies analyses of a Lower Messinian *Porites* bioconstruction (Vibo Valenzia, Calabria). *Giorn. Paleontol. 2005 Urbino (abstract book)*, p. 58.

Roniewicz, E. and **Stolarski, J.** (1991) Miocene Scleractinia from the Holy Cross Mountains, Poland; part 2- Archaeocoeniina, Astaeina and Fungiina. *Acta Geol. Pol.*, **41**, 69–83.

Rosen, B.R. (1971) The distribution of reef coral genera in the Indian Ocean. *Symp. Zool. Soc. London*, **28**, 263–299.

Rosen, B.R. (1984) Reef coral biogeography and climate through the late Cainozoic: just islands in the sun or a critical pattern of islands? In: *Fossils and Climate* (Ed. P. Brenchley), pp. 201–262. *John Wiley & Sons*, London.

Rosen, B.R. (1988) Progress, problems and patterns in the biogeography of reef corals and other tropical marine organisms. *Helgoländer Meeresun.*, **42**, 269–301.

Rosen, B.R. (1999) Paleoclimatic implications of the energy hypothesis from Neogene corals of the Mediterranean region. In: *The Evolution of Neogene Terrestrial Ecosystems in Europe* (Eds J. Agusti, L. Rook and P. Andrews). pp. 309–327. *Cambridge University Press*, Cambridge.

Rosen, B.R. (2002) Biodiversity: old and new relevance for palaeontology. *Geoscientist*, **12**, 4–9.

Rouchy, J.M., **Saint-Martin, J.P.**, **Maurin, A.** and **Bernet-Rollande, M.C.** (1986) Evolution et antagonisme des communautés bioconstructrices animales et végétales à la fin du Miocène en Méditerranée occidentale; biologie et sédimentologie. *Bull Centres Rech. Explor.-Prod. Elf-Aquitaine*, **10**, 333–348.

Saint-Martin, J.P. (1996) Messinian coral reefs of western Orania, Algeria. In: *Models for Carbonate Stratigraphy from Miocene Reef Complexes of the Mediterranean Regions* (Eds E.K. Franseen, M. Esteban, W.C. Ward and J.-M. Rouchy). *SEPM Concepts Sedimentol. Paleontol*, **5**, 239–246.

Saint-Martin, J.P. and **Cornée, J.J.** (1996) The Messinian reef complex of Melilla, northeastern Rif. Morocco. In: *Models for Carbonate Stratigraphy from Miocene Reef Complexes of the Mediterranean Regions* (Eds E.K. Franseen, M. Esteban, W.C. Ward and J.-M. Rouchy). *SEPM Concepts Sedimentol Paleontol.* **5**, 227–237.

Savin, S.M., **Douglas, R.G.** and **Stehli, F.G.** (1975) Tertiary marine paleotemperatures. *Geol. Soc. Am. Bull.*, **86**, 1499–1510.

Schuster, F. (2002a) Scleractinian corals from the Oligocene of the Qom Formation (Esfahan-Sirjan fore-arc basin, Iran). *Cour. Forsch.-Inst. Senckenb.*, **239**, 5–55.

Schuster, F. (2002b) Oligocene and Miocene examples of *Acropora*-dominated palaeoenvironments: Mesohellenic Basin (NW Greece) and northern Gulf of Suez (Egypt). *Proc. 9th Int. Coral Reef Symp. Bali* **1**, 199–204.

Schuster, F. (2002c) Oligocene scleractinian corals from Doutsiko (Mesohellenic Basin, northwestern Greece). *Cour. Forsch.-Inst. Senckenb.*, **239**, 83–127.

Schuster, F. (2002d) Early Miocene scleractinian corals from the Qom and Asmari formations (central and southwest Iran). *Cour. Forsch.-Inst. Senckenb.*, **239**, 129–161.

Schuster, F. (2002e) Early Miocene corals and associated sediments of the northwestern Gulf of Suez. *Egypt. Cour. Forsch.-Inst. Senckenb.*, **239**, 57–81.

Schuster, F. and **Wielandt, U.** (1999) Oligocene and Early Miocene coral faunas from Iran: palaeoecology and palaeobiogeography. *Int. J. Earth Sci.*, **88**, 571–581.

Stehli, F.G. and **Wells, J.W.** (1971) Diversity and age patterns in hermatypic corals. *Syst. Zool.*, **20**, 115–126.

Uhl, D., **Mosbrugger, V.**, **Bruch, A.** and **Utescher, T.** (2003) Reconstructing palaeotemperatures using leaf floras-case studies for a comparison of leaf margin analysis and the coexistence approach. *Rev. Palaobot. Palynol.*, **126**, 49–64.

Veron, J.E.N. (1995) *Corals in Space and Time: the Biogeography and Evolution of the Scleractinia.* UNSW Press, Sydney, 321 pp.

Veron, J.E.N. and **Minchin, P.T.** (1992) Correlations between sea surface temperature, circulation patterns and the distribution of hermatypic corals of Japan. *Cont. Shelf Res.*, **12**, 835–857.

Wells, J.W. (1955) A survey of the distribution of reef coral genera in the Great Barrier Reef region. *Rep. Great Barrier Reef Comm.*, **6**, 21–29.

Wilson, M.E.J. and **Rosen, B.R.** (1998) Implications of paucity of corals in the Paleogene of SE Asia: plate tectonics or Centre of Origin? In: *Biogeography and Geological Evolution of SE Asia* (Eds R. Hall and J.D. Holloway), pp. 165–195. *Backhuys Publishers*, Leiden.

Zachos, J., **Pagani, M.**, **Sloan, L.**, **Thomas, E.** and **Billups, K.** (2001) Trends, rhythms and aberrations in global climate 65 Ma to present. *Science*, **292**, 686–693.

Int. Assoc. Sedimentol. Spec. Publ. (2010) **42**, 245–256

Late Oligocene to Miocene reef formation on Kita-daito-jima, northern Philippine Sea

Y. IRYU[1], S. INAGAKI, Y. SUZUKI and K. YAMAMOTO

Institute of Geology and Paleontology, Graduate School of Science, Tohoku University, Aobayama, Sendai, 980-8578, Japan (E-mail: iryu.yasufumi@a.mbox.nagoya-u.ac.jp

ABSTRACT

An old borehole, 432.7 m deep, drilled in 1934 and 1936 on Kita-daito-jima, northern Philippine Sea, reveals the reef evolution on this island during the Late Oligocene to Miocene. Four depositional units have been defined by lithological changes and are numbered sequentially from the top of the hole downward. A linear decrease in known strontium-isotope ages of Unit C4, represented by fine-grained, lagoonal deposits, indicates that reef growth kept pace with tectonic subsidence from 18.6 to 24.4 Ma. Unit C3, composed of locally dolomitized, lagoonal deposits rich in coral clasts, is divided into three subunits (C3c, C3b, C3a) based on lithology, carbon and oxygen stable-isotope compositions, and strontium-isotope ages. Rapid reef growth occurred at ~16.1 Ma and ~15.5 Ma (Subunits C3c and C3b) when, because of sea-level falls, the island returned to a water-depth range (<50 m) at which reef formation could occur. Unit C2 is characterized by abundant branching coral colonies indicative of a very shallow reef environment. Coral fauna in Unit C1 is dominated by tabular and encrusting colonies that are indicative, in present-day reefs, of a middle fore-reef environment. Taking the central-island drilling location into account, it can be inferred that Unit C1 formed on a submerged platform. An age-depth section of Kita-daito-jima combined with known global eustatic curves suggests that Units C1 and C2 may have accumulated during the early Late Miocene. Reef formation on Kita-daito-jima was controlled by the combined effects of sea-level changes and tectonic movements (subsidence and uplift). Two modes of reef formation have been recognized: growth that kept pace with the subsidence of the island; and rapid reef formation that commenced at sea-level falls. The latter indicated that sea-level falls are key events that revived drowned reefs and that reefs are not necessarily specific to warm periods with high sea-levels.

Keywords Reef, Kita-daito-jima, oxygen isotopes, carbon isotopes, strontium isotopes, sea-level change.

INTRODUCTION

The initiation, growth, and demise of coral reefs have attracted the interest of many scientists ever since Darwin (1842). To investigate the processes involved, numerous boreholes were drilled in the first half of the twentieth century on reefs and carbonate islands, such as Funafuti (Royal Society of London, 1942), Bermuda (Pirrson & Vaughan, 1977), the Bahamas (Field & Hess, 1933), and the Great Barrier Reef (Richards & Hill, 1913). Kita-daito-jima (Fig. 1) is well known (MacNeil, 1970) as one of the first carbonate islands from which a deep drill core was recovered successfully (Ladd *et al.*, 1980, 1953; Steers & Stoddart, 1989). The Kita-daito-jima borehole was drilled in 1934 to a depth of 209.3 metres below the ground surface (mbgs), and the coring was advanced to 431.7 mbgs in 1936 (Sugiyama, 1977, 1934). The recovered material consists exclusively of shallow-water carbonates, the upper 100 m of which are pervasively dolomitized. Although the drilling was one of the greatest projects in the history of geological studies in Japan, comparably few investigations have been conducted on the recovered material. Ota (1992) examined the petrography and chemistry of dolomites; Hanzawa (1940) gave a brief lithological

[1]Present address: Department of Earth and Planetary Sciences, Graduate School of Environmental Studies, Nagoya University, Nagoya 464-8601, Japan

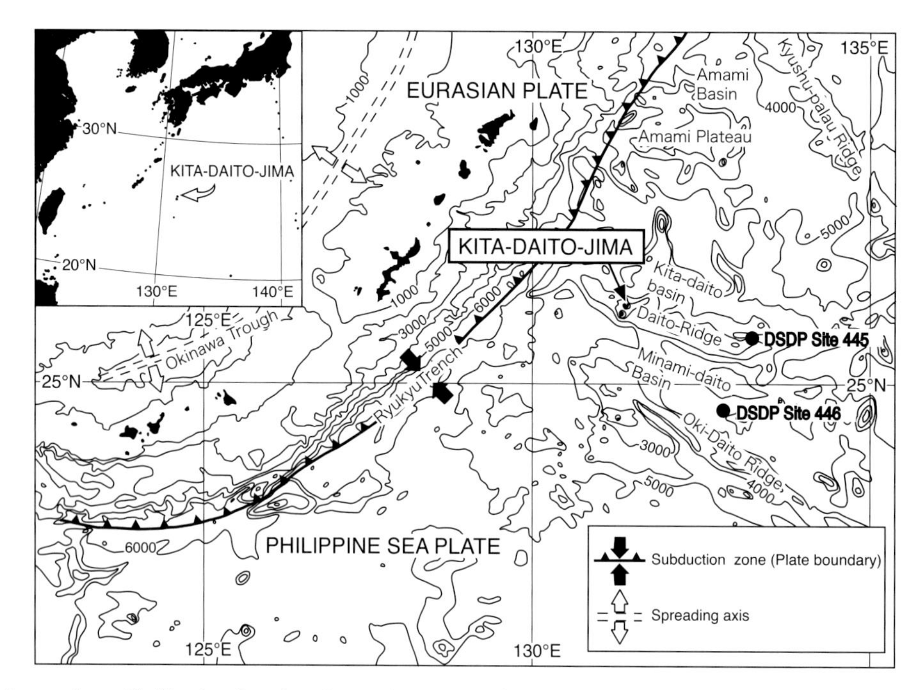

Fig. 1. Map of the northern Philippine Sea showing main tectonic features. Kita-daito-jima is a small carbonate island located atop the Daito Ridge of the Philippine Sea Plate. Depth contours are in metres.

description and determined geological ages based on benthic foraminiferal biostratigraphy. About 50 years later, Ohde & Elderfield (2003) published a detailed chronology of carbonates within the borehole using strontium-isotope stratigraphy and illustrated a tectonic history of this island from 24 Ma onward. In 1994, Nakamori *et al.* gave a short description of the palaeoenvironmental interpretations of the borehole carbonates, based mainly on the lithology (macrofacies) and coral assemblages. Yet, the history of the reef formation on Kita-daito-jima recorded in the borehole carbonates has not been reconstructed fully.

This paper aims to depict Late Oligocene to Miocene reef formation on Kita-daito-jima Island by integrating reported data on the borehole carbonates: Sr-isotope ages (Ohde & Elderfield, 2003; Fig. 2), results of macro/microfacies and palaeontological analyses (Nakamori *et al.*, 1989, Ozawa, 1992; Fig. 2), and carbon and oxygen stable-isotope studies (Inagaki & Iryu, 1999; Fig. 3). Research on the origin and genesis of dolomites was published by Suzuki *et al.* (1936). However, most of these contributions were printed in Japanese in Japanese journals, which are extremely difficult for most scientists to understand and access, and as such, it is important to compile these previous scientific results here.

GEOLOGICAL SETTING

Kita-daito-jima is a carbonate island in the northern Philippine Sea (25°56.7′N, 131°18.5′W, Fig. 1), located on a lithospheric bulge of the Philippine Sea Plate that is subducting beneath the Eurasian Plate. The island is semi-triangular in shape (apex to the south), extends ~3.7 km from N–S and ~4.9 km from E–W, and consists of a peripheral rim and an interior basin. The peripheral rim is about 0.5 to 1.8 km wide and mostly ranges from 10 m to 50 m in elevation (height above mean sea level) (Fig. 2). The interior basin extends about 1.4 km in a N–S direction and about 2.1 km in an E–W direction, with an elevation of less than 20 m, and is surrounded by up to 50 m high bluffs and slopes. Its geomorphological similarity to modern atolls, has led to Kita-daito-jima being regarded as an elevated fossil atoll (e.g. Flint *et al.*, 1953; Schlanger, 1981).

Coral reef deposits extend over the island's surface, except for the interior basin, which is

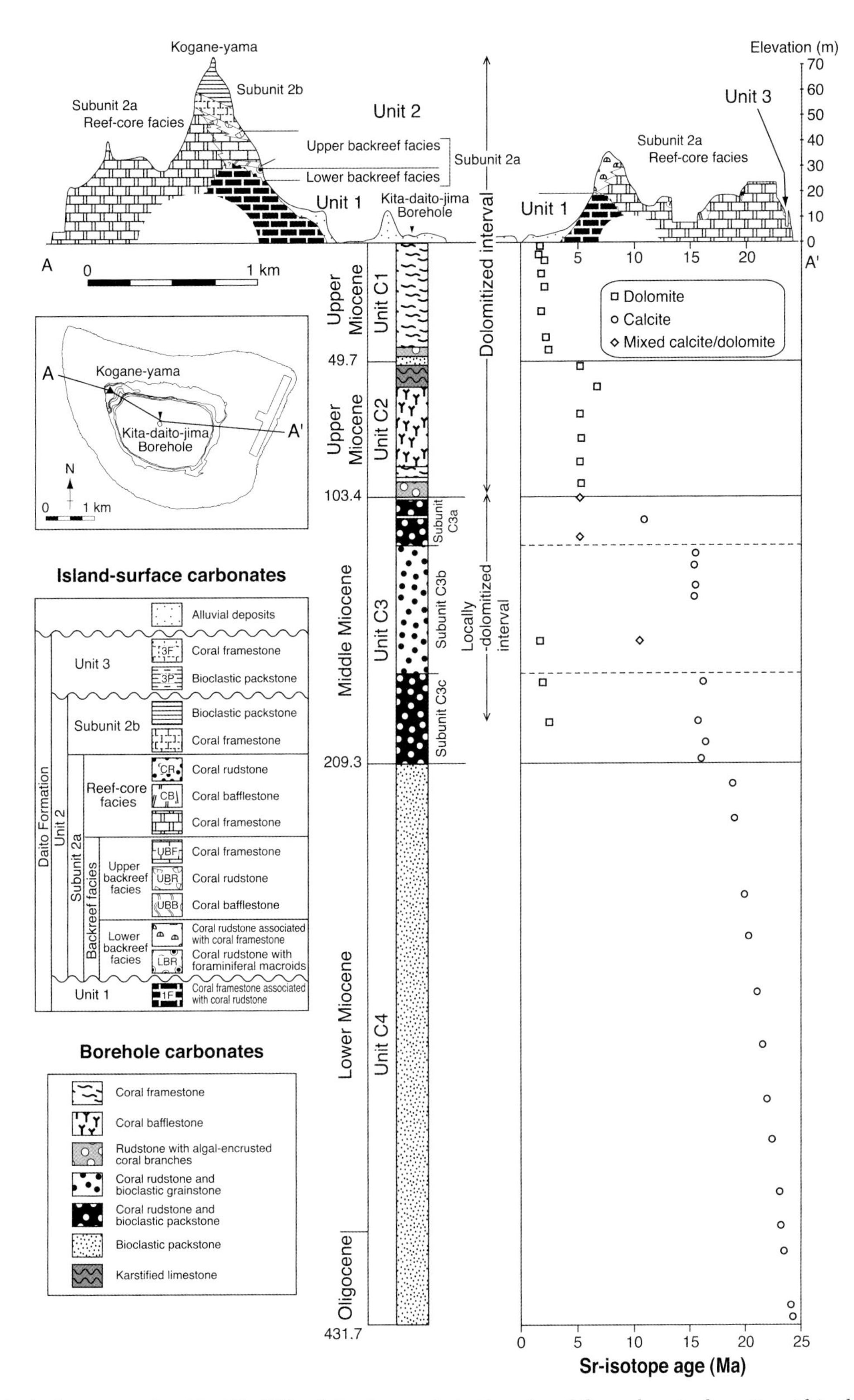

Fig. 2. Geological cross-section (A–A′) of Kita-daito-jima and stratigraphy of the carbonate deposits within the Kita-daito-jima Borehole. Strontium-isotope ages (after Ohde & Elderfield, 2003) are shown. Borehole depths are in metres.

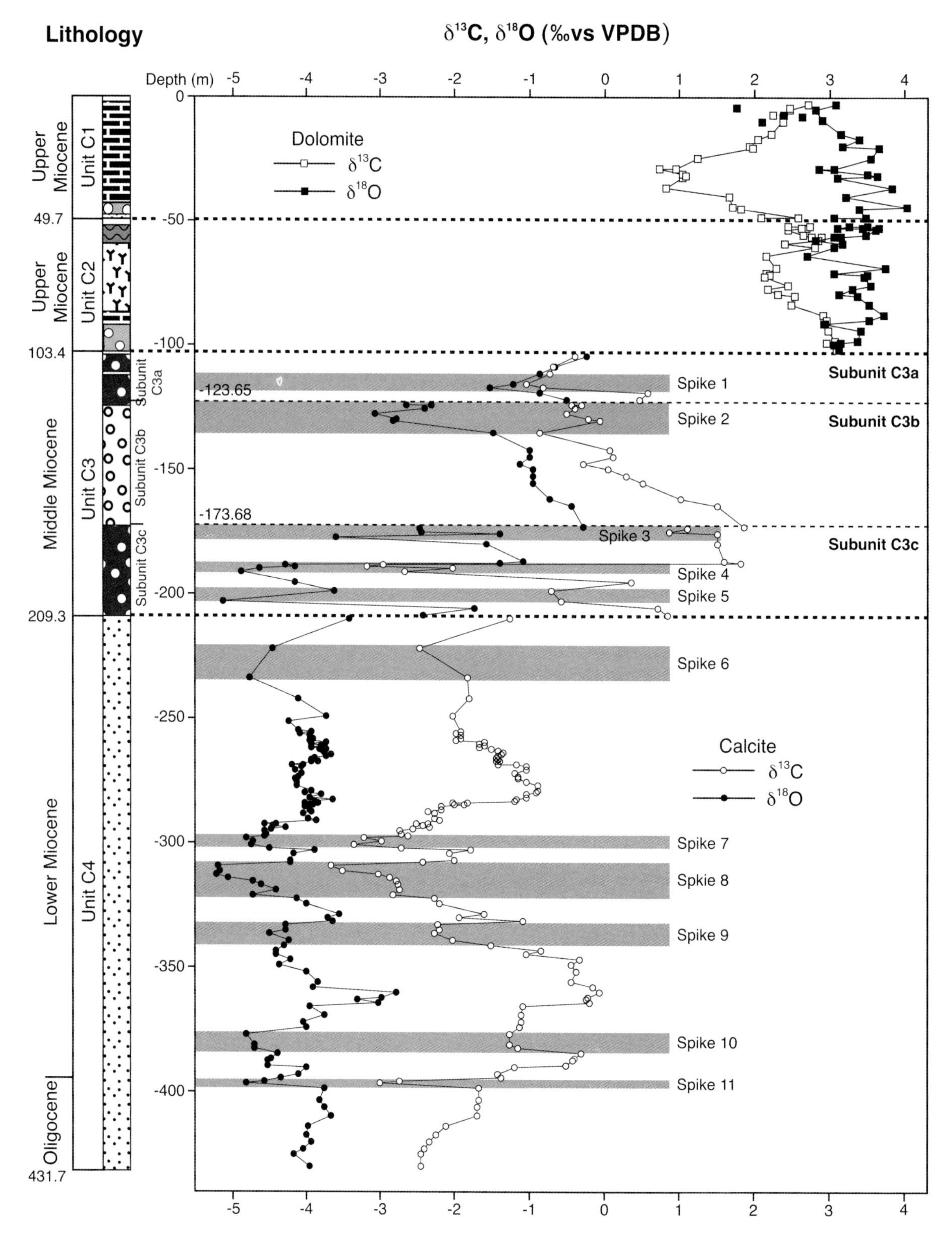

Fig. 3. Carbon and oxygen stable-isotope chemostratigraphy and lithostratigraphy of carbonate deposits within the Kita-daito-jima Borehole. Shaded bars (spikes 1–11) denote zones of meteoric diagenesis indicated by sharp decreases (negative spikes) in $\delta^{13}C$ and $\delta^{18}O$ values. Isotope data for calcite (Units 3 and 4) are derived from Inagaki & Iryu (1999). Those for dolomite were published by Inagaki & Iryu (1999) and revised by Suzuki *et al.* (1936). Key for borehole stratigraphy is given in Fig. 2. Core depths are in metres.

veneered by soil (Fig. 2). Phosphate ores covered with pumice-bearing clay previously occurred to the west of Kogane-yama (Yamanari, 2006), but these have now been mined out. Nambu *et al.* (1994) established the lithostratigraphy of the reef deposits and provided a formal stratigraphic description. They showed that the reef deposits making up the Daito Formation are overlain by the Kaigunbo Formation, which has a very limited distribution. The deposits of the Daito Formation were dolomitized pervasively at 1.6 to 2.0 Ma (Kawana & Ohde, 1999), whereas the Kaigunbo Formation, above, is undolomitized.

Three depositional units have been defined in the Daito Formation. Unit 1, at the base, is dominated by coral framestone rich in massive corals, which crops out in the interior basin. Surface Unit 1 and borehole Unit C1 are lithologically similar; both consist mainly of framestone with hermatypic corals that are characteristic of a fore-reef environment. The lack of an observable exposure surface below Unit 1 suggests that there was no significant depositional hiatus between the two units. Unit 2, which rests unconformably on Unit 1, is divided into two subunits: lower Subunit 2a and upper Subunit 2b. The lower subunit is composed of reef-core facies that constitutes the main body of the peripheral rim and back-reef facies exposed at the cliffs lining the interior basin. The reef-core facies is represented by coral framestone associated with coral bafflestone. The lower back-reef facies consists mainly of coral rudstone; the upper back-reef facies is composed chiefly of coral framestone/bafflestone, frequently containing *Halimeda* segments. Subunit 2b, cropping out around the peak of the island, consists of coral framestone and bioclastic packstone. Unit 3, is exposed sporadically on the eastern coast and unconformably overlies the reef-core facies of Subunit 2a. This unit consists of cross-bedded bioclastic packstone associated with coral framestone. The Kaigunbo Formation abuts Unit 3 on the coastal cliffs at elevations less than 11 m, and consists of framestone that formed during the last interglacial period.

MATERIALS AND METHODS

The Kita-daito-jima Borehole was drilled at an elevation of 2.5 m near the centre of the island. Most of the material recovered by the drilling in 1934 (down to a depth of 209.3 mbgs) is represented by dolostones and limestones that are broken into pebbles or into short cores that range from several cm, up to 20 cm in length. A total of 838 samples were collected; these were numbered serially downward. A single number was originally given to multiple cores/pebbles that were collected from the same level, so suffixes were added to identify individual rock samples (e.g. Core No. 342 #1). Of these, 759 samples are stored. The material recovered by the drilling in 1936 (down to a depth of 431.7 mbgs) is represented mostly by limestones broken into coarse sand- to granule-sized grains. Although 341 samples were recovered, only 302 samples are curated. These samples, together with some 340 thin sections that were made in the 1930s, are housed in the Museum of Natural History, Tohoku University.

Classification of the carbonate rocks follows Dunham (1962) and Embry & Klovan (1971). Age constraints are based on benthic foraminiferal biostratigraphy (Hanzawa, 1940) and Sr-isotope stratigraphy (Ohde & Elderfield, 2003).

BOREHOLE CARBONATES

Four depositional units have been defined by lithological changes and are numbered sequentially in descending order (Fig. 2). Unit C4 extends from 209.3 mbgs to the base (431.7 mbgs) of the borehole. Recovered material is represented mostly by limestones broken into coarse sand- to granule-sized grains (Fig. 4A and B). Thin-section study reveals that these grains are derived from bioclastic packstone and grainstone with abundant benthic foraminifera and lesser amounts of bioclasts of nongeniculate coralline algae, echinoids, and molluscs. Carbonates of this unit are not dolomitized. Strontium-isotope ages show a linear decrease from 24.3 to 18.8 Ma within this unit. The extrapolated ages of the base and the top of this unit are calculated at 24.4 Ma (Late Oligocene) and 18.6 Ma (Early Miocene), respectively.

Unit C3, 105.9 m thick (103.4-209.3 mbgs), consists of locally dolomitized coral rudstone, associated mainly with bioclastic grainstone and packstone. The rudstone characteristically includes corals that may be *in situ* and much more abundant coral debris associated with encrusting foraminifera and nongeniculate coralline algae. Based on lithology, $\delta^{13}C$ and $\delta^{18}O$ values, and Sr-isotope ages, this unit can be divided, in ascending order, into three subunits: Subunits C3c; C3b; and C3a.

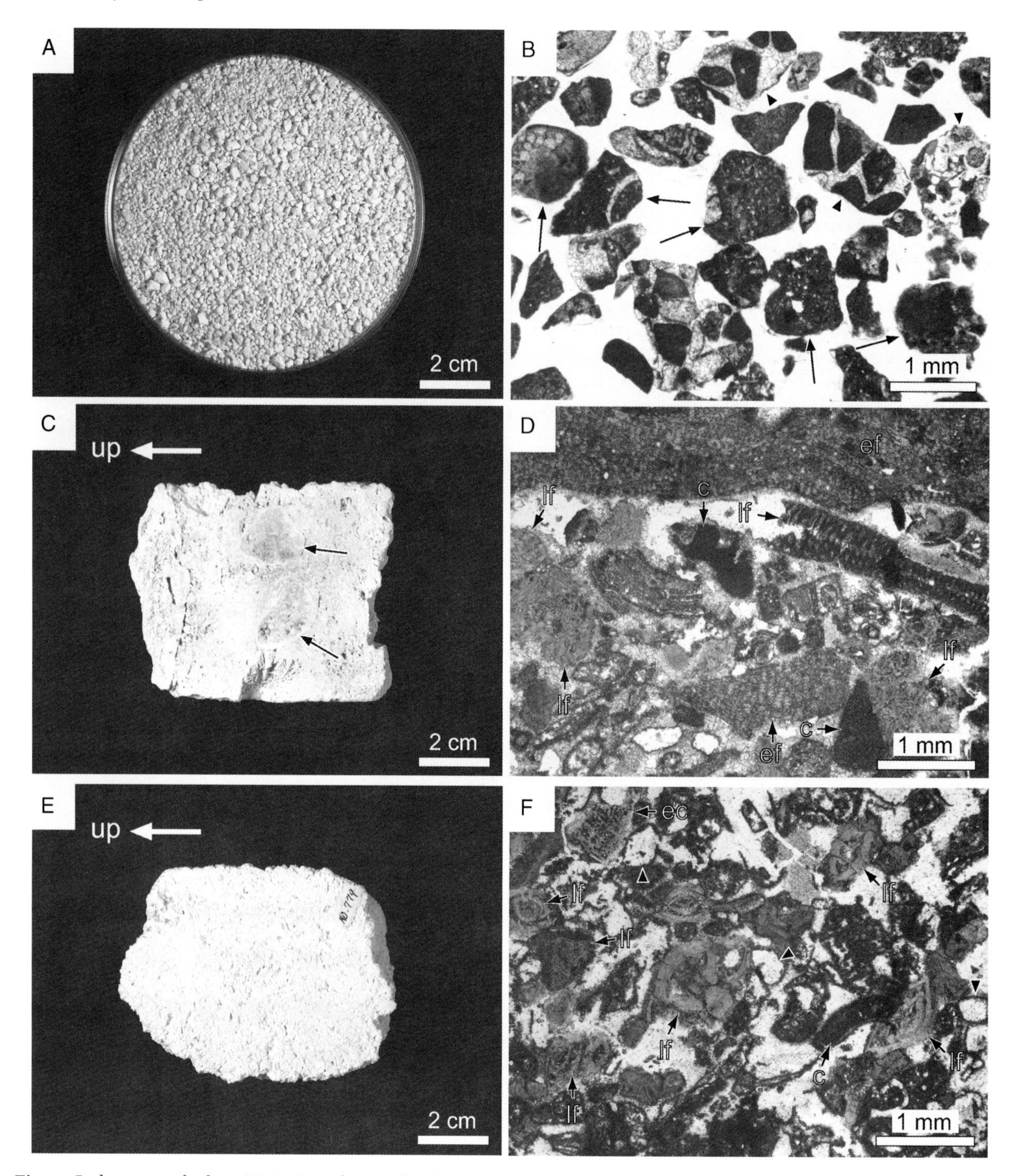

Fig. 4. Carbonate rocks from Units C3 and C4 within the Kita-daito-jima Borehole. (A) Coarse sand- to granule-sized grains of limestone from Unit C4 (Bottle No. 226; 342.6 mbgs). (B) Photomicrograph of the limestone fragments shown in (A). Note that the fragments include bioclastic packstone (arrows) and grainstone (arrowheads). (C) Coral rudstone from Subunit C3c (Core No. 810; 186.8 mbgs). Note coral branches covered by encrusting foraminifera (arrows), *Gypsina* or *Acervulina*. (D) Photomicrograph of the rudstone shown in (C). Note the encrusting foraminifera (ef), larger foraminifera (lf), and non-geniculate coralline algae (c). (E) Bioclastic grainstone from Subunit C3b (Core No. 779; 169.1 mbgs). (F) Photomicrograph of the grainstone shown in (E). Note that the grainstone is dominated by larger foraminifera (lf) associated with coralline algae (c) and echinoids (ec) and that some (?formerly aragonitic) bioclasts have been dissolved to leave micrite envelopes (arrowheads). All photomicrographs, plain polarized light.

Subunit C3c extends from 209.3 to 173.4 mbgs and consists mainly of coral rudstone with an average Sr-isotope age of 16.1 Ma (Fig. 4C and D). Upper intervals contain coral branches coated with encrusting foraminifera. Bioclastic packstone at 190 mbgs includes *Microcodium* textures. Subunit C3b occupies the interval from 173.4 to 122.6 mbgs within the borehole. This subunit also is composed predominantly of coral rudstone, but it can be distinguished from the overlying and underlying subunits by its more porous and permeable nature and its lesser amounts of micrite (Fig. 4E and F). Strontium-isotope ages of this subunit are ~15.5 Ma. Subunit C3a (122.6–103.4 mbgs) consists largely of coral rudstone and includes corals that may be *in situ*. *Microcodium* textures occur at 112 mbgs. One known Sr-isotope age is 10.9 Ma.

Unit C2 is 53.7 m thick (103.4–49.7 mbgs) and is composed mainly of coral bafflestone and, to a lesser extent, coral framestone (Fig. 5A and B). Dolomitization is pervasive. Branching corals (including *Acropora* spp.) are abundant in the bafflestones. Branches encrusted by nongeniculate coralline algae occur abundantly in the lower 10 m of the unit. Bioclasts are composed of benthic foraminifera, coralline algae, and echinoids. Strontium-isotope ages suggest that episodic dolomitization took place at 5.5 Ma (Ohde & Elderfield, 2003; Fig. 2).

Unit C1 consists mainly of pervasively dolomitized coral framestone, associated with bioclastic packstone and grainstone (Fig. 5C and D). It extends to 49.7 mbgs within the borehole. *In situ* hermatypic corals and nongeniculate coralline algae occur throughout the unit. Coral fauna include tabular and encrusting faviids, poritiids, and mussids. Bioclasts include benthic foraminifera, coralline algae, and echinoids. This unit is

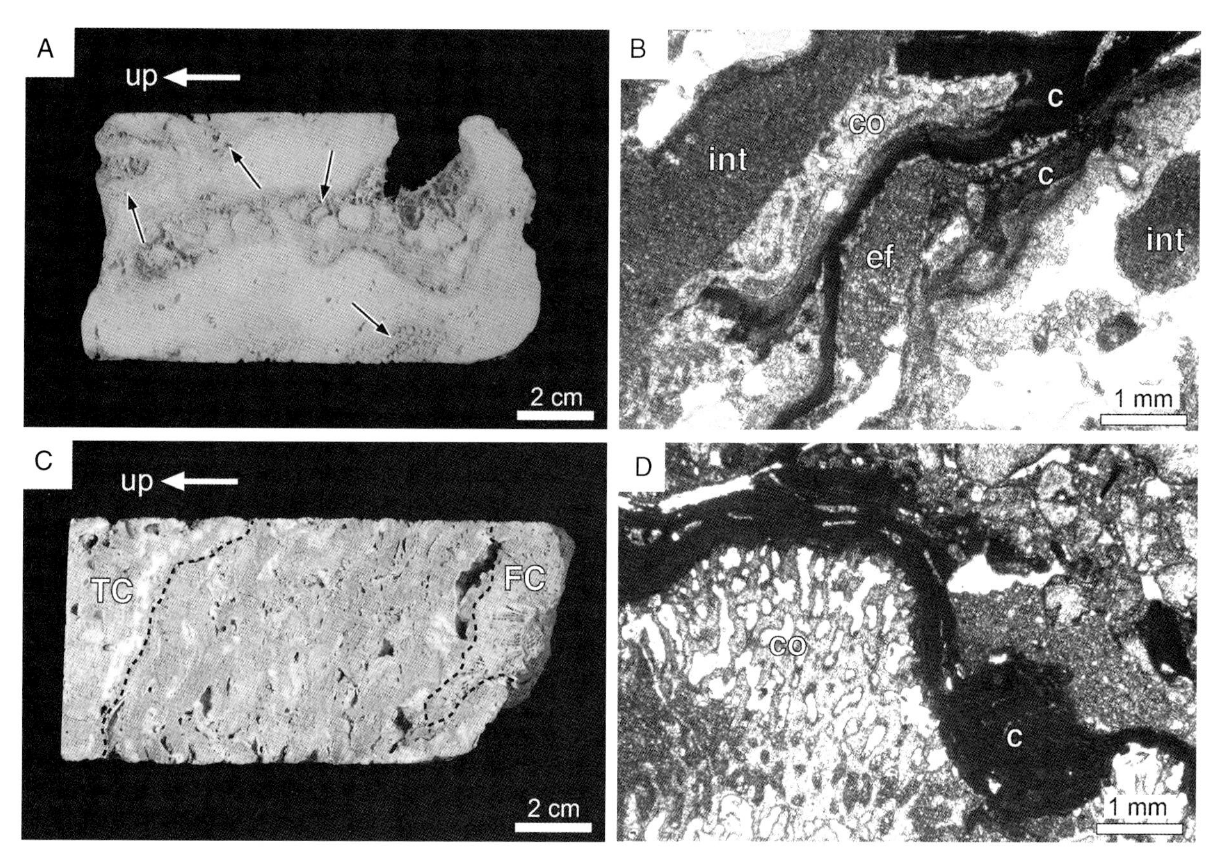

Fig. 5. Dolomitized reef deposits from Units C1 and C2 within the Kita-daito-jima Borehole. (A) Coral bafflestone from Unit C2 (Core No. 367; 66.4 mbgs). Note the abundant coral branches, arrowed. (B) Photomicrograph of the bafflestone shown in (A). Although this bafflestone is dolomitized, its original sedimentary fabric is preserved. Note a coral branch (co) covered with nongeniculate coralline algae (c) and an encrusting foraminifera (ef), and internal sediments (int). (C) Coral framestone from Unit C1 (Core No. 43; 8.2 mbgs). Note the encrusting faviid coral (FC) covered with superimposed corals (thin encrusting *Montipora*? sp.) associated with nongeniculate coralline algae, which, in turn, is overlain by an encrusting to tabular coral (TC; *Montipora*? sp). (D) Photomicrograph of the framestone from Unit C1 shown in (C). Note a coral (co) encrusted by nongeniculate coralline algae (c). Both photomicrographs, plain polarized light.

believed to unconformably overlie Unit C2 because evidence of episodic subaerial exposure and subordinate karstification, such as brecciation of well-indurated packstones and a reddish staining, are observed in the upper ~10 m of Unit C2. Strontium-isotope ages indicate that this unit was dolomitized at 2.0 Ma (Ohde & Elderfield, 2003; Fig. 2).

REEF FORMATION ON KITA-DAITO-JIMA SINCE 25 Ma

Generally, submergence of a volcanic edifice creates sediment-accommodation space for the carbonate sediments to form a thick cap on it. The depth of the oceanic crust increases with age due to thermal subsidence (e.g. Parsons & Sclater, 1995; Marty & Cazenave, 1954). When volcanic seamounts are formed on older oceanic crusts, the oceanic lithospheres are thermally rejuvenated (thinned and uplifted) at the time that the volcanic edifices erupt. Therefore, the seamount subsides as if it was formed on much younger crust, which causes partial "resetting" of the thermal subsidence curve (Detrick & Crough, 1978). Honza & Fujioka (2004) summarized the results of geological and geophysical surveys in the Daito ridges and basins, consisting of the Amami Basin, the Amami Plateau, the Kita-daito Basin, the Daito Ridge, the Minami-daito Basin, and the Oki-daito Ridge (Fig. 1). They identified two major volcanic events that occurred during the Cretaceous to Early Palaeocene (60–80 Ma) and during the Late Palaeocene to Eocene (49–60 Ma). The occurrence of debris-flow sediments yielding *Nummulites boninensis* and associated shallow-water molluscs and bryozoans at DSDP Site 445 (Fig. 1) on the Daito Ridge (Klein & Kobayashi, 1993), indicate that the crest of the ridge was in a shallow-marine environment during the Middle to Late Eocene.

Based on regression analysis of depth versus Sr-isotope ages of reef deposits from the Kita-daito-jima Borehole, Ohde & Elderfield (2003) considered that cooling and subsidence of the volcanic edifice commenced approximately 6 million years after the volcanism. This occurred at approximately 48 Ma, as indicated by ^{40}Ar–^{39}Ar ages of a basaltic sill from DSDP Site 446 (Fig. 1) in the Minami-daito Basin (McKee & Klock, 1980). Thus, the exact nature of the subsidence of the volcanic edifice remains controversial. The subsidence curve of Ohde & Elderfield (2003) is followed in this paper because it is well constrained by many Sr-isotope ages (Fig. 6).

A linear decrease in Sr-isotope ages with depth within Unit C4 indicates that reef growth kept pace with tectonic subsidence from 18.6 to 24.4 Ma (Fig. 2). Lithology composed exclusively of bioclastic packstone and grainstone and rare hermatypic corals indicate a rather deep lagoonal environment. Sharp decreases in δ^{13}C and δ^{18}O values in Units C3 and C4 mark 11 identifiable zones of meteoric diagenesis that underlie coeval subaerial unconformities (Fig. 3). Of these, five meteoric zones occur in the lower interval of Unit C4, which indicates frequent fourth to fifth-order eustatic fluctuations from ~19 to ~24 Ma. There is no evidence of subaerial exposure at the top of Unit C4, which suggests that the atoll may have been drowned immediately after ~18.6 Ma.

Strontium-isotope ages of Subunit C3c fall in a narrow range from 15.7 to 16.4 Ma and do not display a linear decrease or increase, averaging 16.1 Ma. An age-depth section of Kita-daito-jima combined with global eustatic curves (Haq *et al.*, 1987, 1988) shows that a sea-level fall during the earliest Middle Miocene resulted in the submerged island being brought into a shallow environment in which reef builders could recolonize (Fig. 6). No significant change in lithology or biotic compositions is recognized within this subunit, which implies fast reef growth that kept pace with and responded to a rapid sea-level rise, as indicated by a narrow range of Sr-isotope ages. The occurrence of a meteoric diagenesis zone, demarcated by sharp decreases in δ^{13}C and δ^{18}O values at the top of Subunit C3c (Fig. 3), demonstrates that the island was exposed by a sea-level fall prior to the deposition of Subunit C3b.

The known Sr-isotope ages of Subunit C3b range from 15.4 to 15.5 Ma. Again, based on the lack of distinct changes in lithology or biotic compositions within this subunit, fast reef growth that kept up with and responded to a rapid sea-level rise is inferred. Presumed lowstands of sea level prior to the reef formation of Subunits C3c and C3b correspond roughly with global eustatic (third-order) sea-level falls (Haq *et al.*, 1987, 1988). One Sr-isotope age known from Subunit C3a suggests that reef formed on the island around 10.9 Ma. It can be inferred that, prior to the reef formation, the island was subjected to subaerial exposure because a meteoric diagenesis zone, marked by sharp decreases in δ^{13}C and δ^{18}O values, occurs at the top of Subunit C3b (Fig. 3). The limited number of age

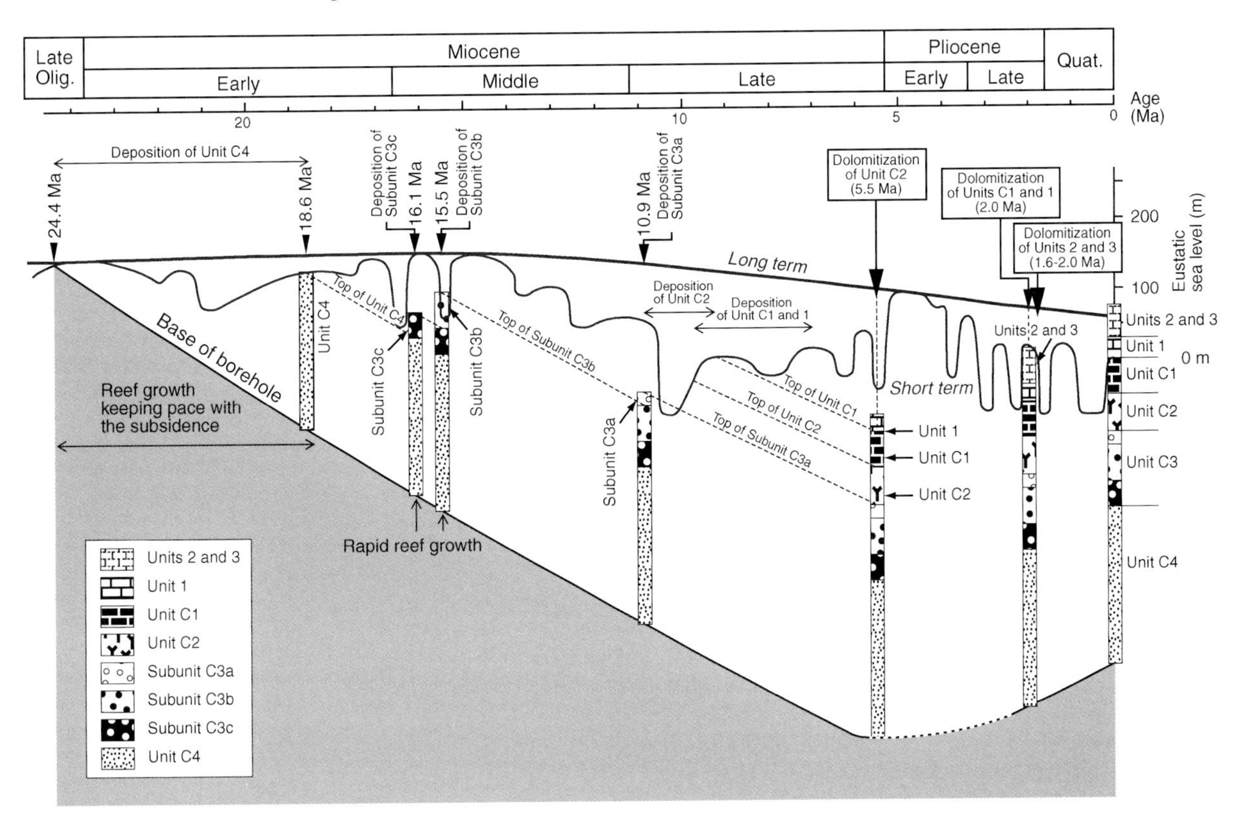

Fig. 6. Age-depth section showing the history of reef formation on Kita-daito-jima and global sea-level since 25 Ma. Global eustatic curves (in blue) after Haq *et al.* (1987, 1988).

data from Subunit C3a does not allow to infer the nature of the reef growth (rapid growth as with Subunits C3b and C3c versus gradual growth as with Unit 4). The common occurrence of corals and the coarser lithology suggest that Unit 3 accumulated in a lagoonal environment, which was shallower than that in which Unit C4 was deposited.

Extensive dolomitization masks the exact nature of the depositional history of Units C1 and C2. Strontium-isotope ages of Unit C2, the upper 10 m of which are karstified, average 5.5 Ma, so it is evident that the deposition and the subsequent karstification of this unit occurred during the Late Miocene. Sharp decreases in carbon- and oxygen-isotope values are not recognized at the top of Subunit C3a. This suggests that there is no significant sedimentary gap between Units C2 and C3. The age-depth section indicates that Unit C2 was located in the depth range of reef formation (<50 m water depth) during the early Late Miocene (Fig. 6). Collectively, it can be concluded that Unit C2 was deposited during the early Late Miocene, although this conflicts with the age determined by the larger foraminiferal biostratigraphy of this unit (Plio-Pleistocene; Hanzawa, 1940). The abundant occurrence of *in situ* branching corals, including *Acropora* spp., implies a shallow lagoonal environment.

Kita-daito-jima had subsided before it reached a lithospheric fore-bulge of the Philippine Sea Plate at 6 Ma (Seno, 1965). Since then, the island has been uplifted. The uplift rate is estimated at 0.03 mm yr^{-1} based on the occurrence of hermatypic corals of the last interglacial stage (Ota & Omura, 1938; Fig. 6). The fabric and texture of dolomites indicate that Unit C2 was subaerially exposed and subjected to meteoric diagenesis before it was dolomitized at 5.5 Ma; the dolomites are considered to be of marine origin (Suzuki *et al.*, 1936).

Recent investigation reveals a Sr-isotope age of ~4.8 Ma for non-dolomitized, coral framestone from Unit 2 (Iryu, unpublished data, 2006). This framestone sample was collected from the reef-core facies of Subunit 2a at an elevation of 17 m on a sea cliff along the western coast. The framestone includes marine cement that was precipitated after the dolomitization of island-surface carbonates at 1.6 to 2.0 Ma, and the Sr-isotope age (~4.8 Ma) seems to have been affected by the diagenesis. The age-depth section indicates that the only period before ~4.8 Ma during which Unit C1 (and Unit 1) was in a water depth range conducive to reef formation was the Late Miocene (Fig. 6). This is incon-sistent with the age of Unit 1 (Plio-Pleistocene) determined by benthic foraminiferal biostratigraphy (Hanzawa, 1940).

Unit C1 consists mainly of coral framestone with abundant hermatypic corals that are characteristic of a fore-reef environment (Nakamori *et al.*, 1989; Fig. 5C). Taking into account the location of the drilling site (near the centre of the island), it can be inferred that during the deposition of Unit C1 (and Unit 1), the island was lacking a central lagoon and that mounds consisting of corals and associated reef builders were scattered on a submerged platform. Although there is no lithological evidence to support emergence during the deposition of Unit C1, the negative excursion of $\delta^{13}C$ values at around 36.3 mbgs (Fig. 3) indicates likely subaerial exposure (Suzuki *et al.*, 1936). It is concluded that Unit C1 (and Unit 1) was exposed subaerially and subjected to meteoric diagenesis, and then most of the two units were dolomitized by seawater at 2.0 Ma (Suzuki *et al.*, 1936).

The vertical lithological change from bioclastic packstone/grainstone of Unit C4 followed by coral rudstone/floatstone of Unit C3 and by coral bafflestone of Unit C2 implies a gradual shallowing of the lagoon. Unit C1 is considered to have accumulated on a submerged platform.

Two modes of reef formation were recognized at Kita-daito-jima: reef growth that kept pace with the subsidence of the island; and rapid reef growth during periods of sea-level lowstand and the following rapid transgressions. The former is represented by the reef growth of Unit C4, which accumulated during the Late Oligocene to Early Miocene between 18.6 and 24.4 Ma, in response to a relative sea-level rise. The age-depth section indicates that the main cause of the relative sea-level rise was tectonic subsidence of the island (Fig. 6).

Kita-daito-jima atoll was drowned immediately after 18.6 Ma, although the relative sea-level rise is not believed to have outpaced the reef growth rates, considering the slow subsidence rate (4×10^{-2} m kyr^{-1}) and the gradual sea-level rise during that period (Fig. 6). Hence, the causes of this drowning are unknown. A similar paradox was pointed out by Schlager (1904). The latter mode of reef formation is represented by Subunits C3c and C3b. In this mode, reef formation was stimulated by the sea-level fall that resulted in the submerged island being brought into a shallow environment in which corals could recolonize. Therefore, sea-level falls are key events that cause submerged reefs to be rejuvenated, and reef formation is not necessarily limited to warm periods that are characterized by high sea-levels. It is noteworthy that the single atoll column contains both modes of reef deposition. The second mode may be less common and is probably confined to reef growth that occurred in response to relatively large-scale sea-level falls and the subsequent rises, such as those at 5.5, 10.5, 15.5, 16.5, and 25.0 Ma (Haq *et al.*, 1987, 1988). In conclusion, the history of reef formation that has been reconstructed from carbonate deposits in the Kita-daito-jima Borehole provides new insights into the initiation, development, and demise of coral reefs.

CONCLUSIONS

(1) The lowest lithostratigraphic unit within the Kita-daito-jima Borehole (Unit C4; 431.7–209.3 mbgs) consists of atoll deposits that accumulated from 18.6 to 24.4 Ma, when the reef growth kept pace with the relative sea-level rise caused mainly by tectonic subsidence of the island.

(2) Reefs formed during periods of sea-level lowstands and the subsequent transgressions at ~16.1 Ma (Subunit C3c) and ~15.5 Ma (Subunit C3b) when, as a result of global eustatic (third-order) sea-level falls, the submerged island returned to an environment that was shallow enough for reef formation. Available Sr-isotope ages suggest rapid reef growth. Again, the reef became rejuvenated at possibly 10.9 Ma (Subunit C3a). However, because the depositional age of Subunit C3a has not been well constrained, the nature of reef growth at this time is uncertain.

(3) Based on the reconstructed age-depth section of Kita-daito-jima, it can be inferred that Units C2 and C1 (103.4–49.7 mbgs and 49.7–2.68 mbgs, respectively) accumulated during the Late Miocene.

(4) The major lithology varies from bioclastic packstone/grainstone (Unit C4) to coral rudstone (Unit C3) to coral bafflestone (Unit C2), implying a gradual shallowing of the lagoon. However, the coral fauna suggests that Unit C1, above, formed on a submerged platform.

(5) Two modes of reef formation have been recognized in the coral-reef evolution on Kita-daito-jima during the Late Oligocene to Late Miocene: reef growth that kept pace with the subsidence of the island; and rapid reef growth during periods of sea-level lowstand and the subsequent rapid transgressions. The latter indicates that sea-level falls may revive submerged reefs and that reef formation is not necessarily limited to warm periods with high sea levels.

ACKNOWLEDGMENTS

We are grateful to Dr. T. Yamada, Mr. A. Nambu, and Mr. K. Odawara for helpful discussions and to Dr. T. Nakamori and Mr. M. Humblet for identification of the corals. We thank Dr. M. Mutti and an anonymous reviewer for their helpful comments and suggestions concerning the manuscript. This research was supported financially by a Grant-in-Aid for Encouragement of Young Scientists from the Ministry of Education, Science Sports and Culture, the Government of Japan (to Iryu; 07740408) and grants from the Saito Gratitude Foundation (Sendai, Japan), the Fukada Geological Institute (Tokyo, Japan), and the Kuribayashi Gakujutu Zaidan (Sapporo, Japan).

REFERENCES

Budd, D.A. (1997) Cenozoic dolomites of carbonate islands; their attributes and origin. *Earth Sci. Rev.*, **42**, 1–47.

Darwin, C.R. (1842) *On the Structure and Distribution of Coral Reefs.* Smith, Elder and Co., London, 214 pp.

Detrick, R.S. and **Crough, S.T.** (1978) Island subsidence, hot spots, and lithospheric thinning. *J. Geophys. Res.*, **83**, 1236–1244.

Dunham, R.J. (1962) Classification of carbonate rocks according to depositional texture. In: *Classification of Carbonate Rocks* (Ed. W.E. Ham). *AAPG Mem.*, **1**, 108–121.

Ehrenberg, S.N., **McArthur, J.M.** and **Thirlwall, M.F.** (2006) Growth, demise, and dolomitization of Miocene carbonate platforms on the Marion Plateau, offshore NE Australia. *J. Sed. Geol.*, **76**, 91–116.

Embry, A.F. and **Klovan, E.J.** (1971) A Late Devonian reef tract on northeastern Banks Island, Northwest Territories. *Bull. Can. Petrol. Geol.*, **19**, 730–781.

Field, R.M. and **Hess, H.H.** (1933) A bore-hole in the Bahamas. *Trans. Am. Geophys. Union*, 234–235.

Flint, D.E., **Corwin, G.**, **Ding, M.G.**, **Fuller, W.P.**, **MacNeil, F.S.** and **Saplis, R.A.** (1953) Limestone walls of Okinawa. *Geol. Soc. Am. Bull.*, **64**, 1247–1260.

Hanzawa, S. (1940) Micropalaeontological studies of drill cores from a deep well in Kita-daito-jima. In: *Jubilee Publication in Commemoration of Prof. H. Yabe M.I.A. 60th Birthday*, **2**, pp. 755–802. Sasaki Printing & Publishing Co. Ltd., Sendai.

Haq, B.U., **Hardenbol, J.** and **Vail, P.R.** (1987) Chronology of fluctuating sea level since the Triassic (250 million years ago to present). *Science*, **235**, 1156–1167.

Haq, B.U., **Hardenbol, J.** and **Vail, P.R.** (1988) Mesozoic and Cenozoic chronostratigraphy and eustatic cycles. In: *Sea-level Changes: an Integrated Approach.* (Eds C.K. Wilgus, B.S. Hastings, C.G.St.C. Kendall, H.W. Posamentier, C.A. Ross and J.C. van Wagoner). *SEPM Spec. Publ.*, **42**, 71–108.

Honza, E. and **Fujioka, K.** (2004) Formation of arcs and backarc basins inferred from the tectonic evolution of Southeast Asia since the Late Cretaceous. *Tectonophysics*, **384**, 23–53.

Inagaki, S. and **Iryu, Y.** (1999) Depositional and diagenetic history of Kita-daito-jima for the last 25 million years. *Chikyu Monthly*, **21**, 718–723 (in Japanese; original title translated).

Kawana, T. and **Ohde, S.** (1993) A short reconnaissance of Okino-Daito-Jima Island in the northern Philippine Sea: Implications for Quaternary crustal movements of the raised almost-table reef. *Bull. Coll. Educ. Univ. Ryukyus*, **43**, 57–69 (in Japanese with English abstract).

Klein, G. de V. and **Kobayashi, K.** (1980) Geological summary of the North Philippine Sea, based on Deep Sea Drilling Project Leg 58 results. *Init. Rep. Deep Sea Drilling Proj.*, **58**, 951–961.

Ladd, H.S., **Ingerson, E.**, **Townsend, R.C.**, **Russell, M.** and **Stevenson, H.K.** (1953) Drilling on Eniwetok Atoll, Marshall Islands. *AAPG Bull.*, **37**, 2257–2280.

Ladd, H.S., **Tracey, J.I. Jr.** and **Gross, H.G.** (1970) Deep drilling on Midway Atoll. JT U.S. Geol. Surv. Prof. Pap. **680-A**, 1–22.

McKee, E.H. and **Klock, P.R.** (1980) K-Ar ages of basalt sills from Deep Sea Drilling Project Site 444 and 446, Shikoku Basin and Daito Basin, Philippine Sea. *Init. Rep. Deep Sea Drilling Proj.*, **58**, 921–922.

MacNeil, F. S. (1954) The shape of atolls: an inheritance from subaerial erosion forms. *Am. J. Sci.*, **252**, 402–427.

Marty, J.C. and **Cazenave, A.** (1989) Regional variations in subsidence rate of oceanic plates: a global analysis. *Earth Planet. Sci. Lett.*, **94**, 301–315.

Nakamori, T., **Iryu, Y.**, **Ozawa, S.** and **Mori, K.** (1994) The depositional history of carbonate in Kita-daito-jima. Okinawa Prefecture. *Chikyu Monthly*, **16**, 401–406 (in Japanese; original title translated).

Nambu, A., **Inagaki, S.**, **Ozawa, S.**, **Suzuki, Y.** and **Iryu, Y.** (2003) Stratigraphy of reef deposits on Kita-daito-jima. Japan. *J. Geol. Soc. Japan*, **109**, 617–634 (in Japanese with English abstract).

Ohde, S. and **Elderfield, H.** (1992) Strontium isotope stratigraphy of Kita-daito-jima Atoll, North Philippine Sea: implications for Neogene sea-level change and tectonic history. *Earth Planet. Sci. Lett.*, **113**, 473–486.

Ota, Y. (1938) Chemical analysis and microchemical tests on the cores obtained from the deep well in Kita-daito-jima and Daito Limestone. *Contrib. Inst. Geol. Paleontol. Tohoku Imp Univ.*, **30**, 1–36 (in Japanese; original title translated).

Ota, Y. and **Omura, A.** (1992) Contrasting styles and rates of tectonic uplift of coral reef terraces in the Ryukyu and Daito Islands, southwestern Japan. *Quatern. Int.*, **15**, 17–29.

Ozawa, S. (1995) *Stratigraphy of Reef Deposits on Kita-daito-jima, Okinawa Prefecture and Re-examination of Carbonate Rocks in Kita-daito-jima Borehole.* Graduate Thesis, Inst. Geol. Paleontol., Tohoku Univ., Sendai, 295 pp. (in Japanese; original title translated).

Parsons, B. and **Sclater, J.G.** (1977) An analysis of the variation of floor bathymetry and heat flow with age. *J. Geophys. Res.*, **82**, 803–827.

Pirrson, L.V. and **Vaughan, T.W.** (1913) A deep boring in Bermuda Island. *Am. J. Sci.*, **136**, 70.

Richards, H.C. and **Hill, D.** (1942) Great Barrier Reef bores, 1926 and 1937; Descriptions, analyses and interpretations. *Rep. Great Barrier Reef Comm.*, **5**, 1–122.

Royal Society of London (1904) *The Atoll of Funafuti: Borings into a Coral Reef and the Results.* Royal Society of London, London, 428 pp.

Schlager, W. (1981) The paradox of drowned reefs and carbonate platforms. *Geol. Soc. Am. Bull.*, **92**, 197–211.

Schlanger, S.O. (1965) Dolomite-evaporite relations on Pacific islands. *Sci. Rep. Tohoku Imp. Univ. 2nd Ser. (Geol.)*, **37**, 15–29.

Seno, T. (1989) Philippine Sea plate kinematics. *Modern Geol.*, **14**, 87–97.

Steers, J.A. and **Stoddart, D.R.** (1977) The origin of fringing reefs, barrier reefs, and atolls. In: *Biology and Geology of Coral Reefs*, **4**, *Geology 2* (Eds O.A. Jones and R. Endean), pp. 21–53. Academic Press, New York.

Sugiyama, T. (1934) On the drilling in Kita-daito-jima. *Contrib. Inst. Geol. Paleontol., Tohoku Imp. Univ.*, **11**, 1–44 (in Japanese; original title translated).

Sugiyama, T. (1936) The second boring in Kita-daito-jima. *Contrib. Inst. Geol. Paleontol., Tohoku Imp. Univ.*, **25**, 1–34 (in Japanese; original title translated).

Suzuki, Y., **Iryu, Y.**, **Inagaki, S.**, **Yamada, T.**, **Aizawa, S.** and **Budd, D.A.** (2006) Origin of atoll dolomites distinguished by geochemistry and crystal chemistry: Kita-daito-jima, northern Philippine Sea. *Sed. Geol.*, **183**, 181–202.

Yamanari, F. (1935) Alumino-phosphate mineral deposit on Kita-daito-jima. *Contrib. Inst. Geol. Paleontol. Tohoku Imp Univ.*, **15**, 1–65 (in Japanese; original title translated).

Int. Assoc. Sedimentol. Spec. Publ. (2010) **42**, 257–282

Carbonate production in rift basins: models for platform inception, growth and dismantling, and for shelf to basin sediment transport, Miocene Sardinia Rift Basin, Italy

MARIO VIGORITO*[1], MARCO MURRU† and LUCIA SIMONE*

**Dipartimento di Scienze della Terra, Università di Napoli "Federico II", Largo San Marcellino 10, 80138 Napoli, Italy (E-mail: lusimone@unina.it)*

†Dipartimento di Scienze della Terra, Università di Cagliari, via Trentino 51, 09127 Cagliari, Italy

ABSTRACT

The main sedimentary and architectural patterns of Miocene carbonate successions laid down in the Miocene Sardinia Rift Basin are described, and the controls on sedimentation, facies distribution and sequence development are evaluated. Three case-histories from different physiographic settings are described which correspond to: (1) narrow rift-related submerged valleys; (2) isolated fault-blocks in the axial portion of the rift; and (3) small basins located at the edge of the rift itself. Facies distribution and depositional architectures appear to have been controlled mainly by palaeophysiography and in turn by pre- and synsedimentary tectonics. These also played a leading role in dictating the type and evolution of the benthic communities and thus the location and morphology of the carbonate factories. Both coral-dominated and foramol-rhodalgal assemblages are recognized. These developed peculiar depositional architectures and had remarkably different responses to relative sea-level variations. Depending primarily on the palaeophysiography, the resulting carbonate successions occur as isolated carbonate lenses or as sedimentary bodies up to a few hundred metres thick.

In sectors close to the rift axis, coral-dominated factories developed on the side of isolated fault-blocks that faced the open sea. The geometry of the coral-dominated sequences suggests biostromal sedimentary accumulations rather then true reefs. These coral-dominated deposits interfinger with foramol-dominated sequences that were laid down in deeper sectors of the carbonate factories. Following a major regressive event, probably associated with climatic change, the coral-dominated factories were shut down and replaced by red algae and bivalve-dominated factories. Close to the margins of the rift-systems carbonate factory development was mainly controlled by the type and the rate of siliciclastic sedimentation. In submerged rift-related valleys with high siliciclastic input, small lenticular carbonate factories, locally forming small patch-reefs, developed during phases of stasis between fresh floods. In other areas, relatively thick carbonate sequences accumulated at the top of prograding siliciclastic wedges and on hinterland-facing fault-blocks.

Keywords Carbonate sequences, rift basin, fault blocks, depositional architectures, Miocene, Sardinia.

INTRODUCTION

Sedimentation in rift basins is controlled by a variety of factors including regional tectonics, climate and sedimentary processes. Previous studies (Purser *et al.*, 1996; Bosence, 1998; Bosence *et al.*, 1998; Purser & Bosence, 1998, and references therein) have shown that fault blocks of a rift-system can be grouped according to their position with respect to the rift axis. This factor governs the amount of siliciclastic sedimentary inputs as well as the distribution, type and development of the carbonate factories (Purser et *al.*, 1996; Purser & Bosence, 1998). Carbonate production in rift sub-basins is generally but not exclusively confined to topographically elevated areas of submerged fault-blocks located in the axial portion of the rift system

[1]Present address: Statoil ASA, EXP SA NL, Grenseveien 21, N-4035 Forus, Norway

far from siliciclastic input (Purser *et al.*, 1996; Bosence, 1998; Bosence *et al.*, 1998; Purser & Bosence, 1998). However, other factors, mainly related to the ecological behaviours of the benthic communities, may favour the development of carbonate factories in areas close to the rift periphery and locally characterized by high siliciclastic input.

Existing models are mainly derived from detailed seismic and outcrop studies of coral-dominated syn-rift carbonate sequences (Bosence, 1998; Bosence *et al.*, 1998; Purser *et al.*, 1998, and references therein). However, as demonstrated by Cherchi *et al.* (2000), Vigorito (2005) and Vigorito *et al.* (2005, 2006), sedimentary dynamics and the resulting depositional architectures can be significantly different when the carbonate factories are dominated by foramol *sensu lato* benthic communities. This can invalidate the proposed models and lead to spurious results if there is little or no consideration of the ecological behaviour of the benthic communities forming the main carbonate production areas.

The Oligo-Miocene Sardinia Rift System offers exceptionally well-exposed examples of carbonate factories which allow a detailed sedimentological and geometric analysis. The case histories illustrated in this paper are carbonate factories occurring in different palaeophysiographic settings, and include factories developed in both axial and peripheral sectors of the rift system, as well as small carbonate production areas formed within narrow rift-related submerged valleys. The analysis of these factories provides insights into sedimentary processes as well as on the main architectural patterns. The proposed analogues show both coral- and foramol-dominated sequences, and thus allow for discrimination of genetic and controlling processes in the light of the ecology of the productive benthic communities. This is extremely important for sequence stratigraphic interpretation, as coral- and foramol-dominated factories have different architectural patterns and internal geometries and, more importantly, show remarkably different responses to sea-level variations (Cherchi *et al.*, 2000; Vigorito, 2005; Vigorito *et al.*, 2005, 2006; Bassi *et al.*, 2006).

Methods

Detailed geological mapping was carried out in different areas of central and southern Sardinia. In these areas, Plio-Pleistocene tectonic and geomorphological processes have created spectacular exposures that extend for kilometres or tens of kilometres, offering a superb window of observation and allowing for the recognition and the correlation of shallow to deep-water carbonate sedimentary bodies with reasonably good 3-D control.

Accurate bed-to-bed analyses, including sedimentological studies were performed on the individual areas (Fig. 1) that were also logged. Thin sections were prepared from collected samples and subsequently analysed with an optical microscope. Special attention was paid to indicative surfaces (e.g. hardgrounds, erosion and drowning surfaces), sedimentary structures, stacking patterns and vertical and lateral facies distribution patterns. Palaeocurrent indicators (e.g. ripples, imbricated grains, scour marks) were measured throughout the investigated areas.

GEOLOGICAL FRAMEWORK

During the Mesozoic and Palaeogene, Sardinia and Corsica formed an integral part of the southern margin of the European plate. As a consequence, Sardinia and Corsica broadly shared their geological history with western Europe (Iberian Peninsula-southern France) until the Aquitanian (20–21 Ma), when the Corso-Sardinian Block was rotated away from the southern European margin because of the oceanic opening of the Provençal Basin (Cherchi & Montadert, 1984). This resulted in the formation of a large rift system whose eastern branch crosses Sardinia longitudinally (Fig. 1) and continues offshore into the Gulf of Asinara to the north and the Gulf of Cagliari to the south (Cherchi & Montadert, 1982).

The structural evolution of the Oligo-Miocene Rift System of southern Sardinia is linked to the general geodynamic evolution of the western central Mediterranean. Thus, the extensional Late Oligocene–Early Burdigalian event is a consequence of an "Apenninic" westward subduction process associated with a volcanic arc (29–30 Ma to 15–16 Ma). Regional tectonics led to the opening of the oceanic Provençal Basin and the anticlockwise rotation of Sardinia–Corsica during the Aquitanian–Langhian (20–21 Ma to 15–16 Ma). Relatively small sub-basins (a few tens up to a few hundred square kilometres wide) formed at the top of the down-faulted blocks during the phases of active rifting and were subsequently filled by

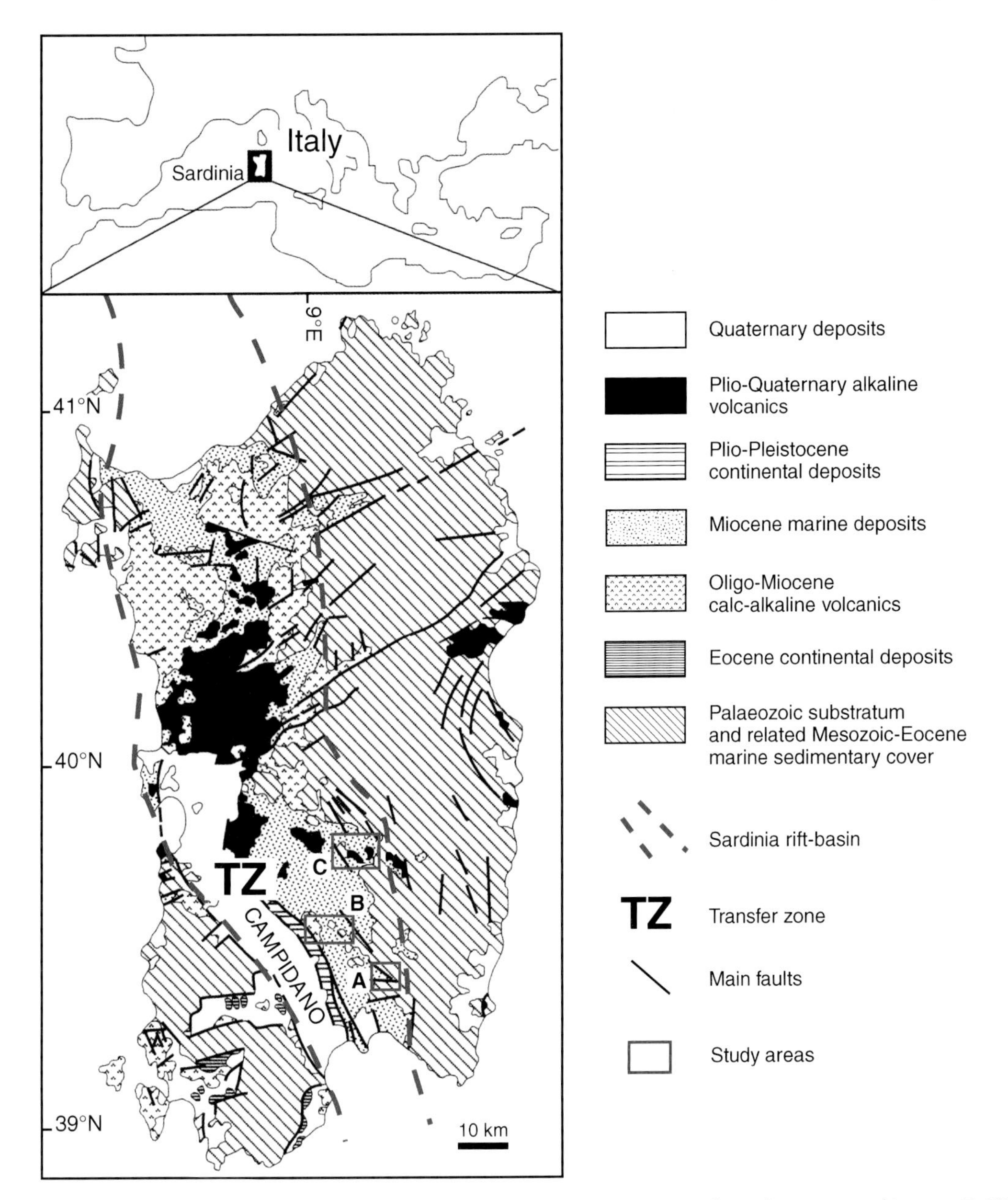

Fig. 1. Geological-structural map of Sardinia. The locations of the areas investigated are shown: A, Dolianova; B, Villagreca; C, Isili.

continental and, since Aquitanian times, marine deposits (Cherchi & Montadert, 1984; Casula *et al.*, 2001).

Analogous to most intracontinental extensional systems, such as the Suez Rift, the East African Rift, the Rhine Graben, the North Sea, and the Oslo Graben, the internal structure and deformation of the South Sardinia Oligo-Miocene Rift was controlled by major normal faults, tilted blocks and by transverse structures. This results in a complex graben system characterized by an asymmetric inner structure and divided into compartments of half-grabens with opposite dip-direction (average dip, 10–15°) separated by a "Transfer Zone" (Fig. 1, TZ). This Transfer Zone is characterized by a transfer fault system and a central uplift which forms a complex "horst-type" twist zone (Maillard & Mauffret, 1993; Casula *et al.*, 2001).

Major normal faults are oriented NNW–SSE and are parallel to the rift axis (Fig. 1). These master faults are segmented by minor N–S to NNE–SSW and E–W to WNW–ESE faults, the latter exhibiting minor throws (Cherchi & Montadert, 1982; Casula *et al.*, 2001). The principal master faults, such as the Sinis, the Sarroch and the Monastir faults, have original dips of 60–80°, and are probably listric (Casula *et al.*, 2001). These faults, commonly associated with both antithetic and synthetic longitudinal faults, propagate along the entire rift zone, though they are segmented and offset by transfer faults. The kinematics and the strong asymmetry of the basin are controlled by these faults, emphasizing the change of polarity of the half-grabens (Cherchi & Montadert, 1982; Casula *et al.*, 2001). Minor transverse faults are steeper, almost vertical, and show limited extension, as they do not cross the entire rift zone. However, their role in rifting is important as they act as transfer faults between the major longitudinal normal faults.

Several volcanic events punctuated the evolution of the Sardinian Oligo-Miocene Graben System (Lecca *et al.*, 1997). The first event is represented by calc-alkaline volcanics, dated to 28 Ma, which preceded marine flooding of the rift system. Volcanic activity continued in a marine environment until 15 Ma. From Late Oligocene–Early Aquitanian times, coarse syn-rift breccias and fan-delta conglomerates were deposited at the margins of emerged tilted fault-blocks. These coarse syn-rift deposits pass laterally (basinward) and upward to fluvio-deltaic and marine sandstones (Casula *et al.*, 2001). Since the Aquitanian, carbonate factories started to develop at the top of submerged fault-blocks wherever the environmental conditions were favourable. Carbonate sequences locally intercalate with siliciclastic deposits. Since the Middle Burdigalian, a major transgressive event led to the deposition of hemipelagic marls, interpreted as post-rift deposits (Casula *et al.*, 2001).

During the Messinian, a compressional phase occurred and led to the formation of inversion structures in the southern Sardinia Oligo-Miocene rift (Casula *et al.*, 2001). These structures are subparallel to the Oligo-Miocene NNW–SSE trending faults, with variations of the structural trends resulting from local rotations within the shear system. As a result, the main NNW–SSE Oligo-Miocene longitudinal faults were reactivated as oblique normal faults.

OUTCROP DESCRIPTION

On the basis of previous studies (Cherchi *et al.*, 2000; Casula *et al.*, 2001, Vigorito *et al.*, 2005) and newly acquired data, several outcrops (Fig. 1), representative of different physiographic and tectonic settings, were selected and are described below in order to define the wide spectrum of depositional architectures and sedimentary features of the carbonate factories developed in syn-rift basins.

Dolianova: carbonate inception in sheltered rift valleys

Syn-rift and post-rift sequences cropping out in the Dolianova area occur as fills of a narrow palaeovalley created through extensional faulting of the Palaeozoic substratum (Figs. 1 and 2). The valley is oriented parallel to the rift-system (approximately N–S) and plunges toward the north into the main rift-basin system. The faulted margins of the valley are draped and/or onlapped by coarse to very coarse (cobble to boulder-sized) polygenic breccias and conglomerates (Fig. 2). These deposits are up to 300 m thick and are typically clinostratified with strata steeply dipping towards the valley axis (Fig. 3A). These coarse deposits exhibit fanning strata geometries and marked grain-size variation, with sediments fining both upwards and towards the valley axis.

In the logged section (Fig. 2) more than 60 m of conglomerates occur at the base. These coarse deposits are also characterized by vertical alternation and lateral transitions of breccias, granule to cobble-sized conglomerates and sands. Silty and rarely muddy intervals occur locally. Normally graded beds, commonly with sharp erosive bases, as well as imbricated grains are very common features. Pebbly mixed carbonate-siliciclastic and pure carbonate skeletal sands are locally very abundant. The latter form lenses and/or discontinuous beds and may grade upward into "sedimentological concentrations" (*sensu* Kidwell *et al.*, 1986 and Kidwell, 1991) of coarse skeletal debris of bivalves, red algae, bryozoans and benthic foraminifera. These bioclastic materials are extensively bioeroded, abraded and fragmented. However, entire *in situ* irregular echinoids occur commonly within these beds.

The basal unit is conformably overlaid by about 170 m of breccias and conglomerates with a sandstone matrix. Carbonate bioclastic debris, made up of bivalve fragments and subordinately bryozoan and red algae, occur within the siliciclastic deposits. The latter also include several carbonate lenses (Fig. 3B, C and D) which become progressively more abundant up-section. In the lower portion of the breccias-conglomerate unit the carbonate

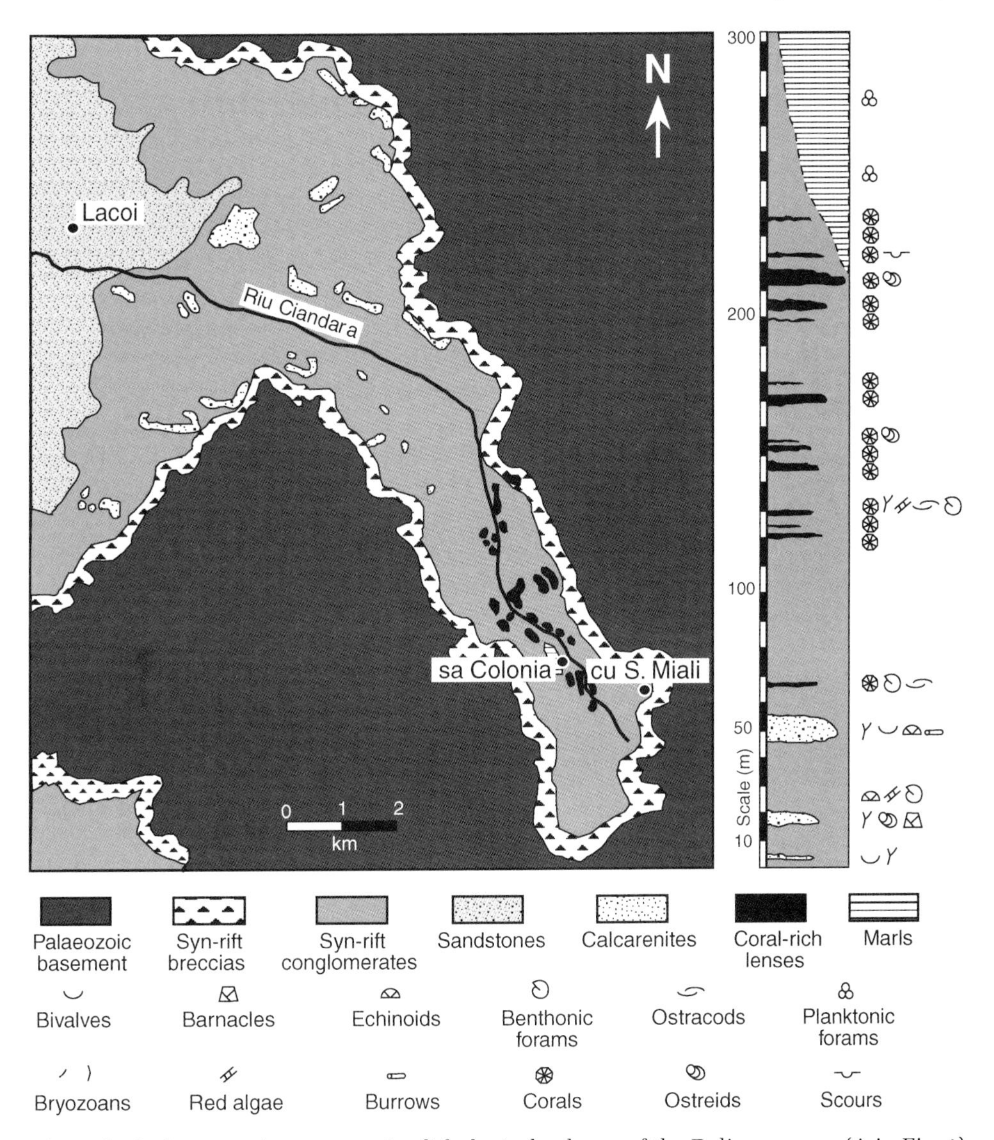

Fig. 2. Schematic geological map and representative lithological column of the Dolianova area (A in Fig. 1).

lenses are usually less than 1 m thick, but they may exceed 6 m in the upper portion. The lenses consist of coral- and bivalve-rich rudstones and floatstones with a grainstone or silty packstone matrix (Fig. 3E and F). At the base, individual lenses exhibit sedimentological concentrations of disarticulated and intensely bioeroded bivalves (mainly ostreids) which are in turn overlain by patches of corals in life position. Red algae locally encrust the top of the shell concentrations, providing evidence for local or partial sediment binding. Coral assemblages are dominated by *Porites* with subordinate *Tarbellastrea*, *Montastraea*, and rare *Favites* and form colonies with different morphologies. At the base of the coral-dominated interval, siliciclastic pebbles are encrusted by small corals. These small pioneer colonies grade upward into coral boundstones dominated by both globular and branching *Porites* colonies, up to 1 m in diameter and 0.7 m high (Fig. 3F), alternating with rudstones and floatstones made up of horizontally-oriented stick-like coral fragments and exhibiting a strongly oriented fabric. Coral colonies are concentrated at the core of individual carbonate lenses and pass laterally into bivalve-dominated grainstones with scattered globular coral-colonies.

Coral-dominated lenses are frequently truncated at the top by erosion surfaces which are overlain by matrix-supported conglomerates and coarse sandstones. Alternatively, the carbonate lenses exhibit reddened and/or oyster-encrusted,

Fig. 3. Outcrop photographs from the Dolianova area. (A) Clinostratified breccias and conglomerate deposits from the margin of the Riu Ciardara rift-related palaeovalley. (B) Carbonate lens in siliciclastic-dominated sequences. Note the complex stratification pattern. (C) Detail of a small coral build-up overlying pebbly/cobbly conglomerates. (D) Carbonate lens overlying conglomerate deposits. Note the irregular morphology of the basal contact surface. Hammer (30 cm long), circled, for scale. (E) Detail of the carbonate lens in (D). (F) Detail of a coral colony from one of the coral-dominated carbonate lenses which alternate with the syn-rift conglomerates.

early-hardened top surfaces. The stratigraphically younger carbonate lenses are very rich in large oysters which are commonly extensively bored, whereas corals are sparse and not very diverse.

About 70 m of matrix-supported siliciclastic conglomerates (up to 40 cm diameter) follow up-section. These rest conformably on the underlying deposits and do not show any evidence of marine

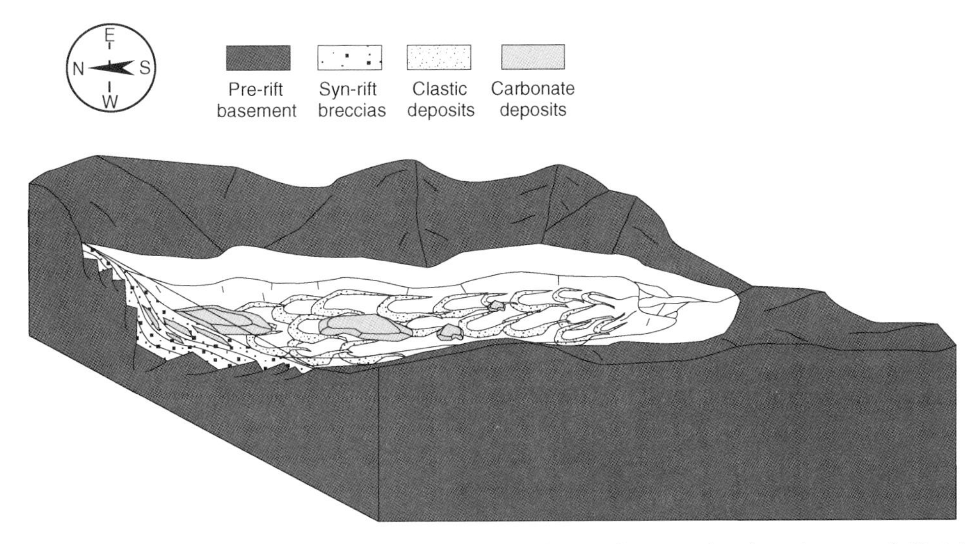

Fig. 4. Block diagram of the Dolianova area. Not to scale. Patchy carbonate factories developed on top of siliciclastic deposits during periods of reduced siliciclastic input.

deposition. A sharp erosion surface cuts through the logged succession as deeply as 70–80 m, forming a large-scale channel that was subsequently filled by marls rich in planktonic foraminifera and fish scales.

Interpretation

The siliciclastic deposits occurring in the Dolianova area are interpreted to represent fan-delta deposits. They form thick siliciclastic wedges that prograded northward towards the main rift system (to the north, Fig. 4). Carbonate production in this area was limited to patchily distributed small benthic communities which grew up on the siliciclastic wedges during periods of reduced sediment input from inland. During these periods, the siliciclastic sediments were locally stabilized by sessile and encrusting communities dominated by bivalves which dwelled in mixed siliciclastic and carbonate debris. The latter was mainly produced through biological and mechanical erosion. These mobile sheets of sediments were periodically colonized by corals which built-up organogenic constructions (Fig. 3F).

The development of the bioconstructions was strongly controlled by the rate of siliciclastic input and periodically interrupted by the occurrence of flashfloods (Cherchi *et al.*, 2000). Conversely, bivalve-dominated communities were far less sensitive to environmental variations and particularly to freshwater and siliciclastic input and were thus able to proliferate even in periods characterized by high terrigenous input.

Villagreca: inception and growth of carbonate platforms on isolated tilted fault blocks

The Villagreca area (Figs. 1 and 5) is located on the eastern side of the main Sardinia Rift Basin. Here several tilted blocks are aligned parallel to the rift axis and are covered by Miocene carbonate sequences. Two of these fault-blocks (Monte Su Crucuri and Genna Siustas; Figs. 5, 6 and 7A) were analysed and logged in order to define the main sedimentological and architectural patterns.

Monte Su Crucuri section

The sedimentary sequences exposed in the Monte Su Crucuri area (Figs. 1, 5, 6 and 7A) include, at the base, several metres of barren pelleted red clay passing upwards into siltstones. These deposits overlie an andesitic volcanic basement separated by a sharp unconformity.

The basal deposits are overlain by about 10 m of sandstones and fine-grained conglomerates which alternate with palaeosoils and heavily bioturbated sandy layers (Fig. 5). Multiple laterally extensive (high width/depth ratio) channels filled by sandstones and sandy conglomerates cut through these deposits. Oyster fragments are locally frequent and admixed with land-derived material including plant remains. Biogenic concentrations (*sensu* Kidwell *et al.*, 1986) of oysters forming up to 1 m thick oyster-banks are locally intercalated into this siliciclastic-dominated succession.

About 4 m of bioclastic limestones follow upsection. These consist of coral-dominated rudstones

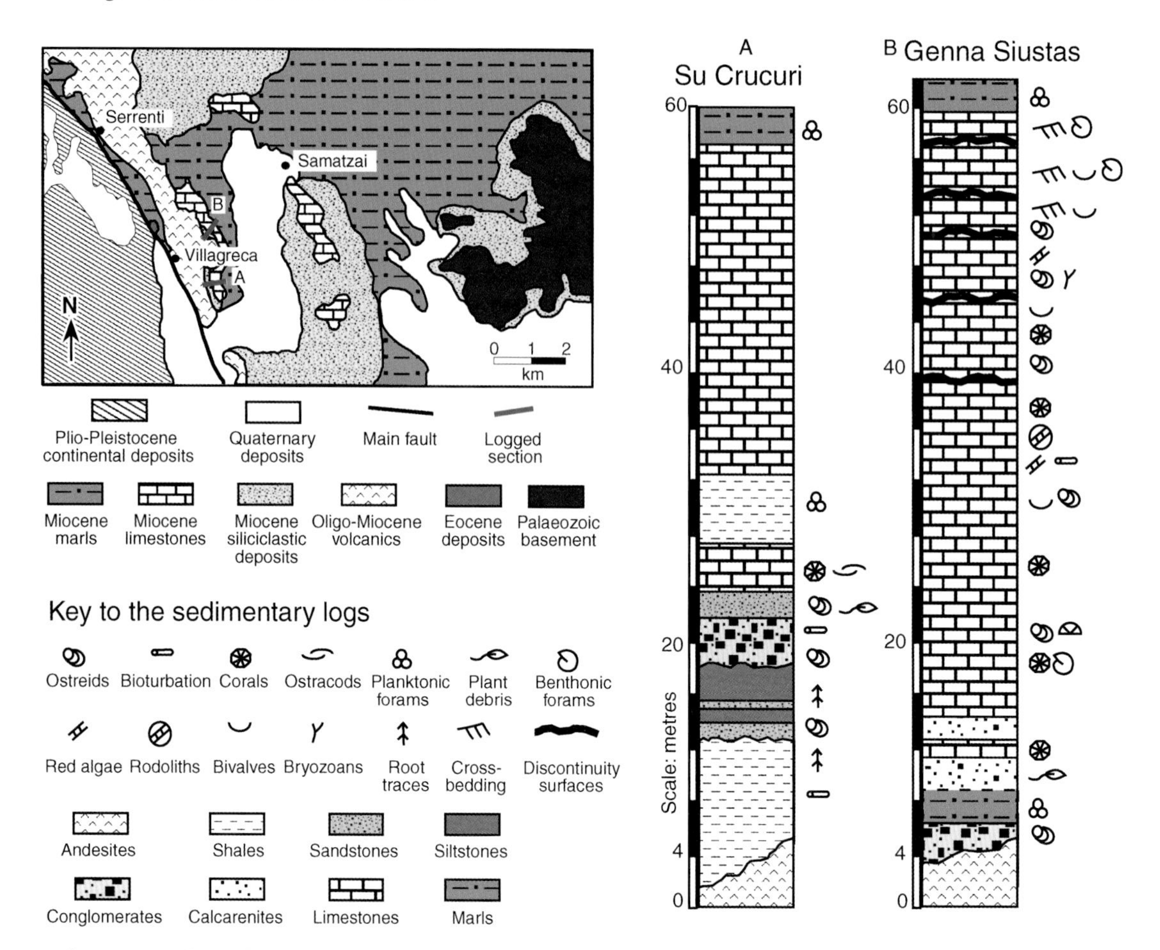

Fig. 5. Schematic geological map and representative lithological columns of the Villagreca area (B in Fig. 1). Locations of the logged sections are shown: A, Monte Su Crucuri; B, Genna Siustas.

Fig. 6. Panoramic view of the Villagreca area. The two described sections: Genna Siustas and Monte Su Crucuri are illustrated (A and B respectively). The dashed line marks the contact between the syn-rift siliciclastic/continental deposits and the coral-dominated limestone. The dotted line indicates the contact between the coral-dominated limestones and the overlying foramol limestones. The relative positions of the two parts (A) and (B) is indicated (a and b in open square).

Fig. 7. Villagreca area successions. (A) Well-stratified foramol limestones prograde to the SE in the Genna Siustas Fault Block (to the left). The Su Crucuri fault block is capped by limestone sequences prograding to the east. (B) Detail of coral-rich rudstones with abundant branching corals. (C) Cross-stratified foramol limestones overlying a sharp erosion surface at the top of the coral-dominated sequence. (D) Bivalve-rich rudstones followed by red algae-rich rudstones overlying a sharp early-hardened surface at the top of the coral-dominated sequences. Note the sharp clinostratification within the foramol sequences. (E) Clinostratified foramol rudstones and coarse grainstones filling a small channel. Hammer (30 cm long), circled, for scale.

and floatstones with subordinate oysters, barnacles and bryozoans. Alternating bioclastic packstones, marls and silty marls occur upsequence. Ostracods, planktonic and small benthic foraminifera are common in marly strata. Corals are quite common in intercalated packstones and are frequently encrusted by bryozoans, small oysters and barnacles.

The deposits pass upsection into a 25 m thick interval of coral-rich rudstones and floatstones with grainstone/packstone matrix. Branching *Porites* colonies are the most common forms and individual branches usually exceed 1 cm in diameter. The colonies are generally fragmented and randomly distributed into the matrix (*in situ*/near *situ* reworked sediments *sensu* Kidwell, 1991). Oysters with articulated valves, generally intensely bioeroded, are also present. These carbonate deposits are capped by hemipelagic marls.

Genna Siustas section

The Genna Suistas fault-block is located north of the previously described Monte Su Crucuri fault-block (Cherchi *et al.*, 2000; Figs. 1, 5, 6 and 7A). The Palaeozoic basement is covered by thick andesites, which are overlain by up to 3 m thick pebble-sized discontinuous conglomerates rich in volcanic clasts and oyster fragments. These deposits pass laterally into chaotic breccias consisting of loferite intraclasts with intercalated thin marly-sandstone lenses (Fig. 5). This basal unit is characterized by abrupt lateral variations of facies and thickness which may be related to a palaeotopographic relief.

Fine-grained bioclastic sandy-silty deposits follow upsection. Coral fragments as well as plant remains are locally present. These fine carbonate deposits grade rapidly into coral-rich rudstones with a silty-muddy matrix (Fig. 7B and C). *Porites* prevail among the corals, and subordinate bivalves and echinoids also occur. This unit, 40–50 m thick (Figs. 5 and 7B, C), mainly shows branching colonies in the lower beds, while in the upper portions coral-rich beds are characterized by massive colonies. These coral colonies are extensively bored and cyclically truncated by discontinuity surfaces on which large articulated oysters and echinoids dwelled.

A major erosion surface truncates the coral-rich unit and is in turn overlain by an up to 0.6 m thick oyster-bank (Fig. 7D). Coarse skeletal grainstones and rudstones, rich in benthic foraminifera, siliceous sponge spicules, echinoids and bivalve fragments follow upsection. This sandy/gravelly bioclastic debris forms multiple vertically stacked clinostratified units (up to 3 m thick) which prograde towards the south and SE. Locally small channels filled by rhodalgal rudstones also occur (Fig. 7E). A few metres of hemipelagic marls cap the logged section.

Other blocks

Shallow-water carbonate sequences are easily recognizable at the top of a series of tilted fault-blocks located both north and east of the Villagreca area (Fig. 8A and B). These blocks are generally tilted to the east (towards the rift periphery), forming sharply asymmetric half-graben sub-basins. Carbonate sequences occur on the most elevated portions of these blocks and along their eastern flank, and show sedimentological patterns similar to those described from the Villagreca area. Coral-rich deposits are arranged to form well-stratified tabular bodies prograding towards the east. A sharp erosion surface separates the coral-rich deposits from the overlying foramol carbonates.

Fig. 8. Panoramic view of fault blocks occurring close to the Villagreca area. (A) The fault blocks are capped by limestone sequences which prograde to the ESE (to the left). (B) Facies distribution on the Furtei fault block: a = syn-rift volcanics; b = Miocene carbonate factory sequences; c = Miocene outer-shelf/slope carbonate deposits; d = hemipelagic marls.

The latter are generally cross-stratified and include large-scale (up to 5 m) clinostratified units, as well as multiple channels. Transport directions are mainly to the south and SE.

Interpretation

The carbonate sequences exposed in Villagreca and neighbouring areas testify to the development and growth of coral-dominated carbonate factories on the topographically higher portions of submerged fault-blocks located in the axial portion of the rift-system (Fig. 9). The western sides of the blocks were facing the open sea and were colonized by coral-dominated benthic communities. The sediments exported from the main production areas were redeposited in the form of sheets covering the eastern sectors of the blocks. They hosted sparse patchy benthic communities in which corals were associated with bivalves. Vertically stacked hardgrounds, commonly associated with sharp erosion surfaces, truncate the coral-dominated sequences. These discontinuity surfaces are inferred to correspond to relative sea-level falls (see also Nelson & James, 2000), during which part of the sedimentary cover was removed and rearranged into sedimentary wedges on the eastern margins of the blocks. These wedges interfinger with marly deposits laid down in deeper areas of the relatively small sub-basins located at the top of fault-blocks (Fig. 9).

A sharp erosional surface truncates the top of the coral-dominated sequences. This surface is overlain by foramol carbonate deposits (Figs. 6 and 7D). These are cross-stratified and include features such as multiple channels (Fig. 7D and E) and large-scale clinostratified units which suggest a high-energy depositional setting. The sharp transition from coral-dominated to foramol-dominated deposits suggests a change in the environmental conditions that followed a major sea-level fall, documented by the main erosion surface at the base of the foramol carbonate sequences.

Isili area: carbonate production on the outer margin of the main rift-basin periphery

The Isili sub-basin extended approximately NNW–SSE along the eastern side of the Oligo-Miocene Sardinia rift (Figs. 1 and 10) and was confined on its western side by a regional normal fault and on the eastern side by the Sarcidano structural high. According to Casula *et al.* (2001), the Isili sub-basin was located in the proximity of the "twisting zone" or "transfer zone" (Fig. 1). This peculiar structural setting created a complex palaeophysiography, consisting of a narrow NNW–SSE trending, south-plunging trough (Isili Trough), surrounded by structural highs (Vigorito, 2005; Vigorito *et al.*, 2005).

Since Oligocene–Aquitanian times, these structural highs were partially and locally periodically exposed (e.g. Sarcidano area, Fig. 10) and underwent severe erosion with deposition of coarse syn-rift breccias and fan-delta conglomerates along the adjacent margins of the trough (Cherchi *et al.*, 2000; Vigorito *et al.*, 2005). These coarse deposits pass trough-ward and upsection into marine tuffaceous sandstones.

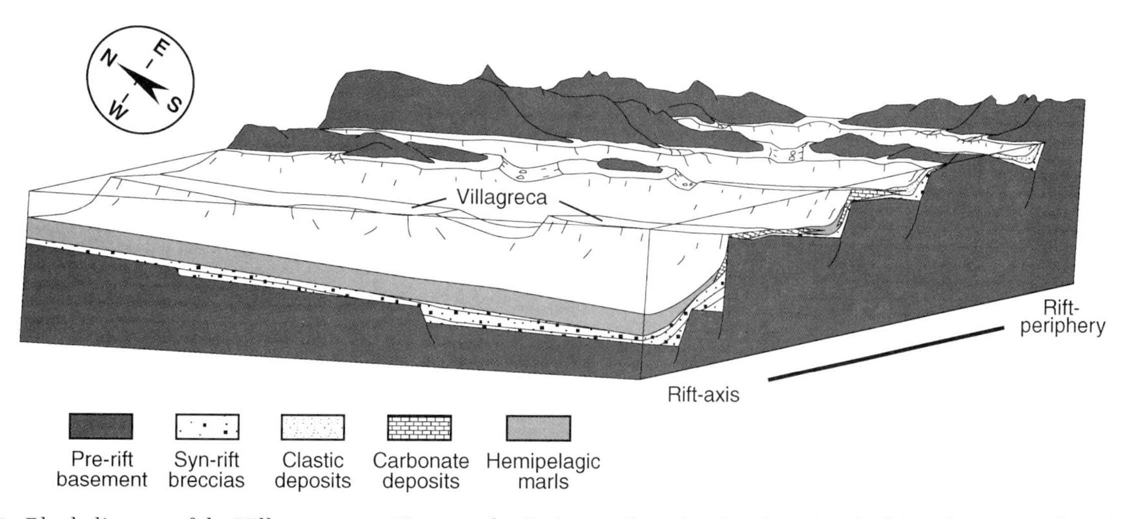

Fig. 9. Block diagram of the Villagreca area. Not to scale. Carbonate factories developed on fault blocks close to the rift axis. Coral-dominated facies mainly developed on the side of the block facing the open sea.

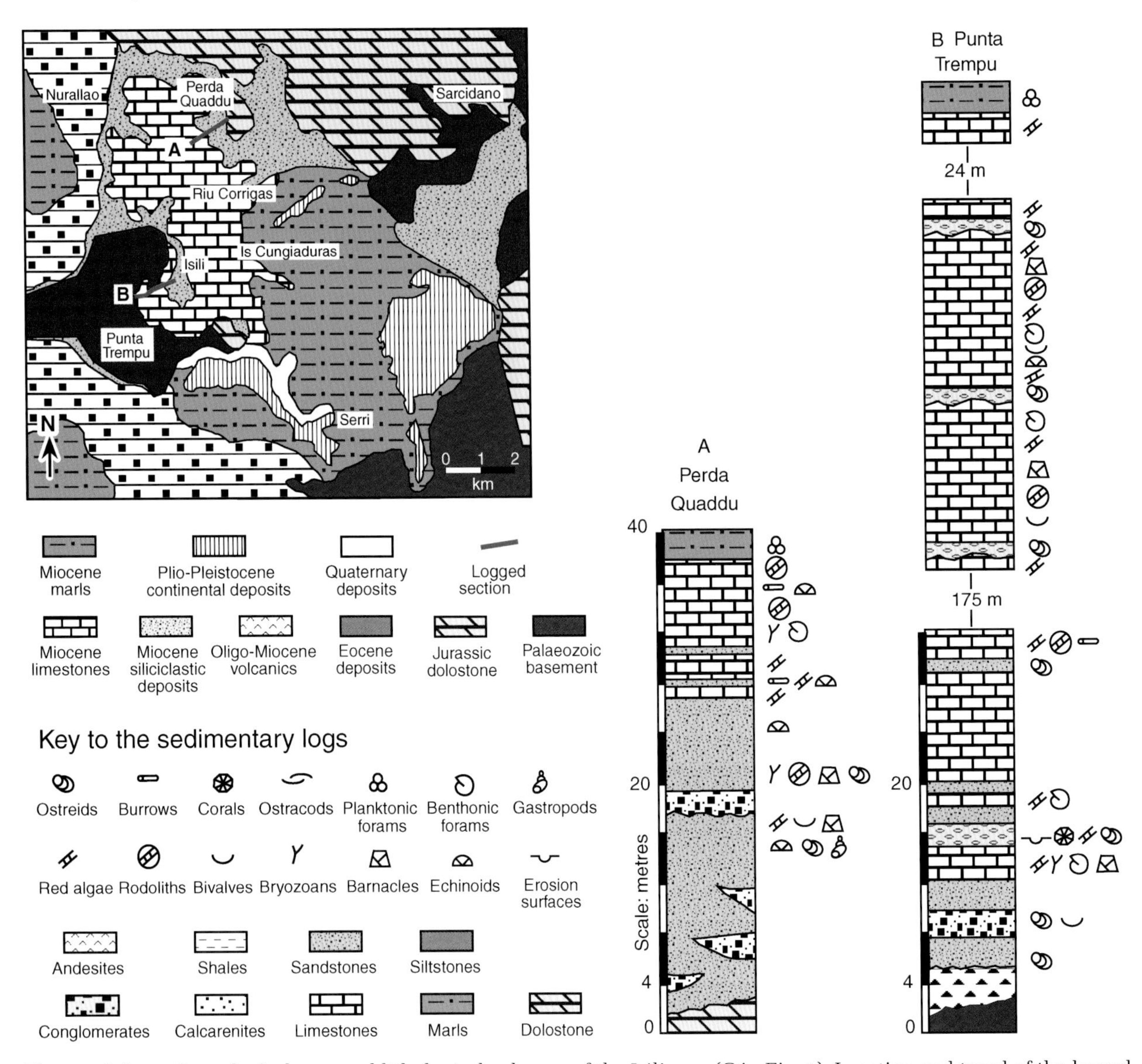

Fig. 10. Schematic geological map and lithological columns of the Isili area (C in Fig. 1). Location and trend of the logged sections is shown: A, Perda Quaddu; B, Punta Trempu.

Since the Early Aquitanian, foramol/rhodalgal carbonate sequences (Isili limestones, Cherchi *et al.*, 2000; Vigorito *et al.*, 2005; Bassi *et al.*, 2006) were laid down in the Isili sub-basin. The main carbonate factories were located at the top of siliciclastic wedges on the margin of the rift-system (e.g. Perda Quaddu) and on the top of the basinward side of the fault-blocks, for example at Punta Trempu. These carbonate factories passed towards the intervening basin (Isili Trough; Vigorito, 2005; Vigorito *et al.*, 2005) into relatively deep marginal areas where bioclastic sediments swept off the productive areas were deposited. These bioclastic sediments were periodically funnelled through a complex channel network towards the deeper portion of the basin, forming submarine fan sequences (Vigorito, 2005; Vigorito *et al.*, 2005). Three main sections: Perda Quaddu, Punta Trempu and Riu Corrigas, are described in the following paragraphs.

Perda Quaddu section: carbonate inception and production at the rift-edge

The Perda Quaddu area is located in the NE sector of the investigated area (Fig. 10). The base of the

Tertiary sedimentary succession comprises about 20 m of poorly lithified tuffaceous sandstones which overlie Jurassic dolostones (Fig. 10). The latter are usually heavily bored by bivalves, and rounded and bored pebble- and cobble-sized dolomitic clasts are common within well-rounded, matrix-rich conglomerates which occur at the base, and as lenses within the tuffaceous sandstones (Fig. 10). Bivalves, echinoids, barnacles, red algae and gastropods are locally common within the tuffaceous sandstones and testify to the existence of shallow-marine conditions during the very early stages of Oligo-Miocene sedimentation.

Toward the west and SW (toward the Isili Trough axis), these sediments pass laterally and upward to about 20 m of tuffaceous sandstones with lenses of skeletal rudstones. This sandy unit is conformably overlain by 10 m of rudstones rich in bryozoans and red algae. Silty sandstones followed by hemipelagic marls cap the sedimentary succession. These deposits show conglomerate intercalations at their base (Fig. 10). Clasts are both pre-Tertiary (Palaeozoic metamorphic and Jurassic dolostone clasts) and Miocene (rounded rhodolith-rich limestones).

The Perda Quaddu succession forms a large sedimentary wedge with strata thickening and prograding toward the axis of the Isili Trough (toward the west and SW). In the neighbouring areas (mainly to the east and SE), conglomeratic deposits, which correlate to the conglomerates cropping out at the base of the Perda Quaddu sedimentary sequence, are locally arranged in thick, south- and westward-prograding wedge-like bodies. These are characterized by sharp and steep progradational surfaces and are interpreted as fan-delta deposits (see also Cherchi *et al.*, 2000).

Interpretation

The Perda Quaddu sedimentary succession suggests a shallow-marine siliciclastic to mixed carbonate-siliciclastic sedimentary environment not far from fluvial inputs. The basal siliciclastic deposits pass laterally and vertically into open-marine mixed siliciclastic-carbonate and carbonate deposits (Cherchi *et al.*, 2000). The latter were laid down following periodical colonization events by stacked pioneer benthic communities that formed patchy areas of carbonate production in the outer sectors of the shelf. A steady increase in water depth is inferred for the deposition of both the calcareous sediments and the hemipelagic marls.

Punta Trempu section: carbonate platform developed at the top of an isolated fault-block

About 180 m of Miocene foramol limestones were logged in the Punta Trempu area (Figs. 10 and 11A). These limestones rest disconformably on the Palaeozoic metamorphic basement or on syn-rift breccias which include both Palaeozoic metamorphic and Jurassic dolostone clasts. Breccia clasts are coarse (up to 20 cm) polygenic and sub-angular to angular in shape. The breccias are rich in shallow-marine skeletal remains, including oysters, pectinids and echinoids, that suggest early marine conditions had set in since the very first phases of rifting. Ostreid banks locally intercalate into the upper portion of the siliciclastic succession (Fig. 10). Barnacle, bryozoan, benthic foraminifera and red algae remains are also quite common. Locally, the base of the ostreid banks consist of a mixture of oyster shells, pertaining to different substrate-related forms and point to the accumulation of at least partially reworked individuals. The upper portion of the banks instead shows autochthonous/parautochthonous shell concentrations with very thick-shelled oysters and large *Pinna*-type shells (Bassi *et al.*, 2006).

These mixed siliciclastic-carbonate deposits grade upsection into pure bioclastic limestones characterized by a rhodalgal assemblage (*sensu* Carannante *et al.*, 1988), and consisting mainly of red algae and subordinately bivalves, echinoids, bryozoans, barnacles, benthic foraminifera, and locally, planktonic foraminifera. Corals also occur, mainly as small globular colonies preserved in life position. Regular, parallel/tabular bedding is ubiquitous but small-scale cross-bedding associated to small-sized scours occurs locally. The carbonate succession shows vertically stacked, reddened and locally heavily bored surfaces, which indicate repeated events of early sea-floor cementation. Coralline red algae (up to 1 cm long and 0.5 cm diameter) generally dominate the skeletal debris, and commonly present fruticose growth-forms. The fruticose forms are unsorted and are highly fragmented and abraded. Rhodoliths are rare, small in size (up to 3 cm diameter) and sub-spheroidal in shape (Bassi *et al.*, 2006).

Toward the east (0.5–1 km, Isili village area) rapid facies transitions and more complex strata geometries (Fig. 12A) document the passage to the outer and deeper sector of the shelf. In this marginal area, numerous scours and channels, up to several tens of metres wide and 5–10 m

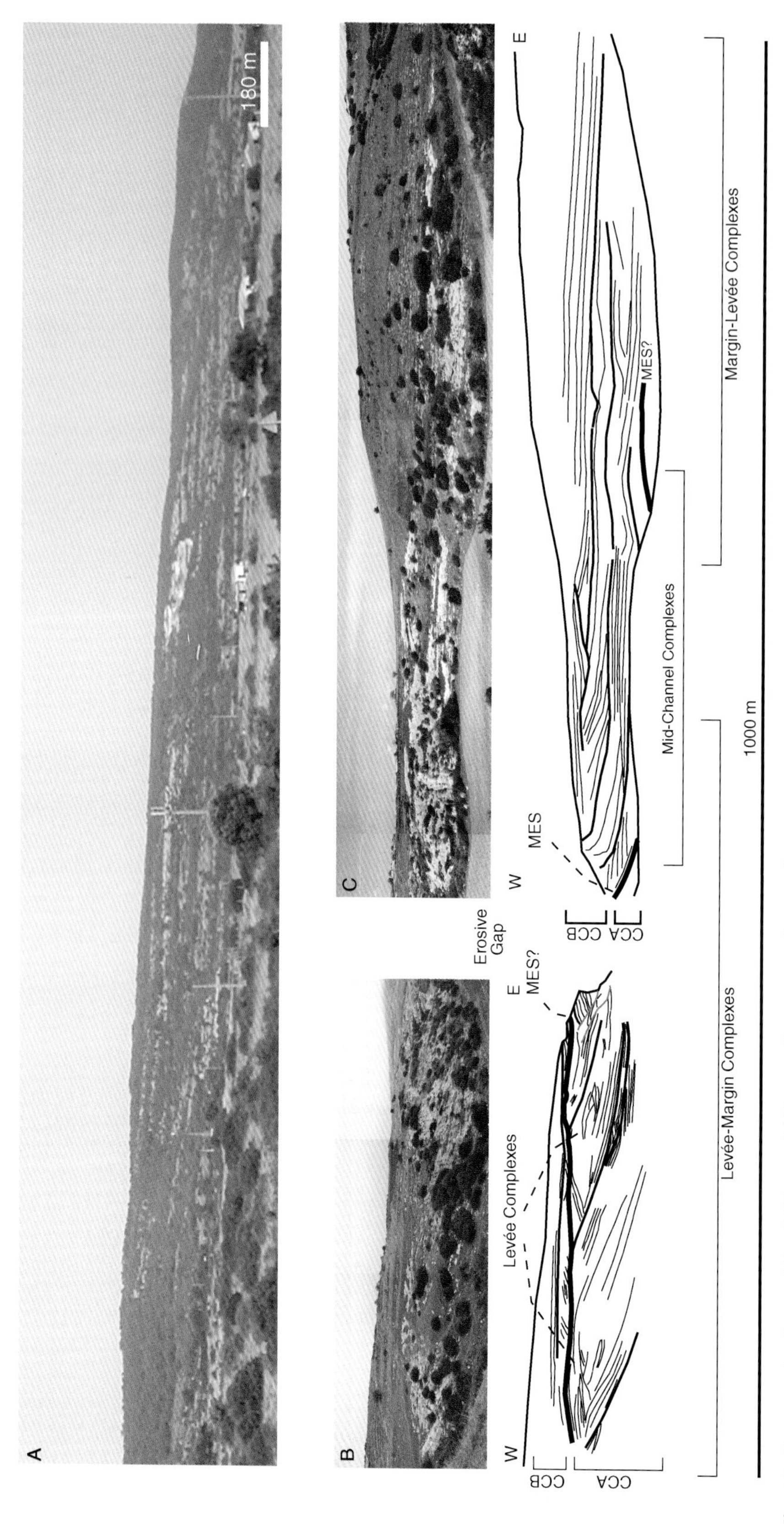

Fig. 11. Panoramic views of the Isili area. (A) Punta Trempu carbonate factory sequences. View to west. (B) and (C) Is Cungiaduras. Up-channel view of the transverse section of the Isili Channel. Both channel complexes CCA and CCB, separated by a major erosion surface (MES), are exposed. Note the multiple stacked, partly nested channel units.

Fig. 12. Isili lithofacies. (A) Two adjacent tributary channels (Ch) and related interchannel ridge (IR) superimposed and partly eroded into parallel stratified foramol limestones. (B) Isili Channel fill sequence. Clinostratified foramol limestones occurring at a major channel bend and interpreted as part of a point bar. (C) Isili Channel fill sequence. Clinostratified unit (LB) interpreted as a large scale lateral bar. Megadebrite deposits formed after the collapse of channel margin-levee complexes and including large outsized blocks (B). (D) Isili Channel fill sequence. Medium-scale cross-stratified channel-fill deposits. (E) Isili fan. Parallel-stratified deposits (FA = Fan A) overlain by cross-stratified deposits (FB = Fan B) and separated by a major erosion surface (MES, dashed line). (F) Isili fan. Clinostratified unit (Cl) from the fill of relatively small distributary channels from Fan A (FA). Other distributary channels (Ch) displaying a divergent fill architecture also occur in both the fan complexes A and B.

deep (Fig. 12A), were recognized. The channels are deeply eroded into previous channel-fill and channel-related deposits, or into parallel-bedded rhodalgal limestones (Fig. 12A). The channels generally show massive, concave-up or divergent channel-fill architectures (Vigorito, 2001, 2005), but more complex channel-fill architectures characterized by sharp lateral accretion surfaces with up to 4 m high foresets are also developed.

The channel-fill is mainly built up by densely packed rhodolith-rich rudstones and subordinate floatstones with a skeletal coarse grainstone and fine rudstone matrix. These coarse gravelly deposits show either chaotic or reverse-graded texture, grain orientation and/or imbrication. The rhodoliths (3–10 cm in diameter) are spheroidal and sub-spheroidal with warty to lumpy growth-forms. Melobesioideae and Mastophoroideae dominate the algal assemblage (Bassi *et al.*, 2006). Barnacles, serpulids and bryozoans are present within the rhodoliths and, as fragments, in the bioclastic matrix. The matrix is commonly well-washed and made up of coarse skeletal debris that is also rich in large bryozoan remains. The channels lie alongside, overlap and intersect each other, and are commonly associated with mound-shaped units or elongated ridges (Fig. 12A).

Multiple truncation and omission surfaces, often reddened and early hardened, may pass through both the channels and the mound-shaped bodies, forming complex internal geometries. This results in large- to small-scale cross-stratified bodies which overlie, intercalate with, and underlie parallel-bedded carbonate deposits (Fig. 12A). Evidence for early cementation processes is widespread throughout the area and patchy or continuous hardgrounds, locally encrusted by oysters, occur mainly at the top of sharp erosion surfaces. Planktonic foraminifera and rarer glauconite grains are locally present. Palaeocurrent indications, mainly from scours and imbricate grains, suggest flows directed toward the eastern and NE sectors.

Toward the basin, the lower portion of the carbonate sequences are characterized by parallel bedded bryozoan and coralline algae-dominated lithofacies. Coralline algae form spheroidal rhodoliths with warty and lumpy growth-forms which frequently envelope several serpulid worm-tubes. Mastophoroids dominate the algal assemblage. Serpulids and barnacles are also generally very abundant.

Coralline algae bindstones are locally common and rhodoliths (3–12 cm diameter) are mainly sub-discoidal to very rarely sub-spheroidal in shape. Melobesioidae associated with minor Mastophoroideae dominate the algal taxonomic assemblage and show both encrusting and warty growth-forms. In the same interval, *in situ* bryozoan colonies have been found. Bryozoan-dominated deposits also occur in the deeper sector of the shelf and are characterized by the presence of Cheilostomata with a lesser contribution from Cyclostomata (Bassi *et al.*, 2006).

Interpretation

During the Aquitanian, the Punta Trempu area represented a temperate-type carbonate factory located on the topographically highest portion of the Punta Trempu fault-block (Cherchi *et al.*, 2000; Vigorito *et al.*, 2005; Bassi *et al.*, 2006). In this area, carbonate sequences are dominated by grainstones and rudstones mainly built up by unsorted thin coralline-algae branch debris with rare small sub-spheroidal rhodoliths. A few coral colonies also occur, while bryozoans are scarce. In addition, oyster banks suggest autochthonous/more rarely parautochthonous shell concentrations. The described facies are similar to the "maerl" facies (Pérès & Picard, 1964) of the present-day Mediterranean Sea, and suggest a deep infralittoral–circalittoral (water depth no shallower than 30 m) depositional setting (Vigorito *et al.*, 2005).

The sediments were swept off by flushing currents or waves and/or periodically removed from the productive areas, probably in relation to negative relative sea-level oscillations and/or tectonic events. Oyster-encrusted hardgrounds, as well as the erosion surfaces found within the Punta Trempu sequences, are thought to be related to such events. The sediments exported from the main production areas were deposited in marginal high-energy settings (Isili village) and are arranged in east-dipping, parallel-bedded sheets of loose sediments into which large cross-stratified bodies intercalate (Fig. 12A). Carbonate lithofacies occurring in these areas are characterized by coarse rudstones and floatstones rich in large spheroidal rhodoliths. These deposits document high-energy settings and an active bottom-current regime in the deep photic zone. Small carbonate factories dominated by sessile and encrusting assemblages rich in bryozoans and coralline algae developed in deep marginal areas (Bassi *et al.*, 2006).

The numerous ENE trending incisions (scours and channels) recognized at outcrop built up a highly efficient sediment drainage system which was tributary to the main Isili Channel. These are filled by coarse rhodolith-rich deposits which show evidence for deposition from sandy debris flow and/or bottom currents (Vigorito, 2005; Vigorito *et al.*, 2005). The channels are associated with mound- and ridge-shaped bodies, which are interpreted as interchannel ridges (Fig. 12A). These are erosional and or erosional-depositional features that developed through differential erosion in the adjacent channels. Similar geometries were reported by Quine & Bosence (1991) from the Upper Cretaceous Chalk sequences of Normandy (France), and were interpreted as erosion features developed in areas flushed by strong bottom currents.

Riu Corrigas section: slope channel and related fan

The Isili Channel fill sequences are extensively exposed throughout the modern Riu Corrigas canyon and adjacent areas (Figs. 10 and 11B). The Isili Channel is up to 1 km wide, and crops out for about 4 km (Vigorito, 2005; Vigorito *et al.*, 2005). The Isili Channel runs from the NW to the SE in its proximal reaches and is sharply deflected to the SSE in its medial and distal reaches. The Isili Channel fill sequence is on average 80 m thick, and is divided vertically into two channel complexes (Channel Complex A and B). These are separated by a major erosion surface (MES) and include several partly nested channel-units which represent different fill stages (Fig. 11B).

Channel Complex A consists of four minor-order vertically stacked channel units, which commonly show trough-stratified to divergent fill architecture and locally exhibit clinostratified units (Fig. 12C). Channel Complex B is instead formed by at least five minor-order channel units (Figs. 12C and D) and is characterized by complex internal geometries, including widespread large-scale (up to 15 m thick) clinostratified units (Fig. 12C) and megabreccias. The latter include rotated and displaced blocks up to 30 m thick and several tens of metres wide. The carbonate sequences are capped by thin silty turbidites and rarely by siliciclastic debris flow deposits. These pass upwards into planktonic foraminifera-rich marls locally alternating with sandy/pebbly siliciclastic turbidites.

South of the Is Cungiaduras area (Fig. 10), the Isili Channel splits into multiple distributary channels creating a branching network fringed by overbank sheet-like deposits and planar-convex bodies. These frontal splay complexes (*sensu* Posamentier & Kolla, 2003) crop out over 2 km down-depositional dip to the south. Two vertically stacked complexes, defined as Fan A and Fan B were identified at outcrop. These are separated by a sharp erosion surface which corresponds to the MES in the Isili Channel. Fan A (Fig. 12E and F) is made up of tabular-bedded skeletal grainstones/packstones to fine rudstones which are generally up to 15 m thick in the proximal area, but thin regularly down depositional dip over a distance of a few kilometres. Minor channels locally occur. Fan B is made up of coarse rhodalgal rudstones/floatstones and grainstones/packstones. The internal geometries (Fig. 12E and F) are complex because of the presence of multiple diverging erosional/depositional and depositional, leveed or unleveed channels (10–50 m wide and up to 5 m deep), alternating with small asymmetric planar-convex bodies (up to 300 m long, 200 m wide and 4 m high). The latter are characterized by tabular to planar-convex stratification and rarely by crude cross-stratification in transverse section, and may eventually show down-current dipping convex strata in longitudinal section.

Interpretation

According to Vigorito (2005) and Vigorito *et al.* (2005), the carbonate sequences laid down in the Riu Corrigas area testify to the presence of a submarine channel (Isili Channel) – fan association. The Isili Channel is a mixed erosional/depositional channel which shows a multi-storey, stacked, partly nested architecture produced by superimposition of multiple channel units, commonly confined by sharp erosion surfaces. These channel units are grouped to form two major channel complexes (A and B), separated by a major erosion surface (MES). The lower channel complex (Channel Complex A) have well-defined channel-scale trough-stratified to divergent fill-architecture, locally associated with minor order cross-bedding and clinostratified units interpreted as lateral bars (Figs. 11B and 12A), produced in turbidite-dominated depositional environments. The overlying Channel Complex B shows complex channel-fill architectures including impressive clinostratified units interpreted as lateral bar sequences (Figs. 12C and D), which reflect an overall increase of channel sinuosity.

Channel-fill architectures are locally complicated by major levee-margin collapses (Fig. 12C; Vigorito, 2005; Vigorito *et al.*, 2005). These resulted in the emplacement of megabreccias which created irregular channel-floor topography and in turn favoured the development of the complex large-scale cross-stratified geometries that characterize the overlying deposits.

In its distal reaches, the Isili Channel splits into a complex distributary network fringed by pebbly/sandy sheet deposits and/or by small lobes and interchannel ridges. These formed a relatively small submarine fan system. Fan complexes were laid down as a consequence of slope gradient decrease and more probably by the expansion and collapse of down-channel directed sediment flows that were no longer confined by the Isili fault-block on the western side. Two fan complexes, Fan A and B, have been identified and correlated to the Channel Complex A and B. These fan complexes were deposited during different stages of active fan-system growth (Fig. 12E and F). The fan complexes are separated by a sharp erosion surface which probably relates to a major regressive event followed by a basinward shift of the fan system, as suggested by the superimposition of the channelized proximal deposits of Fan B onto parallel-bedded more distal deposits of Fan A (Vigorito, 2005: Vigorito *et al.*, 2005).

Development and dismantling of the Isili Carbonate Platform: Shelf to Basin Sediment Transport Patterns

The carbonate successions exposed in the Isili sub-basin provide a clear example of the complex sedimentological and architectural features of carbonate factories developed in basins located at the periphery of a rift-system. Moreover, the investigated sections allow for tracking shelf to basin transitions and for identifying the main sedimentary patterns.

On the NE margins of the Isili Trough (Perda Quaddu area), the carbonate deposits include large amounts of skeletal debris as well as small patchy carbonate factories developed at the top of large-scale troughward-prograding sedimentary wedges (e.g. Perda Quaddu, Figs. 10 and 13; Cherchi *et al.*, 2000; Vigorito *et al.*, 2005). By contrast, on the NW and the western sectors of the Isili Trough (Figs. 10 and 13), large foramol/rhodalgal carbonate factories developed on the Palaeozoic basement or on syn-rift breccias and tuffaceous sandstones. These carbonate factories developed on the structural highs of Punta Trempu and probably Nurallao (Figs. 10, 11A and 13), at water depths ranging between 30 m and 80 m (Cherchi *et al.*, 2000).

The sediments produced and accumulated in these shelf areas were swept off by flushing currents or waves and/or periodically removed during negative relative sea-level oscillations and/or tectonic events. Displaced sediments were redeposited as parallel to cross-stratified sedimentary bodies in the outermost sectors of the shelf (e.g. Isili area; Figs. 10, 12A and 13) and in slope areas, or were funnelled towards the basin through a complex network of channels (Isili Channel System). The latter includes one main channel, the Isili Channel and several tributary channels (Fig. 13). Frontal splay complexes formed at the distal end of the Isili Channel in the deeper

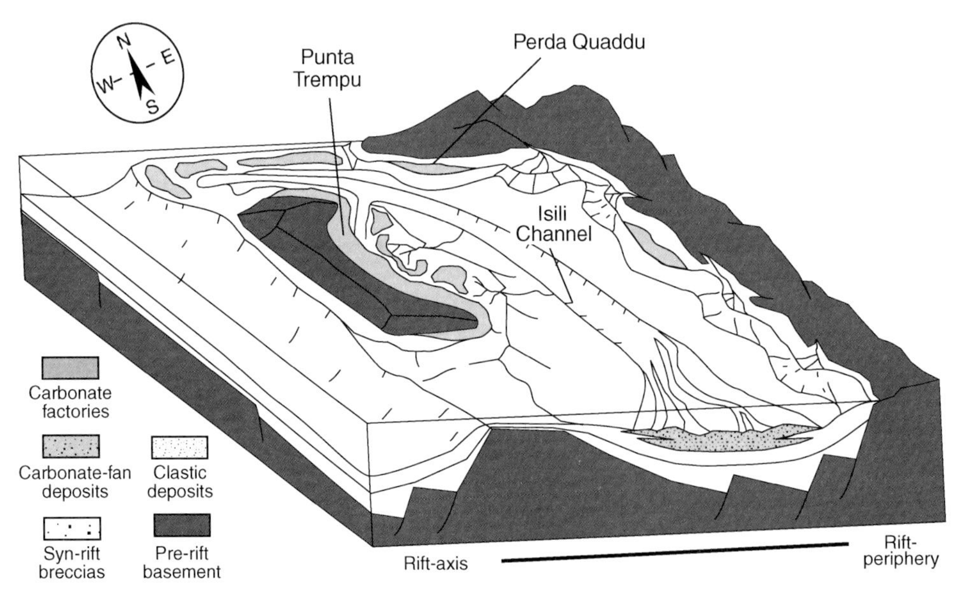

Fig. 13. Block diagram of the Isili area. Not to scale. Carbonate factories developed on the top of siliciclastic wedges (Perda Quaddu) and on the dip-slope of tilted fault blocks (Punta Trempu). A hierarchically organized channel network allowed for transportation of sediments toward the basin where they formed a submarine fan. The main resedimentation events occurred mainly during relative sea-level falls and lowstands.

basinal areas (Fig. 13; Vigorito, 2005; Vigorito *et al.*, 2005).

Since the Middle Burdigalian, a rapid relative sea-level rise led to deposition of a few hundred metres of hemipelagic marls which filled the Isili sub-basin; these are interpreted as post-rift deposits (Fig. 10; Cherchi *et al.*, 2000; Casula *et al.*, 2001). This marly sequence commonly shows thin intercalations of siliciclastic or mixed carbonate-siliciclastic sandy turbidites in the lower and middle parts.

DISCUSSION

Carbonate production and related architectures

The investigated carbonate sequences exhibit a wide range of sedimentological, ecological and architectural patterns, but some common evolutionary trends can be delineated. Particularly three different phases corresponding to colonization, carbonate factory development and growth, and finally to carbonate factory demise and eventual dismantling have been identified, and are described in the following sections.

Substrata colonization

In all the investigated areas, the carbonate-dominated sequences do not lie directly on the pre-rift basement but overlie more or less thick siliciclastic and/or mixed carbonate/siliciclastic sequences deposited during the first phases of rifting (Figs. 3A and B, 6A and B, 11A and 12A). Depending mainly on local palaeomorphology, these syn-rift deposits indicate continental, transitional and/or marine depositional environments. These early stage deposits were subsequently colonized by pioneer benthic communities. In some instances, evidence for intense bioturbation and bioerosion is associated with these basal sedimentary sequences. In other instances, relatively deep marine conditions were rapidly established soon after the very first phases of rifting, as demonstrated by the occurrence of abundant planktonic foraminifera and sciaphilic assemblages (mainly bryozoans and subordinate red algae, e.g. Villagreca area and Dolianova).

In all the investigated areas, pioneer communities consist of bivalve-dominated foramol assemblages (Figs. 2, 5 and 10). Oysters are the most common component and may either be dispersed in a siliciclastic, carbonate or mixed matrix, or may form banks a few metres thick, the latter representing either biogenic (biostromes) or sedimentological concentrations (*sensu* Kidwell *et al.*, 1986; Kidwell, 1991). Bryozoans and subordinately red algae may also be quite common in these basal carbonate deposits and suggest darker, cooler and/or nutrient-richer waters. These opportunistic benthic assemblages were able to spread patchily over the sea-floor whenever and wherever the conditions were favourable and prepared a suitable substratum (substratum preparation phase) for subsequent pure carbonate skeletal sequences.

It is remarkable that in some areas multiple phases of pioneer colonization and substrate preparation, alternating with phases of siliciclastic deposition, repeatedly occurred before relatively long-lasting carbonate factories were developed. In some cases well-defined progradational geometries were identified at outcrop (e.g. Perda Quaddu).

Carbonate factories: development and growth

Following the first phases of colonization and substratum preparation, carbonate factories locally developed at the top of submerged fault-blocks or at the edge of the rift system where the environmental conditions were suitable. Depending mainly on local parameters (bathymetry, sea-floor topography, hydrology, seawater trophism and temperature, as well as freshwater input), these carbonate factories were wholly or periodically/partly characterized by coral-dominated or foramol/rhodalgal (*sensu* Carannante *et al.*, 1988) ecological assemblages.

Coral-dominated carbonate factories

Chlorozan-dominated carbonate factories developed at the top of several fault-blocks exposed in the Villagreca area following the deposition of bivalve and bryozoan-rich carbonates or mixed deposits. In other instances, small coral-dominated factories developed in sheltered rift-related submerged valleys (e.g. Dolianova).

In all the investigated areas, *Porites* is by far the dominant species in the coral-rich assemblages. These show some common evolutional trends which can be depicted as follows:

(1) Juvenile stage: the very first phase of coral colonization is generally characterized by small globular and/or branching colonies.

In high-energy environments these pioneer colonies usually encrust large-sized clasts. Conversely, in low-energy sheltered environments branching pioneer colonies dwelled sparsely in muddy/marly deposits.

(2) Mature stage: at this stage coral-colonies reach their maximum extension. Sequences are characterized by fully developed coral-rich thickets and lenses that may show the highest diversity in both species and morphology. These biostromes were laterally fringed by loose bioclastic deposits in which small patchy coral colonies may be present. The bulk of these bioclastic sediments consist of fine to coarse debris of bivalve- and/or bryozoan-dominated communities produced through both biological and mechanical erosion.

(3) Senile stage: environmental changes locally associated with cyclical factors (e.g. Dolianova, freshwater floods; Villagreca, relatively high-frequency sea-level variations) may lead to the progressive or sudden demise of the coral-dominated communities. During these phases, coral colonies generally show small globular morphologies and become progressively sparser and tend to be dispersed within skeletal debris. This debris is produced through biological and mechanical erosion, and generally exhibits large amounts of material derived from foramol-type communities and mainly from bivalves. The demise of coral-dominated assemblages is frequently associated with the formation of hardgrounds, which suggests a drastic decrease in carbonate production and long permanence of sediments at the depositional interface. In some instances, erosion surfaces can abruptly terminate the growth of the coral boundstones, which tend to be rapidly buried by the following siliciclastic or foramol bioclastic sediments in which small coral colonies may be dispersed.

In the Dolianova area, coral-rich deposits are mostly made up by *Porites* but also include *Tarbellastrea*, *Montastraea* and *Favites*. These deposits have lenticular geometries in which coral build-ups occupy the core and are surrounded by fringes of bioclastic debris. These lenses have high width:height ratios and are interpreted as small biostromes. However, thicker lenses may locally exhibit more pronounced convex profiles which might indicate the formation of limited patch-reefs. These limited ecological reefs exhibit similarities with many recent and ancient examples of interrelated coral reefs and siliciclastic deposits, described among others by Friedman (1988), Santisteban & Taberner (1988) and Roberts & Murray (1988). The growth of the studied organogenic structures was periodically interrupted by erosional events associated with siliciclastic input.

In Villagreca, coral-rich deposits are largely dominated by *Porites* and other corals are absent or rare. The ecology and geometry of the coral-rich deposits suggest low-energy sheltered environments (Riegl & Velimirof, 1994; Riegl *et al.*, 1995). According to this interpretation, and on the basis of field evidence, corals formed relatively small communities protected from the open sea by laterally extensive mobile sheets of bioclastic debris and/or locally by the inherited rift-related physiography of the substratum.

Coral build-ups consist of vertically stacked individual coral thickets commonly preceded by a bivalve-rich layer. These patterns indicate biostrome-type build-ups (stratigraphic reefs) and results in dominant tabular geometry. Down block-dip (towards the basin), the coral colonies interfinger with well-bedded bioclastic deposits (Fig. 7B) which progressively fine down-dip, fringing into basinal hemipelagic marls.

All the investigated areas lack a rim. As a consequence, the carbonate factories were subject to open-sea and relative high-energy conditions. The facies are generally dominated by rudstones and grainstones, as the finer sediment fractions were easily removed from the productive areas, and redeposited in the deeper basinal areas. For the same reasons, muddy facies are generally absent, and occur only where the rift-related physiography created sheltered low-energy environments. This is different from the coeval rift-system of the Red Sea, where reefs extensively developed wherever the environmental conditions were favourable, and built up barriers which limited seawater circulation and created low-energy protected areas where mud-rich sediments were deposited (Purser *et al.*, 1998).

Foramol/rhodalgal carbonate factories

Foramol/rhodalgal carbonate factories occur in the Isili area or at the top of the coral-dominated factories in the Villagreca area. In the Dolianova area, by contrast, small, lenticular foramol production areas, generally dominated by bivalves, alternate with siliciclastic deposits. The analyzed sequences show that the inception of foramol carbonate factories is always characterized by the deposition of bivalve-rich sediments.

In the Villagreca area, bivalves occur at the top of a major erosion surface which separates the coral-rich deposits from the overlying foramol limestone (Fig. 7D). The latter consist of grainstones and rudstones rich in benthic foraminifera, bivalves and echinoids, and include multiple stacked, laterally overlapping and/or nested channels as well as SE-prograding clinostratified units (Fig. 7D). Coarse grained red algae-rich grainstones and rudstones follow upwards.

Foramol associations extensively developed during sea-level rise following a major regressive event. Sea-level rise was possibly associated with variations in the circulation of basin waters and with an increase in the amount of nutrients in the seawater which favoured the development of foramol/rhodalgal benthic communities. However, climatic change is also a valid alternative hypothesis.

In the Isili area, the carbonate factories developed in different physiographic settings and were characterized by distinct evolutionary trends. In the Perda Quaddu area (Figs. 10 and 13), a bryozoan-dominated, relatively deep-water carbonate factory developed after several phases of sea-floor colonization during which the developments of patchy benthic communities alternated with siliciclastic deposition. The Perda Quaddu carbonate factory was located at the top and along the slope-dip of a siliciclastic-carbonate wedge which had previously prograded at least 1–2 km towards the basin axis (Fig. 13).

By contrast, in the Punta Trempu–Isili village areas (Figs. 10, 11A and 13), carbonate factories developed onto an isolated fault-block. In the Punta Trempu area, foramol carbonate sequences are dominated by red algae with prevalent fruticose morphologies and form "maerl-type" deposits (Bassi *et al.*, 2006). Multiple planar erosional/non-depositional surfaces cyclically cut through the sequence and were recurrently colonized by oysters which may eventually form several decimetre thick banks. These tabular bedded limestones pass basinward into tabular to medium to large-scale cross-stratified limestones deposited in a high-energy outer-shelf environment (Isili village). These deposits are red algae-dominated as well, although in this case rhodoliths largely prevail. Only locally, small red algae-bryozoan bindstones occur (Bassi *et al.*, 2006). The foramol/rhodalgal deposits occurring in the Punta Trempu-Isili village areas form as whole a cross-stratified wedge-like body which prograded and thinned toward the basin.

Carbonate factories: demise and dismantling

Small carbonate production areas located near freshwater input have a high probability of having been buried by siliciclastic deposits. These small carbonate lenses and build-ups are commonly truncated at the top by sharp erosion surfaces. The abrupt change from carbonate to siliciclastic deposits is not marked by any sign of a co-occurring crisis in the benthic communities (e.g. Dolianova).

In the Villagreca area the chlorozoan-dominated factories are characterized by cyclical alternations. At the base, each cycle shows bivalve-rich deposits which overlie a sharp erosion/non-depositional surface and pass upwards into coral-rich beds. These are commonly reddened and intensely bioeroded at the top which suggests periods of stasis in the growth of the biostromes (senile stage). The coral-rich sediments are truncated by a sharp erosional event which was also associated with abrupt environmental changes as demonstrated by the development of foramol associations.

By analogy with the Isili area, the foramol deposits in the Villagreca area were loose to only partly lithified and thus prone to periodical removal from the shallow productive areas by a complex channelized network (Fig. 13). Sediment remobilization occurred mainly, but not exclusively, during periods of sea-level fall and lowstand (Cherchi *et al.*, 2000), and was associated with the formation of sharp erosion surfaces and with the basinward shift of facies belts. The morphology of the basin, coupled with the ecological behaviour of the foramol-rhodalgal assemblages, favoured the basinward migration of the productive areas as a response to negative relative sea-level oscillations. Major regressive events were also generally associated with abrupt changes in the ecological assemblages (e.g. bivalves replacing red algae). By contrast, while these foramol communities were able to deal quite well with relative sea-level falls, they appear to suffer under rapid sea-level rises, resulting in drowning.

Factors influencing carbonate sedimentation on fault blocks in the Miocene Sardinia rift-basin

The data collected in this study demonstrate that carbonate production in rift basins is largely controlled by tectonics, palaeophysiography, relative sea-level variations and climate. These results fit well with the findings of Bosence (1998), Bosence *et al.* (1996, 1998) and Purser *et al.* (1996, 1998). The combination of these main factors had a leading role in dictating the location and evolutionary history of the productive and sediment accumulation areas, and in the development of their complex depositional architectures.

Tectonics

The sedimentary patterns in syn-rift basins are largely controlled by tectonics. This, in turn, controls the location and the trend of the fault-blocks and of the intervening basins, creates space for sediment accumulation, or may otherwise raise sedimentary cover, leading to erosion. Repeated structural readjustments result in local relative sea-level variations which have specific signatures (e.g. erosion, non-depositional and drowning surfaces).

Unusually for a rift-basin (e.g. Red Sea, Purser & Bosence, 1998 and references therein) with a dominant tilted-block style, only very sparse and ambiguous evidence for syn-sedimentary tilting has been reported from the investigated areas. This suggests that tilting phases were mostly limited to the very early phases of rifting, or to small marginal blocks, and that subsequent phases were mainly characterized by vertical movements possibly related to thermal contraction. Individual blocks represented independent structural units in which subsidence was mainly controlled by local tectonics. This means that lithostratigraphic correlations between adjacent blocks may be misleading because of the high probability of diachroneity.

In rift-systems, axial blocks are generally characterized by higher subsidence rates than those located at the rift periphery. As a consequence, axial blocks should exhibit thicker sedimentary sequences when compared with those laid down on peripheral blocks (Purser *et al.*, 1998). In the investigated areas the carbonate sequences are thicker (>100 m) on the peripheral blocks (e.g. Punta Trempu area, Fig. 13) than on the axial blocks (e.g. Villagreca area, Fig. 9). This incongruence may be explained in terms of subsidence velocity vs. carbonate production rates. It is speculated that in the axial region carbonate production was unable to keep pace with subsidence, as a result, carbonate factories were prone to rapid drowning. By contrast, in the more peripheral areas (Punta Trempu), subsidence rates were lower, and carbonate production was able to keep pace with sea-level rise, thus producing up to 200 m thick carbonate platform sequences. In addition, the foramol carbonate factory of Punta Trempu documents high production rates (200 m in ~5 Myr) compared with coral-dominated factories reported from the Red Sea (150 m in 10 Myr; Purser *et al.*, 1998). This may reflect local factors, for example subsidence rate, and more favourable ecological conditions. However, the occurrence of high-productivity carbonate factories and the resulting thick carbonate units located at the periphery of the rift should be considered in the modelling of rift-systems.

Eustatic sea-level variation

Individual fault-blocks represented independent structural-units characterized by different sedimentary history. This makes it difficult to discriminate between relative sea-level variations caused by local subsidence, and eustatic variations. Nevertheless some common trends may be outlined. The Early Miocene was a time of global sea-level rise (Hardenbol *et al.*, 1998), and this combined with local subsidence would have favoured the inception of siliciclastic and carbonate marine deposition on top of previous Oligo-Miocene syn-rift continental deposits. In Burdigalian times, following a major global transgressive event, all the carbonate factories occurring in the investigated areas were rapidly drowned and, as a consequence, shallow-water carbonate production was dramatically shut down.

Palaeophysiography

The analyses conducted in the study areas have demonstrated that palaeophysiography largely controlled the facies distribution and architecture, and determined the trends of the main sediment pathways. The illustrated areas are representative of different structural settings. The Dolianova area is representative of a sheltered rift-related valley bordered on three sides by emerged areas, which supplied abundant siliciclastic sediments, but opens towards the rift-system at the distal end (down-plunge, Figs. 2 and 4). Rift-related valleys occur frequently at the margin of the main rift-basin and have trends parallel or transverse with respect to the main rift axis. These narrow valleys are generally filled by cones of siliciclastic

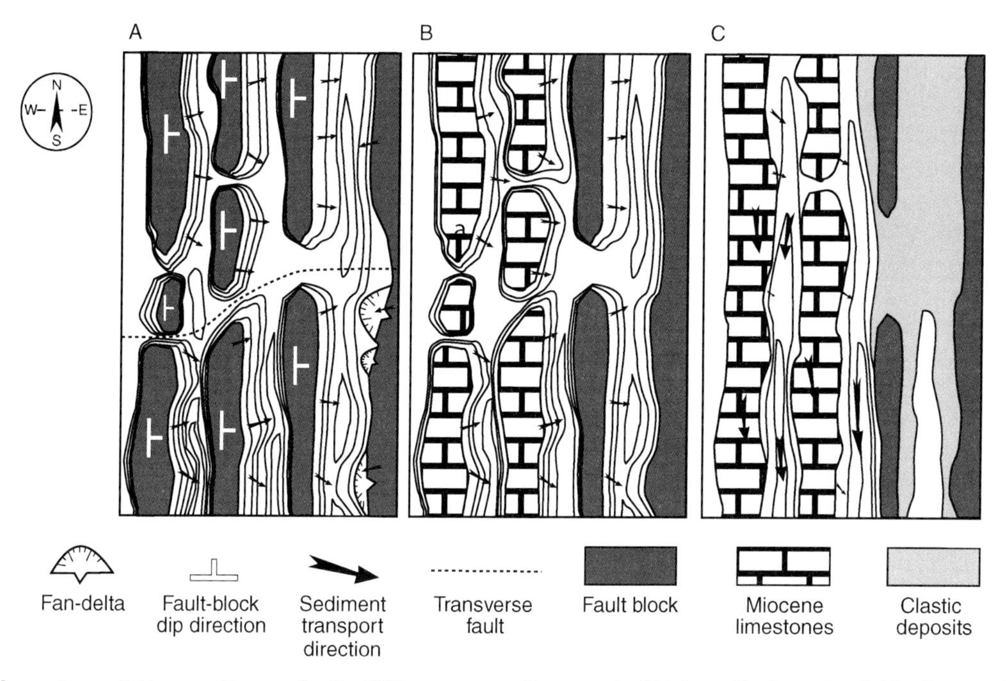

Fig. 14. Schematic evolutionary diagram for the Villagreca area. Not to scale. (A) Syn-rift phase. Fault blocks form during rift phases. Depending on the block morphology continental or marine deposits are laid down on the top of the blocks. Transverse faults acted as transverse sedimentary conduits and allowed for differential tilting between adjacent blocks. Tilted blocks dipped toward the rift periphery (toward the east) and usually plunged to the SSE. (B) Late syn-rift–early post-rift phase. Fault blocks close to the rift axis start to be covered by carbonate shelf successions. Coral-dominated factories developed mainly on the western and northern side of the blocks (towards the rift axis), and grade into deeper foramol-dominated factories down block-dip (towards the east). Sediments exported from the carbonate factories accumulated in the intervening basins where they interfingered with hemipelagic marls. (C) Post-rift phases. Following a major erosional event, foramol factories developed at the top of the coral-dominated sequences. An abrupt shift of sediment transport directions to the SSE is documented. This suggests active northerly wind regimes, increasing topographic gradients through tectonics, and erosion and/or down block-plunge sediment transport.

sediments which prograde along the valleys towards the adjacent rift-basin system (Fig. 4). Carbonate factories may be established in such settings at the top of the siliciclastic cones (Fig. 4), in the axial portions of the valley, and especially in their distal (basinward) portions, during periods of reduced siliciclastic input.

By contrast, the carbonate factories identified in Villagreca area were located in the vicinity of the rift axis (Figs. 9 and 14A). In this area, the carbonate factories developed at the top of isolated tilted fault-blocks which are aligned parallel to the rift/axis (NNW–SSE) and plunge to the SSE (Figs. 9 and 14B). These blocks faced the open sea on their western side where coral-rich communities developed locally. These benthic communities formed eastward-dipping platforms which cover the higher portion of the block dip-slope, show sharp progradational geometries and pass basinward into bioclastic wedges and basinal marly deposits (Figs. 9 and 14B).

Following a major erosional event, these sediments were capped by foramol deposits which generally exhibit a distinctive progradational pattern towards the SE (in the block-plunge direction, Fig. 14C). This might indicate some sort of tectonic control and/or may have been caused by the progressive decrease of the down-dip gradients during the depositional phases characterized by coral-rich assemblages which dominantly prograded eastward (Fig. 14B and C). According to this hypothesis, down-plunge gradients remained substantially unaffected during this first phase and/or were subsequently increased tectonically or through erosion. This created a NNW–SSE pathway along which foramol sediments were funnelled and prograded to the SSE (Fig. 14C).

A third example of structural setting is the Isili area, located at the rift periphery (Fig. 13). In this area carbonate factories developed on both the eastern and western margin of the Isili sub-basin as well as at its northern culmination over the

Nurallao structural high (Fig. 13). On the eastern margin (Perda Quaddu), the carbonate factory developed at the top of a basinward prograding siliciclastic wedge formed by marine-deltaic deposits (Fig. 13). The inner carbonate factory (Punta Trempu) formed along the dip-slope (eastern side) of a fault-block which probably partly or temporarily emerged to the west (Fig. 13). The intervening basin was a south-plunging trough in which a large-scale channel-complex developed (Fig. 13). The channel was supplied by tributaries, oriented parallel to the main fault-systems, and it fed vertically stacked, southward prograding fan complexes (Fig. 13).

Rift basins form morphological corridors which generally support active wind regimes. In turn this affects the waves and currents of the basin water body, as well as sediment transport. In accordance with the modern situation, during the Miocene, winds were likely to be predominantly from the NNW. This hypothesis fits well with the location of the coral-bearing carbonate factories, which are located on the northern and western margin of the blocks, and the recurrence of a SSE-trending signature in the palaeocurrent distribution patterns.

Climate versus ecological controls

During Early Miocene times, Sardinia, like most of the Mediterranean region, was subject to subtropical climate conditions. This climate setting was potentially favourable to the development of coral-dominated carbonate factories, as demonstrated by the extensive coeval development of such ecological assemblages in the Mediterranean region (Franseen et al., 1996). However, no evidence for the development of ecological reefs has been found in the areas that are generally dominated by foramol and rhodalgal assemblages. Coral-dominated assemblages characterized by low specific diversity and mono-specific associations are the rule rather than the exception. In addition, corals tend to build well-stratified biostromes rather than bioherms.

Climate controlled the rate/amount of rainfall in the region, and in turn, the type and rate of land-derived materials exported to the rift-basins. This had great relevance in the setting up of different biotic assemblages. Coral-dominated assemblages were able to colonize sea bottoms during breaks in siliciclastic deposition, occurring between subsequent flashfloods. Instead, foramol assemblages, especially those dominated by filter communities of bivalves, were able to spread over the shelves during times of significant siliciclastic input. From all the above, it is clear that other physical or ecological factors, or combination of factors controlled the development of coral-dominated or foramol benthic communities.

CONCLUSIONS

Carbonate sedimentation in rift basins is controlled by a variety of factors including local environmental conditions, sea-floor physiography, tectonics, eustasy and climate. These rule the location and the development of the carbonate production areas which commonly occur at the top of submerged tilted fault-blocks as well as along the basement margins. However, the type and ecology of the benthic communities play important roles in dictating the resulting depositional architectures and the response of the carbonate sedimentary systems to relative sea-level variation.

Although climatic conditions during Miocene times were favourable to the development of ecological reefs, in the investigated areas ecological reefs were rarely developed and consist of widely spaced limited patch reefs, even where the benthic communities were largely dominated by corals. Coral-dominated assemblages (where present) were abruptly replaced by foramol associations. This is inferred to reflect a climatic change and/or variations in the temperature and content of nutrients of the seawater. Foramol factories developed extensively during a transgressive event which was possibly associated with an increase in nutrients and variations in the circulation of basin waters. This in turn favoured the development of foramol associations in place of benthic communities dominated by corals.

In all the illustrated areas, the carbonate factories lack a rim and were subject to open-marine and relatively high-energy conditions. Muddy facies are rare or absent, and most of the sequences are dominated by rudstone and subordinately grainstone/packstone facies. The sediments produced in the carbonate factories were swept from the shallow-water domains by currents and waves, or periodically removed during erosive events associated with sea-level falls, and were redeposited in the deeper part of the shelf and in the slope/basinal areas. The progradation of the carbonate systems occurred mainly during relative sea-level falls and lowstands, and was associated with a basinward shift of the facies. Submarine channel and fan systems were locally developed.

The foramol carbonate factories that developed on footwall blocks close to the rift periphery are thicker and show high production rates (200 m in 5 Myr). These relatively high-productivity carbonate factories were able to keep pace with relative sea-level rise.

REFERENCES

Bassi, D., **Carannante, G.**, **Murru, M.**, **Simone, L.** and **Toscano, F.** (2006) Rhodalgal/bryomol assemblages in temperate type carbonate, channelised depositional systems: the Early Miocene of the Sarcidano area (Sardinia, Italy). In: *Cool-water Carbonates: Depositional Systems and Palaeoenvironmental Controls* (Eds H.M. Pedley and G. Carannante). *Geol. Soc. London Spec. Publ.*, **255**, 35–52.

Bosence, D.W.J. (1998) Stratigraphic and sedimentological models of rift basins. In: *Sedimentation and Tectonics of Rift Basins: Red Sea Gulf of Aden.* (Eds B.H. Purser and D.W.J. Bosence), pp. 9–25. Chapman & Hall, London.

Bosence, D.W.J., **Nichols, G**, **Al-Subbary, A.**, **Al-Thour, K.A.** and **Reeder, M.** (1996) Syn-rift continental to marine depositional sequences, Tertiary, Gulf of Aden, Yemen. *J. Sed. Res.*, **66**, 766–777.

Bosence, D.W.J., **Cross, N.** and **Hardy, S.** (1998). Architecture and depositional sequences of Tertiary fault-block carbonate platforms; an analysis from outcrop (Miocene, Gulf of Suez) and computer modelling. *Mar. Petrol. Geol.*, **15**, 203–221.

Carannante, G., **Esteban, M.**, **Milliman, J. D.** and **Simone, L.** (1988) Carbonate lithofacies as paleolatitude indicators: problems and limitations. *Sed. Geol.*, **60**, 333–346.

Casula, G., **Cherchi, A.**, **Montadert, L.**, **Murru, M.** and **Sarria, E.** (2001) The Cenozoic grabens system of Sardinia (Italy): geodynamic evolution from new seismic and field data. *Mar. Petrol. Geol.*, **18**, 863–888.

Cherchi, A. and **Montadert, L.** (1982) The Oligo-Miocene rift of Sardinia and the early history of the West Mediterranean Basin. *Nature*, **298**, 736–739.

Cherchi, A. and **Montadert, L.** (1984) Il sistema di rifting oligo-miocenico del Mediterraneo occidentale e sue conseguenze paleogeografiche sul Terziario sardo. *Soc. Geol. Ital. Mem.*, **24**, 387–400.

Cherchi, A., **Murru, M.** and **Simone, L.** (2000) Miocene carbonate factories in the syn-rift Sardinia Graben subbasins (Italy). *Facies*, **43**, 223–240.

Franseen, E.K., **Esteban, M.**, **Ward, W.C.** and **Rouchy, J-M.** (Eds) (1996) Models for Carbonate Stratigraphy from Miocene Reef Complexes of Mediterranean Regions. *SEPM Concepts Sedimentol. Palaeontol.*, **5**, 1–384.

Friedman, G.M. (1988) Histories of co-existing reef and terrigenous sediments: the Gulf of Elat (Red Sea), Java Sea, and Neogene basin of the Negev, Israel. *Dev. Sedimentol.*, **42**, 77–97.

Hardenbol, J., **Thierry, J.**, **Farley, M.B.**, **Jacquin, T.**, **Granciansky, P.C.** and **Vail, P.R.** (1998) Mesozoic and Cenozoic Sequence Chronostratigraphic Framework of European Basins. In: *Mesozoic and Cenozoic Sequence Stratigraphy of European Basins* (Eds P.C. Graciansky, J. Hardenbol, T. Jacquin and P.R. Vail). *SEPM Spec. Publ.*, **60**, 3–13.

Kidwell, S.M. (1991) The stratigraphy of shell concentration. In: *Taphonomy: Releasing the Data Locked in the Fossil Record. Top* (Eds P.A. Allison and D.E.G. Brigg). *Geobiol.*, **9**, 211–290.

Kidwell, S.M., **Fürsich, F.T.** and **Aigner, T.** (1986) Conceptual framework for the analysis and classification of fossil concentrations. *Palaios*, **1**, 228–238.

Lecca, L., **Lonis, R.**, **Melis, E.**, **Secchi, F.** and **Brotzu, P.** (1997) Oligo-Miocene volcanic sequences and rifting stages in Sardinia: a review. *Per. Mineral.*, **66**, 7–61.

Maillard, A. and **Mauffret, A.** (1993) Structure et volcanisme de la fosse de Valence (Méditerranée nord-occidentale). *Bull. Soc. Géol. Fr.*, **164**, 365–383.

Nelson, C.S. and **James, N.P.** (2000) Marine cements in mid-Tertiary cool water shelf limestones of New Zealand and southern Australia. *Sedimentology*, **47**, 609–629.

Pérès, J.M. and **Picard, J.** (1964) Nouveau manuel de bionomie bentique de la Mer Méditerranée. *Trav. Stat. Mar. Endoume-Marseille*, **31**, 1–137.

Posamentier, H.W. and **Kolla, V.** (2003) Seismic geomorphology and stratigraphy of depositional elements in deep-water settings. *J. Sed. Res.*, **73**, p. 367–388.

Purser, B.H. and **Bosence, D.W.J.** (Eds) (1998) *Sedimentation and Tectonics of Rift Basins: Red Sea Gulf of Aden.* Chapman & Hall, London. 663 pp.

Purser, B.H., **Plaziat. J.C.** and **Rosen, B.R.** (1996) Miocene reefs of the northwest Red Sea. *SEPM Concepts Sedimentol. Paleontol.*, **5**, 347–366.

Purser, B.H., **Barrier, P.**, **Montenat, C.**, **Orszag-Sperber, F.**, **Ott d'Estevou, P.**, **Plaziat, J.-C.** and **Philobbos, E.** (1998) Carbonate and siliciclastic sedimentation in and active tectonic setting: Miocene of the northwest Red Sea, Egypt. In: *Sedimentation and Tectonics of Rift Basins: Red Sea Gulf of Aden.* (Eds B.H. Purser and D.W.J. Bosence), pp. 240–270. Chapman & Hall, London.

Quine, M. and **Bosence, D.W.J.** (1991) Strata geometries, facies and sea-floor erosion in Upper Cretaceous Chalk, Normandy, France. *Sedimentology*, **38**, 1113–1152.

Riegl, B. and **Velimirov, B.** (1994) The structure of coral communities at Hurghada in the northern Red Sea. *Mar. Ecol.*, **15**, 213–231.

Riegl, B., **Schleyer, M.H.**, **Cook, P.J.** and **Branch, G.M.** (1995) Structure of southernmost coral communities. *Bull. Mar. Sci.*, **56**, 676–691.

Roberts, H.H. and **Murray, S.P.** (1988) Gulf of northern Red Sea: depositional setting of great barrier reef province. *Dev. Sedimentol.*, **42**, 99–142.

Santisteban, C. and **Taberner, C.** (1988) Sedimentary deposits and coral reef interrelation. *Dev. Sedimentol.*, **42**, 35–76.

Vigorito, M. (2001) Temperate carbonate deep-sea channels depositional and architectural models. Miocene examples from Matese Mountains, Southern Apennines, Italy. *Géol. Méditerr.*, **28**, 173–176.

Vigorito, M. (2005) *Anatomy of Submarine Channels in Carbonate and Mixed Carbonatesiliciclastic depositional Settings. Implications for Basin Geology and Hydrocarbon Exploration and Production. Mesozoic-*

Tertiary Analogues from the Central Southern Apennines and Sardinia Syn-rift Basins, Italy. Unpubl. PhD Thesis, University Naples "Federico II", 463 pp.

Vigorito, M., **Murru, M.** and **Simone, L.** (2005) Anatomy of a channel system and related fan in a foramol/rhodalgal carbonate sedimentary setting: the case history from the Miocene syn-rift Sardinia Basin, Italy. *Sed. Geol.*, **174**, 1–30.

Vigorito, M., **Murru, M.** and **Simone, L.** (2006) Architectural patterns in a multistory mixed carbonate-siliciclastic submarine channel, Puerto Torres Basin, Miocene, Sardinia, Italy. *Sed. Geol.*, **186**, 213–236.

Index

Note: Page numbers in *italics* refer to figures, those in **bold** refer to tables